Y0-BCC-655

Principles of Robotics & Artificial Intelligence

Principles of Robotics & Artificial Intelligence

Editor
Donald R. Franceschetti, PhD

SALEM PRESS

A Division of EBSCO Information Services
Ipswich, Massachusetts

GREY HOUSE PUBLISHING

Cover Image: 3d rendering of human on geometric element technology background, by monsitj (iStock Images)

For information contact Grey House Publishing/Salem Press, 4919 Route 22, PO Box 56, Amenia, NY 12501.

∞ The paper used in these volumes conforms to the American National Standard for Permanence of Paper for Printed Library Materials, Z39.48 1992 (R2009).

Publisher's Cataloging-In-Publication Data
(Prepared by The Donohue Group, Inc.)

Names: Franceschetti, Donald R., 1947- editor.
Title: Principles of robotics & artificial intelligence / editor, Donald R. Franceschetti, PhD.
Other Titles: Principles of robotics and artificial intelligence
Description: [First edition]. | Ipswich, Massachusetts : Salem Press, a division of EBSCO Information Services, Inc. ; Amenia, NY : Grey House Publishing, [2018] | Series: Principles of | Includes bibliographical references and index.
Identifiers: ISBN 9781682179420
Subjects: LCSH: Robotics. | Artificial intelligence.
Classification: LCC TJ211 .P75 2018 | DDC 629.892--dc23

First Printing
Printed in the United States of America

Contents

PUBLISHER'S NOTE

Salem Press is pleased to add *Principles of Robotics & Artificial Intelligence* as the twelfth title in the *Principles of* series that includes *Chemistry, Physics, Astronomy, Computer Science, Physical Science, Biology, Scientific Research, Sustainability, Biotechnology, Programming & Coding* and *Climatology*. This new resource introduces students and researchers to the fundamentals of robotics and artificial intelligence using easy-to-understand language for a solid background and a deeper understanding and appreciation of this important and evolving subject. All of the entries are arranged in an A to Z order, making it easy to find the topic of interest.

Entries related to basic principles and concepts include the following:

- A Summary that provides brief, concrete summary of the topic and how the entry is organized;
- History and Background, to give context for significant achievements in areas related to robotics and artificial intelligence including mathematics, biology, chemistry, physics, medicine, and education;
- Text that gives an explanation of the background and significance of the topic to robotics and artificial intelligence by describing developments such as Siri, facial recognition, augmented and virtual reality, and autonomous cars;
- Applications and Products, Impacts, Concerns, and Future to discuss aspects of the entry that can have sweeping impact on our daily lives, including smart devices, homes, and cities; medical devices; security and privacy; and manufacturing;
- Illustrations that clarify difficult concepts via models, diagrams, and charts of such key topics as Combinatrics, Cyberspace, Digital logic, Grammatology, Neural engineering, Interval, Biomimetics; and Soft robotics; and
- Bibliography lists that relate to the entry.

This reference work begins with a comprehensive introduction to robotics and artificial intelligence, written by volume editor Donald R. Franceschetti, PhD, Professor Emeritus of Physics and Material Science at the University of Memphis.

The book includes helpful appendixes as another valuable resource, including the following:

- Time Line of Machine Learning and Artificial Intelligence, tracing the field back to ancient history;
- A.M. Turing Award Winners, recognizing the work of pioneers and innovators in the field of computer science, robotics, and artificial intelligence;
- Glossary;
- General Bibliography and
- Subject Index.

Salem Press and Grey House Publishing extend their appreciation to all involved in the development and production of this work. The entries have been written by experts in the field. Their names and affiliations follow the Editor's Introduction.

Principles of Robotics & Artificial Intelligence, as well as all Salem Press reference books, is available in print and as an e-book and on the Salem Press online database, at https://online.salempress.com. Please visit www.salempress.com for more information.

Editor's Introduction

Our technologically based civilization may well be poised to undergo a major transition as robotics and artificial intelligence come into their own. This transition is likely to be as earthshaking as the invention of written language or the realization that the earth is not the center of the universe. Artificial intelligence (AI) permits human-made machines to act in an intelligent or purposeful, manner, like humans, as they acquire new knowledge, analyze and solve problems, and much more. AI holds the potential to permit us to extend human culture far beyond what could ever be achieved by a single individual. Robotics permits machines to complete numerous tasks, more accurately and consistently, with less fatigue, and for longer periods of time than human workers are capable of achieving. Some robots are even self-regulating.

Not only are robotics and AI changing the world of work and education, they are also capable of providing new insights into the nature of human activity as well.

The challenges related to understanding how AI and robotics can be integrated successfully into our society have raised several profound questions, ranging from the practical (Will robots replace humans in the workplace? Could inhaling nanoparticles cause humans to become sick?) to the profound (What would it take to make a machine capable of human reasoning? Will "grey goo" destroy mankind?). Advances and improvements to AI and robotics are already underway or on the horizon, so we have chosen to concentrate on some of the important building blocks related to these very different technologies from fluid dynamics and hydraulics. This goal of this essay as well as treatments of principles and terms related to artificial intelligence and robotics in the individual articles that make up this book is to offer a solid framework for a more general discussion. Reading this material will not make you an expert on AI or Robotics but it will enable you to join in the conversation as we all do our best to determine how machines capable of intelligence and independent action should interact with humans.

Historical Background

Much of the current AI literature has its origin in notions derived from symbol processing. Symbols have always held particular power for humans, capable of holding (and sometimes hiding) meaning. Mythology, astrology, numerology, alchemy, and primitive religions have all assigned meanings to an alphabet of "symbols." Getting to the heart of that symbolism is a fascinating study. In the realm of AI, we begin with numbers, from the development of simple algebra to the crisis in mathematical thinking that began in the early nineteenth century, which means we must turn to the Euclid's mathematical treatise, *Elements*, written around 300 BCE. Scholars had long been impressed by Euclidean geometry and the certainty it seemed to provide about figures in the plane. There was only one place where there was less than clarity. It seemed that Euclid's fifth postulate (that through any point in the plane one could draw one and only one straight line parallel to a given line) did not have the same character as the other postulates. Various attempts were made to derive this postulate from the others when finally, it was realized that that Euclid's fifth postulate could be replaced by one stating that no lines could be drawn parallel to the specified line or, alternatively, by one stating that an infinite number of lines could be drawn, distinct from each other but all passing through the point.

The notion that mathematicians were not so much investigating the properties of physical space as the conclusions that could be drawn from a given set of axioms introduced an element of creativity, or, depending on one's point of view, uncertainty, to the study of mathematics.

The Italian mathematician Giuseppe Peano tried to place the emphasis on arithmetic reasoning, which one might assume was even less subject to controversy. He introduced a set of postulates that effectively defined the non-negative integers, in a unique way. The essence of his scheme was the so-called principle of induction: if P(N) is true for the integer N, and P(N) being true implies that P(N+1) is true, then P(N) is true for all N. While seemingly seemingly self-apparent, mathematical logicians distrusted the principle and instead sought to derive a mathematics in which the postulate of induction was not needed. Perhaps the most famous attempt in this direction was the publication of Principia Mathematica, a three-volume treatise by philosophers Bertrand Russell and Alfred North Whitehead. This book was intended to do for mathematics what Isaac Newton's Philosophiæ Naturalis

Principia Mathematica had done in physics. In almost a thousand symbol-infested pages it attempted a logically complete construction of mathematics without reference to the Principle of Induction. Unfortunately, there was a fallacy in the text. In the early 1930's the Austrian (later American) mathematical logician, Kurt Gödel was able to demonstrate that any system of postulates sophisticated enough to allow the multiplication of integers would ultimately lead to undecidable propositions. In a real sense, mathematics was incomplete.

British scholar Alan Turing is probably the name most directly associated with artificial intelligence in the popular mind and rightfully so. It was Turing who turned the general philosophical question "can machines think?' into the far more practical question; what must a human or machine do to solve a problem. Turing's notion of effective procedure requires a recipe or algorithm to transform the statement of the problem into a step by step solution. By tradition one thinks of a Turing machine as implementing its program one step at a time. What makes Turing's contribution so powerful is the existence of a class of universal Turing machine which can them emulate any other Turing machine, so one can feed into a computer a description of that Turing machine, and emulate such a machine precisely until otherwise instructed. Turing announced the existence of the universal Turing machine in 1937 in his first published paper. In the same year. Claude Shannon, at Bel laboratories, published his seminal paper in which he showed that complex switching networks could also be treated by Boolean algebra.

Turing was a singular figure in the history of computation. A homosexual when homosexual orientation was considered abnormal and even criminal, he made himself indispensable to the British War Office as one of the mathematicians responsible for cracking the German "Enigma" code. He did highly imaginative work on embryogenesis as well as some hands-on chemistry and was among the first to advocate that "artificial intelligence" be taken seriously by those in power.

Now. it should be noted that not every computer task requites a Turing machine solution. The simplest computer problems require only that a data base be indexed in some fashion. Thus, the earliest computing machines were simply generalizations of a stack of cards that could be sorted in some fashion. The evolution of computer hard ware and software is an interesting lesson in applied science. Most computers are now of the digital variety, the state of the computer memory being given at any time as a large array of ones and zeros. In the simplest machines the memory arrays are "gates" which allow current flow according to the rules of Boolean algebra as set forth in the mid-Nineteenth century by the English mathematician George Boole. The mathematical functions are instantiated by the physical connections of the gates and are in a sense independent of the mechanism that does the actual computation. Thus, functioning models of a tinker toy compute are sometimes used to teach computer science. As a practical matter gates are nowadays fabricated from semiconducting materials where extremely small sizes can be obtained by photolithography.

Several variations in central processing unit design are worth mentioning. Since the full apparatus of a universal Turing machine is not needed for most applications, the manufacturers of many intelligent devices have devised reduced instruction set codes (RISC's) that are adequate for the purpose intended. At this point the desire for a universal Turing machine comes into conflict with that for an effective telecommunications network. Modern computer terminals are highly networked and may use several different methods to encode the messages they share.

Five Generations of Hardware, Software, and Computer Language

Because computer science is so dependent on advances in computer circuitry and the fabrication of computer components it has been traditional to divide the history of Artificial Intelligence into five generations. The first generation is that I which vacuum tubes are the workhorses of electrical engineering. This might also be considered the heroic age. Like transistors which were to come along later vacuum tubes could either be used as switches or as amplifiers. The artificial intelligence devices of the first generation are those based on vacuum tubes. Mechanical computers are generally relegated to the prehistory of computation.

Computer devices of the first generation relied on vacuum tubes and a lot of them. Now one problem with vacuum tubes was that they were dependent on thermionic emission, the release of electrons from a heated metal surface in vacuum. A vacuum tube-based computer was subject to the burn out of the filament used. Computer designers faced one of two alternatives. The first was run a program which tested every filament needed to check that it had not burned out. The second was to build into the computer an element of redundancy so that the computed result could be used within an acceptable margin of error. First generation computers were large and generally required extensive air conditioning. The amount of programming was minimal because programs had to be written in machine language.

The invention of the transistor in 1947 brought in semiconductor devices and a race to the bottom in the number of devices that could fit into a single computer component. Second generation computers were smaller by far than the computers of the first generation. They were also faster and more reliable. Third generation computers were the first in which integrated circuits replaced individual components. The fourth generation was that in which microprocessors appeared. Computers could then be built around a single microprocessor. Higher level languages grew more abundant and programmers could concentrate on programming rather than the formal structure of computer language. The fifth generation is mainly an effort by Japanese computer manufacturers to take full advantage of developments in artificial intelligence. The Chinese have expressed an interest in taking the lead in the sixth generation of computers, though there will be a great deal of competition for first place.

Nonstandard Logics

Conventional computation follows the conventions of Boolean algebra, a form of integer algebra devised by George Boole in the mid nineteenth century. Some variations that have found their way into engineering practice should be mentioned. The first of these is based on the utility of sometimes it is very useful to use language that is imprecise. How to state that John is a tall man but that others might be taller without getting into irrelevant quantitative detail might involve John having fractional membership in the set of tall people and in the set of not tall people at the same time. The state of a standard computer memory could be described by a set of ones and zeros. The evolution in time of that memory would involve changes in those ones and zeros. Other articles in this volume deal with quantum computation and other variations on this theme.

Traditional Applications of Artificial Intelligence

Theorem proving was among the first applications of AI to be tested. A program called Logic Theorist was set to work rediscovering the Theorem and Proofs that could be derived using the system described in Principia Mathematica. For the most part the theorems were found in the usual sequence but, occasionally Logic Theorist discovered an original proof.

Database Management

The use of computerized storage to maintain extensive databases such as maintained by the Internal Revenue Service, the Department of the Census, and the Armed Forces was a natural application of very low-level database management software. These large databases rise to more practical business software, such that an insurance company could estimate the number of its clients who would pass away from disease in the next year and set its premiums accordingly.

Expert Systems

A related effort was devoted to capturing human expertise. The knowledge accumulated by a physician in a lifetime of medical practice cold be made available to a young practitioner who was willing to ask his or her patients a few questions. With the development of imaging technologies the need for a human questioner could be reduced and the process automated, so that any individual could be examined in effect by the combined knowledge of many specialists.

Natural Language Processing

There is quite a difference between answering a few yes/no questions and normal human communication. To bridge this gap will require appreciable research in computational linguistics, and text processing. Natural language processing remains an area of computer science under active development. Developing a computer program thc can translate say English into German is a relatively modest goal. Developing a program to translate Spanish into the Basque dialect would be a different matter, since most linguists maintain that no native Spaniard has ever mastered the Basque Grammar and Syntax. An even greater challenge is presented by non-alphabetic languages like Chinese.

Important areas of current research are voice synthesis and speech recognition. A voice synthesizer converts able to convert written text into sound. This is not easy in a language like English where a single sound or phoneme can be represented in several different ways. A far more different challenge is present in voice recognition where the computer must be able to discriminate slight differences in speech patterns.

Adaptive Tutoring Systems

Computer tutoring systems are an obvious application of artificial intelligence. Doubleday introduced Tutor Text in the 1960's. A tutor text was a text that required the reader to answer a multiple choice question at the bottom of each page. Depending in the reader's answer he received additional text or was directed to a review selection. Since the 1990's an appreciable amount of Defense Department Funding has been spent on distance tutoring systems, that is systems in which the instructor is physically separated from the student. This was a great equalizer for students who could not study under a qualified instructor because of irregular hours. This is particularly the case for students in the military who may spend long hour in a missile launch capsule or under water in a submarine.

Senses for Artificial Intelligence Applications

All of the traditional senses have been duplicated by electronic sensors. Human vision has a long way to go, but rudimentary electronic retinas have been developed which afford a degree of vision to blind persons. The artificial cochlea can restore the hearing of individuals who have damaged the basilar membrane in their ears through exposure to loud noises. Pressure sensors can provide a sense of touch. Even the chemical senses have met technological substitutes. The sense of smell is registered in regions of the brain. The chemical senses differ appreciably between animal species and subspecies. Thus, most dogs can recognize their owners by scent. An artificial nose has been developed for alcoholic beverages and for use in cheese-making. The human sense of taste is a combination of the two chemical senses of taste and smell.

Remote Sensing and Robotics

Among the traditional reasons for the development of automata that are capable of reporting on environmental conditions at distant sites is the financial cost and hazard to human life that may be encountered there. A great deal can be learned about distant objects by telescopic observation. Some forty years ago, the National Aeronautics and Space administration launched the Pioneer space vehicles which are now about to enter interstellar space. These vehicles have provided numerous insights, some of them quite surprising, into the behavior of the outer planets.

As far as we know, the speed of light, 300 km/sec sets an absolute limit to one event influencing another in the same reference frame. Computer scientists are quick to note that this quantity, which is enormous in terms of the motion of ordinary objects is a mere 30 cm/nanosecond. Thus, computer devices must be less than 30 cm in extent if relativistic effects can be neglected. As a practical matter, this sets a limit to the spatial extent of high precision electronic systems.

Any instrumentation expected to record event over a period of one the or more years must therefore possess a high degree of autonomy.

Scale Effects

Compared to humans, computers can hold far more information in memory, and process that information far more rapidly and in far greater detail. Imagine a human with a mysterious ailment. A computer like IBM's Watson, can compare the biochemical and immunological status of the patient with that of a thousand others in a few seconds. It can then search reports to determine treatment options. Robotic surgery is far better suited to operations on the eyes, ears, nerves and vasculature than using hand held instruments. Advances in the treatment of disease will inevitably follow advances in artificial intelligence. Improvements in public health will likewise follow when the effects of environmental changes are more fully understood.

Search in Artificial Intelligence

Many artificial intelligence applications involve a search for the most appropriate solution. Often the problem can be expressed as finding the best strategy to employ in a game like chess or poker where the space of possible board configurations is very large but finite. Such problems can be related to important problems in full combinatorics, such as the problem of protein folding. The literature is full of examples.

—*Donald R. Franceschetti, PhD*

Bibliography

Dyson, George. *Turing's Cathedral: The Origins of the Digital Universe.* London: Penguin Books, 2013. Print.

Franceschetti, Donald R. *Biographical Encyclopedia of Mathematicians.* New York: Marshall Cavendish, 1999. Print.

Franklin, Stan. *Artificial Minds.* Cambridge, Mass: MIT Press, 2001. Print.

Fischler, Martin A, and Oscar Firschein. *Intelligence: The Eye, the Brain, and the Computer.* Reading (MA): Addison-Wesley, 1987. Print.

Michie, Donald. *Expert Systems in the Micro-Electric Age: Proceedings of the 1979 Aisb Summer School.* Edinburgh: Edinburgh University Press, 1979. Print.

Mishkoff, Henry C. *Understanding Artificial Intelligence.* Indianapolis, Indiana: Howard W. Sams & Company, 1999. Print.

Penrose, Roger. *The Emperor's New Mind: Concerning Computers, Minds and the Laws of Physics.* Oxford University Press, 2016. Print.

ABSTRACTION

SUMMARY

In computer science, abstraction is a strategy for managing the complex details of computer systems. Broadly speaking, it involves simplifying the instructions that a user gives to a computer system in such a way that different systems, provided they have the proper underlying programming, can "fill in the blanks" by supplying the levels of complexity that are missing from the instructions. For example, most modern cultures use a decimal (base 10) positional numeral system, while digital computers read numerals in binary (base 2) format. Rather than requiring users to input binary numbers, in most cases a computer system will have a layer of abstraction that allows it to translate decimal numbers into binary format.

There are several different types of abstraction in computer science. Data abstraction is applied to data structures in order to manipulate bits of data manageably and meaningfully. Control abstraction is similarly applied to actions via control flows and subprograms. Language abstraction, which develops separate classes of languages for different purposes—modeling languages for planning assistance, for instance, or programming languages for writing software, with many different types of programming languages at different levels of abstraction—is one of the fundamental examples of abstraction in modern computer science.

The core concept of abstraction is that it ideally conceals the complex details of the underlying system, much like the desktop of a computer or the graphic menu of a smartphone conceals the complexity involved in organizing and accessing the many programs and files contained therein. Even the simplest controls of a car—the brakes, gas pedal, and steering wheel—in a sense abstract the more complex elements involved in converting the mechanical energy applied to them into the electrical signals and mechanical actions that govern the motions of the car.

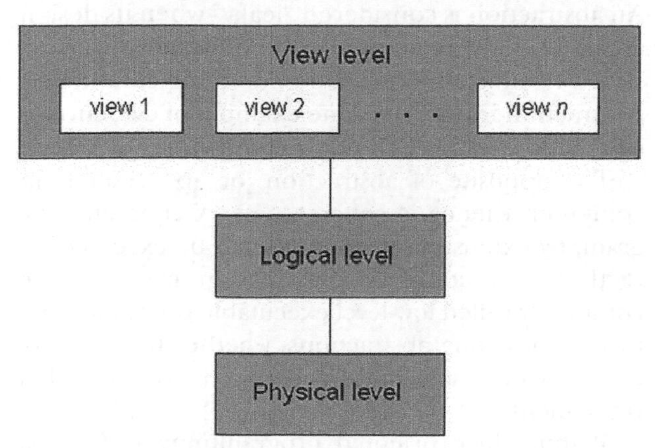

Data abstraction levels of a database system. Doug Bell~commonswiki assumed (based on copyright claims).

BACKGROUND

Even before the modern computing age, mechanical computers such as abacuses and slide rules abstracted, to some degree, the workings of basic and advanced mathematical calculations. Language abstraction has developed alongside computer science as a whole; it has been a necessary part of the field from the beginning, as the essence of computer programming involves translating natural-language commands such as "add two quantities" into a series of computer operations. Any involvement of software at all in this process inherently indicates some degree of abstraction.

The levels of abstraction involved in computer programming can be best demonstrated by an exploration of programming languages, which are grouped into generations according to degree of abstraction. First-generation languages are machine languages, so called because instructions in these languages can be directly executed by a computer's central processing unit (CPU), and are written in binary numerical code. Originally, machine-language instructions were entered into computers directly by setting switches on the machine. Second-generation languages are called assembly languages, designed as shorthand to abstract machine-language instructions into mnemonics in order to make coding and debugging easier.

Third-generation languages, also called high-level programming languages, were first designed in the 1950s. This category includes older, now-obscure and little-used languages such as COBOL and FORTRAN as well as newer, more commonplace languages such as C++ and Java. While different assembly languages are specific to different types of computers, high-level languages were designed to be machine independent, so that a program would not need to be rewritten for every type of computer on the market.

In the late 1970s, the idea was advanced of developing a fourth generation of languages, further abstracted from the machine itself. Some people classify Python and Ruby as fourth-generation rather than third-generation languages. However, third-generation languages have themselves become extremely diverse, blurring this distinction. The category encompasses not just general-purpose programming languages, such as C++, but also domain-specific and scripting languages.

Computer languages are also used for purposes beyond programming. Modeling languages are used in computing, not to write software, but for planning and design purposes. Object-role modeling, for instance, is an approach to data modeling that combines text and graphical symbols in diagrams that model semantics; it is commonly used in data warehouses, the design of web forms, requirements engineering, and the modeling of business rules. A simpler and more universally familiar form of modeling language is the flowchart, a diagram that abstracts an algorithm or process.

PRACTICAL APPLICATIONS

The idea of the algorithm is key to computer science and computer programming. An algorithm is a set of operations, with every step defined in sequence. A cake recipe that defines the specific quantities of ingredients required, the order in which the ingredients are to be mixed, and how long and at what temperature the combined ingredients must be baked is essentially an algorithm for making cake. Algorithms had been discussed in mathematics and logic long before the advent of computer science, and they provide its formal backbone.

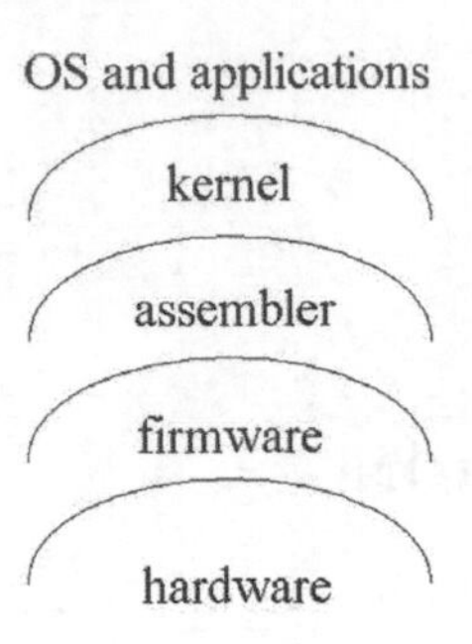

A typical vision of a computer architecture as a series of abstraction layers: hardware, firmware, assembler, kernel, operating system, and applications.

One of the problems with abstraction arises when users need to access a function that is obscured by the interface of a program or some other construct, a dilemma known as "abstraction inversion." The only solution for the user is to use the available functions of the interface to recreate the function. In many cases, the resulting re-implemented function is clunkier, less efficient, and potentially more error prone than the obscured function would be, especially if the user is not familiar enough with the underlying design of the program or construct to know the best implementation to use. A related concept is that of "leaky abstraction," a term coined by software engineer Joel Spolsky, who argued that all abstractions are leaky to some degree. An abstraction is considered "leaky" when its design allows users to be aware of the limitations that resulted from abstracting the underlying complexity. Abstraction inversion is one example of evidence of such leakiness, but it is not the only one.

The opposite of abstraction, or abstractness, in computer science is concreteness. A concrete program, by extension, is one that can be executed directly by the computer. Such programs are more commonly called low-level executable programs. The process of taking abstractions, whether they be programs or data, and making them concrete is called refinement.

Within object-oriented programming (OOP)—a class of high-level programming languages, including

C++ and Common Lisp—"abstraction" also refers to a feature offered by many languages. The objects in OOP are a further enhancement of an earlier concept known as abstract data types; these are entities defined in programs as instances of a class. For example, "OOP" could be defined as an object that is an instance in a class called "abbreviations." Objects are handled very similarly to variables, but they are significantly more complex in their structure—for one, they can contain other objects—and in the way they are handled in compiling.

Another common implementation of abstraction is polymorphism, which is found in both functional programming and OOP. Polymorphism is the ability of a single interface to interact with different types of entities in a program or other construct. In OOP, this is accomplished through either parametric polymorphism, in which code is written so that it can work on an object irrespective of class, or subtype polymorphism, in which code is written to work on objects that are members of any class belonging to a designated superclass.

—Bill Kte'pi, MA

BIBLIOGRAPHY

Abelson, Harold, Gerald Jay Sussman, and Julie Sussman. *Structure and Interpretation of Computer Programs.* 2nd ed, Cambridge: MIT P, 1996. Print.

Brooks, Frederick P., Jr. *The Mythical Man-Month: Essays on Software Engineering.* Anniv. ed. Reading: Addison, 1995. Print.

Goriunova, Olga, ed. *Fun and Software: Exploring Pleasure, Paradox, and Pain in Computing.* New York: Bloomsbury, 2014. Print.

Graham, Ronald L., Donald E. Knuth, and Oren Patashnik. *Concrete Mathematics: A Foundation for Computer Science.* 2nd ed. Reading: Addison, 1994. Print.

McConnell, Steve. *Code Complete: A Practical Handbook of Software Construction.* 2nd ed. Redmond: Microsoft, 2004. Print.

Pólya, George. *How to Solve It: A New Aspect of Mathematical Method.* Expanded Princeton Science Lib. ed. Fwd. John H. Conway. 2004. Princeton: Princeton UP, 2014. Print.

Roberts, Eric S. *Programming Abstractions in C++.* Boston: Pearson, 2014. Print.

Roberts, Eric S. *Programming Abstractions in Java.* Boston: Pearson, 2017. Print.

ADVANCED ENCRYPTION STANDARD (AES)

SUMMARY

Advanced Encryption Standard (AES) is a data encryption standard widely used by many parts of the U.S. government and by private organizations. Data encryption standards such as AES are designed to protect data on computers. AES is a symmetric block cipher algorithm, which means that it encrypts and decrypts information using an algorithm. Since AES was first chosen as the U.S. government's preferred encryption software, hackers have tried to develop ways to break the cipher, but some estimates suggest that it could take billions of years for current technology to break AES encryption. In the future, however, new technology could make AES obsolete.

ORIGINS OF AES

The U.S. government has used encryption to protect classified and other sensitive information for many years. During the 1990s, the U.S. government relied mostly on the Data Encryption Standard (DES) to

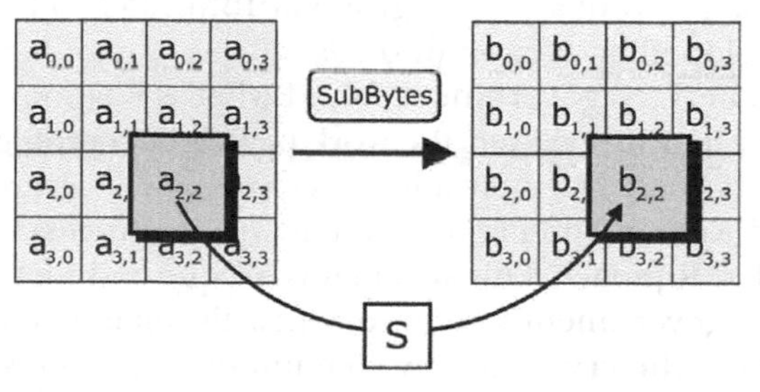

The SubBytes step, one of four stages in a round of AES (wikipedia).

Vincent Rijmen. Coinventor of AES algorithm called Rijndael.

encrypt information. The technology of that encryption code was aging, however, and the government worried that encrypted data could be compromised by hackers. The DES was introduced in 1976 and used a 56-bit key, which was too small for the advances in technology that were happening. Therefore, in 1997, the government began searching for a new, more secure type of encryption software. The new system had to be able to last the government into the twenty-first century, and it had to be simple to implement in software and hardware.

The process for choosing a replacement for the DES was transparent, and the public had the opportunity to comment on the process and the possible choices. The government chose fifteen different encryption systems for evaluation. Different groups and organizations, including the National Security Agency (NSA), had the opportunity to review these fifteen choices and provide recommendations about which one the government should adopt.

Two years after the initial announcement about the search for a replacement for DES, the U.S. government chose five algorithms to research even further. These included encryption software developed by large groups (e.g., a group at IBM) and software developed by a few individuals.

The U.S. government found what is was looking for when it reviewed the work of Belgian cryptographers Joan Daemen and Vincent Rijmen. Daemen and Rijmen had created an encryption process they called Rijndael. This system was unique and met the U.S. government's requirements. Prominent members of the cryptography community tested the software. The government and other organizations found that Rijndael had block encryption implementation; it had 128-, 192-, and 256-bit keys; it could be easily implemented in software, hardware, or firmware; and it could be used around the world. Because of these features, the government and others believed that the use of Rijndael as the AES would be the best choice for government data encryption for at least twenty to thirty years.

REFINING THE USE OF AES

The process of locating and implementing the new encryption code took five years. The National Institute of Standards (NIST) finally approved the AES as Federal Information Processing Standards Publication (FIPS PUB) 197 in November 2001. (FIPS PUBs are issued by NIST after approval by the Secretary of Commerce, and they give guidelines about the standards people in the government should be using.) When the NIST first made its announcement about using AES, it allowed only unclassified information to be encrypted with the software. Then, the NSA did more research into the program and any weaknesses it might have. In 2003—after the NSA gave its approval—the NIST announced that AES could be used to encrypt classified information. The NIST announced that all key lengths could be used for information classified up to SECRET, but TOP SECRET information had to be encrypted using 192- or 256-bit key lengths.

Although AES is an approved encryption standard in the U.S. government, other encryption standards are used. Any encryption standard that has been approved by the NIST must meet requirements similar to those met by AES. The NSA has to approve any encryption algorithms used to protect national security systems or national security information.

According to the U.S. federal government, people should use AES when they are sending sensitive (unclassified) information. This encryption system also can be used to encrypt classified information as long as the correct size of key code is used according to the level of classification. Furthermore, people and organizations outside the federal government can use the AES to protect their own sensitive information. When workers in the federal government use AES, they are supposed to follow strict guidelines to ensure that information is encrypted correctly.

THE FUTURE OF AES

The NIST continues to follow developments with AES and within the field of cryptology to ensure that AES remains the government's best option for encryption. The NIST formally reviews AES (and any other official encryption systems) every five years. The NIST will make other reviews as necessary if any new technological breakthroughs or potential security threats are uncovered.

Although AES is one of the most popular encryption systems on the market today, encryption itself may become obsolete in the future. With current technologies, it would likely take billions of years to break an AES-encrypted message. However, quantum computing is becoming an important area of research, and developments in this field could make AES and other encryption software obsolete. DES, AES's predecessor, can now be broken in a matter of hours, but when it was introduced, it also was considered unbreakable. As technology advances, new ways to encrypt information will have to be developed and tested. Some experts believe that AES will be effective until the 2030s or 2040s, but the span of its usefulness will depend on other developments in technology.

—Elizabeth Mohn

BIBLIOGRAPHY

"Advanced Encryption Standard (AES)." *Techopedia.com.* Janalta Interactive Inc.Web. 31 July 2015. http://www.techopedia.com/definition/1763/advanced-encryption-standard-aes

"AES." *Webopedia.* QuinStreet Inc. Web. 31 July 2015. http://www.webopedia.com/TERM/A/AES.html

National Institute for Standards and Technology. "Announcing the Advanced Encryption Standard (AES): Federal Information Processing Standards Publication 197." NIST, 2001. Web. 31 July 2015. http://csrc.nist.gov/publications/fips/fips197/fips-197.pdf

National Institute for Standards and Technology. "Fact Sheet: CNSS Policy No. 15, Fact Sheet No. 1, National Policy on the Use of the Advanced Encryption Standard (AES) to Protect National Security Systems and National Security Information." NIST, 2003. Web. 31 July 2015. http://csrc.nist.gov/groups/ST/toolkit/documents/aes/CNSS15FS.pdf

Rouse, Margaret "Advanced Encryption Standard (AES)." *TechTarget.* TechTarget. Web. 31 July 2015. http://searchsecurity.techtarget.com/definition/Advanced-Encryption-Standard

Wood, Lamont. "The Clock Is Ticking for Encryption." *Computerworld.* Computerworld, Inc. 21 Mar. 2011. Web. 31 July 2015. http://www.computerworld.com/article/2550008/security0/the-clock-is-ticking-for-encryption.html

AGILE ROBOTICS

SUMMARY

Movement poses a challenge for robot design. Wheels are relatively easy to use but are severely limited in their ability to navigate rough terrain. Agile robotics seeks to mimic animals' biomechanical design to achieve dexterity and expand robots' usefulness in various environments.

ROBOTS THAT CAN WALK

Developing robots that can match humans' and other animals' ability to navigate and manipulate their environment is a serious challenge for scientists and engineers. Wheels offer a relatively simple solution for many robot designs. However, they have severe limitations. A wheeled robot cannot navigate simple stairs, to say nothing of ladders, uneven terrain, or the aftermath of an earthquake. In such scenarios, legs are much more useful. Likewise, tools such as simple pincers are useful for gripping objects, but they do not approach the sophistication and adaptability of a human hand with opposable thumbs. The cross-disciplinary subfield devoted to creating robots that can match the dexterity of living things is known as "agile robotics."

INSPIRED BY BIOLOGY

Agile robotics often takes inspiration from nature. Biomechanics is particularly useful in this respect, combining physics, biology, and chemistry to describe how the structures that make up living things work. For example, biomechanics would describe a running human in terms of how the human body—muscles, bones, circulation—interacts with forces such as gravity and momentum. Analyzing the activities of living beings in these terms allows roboticists to attempt to recreate these processes. This, in turn, often reveals new insights into biomechanics. Evolution has been shaping life for millions of years through a process of high-stakes trial-and-error. Although evolution's "goals" are not necessarily those of scientists and engineers, they often align remarkably well.

Boston Dynamics, a robotics company based in Cambridge, Massachusetts, has developed a prototype robot known as the Cheetah. This robot mimics the four-legged form of its namesake in an attempt to recreate its famous speed. The Cheetah has achieved a land speed of twenty-nine miles per hour—slower than a real cheetah, but faster than any other legged robot to date. Boston Dynamics has another four-legged robot, the LS3, which looks like a sturdy mule and was designed to carry heavy supplies over rough terrain inaccessible to wheeled transport. (The LS3 was designed for military use, but the project was shelved in December 2015 because it was too noisy.) Researchers at the Massachusetts Institute of Technology (MIT) have built a soft robotic fish. There are robots in varying stages of development that mimic snakes' slithering motion or caterpillars' soft-bodied flexibility, to better access cramped spaces.

In nature, such designs help creatures succeed in their niches. Cheetahs are effective hunters because of their extreme speed. Caterpillars' flexibility and strength allow them to climb through a complex world of leaves and branches. Those same traits could be incredibly useful in a disaster situation. A small, autonomous robot that moved like a caterpillar could maneuver through rubble to locate survivors without the need for a human to steer it.

HUMANOID ROBOTS IN A HUMAN WORLD

Humans do not always compare favorably to other animals when it comes to physical challenges. Primates are often much better climbers. Bears are much stronger, cheetahs much faster. Why design anthropomorphic robots if the human body is, in physical terms, relatively unimpressive?

NASA has developed two different robots, Robonauts 1 and 2, that look much like a person in a space suit. This is no accident. The Robonaut is designed to fulfill the same roles as a flesh-and-blood astronaut, particularly for jobs that are too dangerous or dull for humans. Its most remarkable feature is its hands. They are close enough in design and ability to human hands that it can use tools designed for human hands without special modifications.

Consider the weakness of wheels in dealing with stairs. Stairs are a very common feature in the houses and communities that humans have built for themselves. A robot meant to integrate into human society could get around much more easily if it shared a similar body plan. Another reason to create humanoid robots is psychological. Robots that appear more human will be more accepted in health care, customer service, or other jobs that traditionally require human interaction.

Perhaps the hardest part of designing robots that can copy humans' ability to walk on two legs is achieving dynamic balance. To walk on two legs, one must adjust one's balance in real time in response to each step taken. For four-legged robots, this is less of an issue. However, a two-legged robot needs sophisticated sensors and processing power to detect and respond quickly to its own shifting mass. Without this, bipedal robots tend to walk slowly and awkwardly, if they can remain upright at all.

THE FUTURE OF AGILE ROBOTICS

As scientists and engineers work out the major challenges of agile robotics, the array of tasks that can be given to robots will increase markedly. Instead of being limited to tires, treads, or tracks, robots will navigate their environments with the coordination and agility of living beings. They will prove invaluable not just in daily human environments but also in more specialized situations, such as cramped-space disaster relief or expeditions into rugged terrain.

—Kenrick Vezina, MS

Bibliography

Bibby, Joe. "Robonaut: Home." *Robonaut.* NASA, 31 May 2013. Web. 21 Jan. 2016.

Gibbs, Samuel. "Google's Massive Humanoid Robot Can Now Walk and Move without Wires." *Guardian.* Guardian News and Media, 21 Jan. 2015. Web. 21 Jan. 2016.

Murphy, Michael P., and Metin Sitti. "Waalbot: Agile Climbing with Synthetic Fibrillar Dry Adhesives." *2009 IEEE International Conference on Robotics and Automation.* Piscataway: IEEE, 2009. *IEEE Xplore.* Web. 21 Jan. 2016.

Sabbatini, Renato M. E. "Imitation of Life: A History of the First Robots." *Brain & Mind* 9 (1999): n. pag. Web. 21 Jan. 2016.

Schwartz, John. "In the Lab: Robots That Slink and Squirm." *New York Times.* New York Times, 27 Mar. 2007. Web. 21 Jan. 2016.

Wieber, Pierre-Brice, Russ Tedrake, and Scott Kuindersma. "Modeling and Control of Legged Robots." *Handbook of Robotics.* Ed. Bruno Siciliano and Oussama Khatib. 2nd ed. N.p.: Springer, n.d. (forthcoming). *Scott Kuindersma—Harvard University.* Web. 6 Jan. 2016

Algorithm

SUMMARY

An algorithm is a set of steps to be followed in order to solve a particular type of mathematical problem. As such, the concept has been analogized to a recipe for baking a cake; just as the recipe describes a method for accomplishing a goal (baking the cake) by listing each step that must be taken throughout the process, an algorithm is an explanation of how to solve a math problem that describes each step necessary in the calculations. Algorithms make it easier for mathematicians to think of better ways to solve certain types of problems, because looking at the steps needed to reach a solution sometimes helps them to see where an algorithm can be made more efficient by eliminating redundant steps or using different methods of calculation.

Algorithms are also important to computer scientists. For example, without algorithms, a computer would have to be programmed with the exact answer to every set of numbers that an equation could accept in order to solve an equation—an impossible task. By programming the computer with the appropriate algorithm, the computer can follow the instructions needed to solve the problem, regardless of which values are used as inputs.

HISTORY AND BACKGROUND

The word algorithm originally came from the name of a Persian mathematician, Al-Khwarizmi, who lived in the ninth century and wrote a book about the ideas of an earlier mathematician from India, Brahmagupta. At first the word simply referred to the author's description of how to solve equations using Brahmagupta's number system, but as time passed it took on a more general meaning. First it was used to refer to the steps required to solve any mathematical problem, and later it broadened still further to include almost any kind of method for handling a particular situation.

Algorithms are often used in mathematical instruction because they provide students with concrete steps to follow, even before the underlying operations are fully comprehended. There are algorithms for most mathematical operations, including subtraction, addition, multiplication, and division.

For example, a well-known algorithm for performing subtraction is known as the left to right algorithm. As its name suggests, this algorithm requires one to first line up the two numbers one wishes to find the difference between so that the units digits are in one column, the tens digits in another column, and so forth. Next, one begins in the leftmost column and subtracts the lower number from the upper, writing the result below. This step is then repeated for the next column to the right, until the values in the units column have been subtracted from one another. At this point the results from the subtraction of each column, when read left to right, constitute the answer to the problem.

By following these steps, it is possible for a subtraction problem to be solved even by someone still in the process of learning the basics of subtraction. This demonstrates the power of algorithms both for performing calculations and for use as a source of instructional support.

—*Scott Zimmer, MLS, MS*

BIBLIOGRAPHY

Cormen, Thomas H. *Algorithms Unlocked.* Cambridge, MA: MIT P, 2013.

Cormen, Thomas H. *Introduction to Algorithms.* Cambridge, MA: MIT P, 2009.

MacCormick, John. *Nine Algorithms That Changed the Future: The Ingenious Ideas That Drive Today's Computers.* Princeton: Princeton UP, 2012.

Parker, Matt. *Things to Make and Do in the Fourth Dimension: A Mathematician's Journey Through Narcissistic Numbers, Optimal Dating Algorithms, at Least Two Kinds of Infinity, and More.* New York: Farrar, 2014.

Schapire, Robert E., and Yoav Freund. *Boosting: Foundations and Algorithms.* Cambridge, MA: MIT P, 2012.

Steiner, Christopher. *Automate This: How Algorithms Came to Rule Our World.* New York: Penguin, 2012.

Valiant, Leslie. *Probably Approximately Correct: Nature's Algorithms for Learning and Prospering in a Complex World.* New York: Basic, 2013.

ANALYSIS OF VARIANCE (ANOVA)

SUMMARY

Analysis of variance (ANOVA) is a method for testing the statistical significance of any difference in means in three or more groups. The method grew out of British scientist Sir Ronald Aylmer Fisher's investigations in the 1920s on the effect of fertilizers on crop yield. ANOVA is also sometimes called the F-test in his honor.

Conceptually, the method is simple, but in its use, it becomes mathematically complex. There are several types, but the one-way ANOVA and the two-way ANOVA are among the most common. One-way ANOVA compares statistical means in three or more groups without considering any other factor. Two-way ANOVA is used when the subjects are simultaneously divided by two factors, such as patients divided by sex and severity of disease.

English biologist and statistician Ronald Fisher in the 1950s. By Flikr commons via Wikimedia Commons ,

BACKGROUND

In ANOVA, the total variance in subjects in all the data sets combined is considered according to the different sources from which it arises, such as between-group variance and within-group variance (also called "error sum of squares" or "residual sum of squares"). Between-group variance describes the amount of variation among the different data sets. For example, ANOVA may reveal that 50 percent of variation in some medical factor in healthy adults is due to genetic differentials, 30 percent due to age differentials, and the remaining 20 percent due to other factors. Such residual (in this case, the remaining 20 percent) left after the extraction of the factor effects of interest is the within-group variance. The total variance is calculated as the sum of squares total, equal to the sum of squares within plus the sum of squares between.

ANOVA can be used to test a hypothesis. The null hypothesis states that there is no difference between the group means, while the alternative

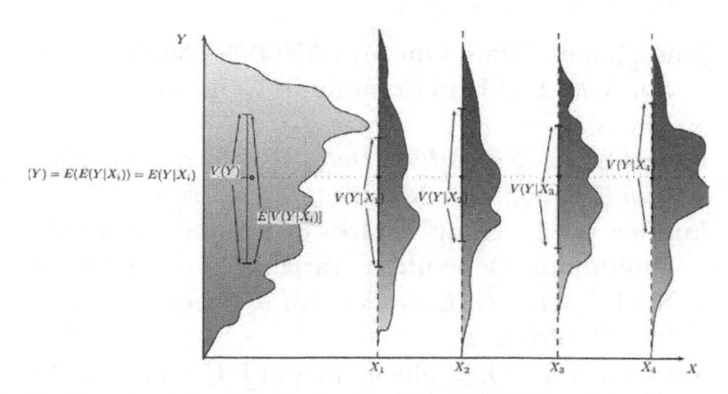

Visual representation of a situation in which an ANOVA analysis will conclude to a very poor fit. By Vanderlindenma (Own work)

hypothesis states that there is a difference (that the null hypothesis is false). If there are genuine differences between the groups, then the between-group variance should be much larger than the within-group variance; if the differences are merely due to random chance, the between-group and within-group variances will be close. Thus, the ratio between the between-group variance (numerator) and the within-group variance (denominator) can be to determine whether the group means are different and therefore prove whether the null hypothesis is true or false. This is what the *F*-test does.

In performing ANOVA, some kind of random sampling is required in order to test the validity of the procedure. The usual ANOVA considers groups on what is called a "nominal basis," that is, without order or quantitative implications. This implies that if one's groups are composed of cases with mild disease, moderate disease, serious disease, and critical cases, the usual ANOVA would ignore this gradient. Further analysis would study the effect of this gradient on the outcome.

CRITERIA

Among the requirements for the validity of ANOVA are

- statistical independence of the observations
- all groups have the same variance (a condition known as "homoscedasticity")
- the distribution of means in the different groups is Gaussian (that is, following a normal distribution, or bell curve)
- for two-way ANOVA, the groups must also have the same sample size

Statistical independence is generally the most important requirement. This is checked using the Durbin-Watson test. Observations made too close together in space or time can violate independence. Serial observations, such as in a time series or repeated measures, also violate the independence requirement and call for repeated-measures ANOVA.

The last criterion is generally fulfilled due to the central limit theorem when the sample size in each group is large. According to the central limit theorem, as sample size increases, the distribution of the sample means or the sample sums approximates normal distribution. Thus, if the number of subjects in the groups is small, one should be alert to the different groups' pattern of distribution of the measurements and of their means. It should be Gaussian. If the distribution is very far from Gaussian or the variances really unequal, another statistical test will be needed for analysis.

The practice of ANOVA is based on means. Any means-based procedure is severely perturbed when outliers are present. Thus, before using ANOVA, there must be no outliers in the data. If there are, do a sensitivity test: examine whether the outliers can be excluded without affecting the conclusion.

The results of ANOVA are presented in an ANOVA table. This contains the sums of squares, their respective degrees of freedom (df; the number of data points in a sample that can vary when estimating a parameter), respective mean squares, and the values of *F* and their statistical significance, given as *p*-values. To obtain the mean squares, the sum of squares is divided by the respective df, and the *F* values are obtained by dividing each factor's mean square by the mean square for the within-group. The *p*-value comes from the *F* distribution under the null hypothesis. Such a table can be found using any statistical software of note.

A problem in the comparison of three or more groups by the criterion *F* is that its statistical significance indicates only that a difference exists. It does not tell exactly which group or groups are different. Further analysis, called "multiple comparisons," is required to identify the groups that have different means.

When no statistical significant difference is found across groups (the null hypothesis is true), there is a tendency to search for a group or even subgroup that stands out as meeting requirements. This post-hoc analysis is permissible so long as it is exploratory in nature. To be sure of its importance, a new study should be conducted on that group or subgroup.

—*Martin P. Holt, MSc*

Bibliography

"Analysis of Variance." *Khan Academy.* Khan Acad., n.d. Web. 11 July 2016.

Doncaster, P., and A. Davey. *Analysis of Variance and Covariance: How to Choose and Construct Models for the Life Sciences.* Cambridge: Cambridge UP, 2007. Print.

Fox, J. *Applied Regression Analysis and Generalized Linear Models.* 3rd ed. Thousand Oaks: Sage, 2016. Print.

Jones, James. "Stats: One-Way ANOVA." *Statistics: Lecture Notes.* Richland Community Coll., n.d. Web. 11 July 2016.

Kabacoff, R. *R in Action: Data Analysis and Graphics with R.* Greenwich: Manning, 2015. Print.

Lunney, G. H. "Using Analysis of Variance with a Dichotomous Dependent Variable: An Empirical Study." *Journal of Educational Measurement* 7 (1970): 263–69. Print.

Streiner, D. L., G. R. Norman, and J. Cairney. *Health Measurement Scales: A Practical Guide to Their Development and Use.* New York: Oxford UP, 2014. Print.

Zhang, J., and X. Liang. "One-Way ANOVA for Functional Data via Globalizing the Pointwise F-test." *Scandinavian Journal of Statistics* 41 (2014): 51–74. Print.

Application Programming Interface (API)

SUMMARY

Application programming interfaces (APIs) are special coding for applications to communicate with one another. They give programs, software, and the designers of the applications the ability to control which interfaces have access to an application without closing it down entirely. APIs are commonly used in a variety of applications, including social media networks, shopping websites, and computer operating systems.

APIs have existed since the early twenty-first century. However, as computing technology has evolved, so has the need for APIs. Online shopping, mobile devices, social networking, and cloud computing all saw major developments in API engineering and usage. Most computer experts believe that future technological developments will require additional ways for applications to communicate with one another.

BACKGROUND

An application is a type of software that allows the user to perform one or more specific tasks. Applications may be used across a variety of computing platforms. They are designed for laptop or desktop computers and are often called desktop applications. Likewise, applications designed for cellular phones and other mobile devices are known as mobile applications.

When in use, applications run inside a device's operating system. An operating system is a type of software that runs the computer's basic tasks. Operating systems are often capable of running multiple applications simultaneously, allowing users to multitask effectively.

Applications exist for a wide variety of purposes. Software engineers have crafted applications that serve as image editors, word processors, calculators, video games, spreadsheets, media players, and more. Most daily computer-related tasks are accomplished with the aid of applications.

APPLICATION

APIs are coding interfaces that allow different applications to exchange information in a controlled manner. Before APIs, applications came in two varieties: open source and closed. Closed applications cannot be communicated with in any way other than directly using the application. The code is secret, and only authorized software engineers have access to it. In contrast, open source applications are completely public. The code is free for users to dissect, modify, or otherwise use as they see fit.

APIs allow software engineers to create a balance between these two extremes. When an API is functioning properly, it allows authorized applications to request and receive information from the original application. The engineer controlling the original application can modify the signature required to request this information at any time, thus immediately modifying which external applications can request information from the original one.

There are two common types of APIs: code libraries and web services APIs. Code libraries operate on a series of predetermined function calls, given either to the public or to specified developers. These function calls are often composed of complicated code, and they are designed to be sent from one application to another. For example, a code library API may have predetermined code designed to fetch and display a certain image, or to compile and display statistics. Web services APIs, however, typically function differently. They specifically send requests through HTTP channels, usually using XML or JSON languages. These APIs are often designed to work in conjunction with a web browser application.

Many of the first APIs were created by Salesforce, a web-based corporation. Salesforce launched its first APIs at the IDG Demo Conference in 2000. It offered the use of its API code to businesses for a fee. Later that year, eBay made its own API available to select partners through the eBay Developers Program. This allowed eBay's auctions to interface with a variety of third-party applications and webpages, increasing the site's popularity.

In 2002, Amazon released its own API, called Amazon Web Services (AWS). AWS allowed third-party websites to display and directly link to Amazon products on their own websites. This increased computer users' exposure to Amazon's products, further increasing the web retailer's sales.

While APIs remained popular with sales-oriented websites, they did not become widespread in other areas of computing until their integration into social media networks. In 2004, the image-hosting website Flickr created an API that allowed users to easily embed photos hosted on their Flickr accounts onto webpages. This allowed users to share their Flickr albums on their social media pages, blogs, and personal websites.

Facebook implemented an API into its platform in August of 2006. This gave developers access to users' data, including their photos, friends, and profile information. The API also allowed third-party websites to link to Facebook; this let users access their profiles from other websites. For example, with a single click, Facebook users were able to share newspaper articles directly from the newspaper's website.

Google developed an API for its popular application Google Maps as a security measure. In the months following Google Maps' release, third-party developers hacked the application to use it for their own means. In response, Google built an extremely secure API to allow it to meet the market's demand to use Google Maps' coding infrastructure without losing control of its application.

While APIs were extremely important to the rise of social media, they were even more important to the rise of mobile applications. As smartphones became more popular, software engineers developed countless applications for use on them. These included location-tracking applications, mobile social networking services, and mobile photo-sharing services.

The cloud computing boom pushed APIs into yet another area of usage. Cloud computing involves connecting to a powerful computer or server, having that computer perform any necessary calculations, and transmitting the results back to the original computer through the Internet. Many cloud computing services require APIs to ensure that only authorized applications are able to take advantage of their code and hardware.

—*Tyler Biscontini*

Bibliography

Barr, Jeff. "API Gateway Update – New Features Simplify API Development." *Amazon Web Services*, 20 Sept. 2016, aws.amazon.com/blogs/aws/api-gateway-update-new-features-simplify-api-development/. Accessed 29 Dec. 2016.

"History of APIs." *API Evangelist*, 20 Dec. 2012, apievangelist.com/2012/12/20/history-of-apis/. Accessed 29 Dec. 2016.

"History of Computers." *University of Rhode Island,* homepage.cs.uri.edu/faculty/wolfe/book/Readings/Reading03.htm. Accessed 29 Dec. 2016.

Orenstein, David. "Application Programming Interface." *Computerworld,* 10 Jan. 2000, www.computerworld.com/article/2593623/app-development/application-programming-interface.html. Accessed 29 Dec. 2016.

Patterson, Michael. "What Is an API, and Why Does It Matter?" *SproutSocial,* 3 Apr. 2015, sproutsocial.com/insights/what-is-an-api. Accessed 29 Dec. 2016.

Roos, Dave. "How to Leverage an API for Conferencing." *HowStuffWorks,* money.howstuffworks.com/business-communications/how-to-leverage-an-api-for-conferencing1.htm. Accessed 29 Dec. 2016.

Wallberg, Ben. "A Brief Introduction to APIs." *University of Maryland Libraries,* 24 Apr. 2014, dssumd.wordpress.com/2014/04/24/a-brief-introduction-to-apis/. Accessed 29 Dec. 2016.

"What Is an Application?" *Goodwill Community Foundation,* www.gcflearnfree.org/computerbasics/understanding-applications/1/. Accessed 29 Dec. 2016.

ARTIFICIAL INTELLIGENCE

SUMMARY

Artificial intelligence is a broad field of study, and definitions of the field vary by discipline. For computer scientists, artificial intelligence refers to the development of programs that exhibit intelligent behavior. The programs can engage in intelligent planning (timing traffic lights), translate natural languages (converting a Chinese website into English), act like an expert (selecting the best wine for dinner), or perform many other tasks. For engineers, artificial intelligence refers to building machines that perform actions often done by humans. The machines can be simple, like a computer vision system embedded in an ATM (automated teller machine); more complex, like a robotic rover sent to Mars; or very complex, like an automated factory that builds an exercise machine with little human intervention. For cognitive scientists, artificial intelligence refers to building models of human intelligence to better understand human behavior. In the early days of artificial intelligence, most models of human intelligence were symbolic and closely related to cognitive psychology and philosophy, the basic idea being that regions of the brain perform complex reasoning by processing symbols. Later, many models of human cognition were developed to mirror the operation of the brain as an electrochemical computer, starting with the simple Perceptron, an artificial neural network described by Marvin Minsky in 1969, graduating to the backpropagation algorithm described by David E. Rumelhart and James L. McClelland in 1986, and culminating in a large number of supervised and nonsupervised learning algorithms.

When defining artificial intelligence, it is important to remember that the programs, machines, and models developed by computer scientists, engineers, and cognitive scientists do not actually have human intelligence; they only exhibit intelligent behavior. This can be difficult to remember because artificially intelligent systems often contain large numbers of facts, such as weather information for New York City; complex reasoning patterns, such as the reasoning needed to prove a geometric theorem from axioms; complex knowledge, such as an understanding of all the rules required to build an automobile; and the ability to learn, such as a neural network learning to recognize cancer cells. Scientists continue to look for better models of the brain and human intelligence.

BACKGROUND AND HISTORY

Although the concept of artificial intelligence probably has existed since antiquity, the term was first used by American scientist John McCarthy at a conference held at Dartmouth College in 1956. In 1955–56, the first artificial intelligence program,

Logic Theorist, had been written in IPL, a programming language, and in 1958, McCarthy invented Lisp, a programming language that improved on IPL. *Syntactic Structures* (1957), a book about the structure of natural language by American linguist Noam Chomsky, made natural language processing into an area of study within artificial intelligence. In the next few years, numerous researchers began to study artificial intelligence, laying the foundation for many later applications, such as general problem solvers, intelligent machines, and expert systems.

In the 1960s, Edward Feigenbaum and other scientists at Stanford University built two early expert systems: DENDRAL, which classified chemicals, and MYCIN, which identified diseases. These early expert systems were cumbersome to modify because they had hard-coded rules. By 1970, the OPS expert system shell, with variable rule sets, had been released by Digital Equipment Corporation as the first commercial expert system shell. In addition to expert systems, neural networks became an important area of artificial intelligence in the 1970s and 1980s. Frank Rosenblatt introduced the Perceptron in 1957, but it was *Perceptrons: An Introduction to Computational Geometry* (1969), by Minsky and Seymour Papert, and the two-volume *Parallel Distributed Processing: Explorations in the Microstructure of Cognition* (1986), by Rumelhart, McClelland, and the PDP Research Group, that really defined the field of neural networks. Development of artificial intelligence has continued, with game theory, speech recognition, robotics, and autonomous agents being some of the best-known examples.

HOW IT WORKS

The first activity of artificial intelligence is to understand how multiple facts interconnect to form knowledge and to represent that knowledge in a machine-understandable form. The next task is to understand and document a reasoning process for arriving at a conclusion. The final component of artificial intelligence is to add, whenever possible, a learning process that enhances the knowledge of a system.

KNOWLEDGE REPRESENTATION. Facts are simple pieces of information that can be seen as either true or false, although in fuzzy logic, there are levels of

Kismet, a robot with rudimentary social skills.

truth. When facts are organized, they become information, and when information is well understood, over time, it becomes knowledge. To use knowledge in artificial intelligence, especially when writing programs, it has to be represented in some concrete fashion. Initially, most of those developing artificial intelligence programs saw knowledge as represented symbolically, and their early knowledge representations were symbolic. Semantic nets, directed graphs of facts with added semantic content, were highly successful representations used in many of the early artificial intelligence programs. Later, the nodes of the semantic nets were expanded to contain more information, and the resulting knowledge representation was referred to as frames. Frame representation of knowledge was very similar to object-oriented data representation, including a theory of inheritance.

Another popular way to represent knowledge in artificial intelligence is as logical expressions. English mathematician George Boole represented knowledge as a Boolean expression in the 1800s. English mathematicians Bertrand Russell and Alfred Whitehead expanded this to quantified expressions in 1910, and French computer scientist Alain Colmerauer incorporated it into logic programming, with the programming language Prolog, in the 1970s. The knowledge of a rule-based expert system is embedded in the if-then rules of the system, and because each if-then rule has a Boolean representation, it can be seen as a form of relational knowledge representation.

Neural networks model the human neural system and use this model to represent knowledge. The brain is an electrochemical system that stores its knowledge in synapses. As electrochemical signals pass through a synapse, they modify it, resulting in the acquisition of knowledge. In the neural network model, synapses are represented by the weights of a weight matrix, and knowledge is added to the system by modifying the weights.

Reasoning. Reasoning is the process of determining new information from known information. Artificial intelligence systems add reasoning soon after they have developed a method of knowledge representation. If knowledge is represented in semantic nets, then most reasoning involves some type of tree search. One popular reasoning technique is to traverse a decision tree, in which the reasoning is represented by a path taken through the tree. Tree searches of general semantic nets can be very time-consuming and have led to many advancements in tree-search algorithms, such as placing bounds on the depth of search and backtracking.

Reasoning in logic programming usually follows an inference technique embodied in first-order predicate calculus. Some inference engines, such as that of Prolog, use a back-chaining technique to reason from a result, such as a geometry theorem, to its antecedents, the axioms, and also show how the reasoning process led to the conclusion. Other inference engines, such as that of the expert system shell CLIPS, use a forward-chaining inference engine to see what facts can be derived from a set of known facts.

Neural networks, such as backpropagation, have an especially simple reasoning algorithm. The knowledge of the neural network is represented as a matrix of synaptic connections, possibly quite sparse. The information to be evaluated by the neural network is represented as an input vector of the appropriate size, and the reasoning process is to multiply the connection matrix by the input vector to obtain the conclusion as an output vector.

Learning. Learning in an artificial intelligence system involves modifying or adding to its knowledge. For both semantic net and logic programming systems, learning is accomplished by adding or modifying the semantic nets or logic rules, respectively. Although much effort has gone into developing learning algorithms for these systems, all of them, to date, have used ad hoc methods and experienced limited success. Neural networks, on the other hand, have been very successful at developing learning algorithms. Backpropagation has a robust supervised learning algorithm in which the system learns from a set of training pairs, using gradient-descent optimization, and numerous unsupervised learning algorithms learn by studying the clustering of the input vectors.

Expert Systems. One of the most successful areas of artificial intelligence is expert systems. Literally thousands of expert systems are being used to help both experts and novices make decisions. For example, in the 1990s, Dell developed a simple expert system that allowed shoppers to configure a computer as they wished. In the 2010s, a visit to the Dell website offers a customer much more than a simple configuration program. Based on the customer's answers to some rather general questions, dozens of small expert systems suggest what computer to buy. The Dell site is not unique in its use of expert systems to guide customer's choices. Insurance companies, automobile companies, and many others use expert systems to assist customers in making decisions.

There are several categories of expert systems, but by far the most popular are the rule-based expert systems. Most rule-based expert systems are created with an expert system shell. The first successful rule-based expert system shell was the OPS 5 of Digital Equipment Corporation (DEC), and the most popular modern systems are CLIPS, developed by the National Aeronautics and Space Administration (NASA) in 1985, and its Java clone, Jess, developed at Sandia National Laboratories in 1995. All rule-based expert systems have a similar architecture, and the shells make it fairly easy to create an expert system as soon as a knowledge engineer gathers the knowledge from a domain expert. The most important component of a rule-based expert system is its knowledge base of rules. Each rule consists of an if-then statement with multiple antecedents, multiple consequences, and possibly a rule certainty factor. The antecedents of a rule are statements that can be true or false and that depend on facts that are either introduced into the system by a user or derived as the result of a rule being fired. For example, a fact could be red-wine and a simple rule could be if (red-wine) then (it-tastes-good). The expert system also has an inference engine that can apply multiple rules in an orderly fashion so that the expert system can draw conclusions by applying its rules to a set of facts introduced by a user. Although it is not absolutely required, most rule-based expert systems have a user-friendly interface and an explanation facility to justify its reasoning.

THEOREM PROVERS. Most theorems in mathematics can be expressed in first-order predicate calculus. For any particular area, such as synthetic geometry or group theory, all provable theorems can be derived from a set of axioms. Mathematicians have written programs to automatically prove theorems since the 1950s. These theorem provers either start with the axioms and apply an inference technique, or start with the theorem and work backward to see how it can be derived from axioms. Resolution, developed in Prolog, is a well-known automated technique that can be used to prove theorems, but there are many others. For Resolution, the user starts with the theorem, converts it to a normal form, and then mechanically builds reverse decision trees to prove the theorem. If a reverse decision tree whose leaf nodes are all axioms is found, then a proof of the theorem has been discovered.

Gödel's incompleteness theorem (proved by Austrian-born American mathematician Kurt Gödel) shows that it may not be possible to automatically prove an arbitrary theorem in systems as complex as the natural numbers. For simpler systems, such as group theory, automated theorem proving works if the user's computer can generate all reverse trees or a suitable subset of trees that can yield a proof in a reasonable amount of time. Efforts have been made to develop theorem provers for higher order logics than first-order predicate calculus, but these have not been very successful.

Computer scientists have spent considerable time trying to develop an automated technique for proving the correctness of programs, that is showing that any valid input to a program produces a valid output. This is generally done by producing a consistent model and mapping the program to the model. The first example of this was given by English mathematician Alan Turing in 1931, by using a simple model now called a Turing machine. A formal system that is rich enough to serve as a model for a typical programming language, such as C++, must support higher order logic to capture the arguments and parameters of subprograms. Lambda calculus, denotational semantics, von Neuman geometries, finite state machines, and other systems have been proposed to provide a model onto which all programs of a language can be mapped. Some of these do capture many programs, but devising a practical automated method of verifying the correctness of programs has proven difficult.

INTELLIGENT TUTOR SYSTEMS. Almost every field of study has many intelligent tutor systems available to assist students in learning. Sometimes the tutor system is integrated into a package. For example, in Microsoft Office, an embedded intelligent helper provides popup help boxes to a user when it detects the need for assistance and full-length tutorials if it detects more help is needed. In addition to the intelligent tutors embedded in programs as part of a context-sensitive help system, there are a vast number of stand-alone tutoring systems in use.

The first stand-alone intelligent tutor was SCHOLAR, developed by J. R. Carbonell in 1970. It used semantic nets to represent knowledge about South American geography, provided a user interface to support asking questions, and was successful enough to demonstrate that it was possible for a computer program to tutor students. At about the same time, the University of Illinois developed its PLATO computer-aided instruction system, which provided a general language for developing intelligent tutors with touch-sensitive screens, one of the most famous of which was a biology tutorial on evolution. Of the thousands of modern intelligent tutors, SHERLOCK, a training environment for electronic troubleshooting, and PUMP, a system designed to help learn algebra, are typical.

ELECTRONIC GAMES. Electronic games have been played since the invention of the cathode-ray tube for television. In the 1980s, games such as Solitaire, Pac-Man, and Pong for personal computers became almost as popular as the stand-alone game platforms. In the 2010s, multiuser Internet games are enjoyed by young and old alike, and game playing on mobile devices has become an important application. In all of these electronic games, the user competes with one or more intelligent agents embedded in the game, and the creation of these intelligent agents uses considerable artificial intelligence. When creating an intelligent agent that will compete with a user or, as in Solitaire, just react to the user, a programmer has to embed the game knowledge into the program. For example, in chess, the programmer would need to capture all possible configurations of a chess board. The programmer also would need to add reasoning procedures to the game; for example, there would have to be procedures to move each individual chess piece on the board. Finally, and most important for game programming, the programmer would need to add one or more strategic decision modules to the program to provide the intelligent agent with a strategy for winning. In many cases, the strategy for winning a game would be driven by probability; for example, the next move might be a pawn, one space forward, because that yields the best probability of winning, but a heuristic strategy is also possible; for example, the next move is a rook because it may trick the opponent into a bad series of moves.

SOCIAL CONTEXT, ETHICS, AND FUTURE PROSPECTS

After artificial intelligence was defined by McCarthy in 1956, it has had a number of ups and downs as a discipline, but the future of artificial intelligence looks good. Almost every commercial program has a help system, and increasingly these help systems have a major artificial intelligence component. Health care is another area that is poised to make major use of artificial intelligence to improve the quality and reliability of the care provided, as well as to reduce its cost by providing expert advice on best practices in health care. Smartphones and other digital devices employ artificial intelligence for an array of applications, syncing the activities and requirements of their users.

Ethical questions have been raised about trying to build a machine that exhibits human intelligence. Many of the early researchers in artificial intelligence were interested in cognitive psychology and built symbolic models of intelligence that were considered unethical by some. Later, many artificial intelligence researchers developed neural models of intelligence that were not always deemed ethical. The social and ethical issues of artificial intelligence are nicely represented by HAL, the Heuristically programmed ALgorithmic computer, in Stanley Kubrick's 1968 film *2001: A Space Odyssey*, which first works well with humans, then acts violently toward them, and is in the end deactivated.

Another important ethical question posed by artificial intelligence is the appropriateness of developing programs to collect information about users of a program. Intelligent agents are often embedded in websites to collect information about those using the site, generally without the permission of those using the website, and many question whether this should be done.

In the mid-to-late 2010s, fully autonomous self-driving cars were developed and tested in the United States. In 2018, an Uber self-driving car hit and killed a pedestrian in Tempe, Arizona. There was a safety driver at the wheel of the car, which was in self-driving mode at the time of the accident. The accident led Uber to suspend its driverless-car testing program. Even before the accident occurred, ethicists raised questions regarding collision avoidance programming, moral and legal responsibility, among others.

As more complex AI is created and imbued with general, humanlike intelligence (instead of concentrated intelligence in a single area, such as Deep Blue and chess), it will run into moral requirements as humans do. According to researchers Nick Bostrom and Eliezer Yudkowsky, if an AI is given "cognitive work" to do that has a social aspect, the AI inherits the social requirements of these interactions. The AI then needs to be imbued with a sense of morality to interact in these situations. If an AI has humanlike intelligence and agency, the Bostrom has also theorized that AI will need to also be considered both persons and moral entities. There is also the potential for the development of superhuman intelligence in AI, which would breed superhuman morality. The questions of intelligence and morality and who is given personhood are some of the most significant issues to be considered contextually as AI advance.

—George M. Whitson III, BS, MS, PhD

Bibliography

Basl, John. "The Ethics of Creating Artificial Consciousness." *American Philosophical Association Newsletters: Philosophy and Computers* 13.1 (2013): 25–30. *Philosophers Index with Full Text.* Web. 25 Feb. 2015.

Berlatsky, Noah. *Artificial Intelligence.* Detroit: Greenhaven, 2011. Print.

Bostrom, Nick. "Ethical Issues in Advanced Artificial Intelligence." *NickBostrom.com.* Nick Bostrom, 2003. Web. 23 Sept. 2016.

Bostom, Nick, and Eliezer Yudkowsky. "The Ethics of Artificial Intelligence." *Machine Intelligence Research Institute.* MIRI, n.d. Web. 23 Sept. 2016.

Giarratano, Joseph, and Peter Riley. *Expert Systems: Principles and Programming.* 4th ed. Boston: Thomson, 2005. Print.

Lee, Timothy B. "Why It's Time for Uber to Get Out of the Self-Driving Car Business." *Ars Technica,* Condé Nast, 27 Mar. 2018, arstechnica.com/cars/2018/03/ubers-self-driving-car-project-is-struggling-the-company-should-sell-it/. Accessed 27 Mar. 2018.

Minsky, Marvin, and Seymour Papert. *Perceptrons: An Introduction to Computational Geometry.* Rev. ed. Boston: MIT P, 1990. Print.

Nyholm, Sven, and Jilles Smids. "The Ethics of Accident-Algorithms for Self-Driving Cars: An Applied Trolley Problem?" *Ethical Theory & Moral Practice,* vol. 19, no. 5, pp. 1275–1289. doi:10.1007/s10677-016-9745-2. *Academic Search Complete,* search.ebscohost.com/login.aspx?direct=true&db=a9h&AN=119139911&site=ehost-live. Accessed 27 Mar. 2018.

Rumelhart, David E., James L. McClelland, and the PDP Research Group. *Parallel Distributed Processing: Explorations in the Microstructure of Cognition.* 1986. Rpt. 2 vols. Boston: MIT P, 1989. Print.

Russell, Stuart, and Peter Norvig. *Artificial Intelligence: A Modern Approach.* 3rd ed. Upper Saddle River: Prentice, 2010. Print.

Shapiro, Stewart, ed. *Encyclopedia of Artificial Intelligence.* 2nd ed. New York: Wiley, 1992. Print.

Augmented Reality

SUMMARY

Augmented reality (AR) refers to any technology that inserts digital interfaces into the real world. For the most part, the technology has included headsets and glasses that people wear to project interfaces onto the physical world, but it can also include cell phones and other devices. In time, AR technology could be used in contact lenses and other small wearable devices.

BASIC PRINCIPLES

Augmented reality is related to, but separate from, virtual reality. Virtual reality attempts to create an entirely different reality that is separate from real life. Augmented reality, however, adds to the real world and does not create a unique world. Users of AR will recognize their surroundings and use the AR technology to enhance what they are experiencing. Both augmented and virtual realities have become better as technology has improved. A number of companies (including large tech companies such as Google) have made investments in augmented reality in the hopes that it will be a major part of the future of technology and will change the way people interact with technology.

In the past, AR was seen primarily as a technology to enhance entertainment (e.g., gaming, communicating, etc.); however, AR has the potential to revolutionize many aspects of life. For example, AR technology could provide medical students with a model of a human heart. It could also help people locate their cars in parking lots. AR technology has already been used in cell phones to help people locate nearby facilities (e.g., banks and restaurants), and future AR technology could inform people about nearby locations, events, and the people they meet and interact with.

HISTORY

The term "augmented reality" was developed in the 1990s, but the fundamental idea for augmented reality was established in the early days computing.

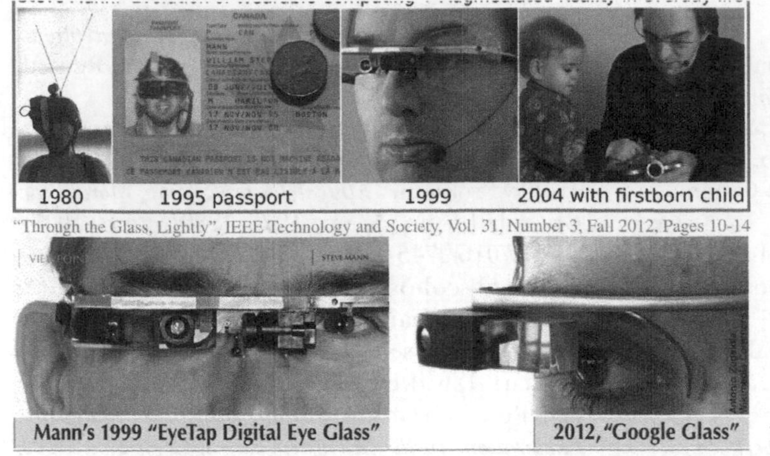

Mann was recognized as "Father of AR" and the "Father of Wearable Computing" (IEEE ISSCC 2000)

Series of self-portraits depicting evolution of wearable computing and AR over 30 year time period, along with Generation-4 Glass (Mann 1999) and Google's Generation-1 Glass. By Glogger (Own work)

Technology for AR developed in the early twenty-first century, but at that time AR was used mostly for gaming technology.

In the early 2010s, technology made it possible for AR headsets to shrink and for graphics used in AR to improve. Google Glass (2012) was one of the first AR devices geared toward the public that was not meant for gaming. Google Glass, created by the large tech company Google, was designed to give users a digital interface they could interact with in ways that were somewhat similar to the way people interacted with smart phones (e.g., taking pictures, looking up directions, etc.). Although Google Glass was not a success, Google and other companies developing similar products believed that eventually wearable technology would become a normal part of everyday life.

During this time, other companies were also interested in revolutionizing AR technology. Patent information released from the AR company Magic Leap (which also received funding from Google) indicated some of the technology the company was working on. One technology reportedly will beam images directly into a wearer's retinas. This design is meant to fool the brain so it cannot tell the difference between light from the outside world and the light coming from an AR device. If this technology works as intended, it could change the way people see the world.

Microsoft's AR company, HoloLens, had plans for technology that was similar to Magic Leap's, though the final products would likely have many differences. HoloLens was working to include "spatial sound" so that the visual images would be accompanied by sounds that seem to be closer or farther away, corresponding with the visuals. For example, a person could see an animal running toward them on the HoloLens glasses, and they would hear corresponding sounds that got louder as the animal got closer.

Other AR companies, such as Leap Motion, have designed AR products to be used in conjunction with technology people already rely on. This company developed AR technology that worked with computers to change the type of display people used. Leap Motion's design allowed people to wear a headset to see the computer display in front of them. They could then use their hands to move the parts of the display seemingly through the air in front of them (though people not wearing the headset would not see the images from the display). Other companies also worked on making AR technology more accessible through mobile phones and other devices that people use frequently.

THE FUTURE OF AR

Although companies such as Magic Leap and HoloLens have plans for the future of AR, the field still faces many obstacles. Developing wearable technology that is small enough, light enough, and powerful enough to provide users with the feeling of reality is one of the biggest obstacles. AR companies are developing new technologies to make AR performance better, but many experts agree that successfully releasing this technology to the public could take years.

Another hurdle for AR technology companies is affordability. The technology they sell has to be priced so that people will purchase it. Since companies are investing so much money in the development of high-tech AR technology, they might not be able to offer affordable AR devices for a number of years. Another problem that AR developers have to manage is the speed and agility of the visual display. Since any slowing of the image or delay in the process could ruin the experience for the AR user, companies have to make sure the technology is incredibly fast and reliable.

AR EdiBear. By Okseduard (Own work)

In the future, AR devices could be shrunk to even small sizes and people could experience AR technology through contact lenses or even bionic eyes. Yet, AR technology still has many challenges to overcome before advanced AR devices become popular and mainstream. Technology experts agree that AR technology will likely play an important role in everyday life in the future.

—Elizabeth Mohn

Bibliography

Altavilla, Dave. "Apple Further Legitimizes Augmented Reality Tech With Acquisition of Metaio." *Forbes.* Forbes.com, LLC. 30 May 2015. Web. 13 Aug. 2015. http://www.forbes.com/sites/davealtavilla/2015/05/30/apple-further-legitimizes-augmented-reality-tech-with-acquistion-of-metaio/

"Augmented Reality." *Webopedia.* Quinstreet Enterprise. Web. 13 Aug. 2051. http://www.webopedia.com/TERM/A/Augmented_Reality.html

Farber, Dan. "The Next Big Thing in Tech: Augmented Reality." *CNET.* CBS Interactive Inc. 7 June 2013. Web. 13 Aug. 2015. http://www.cnet.com/news/the-next-big-thing-in-tech-augmented-reality/

Folger, Tim. "Revealed World." *National Geographic.* National Geographic Society. Web. 13 Aug. 2015. http://ngm.nationalgeographic.com/big-idea/14/augmented-reality

Kofman, Ava. "Dueling Realities." *The Atlantic.* The Atlantic Monthly Group. 9 June 2015. Web. 13 Aug. 2015. http://www.theatlantic.com/technology/archive/2015/06/dueling-realities/395126/

McKalin, Vamien. "Augmented Reality vs. Virtual Reality: What are the differences and similarities?" 6 April 2014. Web. 13 Aug. 2015. *TechTimes.com.* TechTimes.com. http://www.techtimes.com/articles/5078/20140406/augmented-reality-vs-virtual-reality-what-are-the-differences-and-similarities.htm

Vanhemert, Kyle. "Leap Motion's Augmented-Reality Computing Looks Stupid Cool." *Wired.* Condc Nast. 7 July 2015. Web. 13 Aug. 2015. http://www.wired.com/2015/07/leap-motion-glimpse-at-the-augmented-reality-desktop-of-the-future/

Automated Processes and Servomechanisms

SUMMARY

An automated process is a series of sequential steps to be carried out automatically. Servomechanisms are systems, devices, and subassemblies that control the mechanical actions of robots by the use of feedback information from the overall system in operation.

DEFINITION AND BASIC PRINCIPLES

An automated process is any set of tasks that has been combined to be carried out in a sequential order automatically and on command. The tasks are not necessarily physical in nature, although this is the most common circumstance. The execution of the instructions in a computer program represents an automated process, as does the repeated execution of a series of specific welds in a robotic weld cell. The two are often inextricably linked, as the control of the physical process has been given to such digital devices as programmable logic controllers (PLCs) and computers in modern facilities.

Physical regulation and monitoring of mechanical devices such as industrial robots is normally achieved through the incorporation of servomechanisms. A servomechanism is a device that accepts information from the system itself and then uses that information to adjust the system to maintain specific operating conditions. A servomechanism that controls the opening and closing of a valve in a process stream, for example, may use the pressure of the process stream to regulate the degree to which the valve is opened.

The stepper motor is another example of a servomechanism. Given a specific voltage input, the stepper motor turns to an angular position that exactly corresponds to that voltage. Stepper motors are essential components of disk drives in computers, moving the read and write heads to precise data locations on the disk surface.

Another essential component in the functioning of automated processes and servomechanisms is the feedback control systems that provide self-regulation and auto-adjustment of the overall system. Feedback control systems may be pneumatic, hydraulic, mechanical, or electrical in nature. Electrical feedback may be analog in form, although digital electronic feedback methods provide the most versatile method of output sensing for input feedback to digital electronic control systems.

BACKGROUND AND HISTORY

Automation begins with the first artificial construct made to carry out a repetitive task in the place of a person. One early clock mechanism, the water clock, used the automatic and repetitive dropping of a specific amount of water to accurately measure the passage of time. Water-, animal-, and wind-driven mills and threshing floors automated the repetitive action of processes that had been accomplished by humans. In many underdeveloped areas of the world, this repetitive human work is still a common practice.

With the mechanization that accompanied the Industrial Revolution, other means of automatically controlling machinery were developed, including self-regulating pressure valves on steam engines. Modern automation processes began in North America with the establishment of the assembly line as a standard industrial method by Henry Ford. In this method, each worker in his or her position along the assembly line performs a limited set of functions, using only the parts and tools appropriate to that task.

An industrial servomotor. The grey/green cylinder is the brush-type DC motor, the black section at the bottom contains the planetary reduction gear, and the black object on top of the motor is the optical rotary encoder for position feedback. By John Nagle (Own work)

Servomechanism theory was further developed during World War II. The development of the transistor in 1951 enabled the development of electronic control and feedback devices, and hence digital electronics. The field grew rapidly, especially following the development of the microcomputer in 1969. Digital logic and machine control can now be interfaced in an effective manner, such that today's automated systems function with an unprecedented degree of precision and dependability.

HOW IT WORKS

An automated process is a series of repeated, identical operations under the control of a master operation or program. While simple in concept, it is complex in practice and difficult in implementation and execution. The process control operation must be designed in a logical, step-by-step manner that will provide the desired outcome each time the process is cycled. The sequential order of operations must be set so that the outcome of any one step does not prevent or interfere with the successful outcome of any other step in the process. In addition, the physical parameters of the desired outcome must be established and made subject to a monitoring protocol that can then act to correct any variation in the outcome of the process.

A plain analogy is found in the writing and structuring of a simple computer programming function.

The definition of the steps involved in the function must be exact and logical, because the computer, like any other machine, can do only exactly what it is instructed to do. Once the order of instructions and the statement of variables and parameters have been finalized, they will be carried out in exactly the same manner each time the function is called in a program. The function is thus an automated process.

The same holds true for any physical process that has been automated. In a typical weld cell, for example, a set of individual parts are placed in a fixture that holds them in their proper relative orientations. Robotic welding machines may then act upon the setup to carry out a series of programmed welds to join the individual pieces into a single assembly. The series of welds is carried out in exactly the same manner each time the weld cell cycles. The robots that carry out the welds are guided under the control of a master program that defines the position of the welding tips, the motion that it must follow, and the duration of current flow in the welding process for each movement, along with many other variables that describe the overall action that will be followed. Any variation from this programmed pattern of movements and functions will result in an incorrect output.

The control of automated processes is carried out through various intermediate servomechanisms. A servomechanism uses input information from both the controlling program and the output of the process to carry out its function. Direct instruction from the controller defines the basic operation of the servomechanism. The output of the process generally includes monitoring functions that are compared to the desired output. They then provide an input signal to the servomechanism that informs how the operation must be adjusted to maintain the desired output. In the example of a robotic welder, the movement of the welding tip is performed through the action of an angular positioning device. The device may turn through a specific angle according to the voltage that is supplied to the mechanism. An input signal may be provided from a proximity sensor such that when the necessary part is not detected, the welding operation is interrupted and the movement of the mechanism ceases.

The variety of processes that may be automated is practically limitless given the interface of digital electronic control units. Similarly, servomechanisms may be designed to fit any needed parameter or to carry out any desired function.

APPLICATIONS AND PRODUCTS

The applications of process automation and servomechanisms are as varied as modern industry and its products. It is perhaps more productive to think of process automation as a method that can be applied to the performance of repetitive tasks than to dwell on specific applications and products. The commonality of the automation process can be illustrated by examining a number of individual applications, and the products that support them.

"Repetitive tasks" are those tasks that are to be carried out in the same way, in the same circumstances, and for the same purpose a great number of times. The ideal goal of automating such a process is to ensure that the results are consistent each time the process cycle is carried out. In the case of the robotic weld cell described above, the central tasks to be repeated are the formation of welded joints of specified dimensions at the same specific locations over many hundreds or thousands of times. This is a typical operation in the manufacturing of subassemblies in the automobile industry and in other industries in which large numbers of identical fabricated units are produced.

Automation of the process, as described above, requires the identification of a set series of actions to be carried out by industrial robots. In turn, this requires the appropriate industrial robots be designed and constructed in such a way that the actual physical movements necessary for the task can be carried out. Each robot will incorporate a number of servomechanisms that drive the specific movements of parts of the robot according to the control instruction set. They will also incorporate any number of sensors and transducers that will provide input signal information for the self-regulation of the automated process. This input data may be delivered to the control program and compared to specified standards before it is fed back into the process, or it may be delivered directly into the process for immediate use.

Programmable logic controllers (PLCs), first specified by the General Motors Corporation in 1968, have become the standard devices for controlling automated machinery. The PLC is essentially a dedicated computer system that employs a limited-instruction-set programming language. The program of instructions for the automated process is stored in the PLC memory. Execution of the program sends the specified operating parameters to the corresponding machine in such a way that it carries out a set of operations that must otherwise be carried out under the control of a human operator.

A typical use of such methodology is in the various forms of CNC machining. CNC (computer numeric control) refers to the use of reduced-instruction-set computers to control the mechanical operation of machines. CNC lathes and mills are two common applications of the technology. In the traditional use of a lathe, a human operator adjusts all of the working parameters such as spindle rotation speed, feed rate, and depth of cut, through an order of operations that is designed to produce a finished piece to blueprint dimensions. The consistency of pieces produced over time in this manner tends to vary as operator fatigue and distractions affect human performance. In a CNC lathe, however, the order of operations and all of the operating parameters are specified in the control program, and are thus carried out in exactly the same manner for each piece that is produced. Operator error and fatigue do not affect production, and the machinery produces the desired pieces at the same rate throughout the entire working period. Human intervention is required only to maintain the machinery and is not involved in the actual machining process.

Servomechanisms used in automated systems check and monitor system parameters and adjust operating conditions to maintain the desired system output. The principles upon which they operate can range from crude mechanical levers to sophisticated and highly accurate digital electronic-measurement devices. All employ the principle of feedback to control or regulate the corresponding process that is in operation.

In a simple example of a rudimentary application, units of a specific component moving along a production line may in turn move a lever as they pass by. The movement of the lever activates a switch that prevents a warning light from turning on. If the switch is not triggered, the warning light tells an operator that the component has been missed. The lever, switch, and warning light system constitute a crude servomechanism that carries out a specific function in maintaining the proper operation of the system.

In more advanced applications, the dimensions of the product from a machining operation may be tested by accurately calibrated measuring devices before releasing the object from the lathe, mill, or other device. The measurements taken are then compared to the desired measurements, as stored in the PLC memory. Oversize measurements may trigger an action of the machinery to refine the dimensions of the piece to bring it into specified tolerances, while undersize measurements may trigger the rejection of the piece and a warning to maintenance personnel to adjust the working parameters of the device before continued production.

Two of the most important applications of servomechanisms in industrial operations are control of position and control of rotational speed. Both commonly employ digital measurement. Positional control is generally achieved through the use of servomotors, also known as stepper motors. In these devices, the rotor turns to a specific angular position according to the voltage that is supplied to the motor. Modern electronics, using digital devices constructed with integrated circuits, allows extremely fine and precise control of electrical and electronic factors, such as voltage, amperage, and resistance. This, in turn, facilitates extremely precise positional control. Sequential positional control of different servomotors in a machine, such as an industrial robot, permits precise positioning of operating features. In other robotic applications, the same operating principle allows for extremely delicate microsurgery that would not be possible otherwise.

The control of rotational speed is achieved through the same basic principle as the stroboscope. A strobe light flashing on and off at a fixed rate can be used to measure the rate of rotation of an object. When the strobe rate and the rate of rotation are equal, a specific point on the rotating object will always appear at the same location. If the speeds are not matched, that point will appear to move in one direction or the other according to which rate is the faster rate. By attaching a rotating component to a representation of a digital scale, such as the Gray code, sensors can detect both the rate of rotation of the component and its position when it is functioning as part of a servomechanism. Comparison with a digital statement of the desired parameter can then be used by the controlling device to adjust the speed or position, or both, of the component accordingly.

SOCIAL CONTEXT AND FUTURE PROSPECTS

While the vision of a utopian society in which all menial labor is automated, leaving humans free to create new ideas in relative leisure, is still far from reality, the vision becomes more real each time another process is automated. Paradoxically, since the mid-twentieth century, knowledge and technology have changed so rapidly that what is new becomes obsolete almost as quickly as it is developed, seeming to increase rather than decrease the need for human labor.

New products and methods are continually being developed because of automated control. Similarly, existing automated processes can be reautomated using newer technology, newer materials, and modernized capabilities.

Particular areas of growth in automated processes and servomechanisms are found in the biomedical fields. Automated processes greatly increase the number of tests and analyses that can be performed for genetic research and new drug development. Robotic devices become more essential to the success of delicate surgical procedures each day, partly because of the ability of integrated circuits to amplify or reduce electrical signals by factors of hundreds of thousands. Someday, surgeons will be able to perform the most delicate of operations remotely, as normal actions by the surgeon are translated into the miniscule movements of microscopic surgical equipment manipulated through robotics.

Concerns that automated processes will eliminate the role of human workers are unfounded. The nature of work has repeatedly changed to reflect the capabilities of the technology of the time. The introduction of electric street lights, for example, did eliminate the job of lighting gas-fueled streetlamps, but it also created the need for workers to produce the electric lights and to ensure that they were functioning properly. The same sort of reasoning applies to the automation of processes today. Some traditional jobs will disappear, but new types of jobs will be created in their place through automation.

—Richard M. Renneboog, MSc

BIBLIOGRAPHY

Bryan, Luis A., and E. A. Bryan. *Programmable Controllers: Theory and Implementation.* 2nd ed. Atlanta: Industrial Text, 1997. Print.

James, Hubert M. *Theory of Servomechanisms.* New York: McGraw, 1947. Print.

Kirchmer, Mathias. *High Performance through Process Excellence: From Strategy to Execution with Business Process Management.* 2nd ed. Heidelberg: Springer, 2011. Print.

Seal, Anthony M. *Practical Process Control.* Oxford: Butterworth, 1998. Print.

Seames, Warren S. *Computer Numerical Control Concepts and Programming.* 4th ed. Albany: Delmar, 2002. Print.

Smith, Carlos A. *Automated Continuous Process Control.* New York: Wiley, 2002. Print.

AUTONOMOUS CAR

SUMMARY

An autonomous car, also known as a “robotic car” or “driverless car,” is a vehicle designed to operate without the guidance or control of a human driver. Engineers began designing prototypes and control systems for autonomous vehicles as early as the 1920s, but the development of the modern autonomous vehicle began in the late 1980s.

Between 2011 and 2014, fourteen U.S. states proposed or debated legislation regarding the legality of testing autonomous vehicles on public roads. As of November 2014, the only autonomous vehicles used in the United States were prototype and experimental vehicles. Some industry analyses, published since 2010, indicate that autonomous vehicles could become available for public use as early as 2020. Proponents of autonomous car technology believe that driverless vehicles will reduce the incidence of traffic accidents, reduce fuel consumption, alleviate parking issues, and reduce car theft, among other benefits. One of the most significant potential benefits of “fully autonomous” vehicles is to provide independent transportation to disabled individuals who are not able to operate a traditional motor vehicle. Potential complications or problems with autonomous vehicles include the difficulty in assessing liability in the case of accidents and a reduction in the number of driving-related occupations available to workers.

BACKGROUND

Autonomous car technology has its origins in the 1920s, when a few automobile manufacturers,

Google driverless car operating on a testing path. By Flckr user jurvetson (Steve Jurvetson). Trimmed and retouched with PS9 by Mariordo

inspired by science fiction, envisioned futuristic road systems embedded with guidance systems that could be used to power and navigate vehicles through the streets. For instance, the Futurama exhibit at the 1939 New York World's Fair, planned by designer Norman Bel Geddes, envisioned a future where driverless cars would be guided along electrically charged roads.

Until the 1980s, proposals for autonomous vehicles involved modifying roads with the addition of radio, magnetic, or electrical control systems. During the 1980s, automobile manufacturers working with university engineering and computer science programs began designing autonomous vehicles that were self-navigating, rather than relying on modification of road infrastructure. Bundeswehr University in Munich, Germany produced an autonomous vehicle that navigated using cameras and computer vision. Similar designs were developed through collaboration between the U.S. Defense Advanced Research Projects Agency (DARPA) and researchers from Carnegie Mellon University. Early prototypes developed by DARPA used LIDAR, a system that uses lasers to calculate distance and direction. In July 1995, the NavLab program at Carnegie Mellon University produced one of the first successful tests of an autonomous vehicle, known as "No Hands Across America."

The development of American autonomous vehicle technology accelerated quickly between 2004 and 2007 due to a series of research competitions, known as "Grand Challenges," sponsored by DARPA. The 2007 event, called the "Urban Challenge," drew eleven participating teams designing vehicles that could navigate through urban environments while avoiding obstacles and obeying traffic laws; six designs successfully navigated the course. Partnerships formed through the DARPA challenges resulted in the development of autonomous car technology for public use. Carnegie Mellon University and General Motors partnered to create the Autonomous Driving Collaborative Research Lab, while rival automaker Volkswagen partnered with Stanford University on a similar project.

Stanford University artificial intelligence expert Sebastien Thrun, a member of the winning team at the 2005 DARPA Grand Challenge, was a founder of technology company Google's "Self-Driving Car Project" in 2009, which is considered the beginning of the commercial phase of autonomous vehicle development. Thrun and researcher Anthony Levandowski helped to develop "Google Chauffeur," a specialized software program designed to navigate using laser, satellite, and computer vision systems. Other car manufacturers, including Audi, Toyota, Nissan, and Mercedes, also began developing autonomous cars for the consumer market in the early 2010s. In 2011, Nevada became the first state to legalize testing autonomous cars on public roads, followed by Florida, California, the District of Colombia, and Michigan by 2013.

TOPIC TODAY

In May of 2013, the U.S. Department of Transportation's National Highway Traffic Safety Administration (NHTSA) released an updated set of guidelines to help guide legal policy regarding autonomous vehicles. The NHTSA guidelines classify autonomous vehicles based on a five-level scale of automation, from zero, indicating complete driver control, to four, indicating complete automation with no driver control.

Between 2011 and 2016, several major manufacturers released partially automated vehicles for the consumer market, including Tesla, Mercedes-Benz, BMW, and Infiniti. The Mercedes S-Class, which featured automated systems options including parking assistance, lane correction, and a system to detect when the driver may be at risk of fatigue.

Interior of a Google driverless car. By jurvetson

According to a November 2014 article in the *New York Times*, most manufacturers are developing vehicles that will require "able drivers" to sit behind the wheel, even though the vehicle's automatic systems will operate and navigate the car. Google's "second generation" autonomous vehicles are an exception, as the vehicles lack steering wheels or other controls, therefore making human intervention impossible. According to Google, complete automation reduces the possibility that human intervention will lead to driving errors and accidents. Google argues further that fully autonomous vehicles could open the possibility of independent travel to the blind and individuals suffering from a variety of other disabilities that impair the ability to operate a car. In September 2016 Uber launched a test group of automated cars in Pittsburgh. They started with four cars that had two engineers in the front seats to correct errors. The company rushed to be the first to market and plans to add additional cars to the fleet and have them fully automated.

Modern autonomous vehicles utilize laser guidance systems, a modified form of LIDAR, as well as global positioning system (GPS) satellite tracking, visual computational technology, and software that allows for adaptive response to changing traffic conditions. Companies at the forefront of automated car technology are also experimenting with computer software designed to learn from experience, thereby making the vehicle's onboard computer more responsive to driving situations following encounters.

While Google has been optimistic about debuting autonomous cars for public use by 2020, other industry analysts are skeptical about this, given the significant regulatory difficulties that must be overcome before driverless cars can become a viable consumer product. A 2014 poll from Pew Research indicated that approximately 50 percent of Americans are not currently interested in driverless cars. Other surveys have also indicated that a slight majority of consumers are uninterested in owning self-driving vehicles, though a majority of consumers approve of the programs to develop the technology. A survey conducted two years later by New Morning Consult showed a similar wariness of self-driving cars, with 43 percent of registered voters considering autonomous cars unsafe.

Proponents of autonomous vehicles have cited driver safety as one of the chief benefits of automation. The RAND Corporation's 2014 report on autonomous car technology cites research indicating that computer-guided vehicles will reduce the incidence and severity of traffic accidents, congestion, and delays, because computer-guided systems will be more responsive than human drivers and are immune to driving distractions that contribute to a majority of traffic accidents. Research also indicates that autonomous cars will help to conserve fuel, reduce parking congestion, and will allow consumers to be more productive while commuting by freeing them from the job of operating the vehicle. The first fatal accident in an autonomous car happened in July 2016 when a Tesla in automatic mode crashed into a truck. The driver was killed. A second fatal accident involving a Tesla in autonomous mode occurred in early 2018. The first fatal accident involving an autonomous car and a pedestrian occurred in March 2018, when one of Uber's autonomous cars struck and killed a pedestrian in Tempe, Arizona. Uber suspended its road tests after the incident.

In April 2018, the California DMV began issuing road test permits for fully autonomous vehicles. The state had previously allowed road testing only with a human safety operator inside the car.

The most significant issue faced by companies looking to create and sell autonomous vehicles is the issue of liability. Before autonomous cars can become a reality for consumers, state and national lawmakers and the automotive industry must debate and determine the rules and regulations governing responsibility and recourse in the case of automated system failure.

—*Micah Issitt*

Bibliography

Anderson, James M. et al. "Autonomous Vehicle Technology: A Guide for Policymakers." *RAND.* RAND, 2014. Web. 16 Nov. 2014. PDF file.

Bilger, Burkhard. "Auto Correct: Has the Self-Driving Car at Last Arrived?" *New Yorker.* Condé Nast, 25 Nov. 2013. Web. 16 Nov. 2014.

Bogost, Ian. "The Secret History of the Robot Car." *Atlantic.* Atlantic Monthly Group, 14 Oct. 2014. Web. 16 Nov. 2014.

Davies, Alex. "In 20 Years, Most New Cars Won't Have Steering Wheels or Pedals." *Wired.* Condé Nast, 21 July 2014. Web. 16 Nov. 2014.

Moon, Mariella. "Now California's DMV Can Allow Fully Driverless Car Testing." *Engadget,* 3 Apr. 2018, www.engadget.com/2018/04/03/california-fully-driverless-car-testing/. Accessed 3 Apr. 2018.

Ramsey, Mike. "Tesla CEO Musk Sees Fully Autonomous Car Ready in Five or Six Years." *Wall Street Journal.* Dow Jones, 17 Sept. 2014. Web. 16 Nov. 2014.

Smith, Aaron. "U.S. Views of Technology and the Future." *Pew Research.* Pew Foundation, 17 Apr. 2014. Web. 16 Nov. 2014.

Stenquist, Paul. "In Self-Driving Cars, a Potential Lifeline for the Disabled." *New York Times.* New York Times, 7 Nov. 2014. Web. 16 Nov. 2014.

"U.S. Department of Transportation Releases Policy on Autonomous Vehicle Development." *NHTSA.gov.* National Highway Traffic Safety Administration, 30 May 2013. Web. 16 Nov. 2014.

Vanderbilt, Tom. "Autonomous Cars through the Ages." *Wired.* Condé Nast, Feb. 2012. Web. 16 Nov. 2014.

Wakabayashi, Daisuke. "Uber's Self-Driving Cars Were Struggling before Arizona Crash." *The New York Times,* 23 Mar. 2018, www.nytimes.com/2018/03/23/technology/uber-self-driving-cars-arizona.html. Accessed 3 Apr. 2018.

Avatars and Simulation

SUMMARY

Avatars and simulation are elements of virtual reality (VR), which attempts to create immersive worlds for computer users to enter. Simulation is the method by which the real world is imitated or approximated by the images and sounds of a computer. An avatar is the personal manifestation of a particular person. Simulation and VR are used for many applications, from entertainment to business.

VIRTUAL WORLDS

Computer simulation and virtual reality (VR) have existed since the early 1960s. While simulation has been used in manufacturing since the 1980s, avatars and virtual worlds have yet to be widely embraced outside gaming and entertainment. VR uses computerized sounds, images, and even vibrations to model some or all of the sensory input that human beings constantly receive from their surroundings every day. Users can define the rules of how a VR world works in ways that are not possible in everyday life. In the real world, people cannot fly, drink fire, or punch through walls. In VR, however, all of these things are possible, because the rules are defined by human coders, and they can be changed or even deleted. This is why users' avatars can appear in these virtual worlds as almost anything one can imagine—a loaf of bread, a sports car, or a penguin, for example. Many users of virtual worlds are drawn to them because of this type of freedom.

Because a VR simulation does not occur in physical space, people can "meet" regardless of how far apart they are in the real world. Thus, in a company that uses a simulated world for conducting its meetings, staff from Hong Kong and New York can both occupy the same VR room via their avatars. Such virtual meeting spaces allow users to convey nonverbal cues as well as speech. This allows for a greater degree of authenticity than in telephone conferencing.

MECHANICS OF ANIMATION

The animation of avatars in computer simulations often requires more computing power than a single workstation can provide. Studios that produce animated films use render farms to create the smooth and sophisticated effects audiences expect.

Before the rendering stage, a great deal of effort goes into designing how an animated character or avatar will look, how it will move, and how its textures will behave during that movement. For example, a fur-covered avatar that moves swiftly outdoors in the wind should have a furry or hairy texture, with fibers that appear to blow in the wind. All of this must be designed and coordinated by computer animators. Typically, one of the first steps is keyframing, in which animators decide what the starting and ending positions and appearance of the animated object will be. Then they design the movements between the beginning and end by assigning animation variables (avars) to different points on the object. This stage is called "in-betweening," or "tweening." Once avars are assigned, a computer algorithm can automatically change the avar values in coordination with one another. Alternatively, an animator can change "in-between" graphics by hand. When the program is run, the visual representation of the changing avars will appear as an animation.

In general, the more avars specified, the more detailed and realistic that animation will be in its movements. In an animated film, the main characters often have hundreds of avars associated with them. For instance, the 1995 film *Toy Story* used 712 avars for the cowboy Woody. This ensures that the characters' actions are lifelike, since the audience will focus attention on them most of the time. Coding standards for normal expressions and motions have been developed based on muscle movements. The MPEG-4 international standard includes 86 face parameters and 196 body parameters for animating human and humanoid movements. These parameters are encoded into an animation file and can affect the bit rate (data encoded per second) or size of the file.

EDUCATIONAL APPLICATIONS

Simulation has long been a useful method of training in various occupations. Pilots are trained in flight simulators, and driving simulators are used to prepare for licensing exams. Newer applications have included training teachers for the classroom and improving counseling in the military. VR holds the promise of making such vocational simulations much more realistic. As more computing power is added, simulated environments can include stimuli that better approximate the many distractions and detailed surroundings of the typical driving or flying situation, for instance.

VR IN 3-D

Most instances of VR that people have experienced so far have been two-dimensional (2-D), occurring on a computer or movie screen. While entertaining, such experiences do not really capture the concept of VR. Three-dimensional (3-D) VR headsets such as the Oculus Rift may one day facilitate more lifelike business meetings and product planning. They may also offer richer vocational simulations for military and emergency personnel, among others.

—*Scott Zimmer, JD*

BIBLIOGRAPHY

Chan, Melanie. *Virtual Reality: Representations in Contemporary Media.* New York: Bloomsbury, 2014. Print.

Gee, James Paul. *Unified Discourse Analysis: Language, Reality, Virtual Worlds, and Video Games.* New York: Routledge, 2015. Print.

Griffiths, Devin C. *Virtual Ascendance: Video Games and the Remaking of Reality.* Lanham: Rowman, 2013. Print.

Hart, Archibald D., and Sylvia Hart Frejd. *The Digital Invasion: How Technology Is Shaping You and Your Relationships.* Grand Rapids: Baker, 2013. Print.

Kizza, Joseph Migga. *Ethical and Social Issues in the Information Age.* 5th ed. London: Springer, 2013. Print.

Lien, Tracey. "Virtual Reality Isn't Just for Video Games." *Los Angeles Times.* Tribune, 8 Jan. 2015. Web. 23 Mar. 2016.

Parisi, Tony. *Learning Virtual Reality: Developing Immersive Experiences and Applications for Desktop, Web, and Mobile.* Sebastopol: O'Reilly, 2015. Print.

B

Behavioral Neuroscience

SUMMARY

Behavioral neuroscience is the study of the role of the nervous system in human and animal behavior. In approaching behavior from a biological framework, behavioral neuroscience considers not only the roles of the physical structures of the brain and other elements of the nervous system in both normal and abnormal behaviors but also the evolution and development of these roles over time.

Behavioral neuroscience is one of several fields in which biology and psychology intersect. It is distinguished from neuropsychology primarily in that while both are experimental fields of psychology, behavioral neuroscience mostly involves experiments using animals, often those whose biology has some correlation to that of humans, while neuropsychology deals more with human subjects, typically with a narrower focus on the structures and functions of the brain and nervous system. Other related fields include evolutionary psychology, affective neuroscience, social neuroscience, behavioral genetics, and neurobiology.

BRIEF HISTORY

While biological models of behavior are not a modern innovation as such, behavioral neuroscience properly speaking is grounded in both a modern understanding of biology—including evolution, the inheritance of genes, and the functions of hormones and neurotransmitters, just for starters—and the science of psychology. Though philosophers and scientists have discussed the mind since ancient times, psychology as an experimental discipline is a product of the nineteenth century, growing out of the Enlightenment's interests in education and in finding humane treatments for mental illness, both of which were served by seeking a way to better understand the workings of the human mind. In a sense, modern biology and modern psychology developed alongside each other: the basic foundations of each were discovered or formulated concurrently, and in the mid-nineteenth century they were brought together in the famous case of Phineas Gage.

A railroad construction foreman, Gage survived an accident in which an iron rod was driven through his head, destroying most of the left frontal lobe of his brain. It left him blind in the affected eye, but less predictably, it changed his personality for the remainder of his life. Reports of the changes in his personality vary, with those dating from after his death being notably more dramatic than the firsthand accounts written during his life, but they concur in the broad strokes—namely, that Gage's behavior after the accident reflected an absence of social inhibition and an increased degree of impulsivity. Popular culture often depicts Gage as becoming violent, depraved, even psychopathic after his accident; few now believe the change to have been this extreme. Reports from later in his life suggest that he was able to eventually relearn and adapt to social and interpersonal mores.

As significant as any change was the simple fact of Gage's survival, as well as the fact that his mental faculties remained intact with so much brain tissue destroyed. In a sense, the details and degree of Gage's personality change are not as important as the excuse he provided for the scientific community to newly address the "mind-body problem" that had concerned the Western intellectual community for centuries. In modern terms, the mind-body problem is the problem of explaining the relationship between the experiences of the nonphysical mind and those of the physical body within which it resides.

AREAS OF RESEARCH

Since Gage's time, biologists have learned more and more about the brain and related structures, and psychologists have conducted countless experiments relating to the mind and behavior. Research in behavioral neuroscience is conducted by observing the behavior of animals in experimental conditions in which some aspect of the animal's nervous system is measured or altered through various means. (Numerous such means exist, including surgical modifications, psychopharmaceuticals, electrical or magnetic stimulation, and genetic engineering.) One of the classic methods is lesion research, which studies animal subjects that have suffered damage to a particular region of the brain. The infliction of surgical lesions, in which brain or other neural tissue is destroyed through surgical removal, has long been a common research method, while the use of neurotoxins to inflict chemical lesions is a more recent development. An even more recent innovation is the use of special anaesthetics and other methods to induce "temporary lesions" by temporarily disabling neural tissue instead of destroying it entirely.

The study of lesions and their effect on behavior helps inform scientific understanding of which structures in the brain contribute to which of its functions. While early models were based on simple one-to-one correlation, demonstrated by literal maps of the brain that showed where various emotions and mental abilities were believed to be generated or exercised, behavioral neuroscience and other disciplines that study the brain have since shown that its workings are far more complicated than that.

Areas of research in behavioral neuroscience focus on broad categories of behavior that humans have in common with animals, namely needs-motivated behaviors, the senses, movement, memory, learning, sleep, emotion, and the relationships among these areas. Behavioral neuroscientists working with certain species—especially cetaceans, cephalopods, and corvids—may also study consciousness, language, or decisionmaking, but these are more controversial areas of inquiry.

Science has always been informed by philosophy. Only in the modern era have science and philosophy truly been separate disciplines, modern science having descended from what Aristotle called "natural philosophy." That said, although both the philosophy of science and the philosophy of biology are esteemed subfields, the philosophy of neuroscience is more newly developed. Until the 1980s, philosophers in the main proceeded without having integrated much specific understanding of neuroscience into their field—and, by extension, were not equipped to comment or contribute specifically to neuroscience.

This changed with Patricia Churchland's *Neurophilosophy* (1986), which was aimed specifically at bridging the gap. Although Churchland included her own views on the philosophy of neuroscience, she also took the time to include a primer on philosophy for neuroscientists and a primer on neuroscience for philosophers. The interdisciplinary relationship between the fields has continued to develop since, and in the twenty-first century, neuroethics became a prominent field, examining the ethical choices and ramifications of neuroscience. In recent years, neuroethicists have raised critical questions about human cognitive enhancements, the ethics of treating neurological impairments, and animal experimentation in behavioral neuroscience.

—Bill Kte'pi, MA

Bibliography

Bickle, John, ed. *The Oxford Handbook of Philosophy and Neuroscience.* New York: Oxford UP, 2009. Print.

Carlson, Neil R. *Foundations of Behavioral Neuroscience.* 9th ed. Boston: Pearson, 2014. Print.

Churchland, Patricia Smith. *Neurophilosophy: Toward a Unified Science of the Mind-Brain.* Cambridge: MIT P, 1986. Print.

Churchland, Patricia S. *Touching a Nerve: The Self as Brain.* New York: Norton, 2013. Print.

Kean, Sam. *The Tale of the Dueling Neurosurgeons: The History of the Human Brain as Revealed by True Stories of Trauma, Madness, and Recovery.* New York: Little, 2014. Print.

Mele, Alfred R. *Free: Why Science Hasn't Disproved Free Will.* New York: Oxford UP, 2014. Print.

Ramachandran, V. S. *The Tell-Tale Brain: A Neuroscientist's Quest for What Makes Us Human.* New York: Norton, 2011. Print.

Binary Pattern

SUMMARY

The binary number system is the more common name for the base-2 numeral system. Unlike the base-10, or decimal system, which uses digits 0-9 to represent numeric values, the binary number system uses only two symbols, traditionally 0 and 1. Thus the binary representations of base-10 "1, 2, 3" are "1, 10, 11." The binary system is used in computers and other electronics because of its ease of representation with two-state devices such as electrical switches. Throughout history, binary systems have had many uses, including decisionmaking (coin-flipping returns one of two values), divination (the I Ching uses yin/yang system), and encryption (Morse code uses short and long tones), in addition to mathematical applications. Boolean algebra, which became integral to the design of circuitry and computers, was developed by George Boole in 1854 and performs operations with variables assigned the values true or false.

Binary numbers may be manipulated either by conventional arithmetic methods or by using Boolean logical operators in what is usually called a bitwise operation. Bitwise operations are performed by the processor on the individual binary numerals, or bits, of a computer system and are faster and more efficient than arithmetic methods. When computers perform binary operations that result in a 32-bit integer, these operations can be used to form images called binary patterns.

APPLICATION

Despite their underlying simplicity, images formed by binary patterns may appear quite complex. The 32-bit value is key because 32 bits are used in RGBA (Red Green Blue Alpha) color space, in which 8 bits are devoted to the amount of red, green, and blue in an image, with a further 8 bits in the alpha channel reserved for the image's degree of transparency. Images displayed on computers, television screens, and other LCD screens are made up of pixels, each of which is the smallest physical point on the screen. Pixels are not a standard size, however—in comparing two different same-size screens, the one with the highest resolution has the smallest pixels, and thus a greater number of adjustable points.) These pixels are like incredibly tiny dabs of paint adding up to make a coherent image. In a 32-bit RGBA color space, each pixel is associated with 32 bits of information determining its color and transparency.

Binary patterns can form images that take up much less space in memory than if each pixel were encoded individually, by instead storing the formula or series of binary operations used to form the image. Many computer operating systems include basic patterns and tiles that take up a fraction of the space of an image like a photograph, which can't be reconstructed by binary operations. They may be simple repeated shapes, or intricate patterns reminiscent of murals and mandalas.

—Bill Kte'pi, MA

Bibliography

Kolo, Brian. *Binary and Multiclass Classification.* New York: Weatherford, 2010.

Kjaerulff, Uffe, and Anders Madsen. *Bayesian Networks and Influence Diagrams.* New York: Springer, 2014.

Larcher, Gerhard, and Friedrich Pilichshammer, eds. *Applied Algebra and Number Theory.* New York: Cambridge UP, 2015.

Marchand-Maillet, S. *Binary Digital Image Processing.* Waltham, MA: Academic P, 1999.

Nicholas, Patrick. *Scala for Machine Learning.* New York: Packt, 2014.

Reba, Marilyn, and Douglas R. Shier. *Puzzles, Paradoxes, and Problem Solving.* New York: Chapman, 2014.

Stakhov, Alexey, and Scott Anthony Olsen. *The Mathematics of Harmony.* Hackensack: World Scientific, 2009.

Yager, Ronald, and Ali Abbasov, eds. *Soft Computing.* New York: Springer, 2014.

BIOMECHANICAL ENGINEERING

SUMMARY

Biomechanical engineering is a branch of science that applies mechanical engineering principles such as physics and mathematics to biology and medicine. It can be described as the connection between structure and function in living things. Researchers in this field investigate the mechanics and mechanobiology of cells and tissues, tissue engineering, and the physiological systems they comprise. The work also examines the pathogenesis and treatment of diseases using cells and cultures, tissue mechanics, imaging, microscale biosensor fabrication, biofluidics, human motion capture, and computational methods. Real-world applications include the design and evaluation of medical implants, instrumentation, devices, products, and procedures. Biomechanical engineering is a multidisciplinary science, often fostering collaborations and interactions with medical research, surgery, radiology, physics, computer modeling, and other areas of engineering.

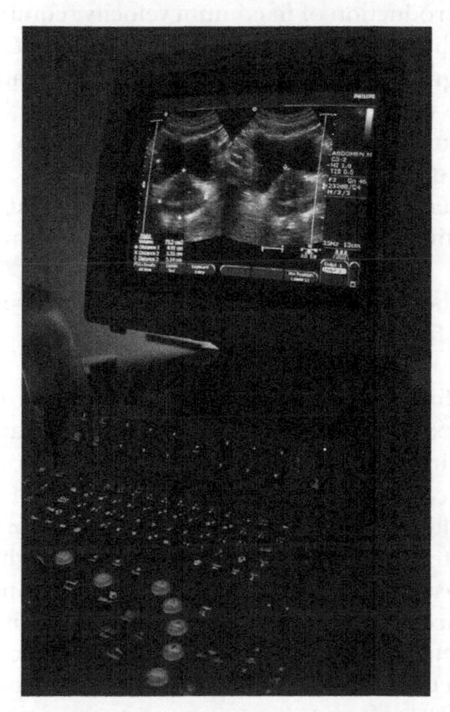

Ultrasound representation of Urinary bladder (black butterfly-like shape) with a hyperplastic prostate. An example of engineering science and medical science working together. By Etan J. Tal (Own work)

DEFINITION AND BASIC PRINCIPLES

Biomechanical engineering applies mechanical engineering principles to biology and medicine. Elements from biology, physiology, chemistry, physics, anatomy, and mathematics are used to describe the impact of physical forces on living organisms. The forces studied can originate from the outside environment or generate within a body or single structure. Forces on a body or structure can influence how it grows, develops, or moves. Better understanding of how a biological organism copes with forces and stresses can lead to improved treatment, advanced diagnosis, and prevention of disease. This integration of multidisciplinary philosophies has lead to significant advances in clinical medicine and device design. Improved understanding guides the creation of artificial organs, joints, implants, and tissues. Biomechanical engineering also has a tremendous influence on the retail industry, as the results of laboratory research guide product design toward more comfortable and efficient merchandise.

BACKGROUND AND HISTORY

The history of biomechanical engineering, as a distinct and defined field of study, is relatively short. However, applying the principles of physics and engineering to biological systems has been developed over centuries. Many overlaps and parallels to complementary areas of biomedical engineering and biomechanics exist, and the terms are often used interchangeably with biomechanical engineering. The mechanical analysis of living organisms was not internationally accepted and recognized until the definition provided by Austrian mathematician Herbert Hatze in 1974: "Biomechanics is the study of the structure and function of biological systems by means of the methods of mechanics."

Greek philosopher Aristotle introduced the term "mechanics" and discussed the movement of living beings around 322 BCE in the first book about biomechanics, *On the Motion of Animals.* Leonardo da Vinci proposed that the human body is subject to the law of mechanics in the 1500s. Italian physicist and mathematician Giovanni Alfonso Borelli, a student of Galileo's, is considered the "father of biomechanics" and developed mathematical models to describe anatomy and human movement mechanically. In the 1890s German zoologist Wilhelm Roux and German surgeon Julius Wolff determined the effects of loading and stress on stem cells in the development of bone architecture and healing. British physiologist Archibald V. Hill and German physiologist Otto Fritz Meyerhof shared the 1922 Nobel Prize for Physiology or Medicine. The prize was divided between them: Hill won "for his discovery relating to the production of heat in the muscle"; Meyerhof won "for his discovery of the fixed relationship between the consumption of oxygen and the metabolism of lactic acid in the muscle."

The first joint replacement was performed on a hip in 1960 and a knee in 1968. The development of imaging, modeling, and computer simulation in the latter half of the twentieth century provided insight into the smallest structures of the body. The relationships between these structures, functions, and the impact of internal and external forces accelerated new research opportunities into diagnostic procedures and effective solutions to disease. In the 1990s, biomechanical engineering programs began to emerge in academic and research institutions around the world, and the field continued to grow in recognition into the twenty-first century.

HOW IT WORKS

Biomechanical engineering science is extremely diverse. However, the basic principle of studying the relationship between biological structures and forces, as well as the important associated reactions of biological structures to technological and environmental materials, exists throughout all disciplines. The biological structures described include all life forms and may include an entire body or organism or even the microstructures of specific tissues or systems. Characterization and quantification of the response of these structures to forces can provide insight into disease process, resulting in better treatments and diagnoses. Research in this field extends beyond the laboratory and can involve observations of mechanics in nature, such as the aerodynamics of bird flight, hydrodynamics of fish, or strength of plant root systems, and how these findings can be modified and applied to human performance and interaction with external forces.

As in biomechanics, biomechanical engineering has basic principles. Equilibrium, as defined by British physicist Sir Isaac Newton, results when the sum of all forces is zero and no change occurs and energy cannot be created or destroyed, only converted from one form to another.

The seven basic principles of biomechanics can be applied or modified to describe the reaction of forces to any living organism.

The lower the center of mass, the larger the base of support; the closer the center of mass to the base of support, and the greater the mass, the more stability increases.

The production of maximum force requires the use of all possible joint movements that contribute to the task's objective.

The production of maximum velocity requires the use of joints in order—from largest to smallest.

The greater the applied impulse, the greater increase in velocity.

Movement usually occurs in the direction opposite that of the applied force.

Angular motion is produced by the application of force acting at some distance from an axis, that is, by torque.

Angular momentum is constant when a body or object is free in the air.

The forces studied can be combinations of internal, external, static, or dynamic, and all are important in the analysis of complex biochemical and biophysical processes. Even the mechanics of a single cell, including growth, cell division, active motion, and contractile mechanisms, can provide insight into mechanisms of stress, damage of structures, and disease processes at the microscopic level. Imaging and computer simulation allow precise measurements and observations to be made of the forces impacting the smallest cells.

APPLICATIONS AND PRODUCTS

Biomechanical engineering advances in modeling and simulation have tremendous potential research and application uses across many health care disciplines. Modeling has resulted in the development of designs for implantable devices to assist with organs or areas of the body that are malfunctioning. The biomechanical relationships between organs and supporting structures allow for improved device design and can assist with planning of surgical and treatment interventions. The materials used for medical and surgical procedures in humans and animals are being evaluated and some redesigned, as biomechanical science is showing that different materials, procedures, and techniques may be better for reducing complications and improving long-term patient health. Evaluating the physical relationship between the cells and structures of the body and foreign implements and interventions can quantify the stresses and forces on the system, which provides more accurate prediction of patient outcomes.

Biomechanical engineering professionals apply their knowledge to develop implantable medical devices that can diagnose, treat, or monitor disease and health conditions and improve the daily living of patients. Devices that are used within the human body are highly regulated by the U.S. Food and Drug Administration (FDA) and other agencies internationally. Pacemakers and defibrillators, also called cardiac resynchronization therapy (CRT) devices, can constantly evaluate a patient's heart and respond to changes in heart rate with electrical stimulation. These devices greatly improve therapeutic outcomes in patients afflicted with congestive heart failure. Patients with arrhythmias experience greater advantages with implantable devices than with pharmaceutical options. Cochlear implants have been designed to be attached to a patient's auditory nerve and can detect sound waves and process them in order to be interpreted by the brain as sound for deaf or hard-of-hearing patients. Patients who have had cataract surgery used to have to wear thick corrective lenses to restore any standard of vision, but with the development of intraocular lenses that can be implanted into the eye, their vision can be restored, often to a better degree than before the cataract developed.

Artificial replacement joints comprise a large portion of medical-implant technology. Patients receive joint replacement when their existing joints no longer function properly or cause significant pain because of arthritis or degeneration. Hundreds of thousands of hip replacements are performed in the United States each year, a number that has grown significantly as the baby boomer portion of the population ages. Artificial joints are normally fastened to the existing bone by cement, but advances in biomechanical engineering have lead to a new process called "bone ingrowth," in which the natural bone grows into the porous surface of the replacement joint. Biomechanical engineering contributes considerable knowledge to the design of the artificial joints, the materials from which they are made, the surgical procedure used, fixation techniques, failure mechanisms, and prediction of the lifetime of the replacement joints.

The development of computer-aided (CAD) design has allowed biomechanical engineers to create complex models of organs and systems that can provide advanced analysis and instant feedback. This information provides insight into the development of designs for artificial organs that align with or improve on the mechanical properties of biological organs.

Biomechanical engineering can provide predictive values to medical professionals, which can help them develop a profile that better forecasts patient outcomes and complications. An example of this is using finite element analysis in the evaluation of aortic-wall stress, which can remove some of the unpredictability of expansion and rupture of an abdominal aortic aneurysm. Biomechanical computational methodology and advances in imaging and processing technology have provided increased predictability for life-threatening events.

Nonmedical applications of biomechanical engineering also exist in any facet of industry that impacts human life. Corporations employ individuals or teams to use engineering principles to translate the scientifically proven principles into commercially viable products or new technological platforms. Biomechanical engineers also design and build experimental testing devices to evaluate a product's performance and safety before it reaches the marketplace, or they suggest more economically efficient design options. Biomechanical engineers also use ergonomic principles to develop new ideas and create new products, such as car seats, backpacks, or even specialized equipment and clothing for elite athletes, military personnel, or astronauts.

SOCIAL CONTEXT AND FUTURE PROSPECTS

The diversity of studying the relationship between living structure and function has opened up vast opportunities in science, health care, and industry. In addition to conventional implant and replacement devices, the demand is growing for implantable tissues for cosmetic surgery, such as breast and tissue implants, as well as implantable devices to aid in weight loss, such as gastric banding.

Reports of biomechanical engineering triumphs and discoveries often appear in the mainstream media, making the general public more aware of the scientific work being done and how it impacts daily life. Sports fans learn about the equipment, training, and rehabilitation techniques designed by biomechanical engineers that allow their favorite athletes to break performance records and return to work sooner after being injured or having surgery. The public is accessing more information about their own health options than ever before, and they are becoming knowledgeable about the range of treatments available to them and the pros and cons of each.

Biomechanical engineering and biotechnology is an area that is experiencing accelerated growth, and billions of dollars are being funneled into research and development annually. This growth is expected to continue.

—April D. Ingram, BSc

Bibliography

"Biomechanics." *Engineering in Medicine and Biology Society*, IEEE, 2018, www.embs.org/about-biomedical-engineering/our-areas-of-research/biomechanics/. Accessed 19 Mar. 2018.

Ethier, C. Ross, and Craig A. Simmons. *Introductory Biomechanics: From Cells to Organisms.* Cambridge, England: Cambridge University Press, 2007. Provides an introduction to biomechanics and also discusses clinical specialties, such as cardiovascular, musculoskeletal, and ophthalmology.

Hall, Susan J. *Basic Biomechanics.* 5th ed. New York: McGraw-Hill, 2006. A good introduction to biomechanics, regardless of one's math skills.

Hamill, Joseph, and Kathleen M. Knutzen. *Biomechanical Basis of Human Movement.* 3d ed. Philadelphia: Lippincott, 2009. Integrates anatomy, physiology, calculus, and physics and provides the fundamental concepts of biomechanics.

Hay, James G., and J. Gavin Reid. *Anatomy, Mechanics, and Human Motion.* 2d ed. Englewood Cliffs, N.J.: Prentice Hall, 1988. A good resource for upper high school students, this text covers basic kinesiology.

Hayenga, Heather N., and Helim Aranda-Espinoza. *Biomaterials Mechanics.* CRC Press, 2017.

Peterson, Donald R., and Joseph D. Bronzino, eds. *Biomechanics: Principles and Applications.* 2d ed. Boca Raton, Fla.: CRC Press, 2008. A collection of twenty articles on various aspects of research in biomechanics.

Prendergast, Patrick, ed. *Biomechanical Engineering: From Biosystems to Implant Technology.* London: Elsevier, 2007. One of the first comprehensive books for biomechanical engineers, written with the student in mind.

Biomechanics

SUMMARY

Biomechanics is the study of the application of mechanical forces to a living organism. It investigates the effects of the relationship between the body and forces applied either from outside or within. In humans, biomechanists study the movements made by the body, how they are performed, and whether the forces produced by the muscles are optimal for the intended result or purpose. Biomechanics integrates the study of anatomy and physiology with physics, mathematics, and engineering principles. It

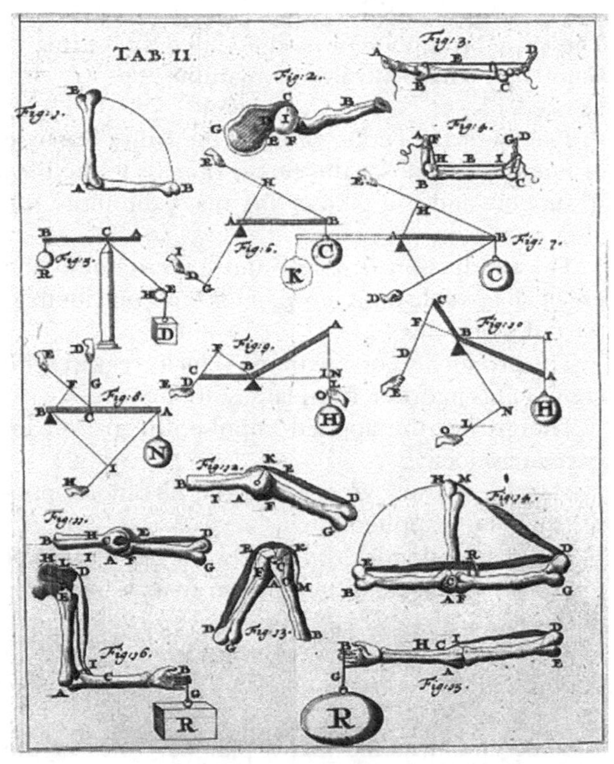

Page of one of the first works of biomechanics (De Motu Animalium of Giovanni Alfonso Borelli). By Giovanni Alfonso Borelli (De Motu Animalium book)

may be considered a subdiscipline of kinesiology as well as a scientific branch of sports medicine.

DEFINITION AND BASIC PRINCIPLES

Biomechanics is a science that closely examines the forces acting on a living system, such as a body, and the effects that are produced by these forces. External forces can be quantified using sophisticated measuring tools and devices. Internal forces can be measured using implanted devices or from model calculations. Forces on a body can result in movement or biological changes to the anatomical tissue. Biomechanical research quantifies the movement of different body parts and the factors that may influence the movement, such as equipment, body alignment, or weight distribution. Research also studies the biological effects of the forces that may affect growth and development or lead to injury. Two distinct branches of mechanics are statics and dynamics. Statics studies systems that are in a constant state of motion or constant state of rest, and dynamics studies systems that are in motion, subject to acceleration or deceleration. A moving body may be described using kinematics or kinetics. Kinematics studies and describes the motion of a body with respect to a specific pattern and speed, which translate into coordination of a display. Kinetics studies the forces associated with a motion, those causing it and resulting from it. Biomechanics combines kinetics and kinematics as they apply to the theory of mechanics and physiology to study the structure and function of living organisms.

BACKGROUND AND HISTORY

Biomechanics has a long history even though the actual term and field of study concerned with mechanical analysis of living organisms was not internationally accepted and recognized until the early 1970s. Definitions provided by early biomechanics specialists James G. Hay in 1971 and Herbert Hatze in 1974 are still accepted. Hatze stated, "Biomechanics is the science which studies structures and functions of biological systems using the knowledge and methods of mechanics."

Highlights throughout history have provided insight into the development of this scientific discipline. The ancient Greek philosopher Aristotle was the first to introduce the term "mechanics," writing about the movement of living beings around 322 BCE. He developed a theory of running techniques and suggested that people could run faster by swinging their arms. In the 1500s, Leonardo da Vinci proposed that the human body is subject to the law of mechanics, and he contributed significantly to the development of anatomy as a modern science. Italian scientist Giovanni Alfonso Borelli, a student of Galileo, is often considered the father of biomechanics. In the mid-1600s, he developed mathematical models to describe anatomy and human movement mechanically. In the late 1600s, English physician and mathematician Sir Isaac Newton formulated mechanical principles and Newtonian laws of motion (inertia, acceleration, and reaction) that became the foundation of biomechanics.

British physiologist A. V. Hill, the 1923 winner of the Nobel Prize in Physiology or Medicine, conducted research to formulate mechanical and structural theories for muscle action. In the 1930s, American anatomy professor Herbert Elftman was able to quantify the internal forces in muscles and joints and developed the force plate to quantify ground reaction. A significant breakthrough in the understanding of muscle action was made by British physiologist Andrew F. Huxley in 1953, when he described his filament theory to explain muscle shortening. Russian physiologist Nicolas Bernstein published a paper in 1967 describing theories for motor coordination and control following his work studying locomotion patterns of children and adults in the Soviet Union.

HOW IT WORKS

The study of human movement is multifaceted, and biomechanics applies mechanical principles to the study of the structure and function of living things. Biomechanics is considered a relatively new field of applied science, and the research being done is of considerable interest to many other disciplines, including zoology, orthopedics, dentistry, physical education, forensics, cardiology, and a host of other medical specialties. Biomechanical analysis for each particular application is very specific; however, the basic principles are the same.

Newton's Laws of Motion. The development of scientific models reduces all things to their basic level to provide an understanding of how things work. This also allows scientists to predict how things will behave in response to forces and stimuli and ultimately to influence this behavior.

Newton's laws describe the conservation of energy and the state of equilibrium. Equilibrium results when the sum of forces is zero and no change occurs, and conservation of energy explains that energy cannot be created or destroyed, only converted from one form to another. Motion occurs in two ways, linear motion in a particular direction or rotational movement around an axis. Biomechanics explores and quantifies the movement and production of force used or required to produce a desired objective.

Seven Principles. Seven basic principles of biomechanics serve as the building blocks for analysis. These can be applied or modified to describe the reaction of forces to any living organism.

The lower the center of mass, the larger the base of support; the closer the center of mass to the base of support and the greater the mass, the more stability increases.

The production of maximum force requires the use of all possible joint movements that contribute to the task's objective.

The production of maximum velocity requires the use of joints in order, from largest to smallest.

The greater the applied impulse, the greater increase in velocity.

Movement usually occurs in the direction opposite that of the applied force.

Angular motion is produced by the application of force acting at some distance from an axis, that is, by torque.

Angular momentum is constant when an athlete or object is free in the air.

Static and dynamic forces play key roles in the complex biochemical and biophysical processes that underlie cell function. The mechanical behavior of individual cells is of interest for many different biologic processes. Single-cell mechanics, including growth, cell division, active motion, and contractile mechanisms, can be quite dynamic and provide insight into mechanisms of stress and damage of structures. Cell mechanics can be involved in processes that lie at the root of many diseases and may provide opportunities as focal points for therapeutic interventions.

APPLICATIONS AND PRODUCTS

Biomechanics studies and quantifies the movement of all living things, from the cellular level to body systems and entire bodies, human and animal. There are many scientific and health disciplines, as well as industries that have applications developed from this knowledge. Research is ongoing in many areas to effectively develop treatment options for clinicians and better products and applications for industry.

Dentistry. Biomechanical principles are relevant in orthodontic and dental science to provide solutions to restore dental health, resolve jaw pain, and manage cosmetic and orthodontic issues. The design of dental implants must incorporate an analysis of load bearing and stress transfer while maintaining the integrity of surrounding tissue and comfortable function for the patient. This work has lead to the development of new materials in dental practices such as reinforced composites rather than metal frameworks.

Forensics. The field of forensic biomechanical analysis has been used to determine mechanisms of injury after traumatic events such as explosions in military situations. This understanding of how parts of the body behave in these events can be used to develop mitigation strategies that will reduce injuries. Accident and injury reconstruction using biomechanics is an emerging field with industrial and legal applications.

Biomechanical Modeling. Biomechanical modeling is a tremendous research field, and it has potential uses across many health care applications. Modeling has resulted in recommendations for prosthetic design and modifications of existing devices. Deformable breast models have demonstrated capabilities for breast cancer diagnosis and treatment. Tremendous growth is occurring in many medical fields that are exploring the biomechanical relationships between organs and supporting structures. These models can assist with planning surgical and treatment interventions and reconstruction and determining optimal loading and boundary constraints during clinical procedures.

Materials. Materials used for medical and surgical procedures in humans and animals are being evaluated and some are being changed as biomechanical science is demonstrating that different materials, procedures, and techniques may be better for reducing complications and improving long-term patient health. Evaluation of the physical relationship between the body and foreign implements can quantify the stresses and forces on the body, allowing for more accurate prediction of patient outcomes and determination of which treatments should be redesigned.

Predictability. Medical professionals are particularly interested in the predictive value that biomechanical profiling can provide for their patients. An example is the unpredictability of expansion and rupture of an abdominal aortic aneurysm. Major progress has been made in determining aortic wall stress using finite element analysis. Improvements in biomechanical computational methodology and advances in imaging and processing technology have provided increased predictive ability for this life-threatening event.

As the need for accurate and efficient evaluation grows, so does the research and development of effective biomechanical tools. Capturing real-time, real-world data, such as with gait analysis and range of motion features, provides immediate opportunities for applications. This real-time data can quantify an injury and over time provide information about the extent that the injury has improved. High-tech devices can translate real-world situations and two-dimensional images into a three-dimensional framework for analysis. Devices, imaging, and modeling tools and software are making tremendous strides and becoming the heart of a highly competitive industry aimed at simplifying the process of analysis and making it less invasive.

SOCIAL CONTEXT AND FUTURE PROSPECTS

Biomechanics has gone from a narrow focus on athletic performance to become a broad-based science, driving multibillion dollar industries to satisfy the needs of consumers who have become more knowledgeable about the relationship between science, health, and athletic performance. Funding for biomechanical research is increasingly available from national health promotion and injury prevention programs, governing bodies for sport, and business and industry. National athletic programs want to ensure that their athletes have the most advanced training methods, performance analysis methods, and equipment to maximize their athletes' performance at global competitions.

Much of the existing and developing technology is focused on increasingly automated and digitized systems to monitor and analyze movement and force. The physiological aspect of movement can be examined at a microscopic level, and instrumented athletic implements such as paddles or bicycle cranks allow real-time data to be collected during an event or performance. Force platforms are being reconfigured as starting blocks and diving platforms to measure reaction forces. These techniques for biomechanical performance analysis have led to revolutionary technique changes in many sports programs and rehabilitation methods.

Advances in biomechanical engineering have led to the development of innovations in equipment, playing surfaces, footwear, and clothing, allowing people to reduce injury and perform beyond previous expectations and records.

Computer modeling and virtual simulation training can provide athletes with realistic training opportunities, while their performance is analyzed and measured for improvement and injury prevention.

—*April D. Ingram, BSc*

Bibliography

Bronzino, Joseph D., and Donald R. Peterson. *Biomechanics: Principles and Practices.* Boca Raton: CRC, 2014. *eBook Collection (EBSCOhost).* Web. 25 Feb. 2015.

Hamill, Joseph, and Kathleen Knutzen. *Biomechanical Basis of Human Movement.* 4th ed. Philadelphia: Lippincott, 2015. Print.

Hatze, H. "The Meaning of the Term 'Biomechanics.'" *Journal of Biomechanics* 7.2 (1974): 89–90. Print.

Hay, James G. *The Biomechanics of Sports Techniques.* 4th ed. Englewood Cliffs: Prentice, 1993. Print.

Kerr, Andrew. *Introductory Biomechanics.* London: Elsevier, 2010. Print.

Peterson, Donald, and Joseph Bronzino. *Biomechanics: Principles and Applications.* Boca Raton: CRC, 2008. Print.

Watkins, James. *Introduction to Biomechanics of Sport and Exercise.* London: Elsevier, 2007. Print.

Biomimetics

SUMMARY

Biomimetics is a branch of science that uses observations in nature to inspire the development of new products or technologies in fields such as medicine, engineering, and architecture. By observing aspects of plant and animal life, researchers can find new ways to perform tasks or develop designs that can be adapted for man-made products. Mimicking biological life through biomimetics has allowed for many scientific innovations.

BACKGROUND

The idea of copying nature is not new. The Greek myth of Icarus, who constructed wings from feathers and wax in an attempt to fly, illustrates that man has long sought to imitate wonders found in the natural world. Centuries after the Greeks told the story of Icarus, Italian Renaissance thinker, artist, and inventor Leonardo da Vinci used birds as models for his intricate drawings of human flight devices. In the early years of the twentieth century, the Wright brothers studied the design of bird wings and incorporated aspects of them into the design of the first airplane. For many years, these attempts to mimic aspects of nature were largely accomplished by individuals and were not a specific discipline of study.

Velcro tape mimics biological examples of multiple hooked structures such as burs. By Ryj (Own work)

That began to change in the 1940s when George de Mestral, a Swiss engineer, was picking some burrs off his trousers and his dog after a walk. Intrigued by the series of small hooks that helped the seed pods cling so tightly to other surfaces, de Mestral began trying to replicate them. He envisioned using this technique in a new clothing fastener. The resulting product—Velcro—was patented in 1958 and eventually became a household word after the National Aeronautics and Space Administration (NASA) used it to help secure loose items in space capsules. Velcro is a hook-and-loop fastener that employs two strips of fabric: one containing tiny hooks and the other containing tiny loops. The hooks and loops cling to one another until the two pieces of fabric are pulled apart.

In the 1960s, at about the same time Velcro was gaining in popularity, an Air Force doctor, Colonel Jack E. Steele, invented the term *bionics*. Steele developed designs for technology that replicated biological functions, such as artificial limbs. Soon other researchers began actively looking for ways to adapt biological forms and functions found in nature to man-made technology.

By the end of the decade, Otto H. Schmitt, an American electrical engineer and professor of biophysics, coined the term *biomimetics*. The word was derived from two Greek words, *bios,* which means "life," and *mimesis,* which means "to imitate." The term was first used in a paper delivered at a conference in Boston in 1969 and first appeared in the dictionary in 1974. Schmitt applied the concept of biomimetics when he studied how frogs planned their leaps between lily pads. He used his findings to develop self-adjusting electronic feedback circuits that are crucial to the function of many types of contemporary electronic equipment.

APPLICATIONS

The biological world has provided scientists and researchers with many ideas to study and copy. Plant life and animal life have evolved over millennia, finding ways to overcome many obstacles and problems and learning to function in various settings. Sometimes the biological inspirations for man-made designs are easy to identify, such as when da Vinci and the Wright brothers studied bird wings. Other times researchers have to seek out the possibilities for using concepts found in nature, as de Mestral did when he used burrs as the inspiration for Velcro. In some cases, moving from natural concept to man-made invention requires a combination of know-how and imagination, such as when Schmitt applied the frog's ability to execute a jump to an improvement for electronic circuits.

The study of biomimetics has applications in many fields. For example, researchers mimicked the brain's neural network when developing patterns for computer programs and networks. Each smaller unit performs its own function, much like a neuron in the brain does. However, these smaller units are interconnected in a way that allows the completion of complex tasks. Other biological applications found in nature include specialized hypodermic needles that copy parts of a mosquito and the use of spider web–like material as a wound dressing.

Nature has helped to improve human travel and propulsion in a number of ways beyond copying bird wings for flight. The skin folds that allow flying squirrels to glide have inspired suits that allow humans to "fly" horizontally without a vehicle. The fins that scuba divers wear to help them swim resemble the webbed feet of water birds such as geese and ducks. The way sharks, tuna, and other large sea creatures move has inspired designs that use the ocean's power for movement. The multi-jointed legs of insects helped scientists find ways to create robotic devices that can move over uneven and difficult landscapes.

Studying how mollusks adhere to surfaces led to the development of new types of adhesives. *Sporosarcina pasteurii,* a bacterium that can use elements found in its natural environment to solidify sand, has helped researchers develop concrete pipes that can self-repair cracks.

The possibilities of biomimicry are almost endless. The natural sensors on insect antennas and the whiskers of many mammals are models for potential electronic sensors. The iridescence of bird and butterfly wings has applications in the study of optics and the development of devices such as computer monitors and windows that are affected by the reflection of light. The wide diversity of nature and its ability to adapt, evolve, and overcome challenges provides much inspiration for solutions and improvements to technologies used in everyday life.

—*Janine Ungvarsky*

Bibliography

Bar-Cohen, Yoseph. "Biomimetics—Using Nature to Inspire Human Innovation." *Bioinspiration & Biomimetics,* vol. 1, 27 Apr. 2006, pp. 1–12, biomimetic.pbworks.com/f/Biomimetics—using+nature+to+inspire+humanBar-Cohen.pdf. Accessed 12 Jan. 2017.

Coldewey, Devin. "Scientists Aspire to Nature's Genius with 'Biomimetic' Research." *NBCNews,* 6 June 2015, www.nbcnews.com/tech/innovation/scientists-attempt-replicate-natures-ingenuity-n365451. Accessed 12 Jan. 2017.

"Dr. Otto H. Schmitt—1978 Inductee." *Minnesota Inventors Hall of Fame,* www.minnesotainventors.org/inductees/otto-h-schmitt.html. Accessed 12 Jan. 2017.

Hwang, Jangsun, et al. "Biomimetics: Forecasting the Future of Science, Engineering, and Medicine." *International Journal of Nanomedicine,* vol. 10, Oct. 2015, pp. 5701–13, www.ncbi.nlm.nih.gov/pmc/articles/PMC4572716/#!po=30.3571. Accessed 12 Jan. 2017.

Mueller, Tom. "Biomimetics: Design by Nature." *National Geographic,* Apr. 2008, ngm.nationalgeographic.com/2008/04/biomimetics/tom-mueller-text. Accessed 12 Jan. 2017.

"Nano-Biomimetics." *Stanford University,* web.stanford.edu/group/mota/education/Physics%2087N%20Final%20Projects/Group%20Gamma/. Accessed 12 Jan. 2017.

Sandhu, Robin. "5 Examples of Biomimetic Technology." *Livewire,* 19 Oc. 2016, www.lifewire.com/examples-of-biomimetic-technology-2495572. Accessed 12 Jan. 2017.

Suddath, Claire. "A Brief History of: Velcro." *Time,* 15 June 2010, content.time.com/time/nation/article/0,8599,1996883,00.html. Accessed 12 Jan. 2017.

Bionics and Biomedical Engineering

SUMMARY

Bionics combines natural biologic systems with engineered devices and electrical mechanisms. An example of bionics is an artificial arm controlled by impulses from the human mind. Construction of bionic arms or similar devices requires the integrative use of medical equipment such as electroencephalograms (EEGs) and magnetic resonance imaging (MRI) machines with mechanically engineered prosthetic arms and legs. Biomedical engineering further melds biomedical and engineering sciences by producing medical equipment, tissue growth, and new pharmaceuticals. An example of biomedical engineering is human insulin production through genetic engineering to treat diabetes.

DEFINITION AND BASIC PRINCIPLES

The fields of biomedical engineering and bionics focus on improving health, particularly after injury or illness, with better rehabilitation, medications, innovative treatments, enhanced diagnostic tools, and preventive medicine.

Bionics has moved nineteenth-century prostheses, such as the wooden leg, into the twenty-first century by using plastic polymers and levers. Bionics integrates circuit boards and wires connecting the nervous system to the modular prosthetic limb. Controlling artificial limb movements with thoughts provides more lifelike function and ability. This mind and prosthetic limb integration is the "bio" portion of bionics; the "nic" portion, taken from the word "electronic," concerns the mechanical engineering that makes it possible for the person using a bionic limb to increase the number and range of limb activity, approaching the function of a real limb.

Biomedical engineering encompasses many medical fields. The principle of adapting engineering techniques and knowledge to human structure and function is a key unifying concept of biomedical engineering. Advances in genetic engineering have produced remarkable bioengineered medications. Recombinant DNA techniques (genetic engineering) have produced synthetic hormones, such as insulin. Bacteria are used as a host for this process; once human-insulin-producing genes are implanted

Velcro was inspired by the tiny hooks found on the surface of burs. By Zephyris (Own work)

in the bacteria, the bacteria's DNA produce human insulin, and the human insulin is harvested to treat diabetics. Before this genetic technique was developed in 1982 to produce human insulin, insulin-dependent diabetics relied on insulin from pigs or cows. Although this insulin was life saving for diabetics, diabetics often developed problems from the pig or cow insulin because they would produce antibodies against the foreign insulin. This problem disappeared with the ability to engineer human insulin using recombinant DNA technology.

BACKGROUND AND HISTORY

In the broad sense, biomedical engineering has existed for millennia. Human beings have always envisioned the integration of humans and technology to increase and enhance human abilities. Prosthetic devices go back many thousands of years: a three-thousand-year-old Egyptian mummy, for example, was found with a wooden big toe tied to its foot. In the fifteenth century, during the Italian Renaissance, Leonardo da Vinci's elegant drawings demonstrated some early ideas on bioengineering, including his helicopter and flying machines, which melded human and machine into one functional unit capable of flight. Other early examples of biomedical engineering include wooden teeth, crutches, and medical equipment, such as stethoscopes.

Electrophysiological studies in the early 1800s produced biomedical engineering information used to better understand human physiology. Engineering principles related to electricity combined with human physiology resulted in better knowledge of the electrical properties of nerves and muscles.

X rays, discovered by Wilhelm Conrad Röntgen in 1895, were an unknown type of radiation (thus the "X" name). When it was accidentally discovered that they could penetrate and destroy tissue, experiments were developed that led to a range of imaging technologies that evolved over the next century. The first formal biomedical engineering training program, established in 1921 at Germany's Oswalt Institute for Physics in Medicine, focused on three main areas: the effects of ionizing radiation, tissue electrical characteristics, and X-ray properties.

In 1948, the Institute of Radio Engineers (later the Institute of Electrical and Electronics Engineers), the American Institute for Electrical Engineering, and the Instrument Society of America held a conference on engineering in biology and medicine. The 1940s and 1950s saw the formation of professional societies related to biomedical engineering, such as the Biophysics Society, and of interest groups within engineering societies. However, research at the time focused on the study of radiation. Electronics and the budding computer era broadened interest and activities toward the end of the 1950s.

James D. Watson and Francis Crick identified the DNA double-helix structure in 1953. This important discovery fostered subsequent experimentation in molecular biology that yielded important information about how DNA and genes code for the expression of traits in all living organisms. The genetic code in DNA was deciphered in 1968, arming researchers with enough information to discover ways that DNA could be recombined to introduce genes from one organism into a different organism, thereby allowing the host to produce a variety of useful products. DNA recombination became one of the most important tools in the field of biomedical engineering, leading to tissue growth as well as new pharmaceuticals.

In 1962, the National Institutes of Health created the National Institute of General Medical Sciences, fostering the development of biomedical engineering programs. This institute funds research in the diagnosis, treatment, and prevention of disease.

Bionics and biomedical engineering span a wide variety of beneficial health-related fields. The common thread is the combination of technology with human applications. Dolly the sheep was cloned in 1996. Cloning produces a genetically identical copy of an existing life-form. Human embryonic cloning presents the potential of therapeutic reproduction of needed organs and tissues, such as kidney replacement for patients with renal failure.

In the twenty-first century, the linking of machines with the mind and sensory perception has provided hearing for deaf people, some sight for the blind, and willful control of prostheses for amputees.

HOW IT WORKS

Restorative bionics integrates prosthetic limbs with electrical connections to neurons, allowing an individual's thoughts to control the artificial limb. Tiny arrays of electrodes attached to the eye's retina connect to the optic nerve, enabling some visual perception for previously blind people. Deaf people hear with electric devices that send signals to auditory nerves, using antennas, magnets, receivers, and electrodes. Researchers are considering bionic skin development using nanotechnology to connect with nerves, enabling skin sensations for burn victims requiring extensive grafting.

Many biomedical devices work inside the human body. Pacemakers, artificial heart valves, stents, and even artificial hearts are some of the bionic devices correcting problems with the cardiovascular system. Pacemakers generate electric signals that improve abnormal heart rates and abnormal heart rhythms. When pulse generators located in the pacemakers sense an abnormal heart rate or rhythm, they produce shocks to restore the normal rate. Stents are inserted into an artery to widen it and open clogged blood vessels. Stents and pacemakers are examples of specialized bionic devices made up of bionic materials compatible with human structure and function.

Cloning. Cloning is a significant area of genetic engineering that allows the replication of a complete living organism by manipulating genes. Dolly the sheep, an all-white Finn Dorset ewe, was cloned from a surrogate mother blackface ewe, which was used as an egg donor and carried the cloned Dolly during gestation (pregnancy). An egg cell from the surrogate was removed and its nucleus (which contains DNA) was replaced with one from a Finn Dorset ewe; the resulting new egg was placed in the blackface ewe's uterus after stimulation with an electric pulse. The electrical pulse stimulated growth and cell duplication. The blackface ewe subsequently gave birth to the all-white Dolly. The newborn all-white Finn Dorset ewe was an identical genetic twin of the Finn Dorset that contributed the new nucleus.

Recombinant DNA. Another significant genetic engineering technique involves recombinant DNA. Human genes transferred to host organisms, such as bacteria, produce products coded for by the transferred genes. Human insulin and human growth hormone can be produced using this technique. Desired genes are removed from human cells and placed in circular bacterial DNA strips called plasmids. Scientists use enzymes to prepare these DNA formulations, ultimately splicing human genes into bacterial plasmids. These plasmids are used as vectors, taken up and reproduced by bacteria. This type of genetic adaptation results in insulin production if the spliced genes were taken from the part of the human genome producing insulin; other cells and substances, coded for by different human genes, can be produced this way. Many biologic medicines are produced using recombinant DNA technology.

APPLICATIONS AND PRODUCTS

Medical Devices. Biomedical engineers produce life-saving medical equipment, including pacemakers, kidney dialysis machines, and artificial hearts. Synthetic limbs, artificial cochleas, and bionic sight chips are among the prosthetic devices that biomedical engineers have developed to enhance mobility, hearing, and vision. Medical monitoring devices, developed by biomedical engineers for use in intensive care units and surgery or by space and deep-sea explorers, monitor vital signs such as heart rate and rhythm, body temperature, and breathing rate.

Equipment and Machinery. Biomedical engineers produce a wide variety of other medical machinery, including laboratory equipment and therapeutic equipment. Therapeutic equipment includes laser devices for eye surgery and insulin pumps (sometimes called artificial pancreases) that both monitor blood sugar levels and deliver the appropriate amount of insulin when it is needed.

Imaging Systems. Medical imaging provides important machinery devised by biomedical engineers. This specialty incorporates sophisticated computers and imaging systems to produce computed tomography (CT), magnetic resonance imaging (MRI), and positron emission tomography (PET) scans. In naming its National Institute of Biomedical Imaging and Bioengineering (NIBIB), the U.S. Department of Health and Human Services emphasized the equal importance and close relatedness of these subspecialties by using both terms in the department's name.

Computer programming provides important circuitry for many biomedical engineering applications, including systems for differential disease diagnosis. Advances in bionics, moreover, rely heavily on computer systems to enhance vision, hearing, and body movements.

Biomaterials. Biomaterials, such as artificial skin and other genetically engineered body tissues, are areas promising dramatic improvements in the treatment of burn victims and individuals needing organ transplants. Bionanotechnology, another subfield of biomedical engineering, promises to enhance the surface of artificial skin by creating microscopic messengers that can create the sensations of touch and pain. Bioengineers interface with the fields of physical therapy, orthopedic surgery, and rehabilitative medicine in the fields of splint development, biomechanics, and wound healing.

Medications. Medicines have long been synthesized artificially in laboratories, but chemically synthesized medicines do not use human genes in their production. Medicines produced by using human genes in recombinant DNA procedures are called biologics and include antibodies, hormones, and cell receptor proteins. Some of these products include human insulin, the hepatitis B vaccine, and human growth hormone.

Bacteria and viruses invading a body are attacked and sometimes neutralized by antibodies produced by the immune system. Diseases such as Crohn's disease, an inflammatory bowel condition, and psoriatic arthritis are conditions exacerbated by inflammatory antibody responses mounted by the affected person's immune system. Genetic antibody production in the form of biologic medications interferes with or attacks mediators associated with Crohn's and arthritis and improves these illnesses by decreasing the severity of attacks or decreasing the frequency of flare-ups.

Cloning and Stem Cells. Cloned human embryos could provide embryonic stem cells. Embryonic stem cells have the potential to grow into a variety of cells, tissues, and organs, such as skin, kidneys, livers, or heart cells. Organ transplantation from genetically identical clones would not encounter the recipient's natural rejection process, which transplantations must overcome. As a result, recipients of genetically identical cells, tissues, and organs would enjoy more successful replacements of key organs and a better quality of life. Human cloning is subject to future research and development, but the promise of genetically identical replacement organs for people with failed hearts, kidneys, livers, or other organs provides hope for enhanced future treatments.

Social Context and Future Prospects

Bionics technologies include artificial hearing, sight, and limbs that respond to nerve impulses. Bionics offers partial vision to the blind and prototype prosthetic arm devices that offer several movements through nerve impulses. The goal of bionics is to better integrate the materials in these artificial devices with human physiology to improve the lives of those with limb loss, blindness, or decreased hearing.

Cloned animals exist but cloning is not a yet a routine process. Technological advances offer rapid DNA analysis along with significantly lower cost genetic analysis. Genetic databases are filled with information on many life-forms, and new DNA sequencing information is added frequently. This basic information that has been collected is like a dictionary, full of words that can be used to form sentences, paragraphs, articles, and books, in that it can be used to create new or modified life-forms.

Biomedical engineering enables human genetic engineering. The stuff of life, genes, can be modified or manipulated with existing genetic techniques. The power to change life raises significant societal concerns and ethical issues. Beneficial results such as optimal organ transplantations and effective medications are the potential of human genetic engineering.

—*Richard P. Capriccioso, BS, MD and Christina Capriccioso*

Bibliography

"Biomedical Engineers." *Occupational Outlook Handbook*. Bureau of Labor Statistics, 17 Dec. 2015, www.bls.gov/ooh/architecture-and-engineering/biomedical-engineers.htm. Accessed 27 Oct. 2016.

Braga, Newton C. *Bionics for the Evil Genius: Twenty-five Build-It-Yourself Projects*. McGraw-Hill, 2006.

Fischman, Josh. "Merging Man and Machine: The Bionic Age." *National Geographic*, vol. 217, no. 1, 2010, pp. 34–53.

Hung, George K. *Biomedical Engineering: Principles of the Bionic Man*. World Scientific, 2010.

Richards-Kortum, Rebecca. *Biomedical Engineering for Global Health*. Cambridge UP, 2010.

Smith, Marquard, and Joanne Morra, editors. *The Prosthetic Impulse: From a Posthuman Present to a Biocultural Future*. MIT, 2007.

Bioplastic

SUMMARY

Bioplastics are nature-based substances that are soft enough to be formed or molded but become hard enough to hold their shape. Unlike conventional plastics, which are derived from nonrenewable fossil fuels like petroleum, bioplastics are made from fats, oils, starches, or microbial life-forms obtained from renewable sources such as corn, soybeans, seaweed, egg whites, and shrimp shells. Petroleum-based plastics decompose so slowly that they are generally considered nonbiodegradable, but most bioplastics will disintegrate relatively quickly under the right conditions.

Bioplastics are commonly used to make items such as disposable food packaging, egg cartons, water bottles, and straws. They can also be used to make grocery store bags, plastic pipes, carpeting, mobile phone cases, and even some medical implant devices.

HISTORY

British inventor Alexander Parkes developed the earliest known synthetic bioplastic in 1856. His polymer, *Parkesine*, was made from a material that came from cellulose. He showed off his discovery at the 1862 Great International Exhibition in London, England. Experimentation with bioplastics continued into the 1900s. American automobile pioneer Henry Ford made car parts out of bioplastics for his Model-T, and he manufactured a car with body parts constructed from soy-based plastics in 1941. In 1926, French chemist Maurice Lemoigne discovered that *Bacillus megaterium*, a type of bacterium, produced a form of non-synthetic plastic known as *polyhydroxybutyrate* (PHB).

Over the next several decades, researchers attempted to identify cheaper, renewable raw materials to keep up with the growing demand for plastic products, but most plastics were still made from petroleum-based polymers. That began to change in the 1980s as concerns about the pollution generated by nonbiodegradable plastics increased and more scientists began to look into bioplastics.

Knives, forks, and spoons made from a biodegradable starch-polyester material. By Scott Bauer

TYPES OF BIOPLASTICS

There are several types of bioplastics. The most common bioplastics made in the twenty-first century are plant-starch based. *Celluloids* are a common form of bioplastics made from plant cellulose, much like the material used by Parkes in the 1860s. PHB, the bacteria-produced plastic identified by Lemoigne, is growing in popularity as well. *Polylactic acid* (PLA) is another common bioplastic produced from corn sugar.

Polyhydroxyalkanoates (PHA) are a form of bioplastic generated by a bacterial fermentation process, much like PHB. For mass production, the microbes are provided within an environment that encourages the production of PHA, which is then harvested and processed. The petroleum-based plastic polyethylene can also be made as a bioplastic using ethylene generated from natural sources like sugarcane.

BENEFITS

Bioplastics provide some environmental benefits over petroleum-based polymers. Most bioplastics break down quickly with the help of natural microorganisms. This could help alleviate some of the problems caused by discarding large amounts of plastic products, which include water pollution and overstuffed landfills. Plastic store bags and food packages that dissolve when exposed to the elements can also help curtail unsightly litter.

Packaging blister made by bioplastics (Celluloseacetat). By Christian Gahle, nova-Institut GmbH (Work by Christian Gahle, nova-Institut GmbH)

This protects the natural environment as well as the health of the animals that live in it.

The use of bioplastics decreases the need for fossil fuels to produce plastic items. The renewable nature of the materials used to make bioplastics offers another advantage over plastics produced from nonrenewable fossil fuels. Also, the process used to make bioplastics generally has a lower carbon footprint than the process used to make traditional plastics.

Additionally, bioplastics are often softer than synthetic plastics, with a feel that may be more appealing to consumers. They are easier to print on, they are less likely to alter the flavor of any food they contain, and they are more transparent than plastics made from petroleum. Some also allow water vapor to pass through more easily, which can be an advantage in some applications, such as wrapping fresh bakery items. Researchers are also developing specialized bioplastics. One example is a bioplastic made from egg whites that has antibacterial properties and could have many applications in the medical and food industries.

Bioplastics are also free of *bisphenol A* (BPA), a chemical that many believe has a detrimental effect on human health when used in plastics that store food.

CONCERNS

Most but not all bioplastics are biodegradable, and some, while technically considered biodegradable, require a specific set of circumstances for decomposition to occur. In order to decompose, most bioplastics need an environment with moisture and oxygen, such as a compost pile. A bioplastic item placed in an airtight landfill will become as nonbiodegradable as its petroleum-based counterpart. This is also a concern for products that are made of multiple types of plastic, such as ink pens, which might have a biodegradable casing but contain a traditional plastic tube that holds the ink.

Some researchers are also raising concerns about by-products from the biodegradation process that are showing up in water and soil. One problem is the amount of *methane*—a greenhouse gas that raises the temperature of the atmosphere—produced when bioplastics are placed in landfills.

Another issue is the cost of producing bioplastics. Compared to traditional plastics, the bioplastic PLA costs about 20 percent more to produce. PHA, which is slightly more biodegradable than PLA, costs twice as much to produce as a petroleum-based plastic.

—*Janine Ungvarsky*

Bibliography

"Bioplastics Made From Egg Whites May Soon Make Conventional Plastic History." *Business Standard.* Business Standard Private Ltd. 19 Apr. 2015. Web. 6 Feb. 2016. http://www.business-standard.com/article/news-ani/bioplastics-made-from-egg-whites-may-soon-make-conventional-plastic-history-115041900294_1.html

Chen, Ying Jian. "Bioplastics and Their Role in Achieving Global Sustainability." *Journal of Chemical and Pharmaceutical Research* 6.1(2014): 226–231. Web. 6 Feb. 2016. http://jocpr.com/vol6-iss1-2014/JCPR-2014-6-1-226-231.pdf

Dell, Kristina. "The Promise and Pitfalls of Bioplastic." *Time.* Time Inc. 3 May 2010. Web. 6 Feb. 2016. http://content.time.com/time/magazine/article/0,9171,1983894,00.html

DiGregorio, Barry E. "Biobased Performance Bioplastic: Mirel." *Chemistry & Biology* 16.1. (2009): 1–2. *ScienceDirect.* Web. 6 Feb. 2016. http://www.sciencedirect.com/science/article/pii/S1074552109000076

Jain, Roopesh, and Archana Tiwari. "Biosynthesis of Planet Friendly Bioplastics Using Renewable Carbon Source." *Journal of Environmental Health Science and Engineering* 13.11 (2015). *PMC.* Web. 6 Feb. 2016. http://www.ncbi.nlm.nih.gov/pmc/articles/PMC4340490/

"Lifecycle of a Plastic Product." *American Chemistry Council.* American Chemistry Council Inc. Web. 11 Feb. 2016. https://plastics.americanchemistry.com/Education-Resources/Plastics-101/Lifecycle-of-a-Plastic-Product.html

Phillips, Anna Lena. "Bioplastics Boom." *American Scientist.* The Scientific Research Society. Web. 6 Feb. 2016. http://www.americanscientist.org/issues/pub/bioplastics-boom

Vidal, John. "'Sustainable' Bio-plastic Can Damage Environment." *Guardian.* Guardian News and Media Limited. 25 Apr. 2008. Web. 6 Feb 2016. http://www.theguardian.com/environment/2008/apr/26/waste.pollution

Bioprocess Engineering

SUMMARY

Bioprocess engineering is an interdisciplinary science that combines the disciplines of biology and engineering. It is associated primarily with the commercial exploitation of living things on a large scale. The objective of bioprocess engineering is to optimize either growth of organisms or the generation of target products. This is achieved mainly by the construction of controllable apparatuses. Both government agencies and private companies invest heavily in research within this area of applied science. Many traditional bioprocess engineering approaches (such as antibiotic production by microorganisms) have been advanced by techniques of genetic engineering and molecular biology.

DEFINITION AND BASIC PRINCIPLES

Bioprocess engineering is the use of engineering devices (such as bioreactors) in biological processes carried out by microbial, plant, and animal cells in order to improve or analyze these processes. Large-scale manufacturing involving biological processes requires substantial engineering work. Throughout history, engineering has helped develop many bioprocesses, such as the production of antibiotics, biofuels, vaccines, and enzymes on an industrial scale. Bioprocess engineering plays a role in many industries, including the food, microbiological, pharmaceutical, biotechnological, and chemical industries.

BACKGROUND AND HISTORY

People have been using bioprocessing for making bread, cheese, beer, and wine—all fermented foods—for thousands of years. Brewing was one of the first applications of bioprocess engineering. However, it was not until the nineteenth century that the scientific basis of fermentation was established, with the studies of French scientist Louis Pasteur, who discovered the microbial nature of beer brewing and wine making.

During the early part of the twentieth century, large-scale methods for treating wastewater were developed. Considerable growth in this field occurred toward the middle of the century, when the bioprocess for large-scale production of the antibiotic penicillin was developed. The World War II goal of industrial-scale production of penicillin led to the development of fermenters by engineers working together with biologists from the pharmaceutical company Pfizer. The fungus *Penicillium* grows and produces antibiotics much more effectively under controlled conditions inside a fermenter.

Later progress in bioprocess engineering has followed the development of genetic engineering, which raises the possibility of making new products from genetically modified microorganisms and plants grown in bioreactors. Just as past developments in bioprocess engineering have required contributions from a wide range of disciplines, including microbiology, genetics, biochemistry, chemistry, engineering, mathematics, and computer science, future developments are likely to require cooperation among scientists in multiple specialties.

HOW IT WORKS

Living cells may be used to generate a number of useful products: food and food ingredients (such as cheese, bread, and wine), antibiotics, biofuels, chemicals (enzymes), and human health care products such as insulin. Organisms are also used to destroy or break down harmful wastes, such as those created by the 2010 oil spill in the Gulf of Mexico, or to reduce pollution.

A good example of how bioprocess engineering works is the development of a bioprocess using bacteria for industrial production of the human hormone insulin. Without insulin, which regulates blood sugar levels, the body cannot use or store glucose properly. The inability of the body to make sufficient insulin causes diabetes. In the 1970's, the U.S. company Genentech developed a bioprocess for insulin production using genetically modified bacterial cells.

The initial stages involve genetic manipulation (in this case, transferring a human gene into bacterial DNA). Genetic manipulation is done in laboratories by scientists trained in molecular biology or biochemistry. After creating a genetically engineered bacterium, scientists grow it in a small tubes or flasks and study its growth characteristics and insulin production.

Once the bacterial growth and insulin production characteristics have been identified, scientists increase the scale of the bioprocess. They use or build small bioreactors (1-10 liters) that can monitor temperature, pH (acidity-alkalinity), oxygen concentration, and other process characteristics. The goal of this scale-up is to optimize bacterial growth and insulin production.

The next step is another scale-up, this time to a pilot-scale bioreactor. These bioreactors can be as large as 1,000 liters and are designed and built by engineers to study the response of bacterial cells to large-scale production. During a scale-up, decreased product yields are often experienced because the conditions in the large-scale bioreactors (temperature, pH, aeration, and nutrient supply) differ from those in small, laboratory-scale systems. If the pilot-scale bioreactors work efficiently, engineers will design industrial-scale bioreactors and supporting facilities (air supply, sterilization, and process-control equipment).

All these stages are part of upstream processing. An important part of bioprocess engineering is the product recovery process, or so-called downstream processing. Product recovery from cells often can be very difficult. It involves laboratory procedures such as mechanical breakage, centrifugation, filtration, chromatography, crystallization, and drying. The final step in bioprocess engineering is testing of the recovered product, in which animals are often used.

APPLICATIONS AND PRODUCTS

A wide range of products and applications of bioprocess engineering are familiar, everyday items.

Foods, Beverages, Food Additives, and Supplements. Living organisms play a major role in the production of food. Foods, beverages, additives, and supplements traditionally made by bioprocess engineering include dairy products (cheeses, sour cream, yogurt, and kefir), alcoholic beverages (beer, wines, and distilled spirits), plant products (soy sauce, tofu, sauerkraut), and food additives and supplements (flavors, proteins, vitamins, and carotenoids).

Traditional fermenters with microorganisms are used to obtain products in most of these applications. A typical industrial fermenter is constructed from stainless steel. Mixing of the microbial culture in fermenters is achieved by mechanical stirring, often with baffles. Airlift bioreactors have also been applied in the manufacturing of food products such as crude proteins synthesized by microorganisms. Mixing and liquid circulation in these bioreactors are induced by movement of an injected gas (such as air).

Biofuels. Bioprocess engineering is used in the production of biofuels, including ethanol (bioethanol), oil (biodiesel), butanol, biohydrogen, and biogas (methane). These biofuels are produced by the action of microorganisms in bioreactors, some of which use attached (immobilized) microorganisms. Cells, when immobilized in matrices such as agar, polyurethane, or glass beads, stabilize their growth and increase their physiological functions. Many microorganisms exist naturally in a state similar to immobilization, either on the surface of soil particles or in symbiosis with other organisms.

Environmental Applications. Bioprocess engineering plays an important role in removing pollution from the environment. It is used in treatment of wastewater and solid wastes, soil bioremediation, and mineral recovery. Environmental applications are based on the ability of organisms to use pollutants or other compounds as their food sources. One of the most important and widely used environmental applications is the treatment of wastewater by microorganisms. Microbes eat organic and inorganic compounds in wastewater and clean it at the same time. In this application, microorganisms are placed inside bioreactors (known as digesters) specifically designed by engineers. Engineers have also developed biofilters, bioreactors for removing pollutants from the air. Biofilters are used to remove pollutants, odors, and dust from air by the action of microorganisms. In addition, the mining industry uses bioprocess engineering for extracting minerals such as copper and uranium through the use of bacteria. Microbial leaching uses leaching dumps or tank bioreactors designed by engineers.

Enzymes. Enzymes are used in the health, food, laundry, pulp and paper, and textile industries. They are produced mainly from fungi and bacteria using bioprocess engineering. One of these enzymes is glucose isomerase, important in the production of fructose syrup. Genetic manipulation provides the means to produce many different enzymes, including those not normally synthesized by microorganisms. Fermenters for enzyme production are usually up to 100,000 liters in volume, although very expensive enzymes may be produced in smaller bioreactors, usually with immobilized cells.

Antibiotics and Other Health Care Products. Most antibiotics are produced by fungi and bacteria. Industrial production of antibiotics usually occurs in fermenters (stirred tanks) of 40,000- to 200,000-liter capacity. The bioprocess for antibiotics was developed by engineers during World War II, although it has undergone some changes since the 1980's. Various food sources, including glucose and sucrose, have been adopted for antibiotic production by microorganisms. The modern bioprocess is highly efficient (90 percent). Process variables such as pH and aeration are controlled by computer, and nutrients are fed continuously to sustain maximum antibiotic production. Product recovery is also based on continuous extraction.

The other major health care products produced with the help of bioprocess engineering are steroids, bacterial vaccines, gene therapy vectors, and therapeutic proteins such as interferon, growth hormone, and insulin. Steroids are important hormones that are manufactured by the process of biotransformation, in which microorganisms are used to chemically modify an inexpensive material to create a desired product. Health care products are produced in traditional fermenters.

Biomass Production. Biomass is used as a fuel source, as a source of protein for human food or animal feed, and as a component in agricultural pesticides or fertilizer. Baker's yeast biomass is a major product of bioprocess engineering. It is required for making bread and other baked goods, beer, wine, and ethanol. Yeast is produced in large aerated fermenters of up to 200,000 liters. Molasses is used as a nutrient source for the cells. Yeast is recovered from the fermentation liquid by centrifugation and then is dried. People also use the biomass of algae. Algae are a source of animal feed, plant fertilizer, chemicals, and biofuels. Algal biomass is produced in open ponds, in tubular glass, or in plastic bioreactors.

Animal and Plant Cell Cultures. Bioprocess engineering incorporating animal cell culture is used primarily for the production of health care products such as viral vaccines or antibodies in traditional fermenters or bioreactors with immobilized cells. Antibodies, for example, are produced in bioreactors with hollow-fiber immobilized animal cells. Plant cell culture is also an important target of bioprocess engineering. However, only a few processes have been successfully developed. One successful process is the production of the pigment shikonin in Japan. Shikonin is used as a dye for coloring food and has applications as an anti-inflammatory agent.

Chemicals. There is an on-going trend in the chemical industry to use bioprocess engineering instead of pure chemistry for production of a variety of chemicals such as amino acids, polymers, and organic acids (citric, acetic, and lactic). Some of these chemicals (citric and lactic acids) are used as food preservatives. Many chemicals are produced in traditional fermenters by the action of microbes.

SOCIAL CONTEXT AND FUTURE PROSPECTS

The role of bioprocess engineering in industry is likely to expand because scientists are increasingly able to manipulate organisms to expand the range and yields of products and processes. Developments in this field continue rapidly.

Bioprocess engineering can potentially be the answer to several problems faced by humankind. One such problem is global warming, which is caused by rising levels of carbon dioxide and other greenhouse gases. A suggested method of addressing this issue is carbon dioxide removal, or sequestration, based on bioprocess engineering. This bioprocess uses microalgae (microscopic algae) in photobioreactors to capture the carbon dioxide that is discharged into the atmosphere by power plants and other industrial facilities. Photobioreactors are various types of closed systems made of an array of transparent tubes in which microalgae are cultivated and monitored under illumination.

The health care industry is another area where bioprocess engineers are likely to be active. For example, if pharmaceutical applications are found for stem cells, a bioprocess must be developed to produce a reliable, plentiful source of stem cells so that these drugs can be produced on a large scale. The process for growing and harvesting cells must be standardized so that the cells have the same characteristics and behave in a predictable manner. Bioprocess engineers must take these processes from laboratory procedures to industrial protocols.

In general, the future of bioprocess engineering is bright, although questions and concerns, primarily about using genetically modified organisms, have arisen. Public education in such a complex area of science is very important to avoid public mistrust of bioprocess engineering, which is very beneficial in most applications.

—Sergei A. Markov, PhD

Bibliography

Bailey, James E., and David F. Ollis. *Biochemical Engineering Fundamentals.* 2d ed. New York: McGraw-Hill, 2006. Covers all aspects of biochemical engineering in an understandable manner.

Bougaze, David, Thomas R. Jewell, and Rodolfo G. Buiser. *Biotechnology. Demystifying the Concepts.* San Francisco: Benjamin/Cummings, 2000. Classical book on biotechnology and bioprocessing.

Doran, Pauline M. *Bioprocess Engineering Principles.* London: Academic Press, 2009. A solid, basic textbook for students entering the field.

Glazer, Alexander N., and Hiroshi Nikaido. *Microbial Biotechnology: Fundamentals of Applied Microbiology.* New York: Cambridge University Press, 2007. In-depth analysis of the application of microorganisms in bioprocessing.

Heinzle, Elmar, Arno P. Biwer, and Charles L. Cooney. *Development of Sustainable Bioprocesses: Modeling and Assessment.* Hoboken, N.J.: John Wiley & Sons, 2007. Looks at making bioprocesses sustainable by improving them. Includes case studies on citric acid, biopolymers, antibiotics, and biopharmaceuticals.

Nebel, Bernard J., and Richard T. Wright. *Environmental Science: Towards a Sustainable Future.* 10th ed. Englewood Cliffs: Prentice Hall, 2008. Describes several bioprocesses used in waste treatment and pollution control.

Yang, Shang-Tian. *Bioprocessing for Value-Added Products from Renewable Resources: New Technologies and Applications.* Amsterdam: Elsevier, 2007. Reviews the techniques for producing products through bioprocesses and lists suitable organisms, including bacteria and algae, and describes their characteristics.

C

C

SUMMARY

C is a programming language that is used to create computer programs. At the most basic level, computer programming involves writing code that tells a computer what to do. Computers are not very useful without codes that tell them which operations to perform. Programmers use different languages, including the C programming language, to tell computers what to do. Although the C programming language has been used since the 1970s, programmers still learn C today. Many advanced programming languages have been developed over the years, but some programmers find instances where they prefer to use C for specific functions.

HISTORY OF C

A computer scientist named Dennis Ritchie was working at Bell Laboratories in the 1970s when he started to develop C. The C programming language was based on other programming languages, but it was unique because it was simpler. In 1978, Ritchie and Brian Kernighan wrote a book that introduced the fundamental elements of the C programming language. The version of C that Ritchie and Kernighan wrote about was called the K&R (for Kernighan and Ritchie) version of C.

C became an important programming language during the 1980s, and an improved version of the program was eventually adopted by the American National Standards Institute (ANSI). This improved version of the program became known as ANSI C. The ANSI adopted one version of C because several versions existed and accepting one version (or other very similar versions) created a standard in the industry. Having a standard version of C allowed people to write programs that were compatible with one another, and this encouraged people to use the language and develop software with it.

Some programmers who use the language use "nonstandard dialects" of C that are somewhat different from the K&R and ANSI C versions. However, these dialects (much like dialects of a spoken language) are based on the same basic ideas and functions. Many of these dialects can be translated into Standard C (another name for ANSI C) using Standard C translators.

C remained an important and popular programming language in the twenty-first century, though it had undergone three major revisions and faced competition from other languages. Even parts of some computer operating systems, such as Windows and Linux, were written using C. Although the language is still used today, it is difficult for programmers to find jobs in which they only program in C. Many C programmers also use C++ (which was based on C), Java, and similar languages.

WORKING WITH C

C is a high-level computer programming language, which means that it is often used to develop new software programs (e.g., computer operating systems, word processing software, etc.). Even though it is a high-level language, C was designed to be somewhat simplistic. Because the language is so simple, it is very flexible and allows programmers to develop many types of programs and commands.

Ken Thompson (left) with Dennis Ritchie (right), the inventor of the C programming language.

C is different from some programming languages because it has to be put through a compiler program. A programmer must write the C program in a text editor, and the finished program has to be passed through a compiler. A compiler is a computer program that translates the C language into machine code. The compiler creates an executable file that will run on a computer.

The C programming language contains functions and variables. These functions and variables tell a computer what to do. Programmers use the functions and variables to string together different commands. Together, these commands form a computer program. C also uses some words and characters for special purposes. Key words are grouped into categories (e.g., labels, operations, and storage class) and can be used in the code only for one unique purpose. C also has a number of operations that are represented by one or more symbols.

ADVANTAGES AND DRAWBACKS OF C

Using the C language for programming has a few advantages. One of the advantages is that C is used in many places. Some operating systems include C programming, and many web-based programs use the language. C is also fairly portable, which means it can be used on different machines without having to be recompiled. Nevertheless, other programs—such as Java—are even more portable. Another advantage of the C language is that it is simplistic, though this does not necessarily mean that the language is simple to learn or use. The simplicity of C makes it flexible and gives programmers many options when using it. A third advantage of the C language is that it has been used in different forms since the 1970s.

There are some drawbacks to using C. One of the major disadvantages of the language is that some of the functions can make it possible for malware to penetrate programs written using C. When C was developed, programmers did not anticipate people creating malware that could hack C programs. Due to this threat, programmers currently avoid some C functions. Another drawback of C is that it is easy to program errors and bugs into the code. C is a concise and simple language, but it does not contain the functions that other languages have to correct errors in code. Many C programmers learn to check their work periodically to ensure that it does not contain bugs, but it is still difficult to make C free of errors. Because of this, C can sometimes be more difficult for beginning programmers, even though it is a simple language.

The C Programming Language 1st edition cover.

—*Elizabeth Mohn*

Bibliography

"The Basics: How Programming Works." *Microsoft Developer Network.* Microsoft. Web. 11 Aug. 2015. https://msdn.microsoft.com/en-us/library/ms172579(v=vs.90).aspx

Bolton, David. "Is C Still Relevant in the 21st Century?" *Dice.* Dice. 6 Dec. 2014. Web. 11 Aug. 2015. http://insights.dice.com/2014/12/08/c-still-relevant-21st-century/

Burgess, Mark. "C Programming Tutorial." *Mark Burgess.* Mark Burgess. Web. 11 Aug. 2015. http://markburgess.org/CTutorial/C-Tut-4.02.pdf

"C Language Reference." *Microsoft Developer Network.* Microsoft. Web. 11 Aug. 2015. https://msdn.microsoft.com/en-us/library/fw5abdx6.aspx

"The C Programming Language." *College of Engineering & Computer Science: University of Michigan, Dearborn.* The Regents of the University of Michigan. Web. 11 Aug. 2015. http://groups.engin.umd.umich.edu/CIS/course.des/cis400/c/c.html

"C Programming Language." *DI Management Services.* DI Management Services Pty Limited. Web. 11 Aug. 2015. http://www.di-mgt.com.au/cprog.html

C++

SUMMARY

C++ is an object-oriented programming language based on the C programming language. Pronounced "C plus plus," C++ was developed in the late 1970s, though it originally was given a different name. C++ standards later followed. Some of the main features of C++ are its power, portability, and simplicity. C++ is one of the most popular programming languages in use today.

AT&T Bell Labs, Holmdel, N.J. By derivative work: Lee Beaumont (Bell_Labs_Holmdel,_The_Oval2.png) , via Wikimedi

APPLICATIONS

The C++ programming language was developed by computer scientist Bjarne Stroustrup in 1979. Stroustrup was working with a language known as Simula, which many consider the first object-oriented programming language. Object-oriented programming is a type of programming that defines both the data types and the types of functions of a data structure, thereby making the data structure an object. A data type is a classification of information. A function is a named procedure that performs an action. A data structure is a format for organizing information. An object is an item that can be manipulated.

Stroustrup discovered that object-oriented programming could be used for software development, but Simula was not fast enough. So Stroustrup set out to take another existing language, the C language, and incorporate object-oriented programming into it. C with Classes was soon born. This new language was so named because it combined the feature of classes with all of the features of the C language. A class is a category of objects. Furthermore, a class defines the commonalities of its objects.

Photo of Bjarne Stroustrup, creator of the programming language C++.

In 1983, C with Classes was renamed C++. In 1985, C++ was released for commercial use. That same year, Stroustrup published the book *The C++ Programming Language*, which is a reference guide for C++. *The Annotated C++ Reference Manual* followed in 1990. C++ standards also were developed, including one in 1998 (informally called C++98), another in 2005 (informally referred to as C++0x), and one in 2011 (informally named C++11).

C++ is known for its power, simplicity, and portability, among other features. Its power can be seen in its wide range of data types, functions, and statements. A statement is an instruction that performs an action. Some of the statements that C++ includes are control statements and decisionmaking statements. As for simplicity, C++ programs are able to be written in simple language that can be easily understood. The portability of C++ is evident because it can be used on a variety of platforms.

C++ is a widely used programming language. Many programs have been written in C++. These include Adobe applications, Microsoft applications, parts of the Mac OS/X operating system, and MongoDB databases. According to Katie Bouwkamp of a programming boot camp called Coding Dojo, as of 2016, C++ was among the most in-demand programming languages based on the number of programming jobs available. Only five other programming languages had more programming jobs available: SQL, Java, JavaScript, C#, and Python, in that order.

—*Michael Mazzei*

BIBLIOGRAPHY

Beal, Vangie. "OOP—Object Oriented Programming." *Webopedia*, www.webopedia.com/TERM/O/object_oriented_programming_OOP.html. Accessed 18 Jan. 2017.

Bouwkamp, Katie. "The 9 Most In-Demand Programming Languages of 2016." *Coding Dojo Blog*, 27 Jan. 2016, www.codingdojo.com/blog/9-most-in-demand-programming-languages-of-2016/. Accessed 18 Jan. 2017.

"A Brief Description." *cplusplus.com*, www.cplusplus.com/info/description/. Accessed 18 Jan. 2017.

"C/C++." *TechTerms*, techterms.com/definition/cplusplus. Accessed 18 Jan. 2017.

"C++ Programming Language." *Techopedia*, www.techopedia.com/definition/26184/c-programming-language. Accessed 18 Jan. 2017.

"Features of C++." *Sitesbay*, www.sitesbay.com/cpp/features-of-cpp. Accessed 18 Jan. 2017.

"History of C++." *cplusplus.com*, www.cplusplus.com/info/history/. Accessed 18 Jan. 2017.

"History of the C++ Language." *CodingUnit Programming Tutorials*, www.codingunit.com/cplusplus-tutorial-history-of-the-cplusplus-language. Accessed 18 Jan. 2017.

CENTRAL LIMIT THEOREM

SUMMARY

The *central limit theorem* is a concept in statistics that states that the distribution of the statistical mean in a sample of the population will approach normal distribution as the sample size gets larger. In other words, even if data obtained from independent random samples seems skewed to one particular side, when the sample size is increased, eventually the data will reflect the probable statistical average. The central limit theorem is a fundamental pillar of statistics and allows researchers to arrive at conclusions about entire populations by examining data from smaller sample sizes. Statisticians disagree on what constitutes a large enough sample size for the central limit theorem to provide valid results. In general, sample sizes of thirty or more are considered sufficient, although some researchers believe samples should be more than forty or fifty.

BACKGROUND

Statistics is a mathematical science that collects numerical data samples and analyzes those samples to determine the probability that they represent a larger whole. If a statistician wanted to discover the average height of every person in the United States, it would be a near impossible task to measure more than three hundred million people. Therefore, a researcher would measure a smaller

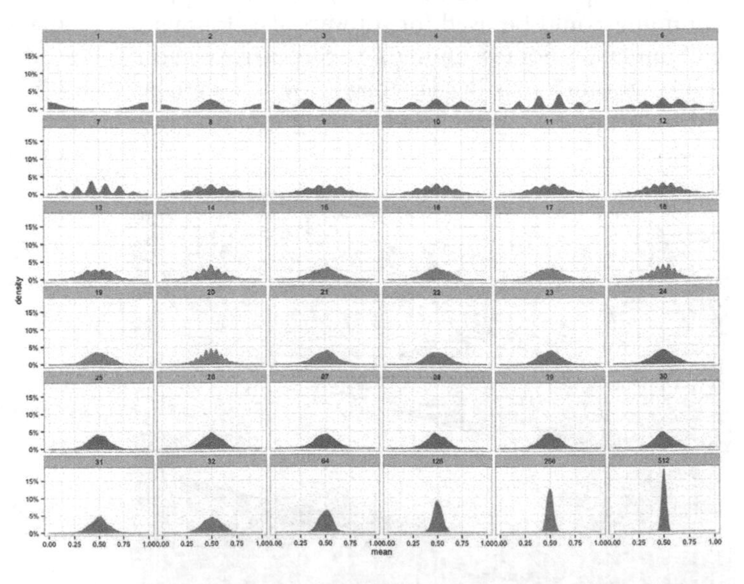

This chart shows the mean of binomial distributions of different sample sizes simulations to show as it increases the sample, it converges toward the real mean. By Daniel Resende

sample size of people chosen at random so as not to unintentionally influence the results. Taking the sum of all the heights in the example and dividing it by the number of people sampled would reveal the statistical mean, or average.

Because the statistician is only measuring a segment of the population, there are several variable factors that need to be taken into consideration. The variance measures the distance each number in a data set is from the mean. Mathematically, variance is determined by taking the distance each point is from the mean and squaring that number, or multiplying the number by itself. The variance is the average of those results. The standard deviation is a measure of the dispersion of the data set from the mean. This means the further the data points are from the mean, the higher the standard deviation will be. Standard deviation is measured as the square root of the variance. For example, if a sample size of four people reveals their heights to be 50 inches, 66 inches, 74 inches, and 45 inches, then the mean would be 58.75 inches. The variance in this case would be 137.82, and the standard deviation would be 11.74.

Normal distribution is the probability distribution of data points in a symmetrical manner with most of the points situated around the mean. This can be illustrated in the common bell curve that is a graphic with a rounded peak in the center that tapers away at either end. In a graph representing normal distribution, the mean is represented by the central peak of the curve, while the standard deviation determines the peak's height.

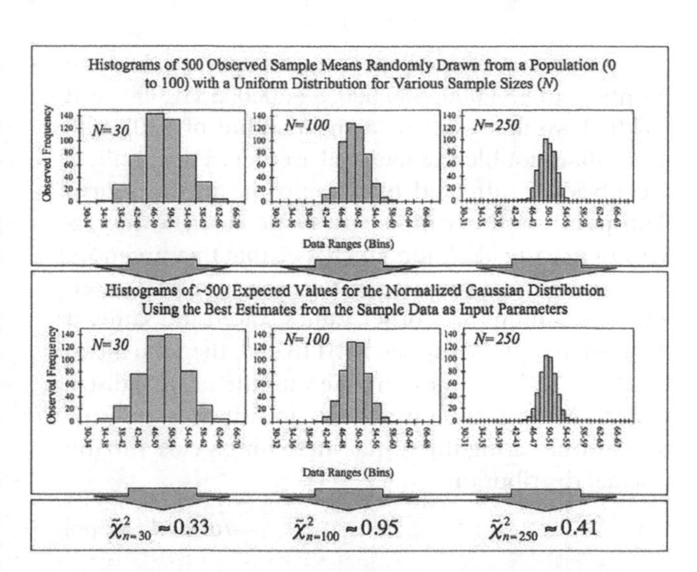

This figure demonstrates the central limit theorem. It illustrates that increasing sample sizes result in sample means which are more closely distributed about the population mean. It also compares the observed distributions with the distributions that were estimated. By Gerbem (Own work)

PRACTICAL APPLICATIONS

Using height as an example is fairly straightforward, as most people in a population tend to be at or close to average. The central limit theorem comes into play when the data from a sample size does not fit the normal distribution and seems to misrepresent the statistical probability of the data. In an analysis of average height, recording the measurements of eight people may be able to yield reliable data that can lead to an accurate result. If a statistician is trying to measure wealth by recording the incomes of eight people, however, a disparity in one respondent's income may skew the results. For example, seven of the eight people may earn between $30,000 and $100,000, but if the eighth person is a millionaire, then the statistical mean would calculate to far more than the second wealthiest respondent. The central limit theorem holds that if the sample size is increased, then the results will move closer to normal distribution, and a more accurate depiction of the average household wealth. Many statisticians say that a sample size of thirty or more is enough to achieve accurate results; however, some insist on a size of more than forty or fifty. In cases where the data points are unusually irregular, a statistician may need to utilize a larger sample size.

In practice, statisticians often take repeated random samples of the same size from the same population. These individual samples are then averaged together to get a data point. The process is repeated a number of times to arrive at a data set. If, in the wealth example, the sample size is three people, then the incomes of three people would be recorded, averaged to find the mean, and that figure would become a data point. Therefore, if a random sampling of three people asked for their income, and they responded with $21,000, $36,000, and $44,000, then their mean income would be $33,667.

For comparison's sake, assume the average salary in the United States was $45,000. If the survey was done correctly, the normal distribution should be near this figure. A group of five data points that yielded income numbers of $33,000, $39,000, $44,000, $351,000, and $52,000 would result in a mean value of $103,800, more than double the national average. The numbers are obviously affected by the income in the fourth data point. If the sample size is increased to ten respondents, the likelihood increases that they are more representative of the true average salary in the United States. Assuming the other values stayed the same, if the fourth figure dropped to $118,000, then the mean would be $57,200, more in line with the normal distribution. Moving the sample size to fifteen, twenty, or thirty would bring the results increasingly closer to the normal distribution.

—Richard Sheposh

BIBLIOGRAPHY

Adams, William J. *The Life and Times of the Central Limit Theorem.* 2nd ed., American Mathematical Society, 2009.

Annis, Charles. "Central Limit Theorem." *Statistical Engineering,* www.statisticalengineering.com/central_limit_theorem.htm. Accessed 24 Jan. 2017.

"Central Limit Theorem." *Boston University School of Public Health,* 24 July 2016, sphweb.bumc.bu.edu/otlt/MPH-Modules/BS/BS704_Probability/BS704_Probability12.html. Accessed 25 Jan. 2017.

Deviant, S. "Central Limit Theorem." *The Practically Cheating Statistics Handbook.* 3rd ed., CreateSpace Independent Publishing, 2010, pp. 88–97.

Dunn, Casey. "As 'Normal' as Rabbits' Weights and Dragons' Wings." *New York Times,* 23 Sept. 2013, www.nytimes.com/2013/09/24/science/as-normal-as-rabbits-weights-and-dragons-wings.html. Accessed 24 Jan. 2017.

Martz, Eston. "How the Central Limit Theorem Works." *The Minitab Blog,* 15 Apr. 2011, blog.minitab.com/blog/understanding-statistics/how-the-central-limit-theorem-works. Accessed 24 Jan. 2017.

Nedrich, Matt. "An Introduction to the Central Limit Theorem." *Atomic Object,* 15 Feb. 2015, spin.atomicobject.com/2015/02/12/central-limit-theorem-intro/. Accessed 24 Jan. 2017.

Paret, Michelle, and Eston Martz. "Tumbling Dice & Birthdays: Understanding the Central Limit Theorem." *Minitab,* Aug. 2009, www.minitab.com/uploadedFiles/Content/News/Published_Articles/CentralLimitTheorem.pdf. Accessed 25 Jan. 2017.

CHARLES BABBAGE'S DIFFERENCE AND ANALYTICAL ENGINES

SUMMARY

Charles Babbage invented the Analytical Engine in the mid-nineteenth century. In computer history, Babbage's Analytical Engine is considered to be one of the first computers. For a variety of reasons, the machine never came to fruition as an engine that was actually built, not least of which that it was thought to be a grandiose idea at that time. Instead, the Analytical Engine remained a complex design that preempted the creation of other general purpose computers by a century. The Analytical Engine was designed as a mechanical type of modern computer, without being electronic. Some of the internal functions are remarkably similar to microprocessors and programmable computers of the modern age. The London Science Museum has exhibits of parts of Babbage's machinery.

BRIEF HISTORY

CHARLES BABBAGE. Charles Babbage (1791–1871) rose to prominence as an English mathematician and mechanical engineer who invented the Analytical Engine. Babbage was also known as a philosopher and as someone proficient in code-breaking. He studied mathematics at Cambridge University and subsequently gained the position of professor of mathematics there. Babbage is considered inventor of the first mechanical computer.

Ada Lovelace. Augusta Ada King (1815–52), daughter of the poet Lord Byron, was the Countess of Lovelace. Ada Lovelace was an English writer who described Babbage's Analytical Engine. Her translation of Luigi Menabrea's Italian essay on the Analytical Engine was a significant step in computer history. She wrote detailed annotations that comprised a method of calculating Bernoulli numbers. This first algorithm method appears as an early type of computer programming. Lovelace is also recognized as having seen beyond Babbage's focus on the mathematical calculation capacities of the Analytical Engine, perceiving the possibility of computers to do even more than that.

Difference and Analytical Engines. The Difference Engine was Charles Babbage's first computer design, which he began working on in 1821. The primary function of the Difference Engine was as a calculator, able only to work with the addition principle of mathematics via the differences method, rather than performing more complex arithmetical calculations. The Analytical Engine, on the other hand, has an increased capacity. It does not function merely to calculate arithmetic in a mechanized way, but rather acts as general-purpose type of computer. The Analytical Engine went through numerous experimentations and progressions until it reached its completed design in 1840.

Numerous obstacles prevented the Analytical Engine from being built. The machine was designed in an extremely complex way, even though the basis of the conceptualization was logical. Production would have been very expensive, and Babbage lacked these funds or the project foundation and management abilities to actualize seeing the creation move from idea to actualization. Moreover, had Parliament been prepared to finance the construction of the Analytical Engine, this would have been advantageous. This was not to be, however, as there were many other projects advocating for funds, and it was challenging to raise awareness and determine the value of the Analytical Engine project.

APPLICATION AND FUNCTION

The Analytical Engine was created as a machine that could process arithmetical calculations. Within the Victorian era context, Babbage's invention was part of the industry of machines. His mathematical genius took the concept of a machine being able to do physical things to a higher level, that of mental type calculations. His fascination with numbers and the desire for accuracy in computing tables of sums led to the calculating functionality of his experiments. Printing out the tables was a further step in avoiding errors by human hand. Babbage's desire was that his Analytical Engine was not restricted to one method of calculation, but that the possibilities of calculating and computing mathematical numbers would be limitless. A profound innovation was the idea of the machine's user controlling what data he or she wished to put in and to assess and ascertain outcome. Generally, machines had functioned to do a specific task, rather than being controlled by the person using it in a general way.

Methodology. The Analytical Engine was designed to utilize punch cards, an idea generated from their usage in French Jacquard textile weaving looms. These were to be used for data input to deliver instructions, in addition to allowing running of the operational program. A system of output devices was also put in place, foreseeing a similar mechanism in electronic computers of the future. Babbage devised internal functional systems that appear to be uniquely modern, albeit that he thought of these ideas in the middle of the nineteenth century. The Analytical Engine contained a data storage capacity for memory purposes, as well as a unit of arithmetic. Punch cards remained popular in the twentieth century for data storage and input and for processing purposes. These remained in use in early digital computers.

Similarities to modern digital computers: The Analytical Engine possessed a storage space, a "store" and a "mill," as two units with different functional services. The store held numbers and intermediate math results, whereas the mill processed the arithmetic. These correspond to the memory bank and central processor with which later computers are known. The notion of data input and output was also part of the mechanism and function of the Analytical Engine. Hardcopy printing out from the engine was also in place. Although Babbage had not conceived the terminology associated with modern computers, the engine's capacity to perform modern-day functions was apparent. These include aspects related to conditional branching, looping, and microprogramming. Latching, iteration, and polling, as well as parallel processing, are further capabilities inherent in his machine. While Babbage intended the user of the Analytical Engine to be able to perform generalized actions, the notion of the specific was to be focused on later. This would manifest in future developments as the kind of software with which modern day computer users are familiar.

Charles Babbage did not live to see his inventions constructed in. A certain part of his first Difference Engine was completed. Construction of the Analytical Engine was not accomplished. At the time of his death, he had been working on an experimental section of the engine. His legacy has become increasingly evident as the forerunner of computer engines. His drawing designs are studied, as are the pieces of equipment that he left behind. The design of the Analytical Engine is perceived as revolutionary, particularly his conceptualization of the memory storage, central processing unit, and programmable performance capabilities.

—Leah Jacob, MA

BIBLIOGRAPHY

"Ada Lovelace." ***History Mesh.*** History Mesh, n.d. Web. 15 June 2016.

"The Analytical Engine." ***History Mesh.*** History Mesh, n.d. Web. 15 June 2016.

"The Analytical Engine of Charles Babbage." History-computer.com. ***History-computer.com,*** n.d. Web. 15 June 2016.

"The Babbage Engine." ***Computer History Museum.*** Computer History Museum, n.d. Web. 15 June 2016.

"The Babbage Engine. A Brief History." ***Computer History Museum.*** Computer History Museum, n.d. Web. 15 June 2016.

"Charles Babbage (1791-1871)." ***BBC.*** BBC, 2014. Web. 15 June 2016.

"Charles Babbage." ***History Mesh.*** History Mesh, n.d. Web. 15 June 2016.

"Charles Babbage's Analytical Engine." Online video clip. ***YouTube.*** YouTube, 12 Aug. 2012. Web. 15 June 2016.

"The Engines." ***Computer History Museum.*** Computer History Museum, n.d. Web. 15 June 2016.

Menabrea, L.F. "Sketch of The Analytical Engine Invented by Charles Babbage." ***Fourmilab.***Fourmilab, n.d. Web. 15 June 2016

Park, Edwards. "What a Difference the Difference Engine Made From Charles Babbage's Calculator Emerged Today's Computer." ***Smithsonian.com.*** Smithsonian Magazine, Feb. 1996. Web. 15 June 2016.

"Punch Cards." ***History Mesh.*** History Mesh, n.d. Web. 15 June 2016.

Walker, John. "The Analytical Engine. The First Computer." ***Fourmilab.*** Fourmilab, n.d. Web. 15 June 2016.

"Who Was Charles Babbage?" ***University of Minnesota.*** Regents of the University of Minnesota, 9 Feb. 2016. Web. 15 June 2016.

CLIENT-SERVER ARCHITECTURE

SUMMARY

Client-server architecture is a computing relationship used to share files across one or more computers, known as clients, and a central computer system, known as a server.

Client-server architecture sometimes is called a networking computing model because a network is used to relay all requests and services between a client and server. A *network* is a group of two or more connected computer systems or devices. Client-server architecture also is known as producer-consumer architecture, in which a server is the producer and the client is the consumer. In this case, the producer provides services that are in demand by the consumer.

WHAT IS CLIENT-SERVER ARCHITECTURE?

Essentially, a client-server architecture relationship is one in which a client links to a server, typically by way of an Internet connection, to complete specific tasks. Some of these tasks include access to applications, access to the server's processing power, database management, file sharing, printing, and storing files.

Architecture refers to the hardware and software on a computer. *Hardware* refers to physical items on a computer such as the display screen, keyboard, and disk drive, and *software* includes the programs, instructions, and data stored on a computer.

A *client* is considered an application, or program, that runs on a range of devices such as a personal computer (PC), laptop, tablet, and cell phone or smartphone. These devices must have the capability to connect to the Internet to link to a server. Once clients link to a server, they can obtain files, processing power, and more. An e-mail client, an application that enables a person to send and receive e-mails on a computer, is an example of a client in a client-server architecture relationship.

A *server* is a computer that manages certain network tasks and contains any files the computers in the network may need to access. It can be thought of as a central hub. However, a server can only communicate with so many computers at once. If the server is overloaded with connections, the connections will slow down or stop. This means that computers in the network will lose access to the server. Additionally, each computer in the network has just one path to communicate with the server. This path is limited by *bandwidth*, which is how much data can be transferred along the connection at one time. The limited bandwidth of a single connection causes a hard limit to how quickly files can be transferred between a central server and a computer.

Servers enable clients to do a particular task and are dedicated to complete only one specific task. (Multiprocessing operating systems, however, can complete multiple tasks at once, referring services as needed.) Because of this, several types of servers exist. *File servers* manage disk drives and store files. Most users on a network can access and save files to a file server. *Print servers* allow clients to print to one or several printers. *Network servers* manage traffic, or activity, on the network. *Database servers* process information on the database to quickly locate requests. A *database* is a collection of organized information.

A server can manage several clients at the same time, and a client can connect to several servers simultaneously to access different services.

An example of client-server architecture is the Internet: A Web server provides clients with Web pages. For example, when a person accesses his or her bank account from a computer, a client program sends a request to the server at the bank. This request is forwarded on until the correct banking information is retrieved from a database. The bank account information requested is then sent back to the person's client program and is displayed on the computer to the person.

PEER-TO-PEER (P2P) ARCHITECTURE

Peer-to-peer (P2P) architecture is another type of network architecture used to share files across computers. A P2P network is an alternative to client-server architecture.

In a P2P network, there is no central server. Instead, every computer in a network uses specialized software to connect itself to every other computer in the network at the same time. The P2P network software breaks files into tiny data packets. These packets are sent through the network to the computer asking for the file and then are reassembled into a copy of the original file.

P2P networks are most useful when more than one computer in the network contains a file. In this

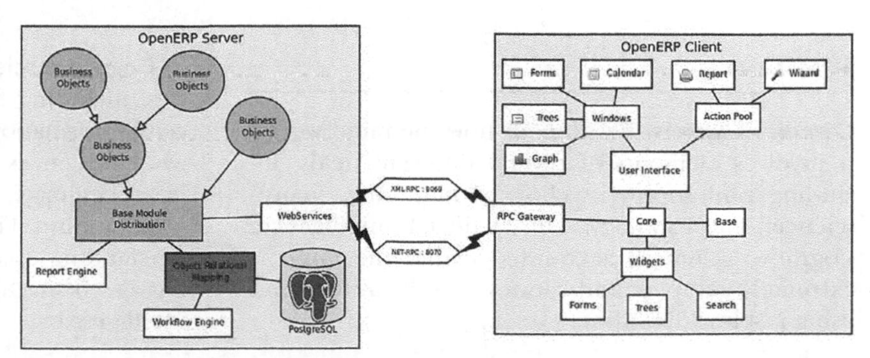

Software architecture of OpenERP. Nicos interests at en.wikipedia

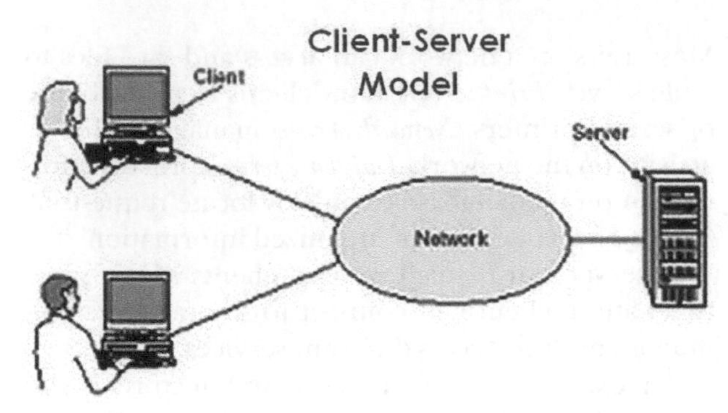

This is a Client-Server Model. By Bp2010.hprastiawan (Own work)

circumstance, multiple computers will send packets of data to the computer asking for the file over their own connections, providing an extremely fast download rate. The process of providing a file for download over a P2P network is called *seeding*, while the process of downloading a file over a P2P network is called *leeching*.

In a client-server architecture relationship, if a central server fails, the entire network is taken offline until the server can be repaired. The files on the central server may not always be recoverable, leading to permanent data loss. However, on a P2P network, there is no central server to fail. Should one computer be removed from the network, the network has lost only a single leeching *node*, or peer. In a large P2P network, most files are distributed across a large number of computers to increase download speeds; therefore, files are not lost if a single computer is removed from the network.

P2P networks are used for many purposes. For example, the music streaming service Spotify uses a P2P network to increase download speeds. P2P networks also are used by research facilities and video game distributers to download extremely large files faster than other server networks.

—Angela Harmon

BIBLIOGRAPHY

Beal, Vangie. "Client-Server Architecture." *Webopedia*. Quinstreet Enterprise, n.d. Web. 7 Oct. 2015. http://www.webopedia.com/TERM/C/client_server_architecture.html.

"Client/Server Architecture." *Techopedia*. Janalta Interactive, 2015. Web. 7 Oct. 2015. https://www.techopedia.com/definition/438/clientserver-architecture.

"Client/Server (Client/Server Model, Client/Server Architecture) Definition." *TechTarget*. TechTarget, 2015. Web. 7 Oct. 2015. http://searchnetworking.techtarget.com/definition/client-server.

Kayne, R. "What Is P2P?" *WiseGeek*. Conjecture, 2015. Web. 7 Oct. 2015. http://www.wisegeek.com/what-is-p2p.htm#comments

Tyson, Jeff. "How Internet Infrastructure Works." *HowStuffWorks*. InfoSpace, 2015. Web. 7 Oct. 2015. http://computer.howstuffworks.com/internet/basics/internet-infrastructure.htm.

"What Is Peer-to-Peer (P2P)?" *GigaTribe*. GigaTribe, n.d. Web. 7 Oct. 2015. https://www.gigatribe.com/en/help-p2p-intro.

COGNITIVE SCIENCE

SUMMARY

Cognitive science is the study of how the mind works. It involves elements of several different fields, including philosophy, psychology, linguistics, neuroscience, anthropology, and artificial intelligence. Cognitive scientists acknowledge that the mind is extremely complex and cannot be fully understood using just one discipline.

Cognitive science covers several different categories, including how people gather, retain, and perceive information, what triggers emotional responses, what influences different behaviors, and how people learn language. It also includes subjects beyond the human mind. For example, it has increased the understanding of animal behaviors and intelligence. It has also been the basis for advancements in artificial intelligence.

Studies in this field have led to major breakthroughs. They have helped identify and address mental health problems and led to new methods of psychological therapy. They have changed the field of linguistics. Discoveries regarding how the brain computes information have led to advances in computer technology and artificial intelligence.

BRIEF HISTORY

Efforts to study the brain and the mind date back to ancient Greek philosophers, but the interdisciplinary approach that makes up modern cognitive science did not take form until the mid-twentieth century. In the 1940s, Warren McCulloch and Walter Pitts developed models for computer networks based on what was known about biological neural patterns at the time. Using artificial models to replicate brain activity was a crucial step in understanding how the brain's physical attributes led to abstract results such as thoughts and decisions.

In the 1950s, J.C.R. Licklider began testing theories on brain activity at the Massachusetts Institute of Technology (MIT). Using computers to simulate a human brain's processes, Licklider researched psychology, laying the groundwork for what would become cognitive science. These studies converged with new developments in the linguistics field.

Linguistics was initially designed to study the history of languages, but since the turn of the twentieth century, it underwent a shift; it now examines the structures of languages and how they changed over time. This new emphasis revealed some common trends throughout many languages, leading linguists to theorize that certain elements of communication were integral to human development. By the 1950s, a clear connection between patterns in linguistics and in the human mind was commonly accepted.

Previously, concepts typically associated with the mind, such as ideas, emotions, and opinions, were considered too abstract and complex to connect with biological concepts such as the brain's components and activity. But mid-twentieth-century discoveries in computer science, psychology, and linguistics overlapped in a way that captured researchers' attention, particularly those at MIT.

The term "cognitive science" did not emerge until 1973. The 1970s saw the field undergo significant growth: the Cognitive Science Society launched, along with the periodical journal *Cognitive Science*. In 1982, Vassar College became the first school to offer Cognitive Science as an undergraduate degree. Others soon followed, and it became a global science.

In the late twentieth century, cognitive science made great strides in artificial intelligence (AI). Two major forms of AI emerged. One used programming designed to emulate decisionmaking processes and respond to specific situations. Later, cognitive scientists focused on creating AIs based on neural networks that were inspired by the complex layout of the brain. Both symbolic and connectionist models of AI had different specialties.

IMPACT

Because of the many disciplines included in cognitive science, it has made an impact in several other fields. Since the formation of cognitive science, the field of linguistics has become intertwined with the mind. It has been used to find the most effective ways of teaching languages to both children and adults. It has also been applied to translation projects, negotiation situations, and computer programming.

Artificial intelligence (AI) was used since the infancy of cognitive science to create models of brain activity. With increased understanding of the human brain came more knowledge of how to program AI. The field advanced at an extremely fast rate in the late twentieth century, and only sped up in the twenty-first. AI software is prevalent in many forms in modern society. Search engines, social media, and online advertising use AI programs, as do cell phones. It has made an impact on the workforce and production, in the form of autonomous machines in manufacturing plants and software that can track data and inventory. Machines with AI help with medical procedures. AI is also changing how the military operates; the military now uses remotely operated drones capable of performing reconnaissance or carrying out precise strikes without risking human lives.

With society's increasing reliance on AI, the possibilities of programming morality and increased adaptability have come to the forefront. While AI can rapidly carry out several complex calculations and computations, the technology has not yet reached the point where it can emulate the ethics-based choices humans make. It also has difficulty reacting quickly to unpredictable situations. Cognitive science plays the important role of learning more about both the human brain and AI to overcome these limitations.

The research of cognitive science has improved not only understanding of the human mind, but also of animal minds. While a general consensus has existed for many decades that a few particular species have some degree of intelligence, most animal behavior was considered instinctive rather than learned. However, until recently, most AI tests were rather rigid in their premises and standards. Modern tests, taking into account how human brains respond differently to different stimuli, are tailored to different animals' environments and routines. These tests have demonstrated that many animal species are capable of learned behaviors and problem solving.

Advances in cognitive science have improved the treatment and perception of mental health. Until the twentieth century, most mental illnesses had no known scientific cause, and disorders like depression were often treated as simple bad moods. Mental disorders did not display visible or tangible signs, which made it difficult for society to consider them something as real and dangerous as physical diseases. Advances in cognitive science helped provide those concrete signs of some illnesses. Scientists were able to detect different patterns in brain activity, and certain genetic disorders could be identified in DNA.

Cognitive science's study of brain activity and how it is expressed in behavior has been essential to finding treatments and medication for some mental illnesses. This includes the development of Cognitive Behavioral Therapy. This form of therapy centers on perception, helping patients overcome mental blocks and stress by attempting to cast the same information in a different light.

—Reed Kalso

Bibliography

"About Cognitive Linguistics." *ICLA*, RLINK" http://www.cognitivelinguistics.org/en/about-cognitive-linguistics.%20Accessed%2019%20Dec.%202016" www.cognitivelinguistics.org/en/about-cognitive-linguistics. Accessed 19 Dec. 2016.

"About." *Cognitive Science Society*, www.cognitivesciencesociety.org/description/. Accessed 18 Dec. 2016.

"Cognitive Behavioral Therapy." *Mayo Clinic*, 23 Feb. 2016, www.mayoclinic.org/tests-procedures/cognitive-behavioral-therapy/home/ovc-20186868. Accessed 17 Dec. 2016.

De Waal, Frans. "To Each Animal Its Own Cognition." *The Scientist*, 1 May 2016, www.the-scientist.com/?articles.view/articleNo/45884/title/To-Each-Animal-Its-Own-Cognition/. Accessed 20 Dec. 2016.

Estienne, Sophie. "Artificial Intelligence Creeps into Daily Life." *Phys. Org*, 15 Dec. 2016, phys.org/news/2016-12-artificial-intelligence-daily-life.html. Accessed 21 Dec. 2016.

"The Evolution of AI: Can Morality Be Programmed?" *Future of Life*, 6 July 2016, futureoflife.org/2016/07/06/evolution-of-ai/. Accessed 20 Dec. 2016.

"History & Timeline." *Brain & Cognitive Sciences*, https://bcs.mit.edu/about-bcs/history-timeline. Accessed 18 Dec. 2016.

Miller, George. "The Cognitive Revolution: A Historical Perspective." *TRENDS in Cognitive Sciences*, vol. 7, no.3, Mar. 2003, pp. 141–144.

"What Is Cognitive Science?" *ICOGSCI*, 2012, www.cogs.indiana.edu/icogsci/what.shtml. Accessed 18 Dec. 2016.

Combinatorics

SUMMARY

Combinatorics is a branch of mathematics that determines the number of ways that something can be done. In other words, it is the mathematics of counting. Combinatorics has numerous applications to probability, computer science, and experimental design. For example, the probability of a favorable outcome can be determined by dividing the number of favorable outcomes by the total number of outcomes. The number of ways that a set of outcomes A can occur may be represented by the symbol $|A|$.

One of the most basic methods in combinatorics is the Addition Principle. The Addition Principle states that if A and B cannot both occur, then the number of ways A or B can occur is computed by $|A \cup B| = |A| + |B|$.

A generalization of the Addition Principle is the Principle of Inclusion and Exclusion. This states that the number of ways that A or B can occur is given by $|A \cup B| = |A|+|B|-|A \cap B|$,where $|AcapB|$ is the number of ways both A and B can occur.

Another basic method is the Multiplication Principle. The Multiplication Principle states that if outcome of A does not affect the outcome of B, then the combination of A and B is computed by $|A||B|$. As an example, consider ordering a meal at a restaurant. This boils down to a series of choices (drink, appetizer, entrée, side, and dessert), none of which affects any other choice. So if there are five choices for drink, three for appetizer, ten for entrée, seven for side, and three for dessert, then the number of meals that can ordered is $5(3)(10)(7)(3) = 3150$.

The Multiplication Principle allows us to count a number of things. For example, suppose that we want to give out k different trophies to n different people in such a way that each person can receive multiple trophies or go home with nothing. There are n choices for the first trophy, n choices for the second, and so on. Hence, the Multiplication Principle gives the number of distributions as $n^k = n(n)\dots(n)$.

Consider the problem of lining up n people. There are n ways to place the first person in line. Regardless of who is selected, there are n–1 ways to place the second n–2 for the third, and so on. Thus, the Multiplication Principle shows that the number of ways to line up the people is $n! = n(n-1)(n-2)\dots(2)(1)$. The symbol $n!$ is read " n factorial."

Using a similar technique, we can count the number of ways to select n officers of different rank from a group of n people. Again, there are n choices for the first individual, n–1 choices for the second, and so on, with n–k+1 choices for the last individual. So the number of ways to select and rank the individuals is given by $P(n,k) = n(n-1)\dots(n-k+1) = \frac{n!}{(n-k)!}$.

Suppose that instead we want to determine the number of ways to select a committee of k (where every member has the same rank) from a group of n people. To do this, we can simply divide $P(n, k)$ by $k!$, the number of ways to rank the n individuals. This results in $C(n,k) = \frac{n!}{k!(n-k)!}$.

As example, consider the problem of drawing a "three of a kind" in five-card poker. There are $C(13, 3)$ ways to select ranks for the hand. There are 3 ways to select which of these ranks to triple. There are $C(4, 3)$ ways to select suits for the triple. There are 4^2 ways to select suits for the other two cards. Thus, the Multiplication Principle yields the number of acceptable hands as $C(13, 3)* C(4, 3)* 4^2 = 18304$.

—*Robert A. Beeler, PhD*

BIBLIOGRAPHY

Benjamin, Arthur T., and Jennifer J. Quinn. *Proofs That Really Count—The Art of Combinatorial Proof.* Washington, DC: Mathematical Assoc. of America, 2003.

DeGroot, Morris H., and Mark J. Schervish. *Probability and Statistics.* 4th ed. Upper Saddle River, NJ: Pearson, 2011.

Hollos, Stefan, J. Richard Hollos. *Combinatorics Problems and Solutions.* Longmont, CO:Abrazol, 2013.

Martin, George E. *Counting: The Art of Enumerative Combinatorics.* New York, NY: Springer, 2001.

Tucker, Alan. *Applied Combinatorics.* Hoboken, NJ: Wiley, 2012.

COMPUTED TOMOGRAPHY

SUMMARY

Computed tomography (CT) is an imaging modality that relies on the use of X rays and computer algorithms to provide high-quality image data. CT scanners are an integral part of medical health care in the developed world. Physicians rely on CT scanners to acquire important anatomical information on patients. CT images provide a three-dimensional view of the body that is both qualitative and quantitative in nature. CT scans are often used in the diagnosis of cancer, stroke, bone disorders, lung disease, atherosclerosis, heart problems, inflammation, and a range of other diseases and physical ailments, such as a herniated disc and digestive problems.

DEFINITION AND BASIC PRINCIPLES

Computed tomography (CT; also known as computer-aided tomography, or CAT) is an imaging modality that uses X rays and computational and mathematical processes to generate detailed image data of a scanned subject. X rays are generated through an evacuated tube containing a cathode, an anode, and a target material. The high voltage traveling through the tube accelerates electrons from the cathode toward the anode. This is very similar to a light bulb, with the addition of a target that generates the X rays and directs them perpendicularly to the tube. As electrons interact with the target material, small packets of energy, called photons, are produced. The photons have energies ranging from 50 to 120 kilovolts, which is characteristic of X-ray photons. The interaction of X-ray photons with a person's body produces a planar image with varying contrast depending on the density of the tissue being imaged. Bone has a relatively high density and is more readily absorbed by X rays, resulting in a bright image on the X-ray film. Less dense tissues, such as lungs, do not absorb X rays as readily, and as a result, the image produced is only slightly exposed and therefore dark.

In computed tomography, X rays are directed toward the subject in a rotational manner, which generates orthogonal, two-dimensional images. The X-ray tube and the X-ray detector are placed on a rotating gantry, which allows the X rays to be detected at every possible gantry angle. The resulting two-dimensional images are processed through computer algorithms, and a three-dimensional image of the subject is constructed. Because computed tomography relies on the use of ionizing radiation, it has an associated risk of inducing cancer in the patient. The radiation dose obtained from CT procedures varies considerably depending on the size of the patient and the type of imaging performed.

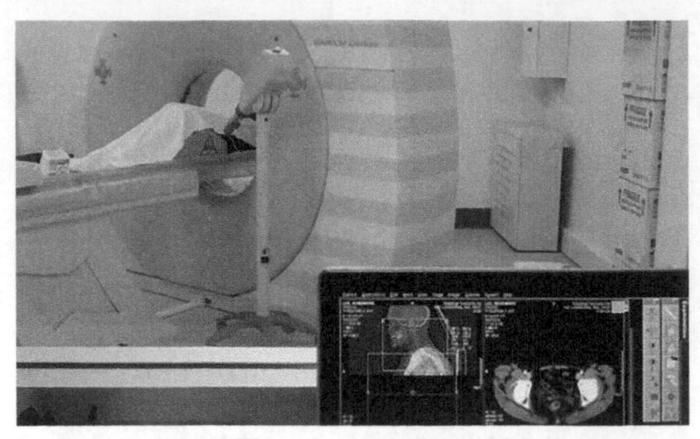

A patient is receiving a CT scan for cancer. Outside of the scanning room is an imaging computer that reveals a 2D image of the body's interior. By rosiescancerfund.com

BACKGROUND AND HISTORY

X rays were first discovered by Wilhelm Conrad Röntgen in November, 1895, during his experiments with cathode-ray tubes. He noticed that when an electrical discharge passed through these tubes, a certain kind of light was emitted that could pass through solid objects. Over the course of his experimentations with this light, Röntgen began to refer to it as an "X ray," as *x* is the mathematical term for an unknown. By the early 1900s, medical use of X rays was widespread in society. They were also used to entertain people by providing them with photographs of their bodies or other items. In 1901, Röntgen was awarded the first Nobel Prize in Physics for his discovery of X rays. During World War II, X rays were frequently used on injured soldiers to locate bullets or bone fractures. The use of X rays in medicine increased drastically by the mid-twentieth century.

The development of computer technology in the 1970s made it possible to invent computed tomography. In 1972, British engineer Godfrey Hounsfield and South African physicist Allan Cormack, who was working at Tufts University in Massachusetts, independently developed computed tomography. Both scientists were awarded the 1979 Nobel Prize in Physiology or Medicine for their discovery.

HOW IT WORKS

The first generation of CT scanners was built by Electric and Musical Industries (EMI) in 1971. The CT scanner consisted of a narrow X-ray beam (pencil beam) and a single detector. The X-ray tube moved linearly across the patient and subsequently rotated to acquire data at the next gantry angle. The process of data acquisition was lengthy, taking several minutes for a single CT slice. By the third generation of CT scanners, numerous detectors were placed on an arc across from the X-ray source. The X-ray beam in these scanners is wide (fan beam) and is covered by the entire area of detectors. At any single X-ray emission, the entire subject is in the field of view of the detectors, and therefore linear movement of the X-ray source is eliminated. The X-ray tube and detectors remain stationary, while the entire apparatus rotates about the patient, resulting in a drastic reduction in scan times. Most medical CT scanners in the world are of the third-generation type.

Mode of Operation. The process of CT image acquisition begins with the X-ray beam. X-ray photons are generated when high-energy electrons bombard a target material, such as tungsten, that is placed in the X-ray tube. At the atomic level, electrons interact with the atoms of the target material through two processes to generate X rays. One mode of interaction is when the incoming electron knocks another electron from its orbital in the target atom. Another electron from within the atom fills the vacancy, and as a result, an X-ray photon is emitted. Another mode of interaction occurs when an incoming electron interacts with the nucleus of the target atom. The electron is scattered by the strong electric field in the nucleus of the target atom, and as a result, an X-ray photon is emitted. Both modes of interaction are very inefficient, resulting in a considerable amount of energy that is dissipated as heat. Cooling methods need to be considered in the design of X-ray machines to prevent overheating in the X-ray tube. The resulting X-ray beam has a continuous energy spectrum, ranging from low-energy photons to the highest energy photons, which corresponds with the X-ray tube potential. However, since low-energy photons increase the dose to the body and do not contribute to image quality, they are filtered from the X-ray beam.

Once a useful filtered X-ray beam is generated, the beam is directed toward the subject, while an image receptor (film or detector) is placed in the beam direction past the subject to collect the X-ray signal and provide an image of the subject. As X rays interact with the body's tissues, the X-ray beam is attenuated by different degrees, depending on the density of the material. High-density materials, such as bone, attenuate the beam drastically and result in a brighter X-ray image. Low-density materials, such as lungs, cause minimal attenuation of the X-ray beam and appear dark on the X-ray image because most of the X rays strike the detector.

Image Acquisition. In CT, X-ray images of the subject are taken from many angles and reconstructed into a three-dimensional image that provides an excellent view of the scanned subject. At each angle, X-ray detectors measure the X-ray beam intensities, which are characteristic of the attenuation coefficients of the material through which the X-ray beam passes. Generating an image from the acquired detector measurements involves determining the attenuation coefficients of each pixel within the image matrix and using mathematical algorithms to reconstruct the raw image data into cross-section CT image data.

APPLICATIONS AND PRODUCTS

The power of computed tomography to provide detailed visual and quantitative information on the object being scanned has made it useful in many fields and suitable for numerous applications. Aside from disease diagnosis, CT is also commonly used as a real-time guide for surgeons to accurately locate their target within the human body. It is also used in the manufacturing industry for nondestructive evaluation of manufactured products.

Disease Diagnosis. The most common use for CT is in radiological clinics, where it is used as an initial procedure to evaluate specific patients' complaints or symptoms, thereby providing information for a diagnosis, and to assess surgical or treatment options. Radiologists, medical professionals specialized in reading and analyzing patient CT data, look for foreign bodies such as stones, cancers, and fluid-filled cavities that are revealed by the images. Radiologists can also analyze CT images for size and volume of body organs and detect abnormalities that suggest diseases and conditions involving changes in tissue density or size, such as pancreatitis, bowel disease, aneurysms, blood clots, infections, tuberculosis, narrowing of blood vessels, damaged organs, and osteoporosis.

In addition to disease diagnosis, CT has been used by private radiological clinics to provide full-body scans to symptom-free people who desire to obtain a CT image of their bodies to ascertain their health and to detect any conditions or abnormalities that might indicate a developing problem. However, the use of CT imaging for screenings in the absence of symptoms is controversial because the X-ray radiation used in CT has an associated risk of cancer. The dose of radiation deposited by a CT scan is between fifty and two hundred times the dose deposited by a conventional X-ray image. Although the association between CT imaging and cancer induction is not well established, its casual use remains an area of considerable debate.

Guided Biopsy. A biopsy is a time-consuming, sometimes inaccurate, and invasive procedure. Traditionally, doctors obtained a biopsy (sample) of the tissue of interest by inserting a needle into a patient at the approximate location of the target tissue. However, real-time CT imaging allows doctors to observe the location of the biopsy needle within the patient. Therefore, doctors can obtain a more accurate tissue biopsy in a relatively short time and without using invasive procedures.

CT Microscopy. The resolution of clinical CT scanners is limited by practical scan times for the patients and the size of X-ray detectors used. In the case of small animals, higher resolutions can be obtained by use of smaller detectors and longer scan time. The field of microCT, or CT microscopy, has rapidly developed in the early twenty-first century as a way of studying disease pathology in animal models of human disease. Numerous disorders can be modeled in small animals, such as rats and mice, to obtain a better understanding of the disease biology or assess the efficacy of emerging treatments or drugs. Traditionally, animal studies would involve killing the animal at a specific time point and processing the tissue for viewing under the microscope. However, the development of CT microscopy has allowed scientists and researchers to investigate disease pathology and treatment efficacy at very high resolutions while the animal is alive, reaching one-fifth of the resolution of a light microscope.

Nondestructive Evaluation. Computed tomography has gained wide use in numerous manufacturing industries for nondestructive evaluation of composite materials. Nondestructive evaluation is used to inspect specimens and ensure the integrity of manufactured products, either through sampling or through continuous evaluation of each product. CT requirements for nondestructive testing differ from those for medical imaging. For medical imaging, scan times have to be short, exposure to radiation has to be minimal, and patient comfort throughout the procedure must be taken into consideration. For nondestructive evaluation, patient comfort and exposure to radiation are no longer important issues. However, keeping scan times short is advantageous, especially for large-scale industries. Furthermore, the X-ray energy for scanning industrial samples can vary significantly from the energy used for patient imaging, since patient composition is primarily water and industrial samples can have a wide range of compositions and associated densities. Engineers have custom-designed CT scanners for specific materials, including plastics, metals, wood, fibers, glass, soil, concrete, rocks, and various composites. The capability of CT to provide excellent qualitative image data and accurate quantitative data on the density of the specimen has made it a powerful tool for nondestructive evaluation. CT is used in the aerospace industry to ensure the integrity of various airplane components and in the automotive industry to evaluate the structure of wheels and tires. In addition to industrial applications, CT is commonly used in research centers to further their imaging and analytical power.

SOCIAL CONTEXT AND FUTURE PROSPECTS

The numbers of CT scanners and scans being performed have risen considerably since the 1980s. By the early twenty-first century, there were more than 6,000 scanners in the United States, and more than 72 million CT scans were performed in the country in 2007. The rapid and wide acceptance of CT scanners in health care institutions has sparked controversy in the media and among health care practitioners regarding the radiation doses being delivered through the scans. Risk of cancer induction rises with increased exposure to radiation and this risk has to be carefully weighed against the benefits of a CT scan. The issue of cancer induction is more alarming when CT procedures are performed on young children or infants—a 2013 Australian study found that the incidence of cancer in people who had been exposed to CT scans in childhood or adolescence was 24 percent greater than in the unexposed population. Studies have recommended that CT scanning of children should not be performed using the same protocols as used for adults because the children are generally more sensitive to radiation than adults. More strict federal and state regulations are being instituted to better control the use of CT in health care.

—Ayman Oweida, BSc, MSc

BIBLIOGRAPHY

Aichinger, H., et al. *Radiation Exposure and Image Quality in X-ray Diagnostic Radiology: Physical Principles and Clinical Applications.* Springer, 2003.

Bossi, R. H., et al. "Computed Tomography." *Non-destructive Testing of Fibre-reinforced Plastics Composites.* Edited by John Summerscales, Elsevier Science, 1990.

Brenner, David, and Eric Hall. "Computed Tomography: An Increasing Source of Radiation Exposure." *New England Journal of Medicine,* vol. 357, no. 22, 2007, pp. 2277–284.

"Do You Really Need a Scan?" *Consumer Reports on Health,* vol. 27, no. 3, 2015, pp., 1–5.

Hsieh, Jiang. *Computed Tomography: Principles, Design, Artifacts, and Recent Advances.* 2nd ed., International Society for Optical Engineering, 2009.

Mathews, J. D., et al. "Cancer Risk in 680,000 People Exposed to Computed Tomography Scans in Childhood or Adolescence: Data Linkage Study of 11 Million Australians." *BMJ,* vol. 346, 2013, p. f2360.

Otani, Jun, and Yuzo Obara, editors. *X-ray CT for Geomaterials: Soils, Concrete, Rocks.* Balkema, 2004.

"Radiation-Emitting Products: Computed Tomography (CT)." *U.S. Food and Drug Administration.* U.S. Food and Drug Administration, 7 Aug. 2014, www.fda.gov/Radiation-EmittingProducts/RadiationEmittingProductsandProcedures/MedicalImaging/MedicalX-Rays/ucm115317.htm. Accessed 3 Mar. 2015.

COMPUTER ENGINEERING

SUMMARY

Computer engineering refers to the field of designing hardware and software components that interact to maximize the speed and processing capabilities of the central processing unit (CPU), memory, and the peripheral devices, which include the keyboard, monitor, disk drives, mouse, and printer. Because the first computers were based on the use of on-and-off mechanical switches to control electrical circuits, computer hardware is still based on the binary number system. Computer engineering involves the development of operating systems that are able to interact with compilers that translate the software programs written by humans into the machine instructions that depend on the binary number system to control electrical logic circuits and communication ports to access the Internet.

DEFINITION AND BASIC PRINCIPLES

Much of the work within the field of computer engineering focuses on the optimization of computer hardware, which is the general term that describes the electronic and mechanical devices that make it possible for a computer user (client) to utilize the power of a computer. These physical devices are based on binary logic. Humans use the decimal system for numbers, instead of the base two-number system of binary logic, and naturally humans communicate with words. A great deal of interface activity is necessary to bridge this communication gap, and computer engineering involves additional types of software (programs) that function as intermediate interfaces to translate human instructions into hardware activity. Examples of these types of software

Computer components. By User Mike1024 on en.wikipedia (Photo taken by Mike1024.)

include operating systems, drivers, browsers, compilers, and linkers.

Computer hardware and software generally can be arranged in a series of hierarchical levels, with the lowest level of software being the machine language, consisting of numbers and operands that the processor executes. Assembly language is the next level, and it uses instruction mnemonics, which are machine-specific instructions used to communicate with the operating system and hardware. Each instruction written in assembly language corresponds to one instruction written in machine code, and these instructions are used directly by the processor. Assembly language is also used to optimize the runtime execution of application programs. At the next level is the operating system, which is a computer program written so that it can manage resources, such as disk drives and printers, and can also function as an interface between a computer user and the various pieces of hardware. The highest level includes applications that humans use on a daily basis. These are considered the highest level because they consist of statements written in English and are very close to human language.

BACKGROUND AND HISTORY

The first computers used vacuum tubes and mechanical relays to indicate the switch positions of *on* or *off* as the logic units corresponding to the binary digits of 0 or 1, and it was necessary to reconfigure them each time a new task was approached. They were large enough to occupy entire rooms and required huge amounts of electricity and cooling. In the 1930s the Atanasoff-Berry Computer (ABC) was created at Iowa State University to solve simultaneous numerical equations, and it was followed by the Electronic Numerical Integrator And Computer (ENIAC), developed by the military for mathematical operations.

The transistor was invented in 1947 by John Bardeen, Walter Brattain, and William Shockley, which led to the use of large transistors as the logic units in the 1950s. The integrated circuit chip was invented by Jack St. Clair Kilby and Robert Norton Noyce in 1958 and caused integrated circuits to come into usage in the 1960s. These early integrated circuits were still quite large and included transistors, diodes, capacitors, and transistors. Modern silicon chips can hold these components and as many as 55 million transistors. Silicon chips are called microprocessors, because each microprocessor can hold these logic units within just over a square inch of space.

HOW IT WORKS

Hardware. The hardware, or physical components, of a computer can be classified according to their general uses of input, output, processing, and storage. Typical input devices include the mouse, keyboard, scanner, and microphone that facilitate communication of information between the human user and the computer. The operation of each of these peripheral devices requires special software called a driver, which is a type of controller, that is able to translate the input data into a form that can be communicated to the operating system and controls input and output peripheral devices.

A read-only memory (ROM) chip contains instructions for the basic input-output system (BIOS) that all the peripheral devices use to interact with the CPU. This process is especially important when a user first turns on a computer for the boot process.

When a computer is turned on, it first activates the BIOS, which is software that facilitates the interactions between the operating system, hardware, and peripherals. The BIOS accomplishes this interaction by first running the power-on self test (POST), which is a set of routines that are always available at a specific memory address in the read-only memory. These routines communicate with the keyboard, monitor, disk drives, printer, and communication ports to access the Internet. The BIOS also controls the time-of-day clock. These tasks completed by the BIOS are sometimes referred to as booting up (from the old expression "lift itself up by its own bootstraps"). The last instruction within the BIOS is to start reading the operating system from either a boot disk in a diskette drive or the hard drive. When shutting down a computer there are also steps that are followed to allow for settings to be stored and network connections to be terminated.

The CPU allows instructions and data to be stored in memory locations, called registers, which facilitate the processing of information as it is exchanged between the control unit, arithmetic-logic unit, and any peripheral devices. The processor interacts continuously with storage locations, which can be classified as either volatile or nonvolatile types of memory. Volatile memory is erased when the computer is turned off and consists of main memory, called random-access memory (RAM) and cache memory. The fundamental unit of volatile memory is the flip-flop, which can store a value of 0 or 1 when the computer is on. This value can be flipped when the computer needs to change it. If a series of 8 flip-flops is hooked together, an 8-bit number can be stored in a register. Registers can store only a small amount of data on a temporary basis while the computer is actually on. Therefore, the RAM is needed for larger amounts of information. However, it takes longer to access data stored in the RAM because it is outside the processor and needs to be retrieved, causing a lag time. Another type of memory, called cache, is located in the processor and can be considered an intermediate type of memory between registers and main memory.

Nonvolatile memory is not erased when a computer is turned off. It consists of hard disks that make up the hard drive or flash memory. Although additional nonvolatile memory can be purchased for less money than volatile memory, it is slower. The hard drive consists of several circular discs called platters that are made from aluminum, glass, or ceramics and covered by a magnetic material so that they can develop a magnetic charge. There are read and write heads made of copper so that a magnetic field develops that is able to read or write data when interacting with the platters. A spindle motor causes the platters to spin at a constant rate, and either a stepper motor or voice coil is used as the head actuator to initiate interaction with the platters.

The control unit of the CPU manages the circuits for completing operations of the arithmetic-logic unit. The arithmetic-logic unit of the CPU also contains circuits for completing the logical operations, in addition to data operations, causing the CPU essentially to function as the brain of the computer. The CPU is located physically on the motherboard. The motherboard is a flat board that contains all the chips needed to run a computer, including the CPU, BIOS, and RAM, as well as expansion slots and power-supply connectors. A set of wires, called a bus, etched into the motherboard connects these components. Expansion slots are empty places on the motherboard that allow upgrades or the insertion of expansion cards for various video and voice controllers, memory expansion cards, fax boards, and modems without having to reconfigure the entire computer. The motherboard is the main circuit board for the entire computer.

Circuit Design and Connectivity. Most makers of processor chips use the transistor-transistor logic (TTL) because this type of logic gate allows for the output from one gate to be used directly as the input for another gate without additional electronic input, which maximizes possible data transmission while minimizing electronic complications. The TTL makes this possible because any value less than 0.5 volt is recognized as the logic value of 0, while any value that exceeds 2.7 volts indicates the logic value of 1. The processor chips interact with external computer devices via connectivity locations called ports. One of the most important of these ports is called the Universal Serial Bus (USB) port, which is a high-speed, serial, daisy-chainable port used to connect keyboards, printers, mice, external disk drives, and additional input and output devices.

Software. The operating system consists of software programs that function as an interface between the user and the hardware components. The operating system also assists the output devices of printers and monitors. Most of the operating systems being used also have an application programming interface (API), which includes graphics and facilitates use. APIs are written in high-level languages (using statements approximating human language), such as C++, Java, and Visual Basic.

APPLICATIONS AND PRODUCTS

STAND-ALONE COMPUTERS. Most computer users rely on relatively small computers such as laptops and personal computers (microcomputers). Companies that manufacture these relatively inexpensive computers have come into existence only since the early 1980s and have transformed the lives of average Americans by making computer usage a part of everyday life. Before microcomputers came into such wide usage, the workstation was the most accessible smaller-size computer. It is still used primarily by small and medium-size businesses that need the additional memory and speed capabilities. Larger organizations such as universities use mainframe computers to handle their larger power requirements. Mainframes generally occupy an entire room. The most powerful computers are referred to as supercomputers, and they are so expensive that often several universities will share them for scientific and computational activities. The military uses them as well. They often require the space of several rooms.

INTER-NETWORK SERVICE ARCHITECTURE, INTERFACES, AND INTER-NETWORK INTERFACES. A network consists of two or more computers connected together in order to share resources. The first networks used coaxial cable, but now wireless technologies allow computer devices to communicate without the need to be physically connected by a coaxial cable. The Internet has been a computer-engineering application that has transformed the way people live. Connecting to the Internet first involved the same analog transmission used by the plain old telephone service (POTS), but connections have evolved to the use of fiber-optic technology and wireless connections. Laptop computers, personal digital assistants (PDAs), cell phones, smartphones, RFID (radio frequency identification), iPods, iPads, and Global Positioning Systems (GPS) are able to communicate, and their developments have been made possible by the implementation of the fundamental architectural model for inter-network service connections called the Open Systems Interconnection (OSI) model. OSI is the layered architectural model for connecting networks. It was developed in 1977 and is used to make troubleshooting easier so that if a component fails on one computer, a new, similar component can be used to fix the problem, even if the component was manufactured by a different company.

OSI's seven layers are the application, presentation, session, transport, network, data link, and physical layers.

The application and presentation layers work together. The application layer synchronizes applications and services in use by a person on an individual computer with the applications and services shared with others via a server. The services include e-mail, the World Wide Web, and financial transactions. One of the primary functions of the presentation layer is the conversion of data in a native format such as extended binary coded decimal interchange code (EBCDIC) into a standard format such as American Standard Code for Information Interchange (ASCII).

As its name implies, the primary purpose of the session layer is to control the dialog sessions between two devices. The network file system (NFS) and structured query language (SQL) are examples of tools used in this layer.

The transport layer controls the connection-oriented flow of data by sending acknowledgments to data senders once the recipient has received data and also makes sure that segments of data are retransmitted if they do not reach the intended recipient. A router is one of the fundamental devices that works in this layer, and it is used to connect two or more networks together physically by providing a high degree of security and traffic control.

The data link layer translates the data transmitted by the components of the physical layer, and it is within the physical layer that the most dramatic technological advances made possible by computer engineering have had the greatest impact. The coaxial cable originally used has been replaced by fiber-optic technology and wireless connections.

Fiber Distributed Data Interface (FDDI) is the fiber-optic technology that is a high-speed method of networking, composed of two concentric rings of fiber-optic cable, allowing it to transmit over a longer distance but at greater expense. Fiber-optic cable uses glass threads or plastic fibers to transmit data. Each cable contains a bundle of threads that work by reflecting light waves. They have much greater bandwidth than the traditional coaxial cables and can carry data at a much faster rate because light travels faster than electrical signals. Fiber-optic cables are also not susceptible to electromagnetic interference and weigh less than coaxial cables.

Wireless networks use radio or infrared signals for cell phones and a rapidly growing variety of handheld devices, including iPhones, iPads, and tablets. New technologies using Bluetooth and Wi-Fi for mobile connections are leading to the next phase of internetwork communications with the Internet called cloud computing. Cloud computing is basically a wireless Internet application where servers supply resources, software, and other information to users on demand and for a fee.

Smartphones that use Google Android operating system for mobile phones have a PlayStation emulator called PSX4Droid that allows PlayStation games to be played on these phones. Besides making games easily accessible, cloud computing is also making it easier and cheaper to do business all around the world with applications such as Go To Meeting, which is one example of video conferencing technology used by businesses.

SOCIAL CONTEXT AND FUTURE PROSPECTS

The use of cookies, a file stored on an Internet user's computer that contains information about that user, an identification code or customized preferences, that is recognized by the servers of the websites the user visits, as a tool to increase online sales by allowing e-businesses to monitor the preferences of customers as they access different web pages is becoming more prevalent. However, this constant online monitoring also raises privacy issues. In the future there is the chance that one's privacy could be invaded with regard to medical or criminal records and all manner of financial information. In addition, wireless technologies are projected to increase the usage of smartphones, the iPad, and many new consumer gadgets, which can access countless e-business and social networking websites. Thus, the rapid growth of Internet applications that facilitate communication and financial transactions will continue to be accompanied by increasing rates of identity theft and other cyber crimes, ensuring that network security will continue to be an important application of computer engineering.

—Jeanne L. Kuhler, BS, MS, PhD

BIBLIOGRAPHY

Das, Sumitabha. *Your UNIX: The Ultimate Guide.* New York: McGraw, 2005. Print.

Dhillon, Gurphreet. "Dimensions of Power and IS Implementation." *Information and Management* 41.5 (2004): 635–44. Print.

Irvine, Kip R. *Assembly Language for Intel-Based Computers.* 5th ed. Upper Saddle River: Prentice, 2006. Print.

Kerns, David V., Jr., and J. David Irwin. *Essentials of Electrical and Computer Engineering* 2nd ed. Upper Saddle River: Prentice, 2004. Print.

Limoncelli, Tom, Strata R. Chalup, and Christina J. Hogan. *The Practice of Cloud System Administration: Designing and Operating Large Distributed Systems.* Upper Saddle River: Addison, 2015. Print.

Magee, Jeff, and Jeff Kramer. *Concurrency: State Models and Java Programming.* Hoboken: Wiley, 2006. Print.

Silvester, P. P., and D. A. Lowther. *Computer Engineering Circuits, Programs, and Data.* New York: Oxford UP, 1989. Print.

Sommerville, Ian. *Software Engineering.* 9th ed. Boston: Addison, 2010. Print.

COMPUTER LANGUAGES, COMPILERS, AND TOOLS

SUMMARY

Computer languages are used to provide the instructions for computers and other digital devices based on formal protocols. Low-level languages, or machine code, were initially written using the binary digits needed by the computer hardware, but since the 1960s, languages have evolved from early procedural languages to object-oriented high-level languages, which are more similar to English. There are many of these high-level languages, with their own unique capabilities and limitations, and most require

```
// class declaration
public class ProgrammingExample {

    // method declaration
    public void sayHello() {

        // method output
        System.out.println("Hello World!");
    }
}
```

An example of source code written in the Java programming language, which will print the message "Hello World!" to the standard output when it is compiled and executed. By Bobbygammill (Using SublimeText)

some type of compiler or other intermediate translator to communicate with the computer hardware. The popularity of the Internet has created the need to develop numerous applications and tools designed to share data across the Internet.

DEFINITION AND BASIC PRINCIPLES

The traditional process of using a computer language to write a program has generally involved the initial design of the program using a flowchart based on the purpose and desired output of the program, followed by typing the actual instructions for the computer (the code) into a file using a text editor, and then saving this code in a file (the source code file). A text editor is used because it does not have the formatting features of a word processor. An intermediate tool called a compiler then has been used to convert this source code into a format that can be run (executed) by a computer.

However, in the 2010s, a tool much faster and more efficient than compilers, called interpreters, gained prominence and replaced most compilers. Larger, more complex programs have evolved that have required an additional step to link external files. This process is called linking and it joins the main, executable program created by the compiler to other necessary programs. Finally, the executable program is run and its output is displayed on the computer monitor, printed, or saved to another digital file. If errors are found, the process of debugging is followed to go back through the code to make corrections.

BACKGROUND AND HISTORY

Early computers such as ENIAC (Electronic Numerical Integrator and Computer), the first general-purpose computer, were based on the use of switches that could be turned on or off. Thus, the binary digits of 0 and 1 were used to write machine code. In addition to being tedious for a programmer, the code had to be rewritten if used on a different type of machine, and it certainly could not be used to transmit data across the Internet, where different computers all over the world require access to the same code.

Assembly language evolved by using mnemonics (alphabetic abbreviations) for code instead of the binary digits. Because these alphabetic abbreviations of assembly language no longer used the binary digits, additional programs were developed to act as intermediaries between the human programmers writing the code and the computer itself. These additional programs were called compilers, and this process was initially known as compiling the code. This compilation process was still machine and vendor dependent, however, meaning, for example, that there were several types of compilers that were used to compile code written in one language. This was expensive and made communication of computer applications difficult.

The evolution of computer languages from the 1950's has accompanied technological advances that have allowed languages to become increasingly powerful, yet easier for programmers to use. FORTRAN and COBOL languages led the way for programmers to develop scientific and business application programs, respectively, and were dependent on a command-line user interface, which required a user to type in a command to complete a specific task. Several other languages were developed, including Basic, Pascal, PL/I, Ada, Lisp, Prolog, and Smalltalk, but each of these had limited versatility and various problems. The C and C++ languages of the 1970's and 1980's, respectively, have emerged as being the most useful and powerful languages and are still in use. These were followed by development tools written in the Java and Visual Basic languages, including integrated development environments with editors, designers, debuggers, and compilers all built into a single software package.

HOW IT WORKS

BIOS and Operating System. The programs within the BIOS are the first and last programs to execute whenever a computer device is turned on or off. These programs interact directly with the operating system (OS). The early mainframe computers that were used in the 1960's and 1970's depended on several different operating systems, most of which are no longer in usage, except for UNIX and DOS. DOS (Disk Operating System) was used on the initial microcomputers of the 1980's and early 1990's, and it is still used for certain command-line specific instructions.

Graphical User Interfaces (GUIs). Microsoft dominates the PC market with its many updated operating systems, which are very user-friendly with GUIs. These operating systems consist of computer programs and software that act as the management system for all of the computer's resources, including the various application programs most taken for granted, such as Word (for documents), Excel (for mathematical and spreadsheet operations), and Access (for database functions). Each of these applications is actually a program itself, and there are many more that are also available.

Since the 1980's, many programming innovations increasingly have been built to involve the client-server model, with less emphasis on large main frames and more emphasis on the GUIs for smaller microcomputers and handheld devices that allow consumers to have deep color displays with high resolution and voice and sound capabilities. However, these initial GUIs on client computers required additional upgrades and maintenance to be able to interact effectively with servers.

World Wide Web. The creation of the World Wide Web provided an improvement for clients to be able to access information, and this involvement of the Internet led to the creation of new programming languages and tools. The browser was developed to allow an end user (client) to be able to access Web information, and hypertext markup language (HTML) was developed to display Web pages. Because the client computer was manufactured by many different companies, the Java language was developed to include applets, which are mini-programs embedded into Web pages that could be displayed on any type of client computer. This was made possible by a special type of compiler-like tool called the Java Virtual Machine, which translated byte code. Java remains the primary computer language of the Internet.

APPLICATIONS AND PRODUCTS

FORTRAN and COBOL. FORTRAN was developed by a team of programmers at IBM and was first released in 1957 to be used primarily for highly numerical and scientific applications. It derived its name from formula translation. Initially, it used punched cards for input, because the text editors were not available in the 1950's. It has evolved but still continues to be used primarily in many engineering and scientific programs, including almost all programs written for geology research. Several updated versions have been released. FORTRAN77, released in 1980, had the most significant language improvements, and FORTRAN2000, released in 2002, is the most recent. COBOL (Common Business-Oriented Language) was released in 1959 with the goal of being used primarily for tracking retail sales, payroll, inventory control, and many other accounting-related activities. It is still used for most of these business-oriented tasks.

C and C++. The C computer language was the predecessor to the C++ language. Programs written in C were procedural and based on the usage of functions, which are small programming units. As programs grew in complexity, more functions were added to a C program. The problem was that eventually it became necessary to redesign the entire program, because trying to connect all of the functions, which added one right after the other, in a procedural way, was too difficult. C++ was created in the 1980's based on the idea of objects grouped into classes as the building blocks of the programs, which meant that the order did not have to be procedural anymore. Object-oriented programming made developing complex programs much more efficient and is the current standard.

Microsoft.NET. In June 2000, Microsoft introduced a suite of languages and tools named Microsoft.NET along with its new language called Visual C#. Microsoft.NET is a software infrastructure that consists of many programs that allow a user to write programs for a range of new applications such as server components and Web applications by using new tools. Although programs written in Java can be run on any machine, as long as the entire program is written in Java, Microsoft.NET allows various programs to be run on the Windows OS. Additional advantages of Microsoft.NET involve its use of Visual C#. Visual C# provides services to help Web pages already in existence, and C# can be integrated with the Visual Basic and Visual C++ languages, which facilitate the work of Web programmers by allowing them to update existing Web applications, rather than having to rewrite them.

The Microsoft.NET framework uses a common type system (CTS) tool to compile programs written in Cobol.NET, PerlNET, Visual Basic.NET, Jscript, Visual C++, Fortran.NET, and Visual C# into an intermediate language. This common intermediate language (CIL) can then be compiled to a common language runtime (CLR). The result is that the .NET programming environment promotes interoperability to allow programs originally written in different languages to be executed on a variety of operating systems and computer devices. This interoperability is crucial for sharing data and communication across the Internet.

SOCIAL CONTEXT AND FUTURE PROSPECTS

The Internet continues to bring the world together at a rapid pace, which has both positive and negative ramifications. Clearly, consumers have much easier access to many services, such as online education, telemedicine, and free search tools to locate doctors, learn more about any topic, comparison shop and purchase, and immediately access software, movies, pictures, and music. However, along with this increase in electronic commerce involving credit card purchases, bank accounts, and additional financial transactions has been the increase of cybercrime. Thousands of dollars are lost each year to various Internet scams and hackers being able to access private financial information. Some programmers even use computer languages to produce viruses and other destructive programs for purely malicious purposes, which have a negative impact on computer security.

Modern computer programs follow standard engineering principles to solve problems involving detail-driven applications ranging from radiation therapy and many medical devices to online banking, auctions, and stock trading. These online application programs require the special characteristics of Web-enabled software such as security and portability, which has given rise to the development of additional tools and will continue to produce the need for increased security features within computer languages and tools.

—Jeanne L. Kuhler, BS, MS, PhD

Bibliography

Bouwkamp, Katie. "The 9 Most In-Demand Programming Languages of 2016." *Coding Dojo*, 27 Jan. 2016, www.codingdojo.com/blog/9-most-in-demand-programming-languages-of-2016/. Accessed 24 Oct. 2016.

Das, Sumitabha. *Your UNIX: The Ultimate Guide.* 2d ed. Boston: McGraw-Hill, 2006.

Dhillon, Gupreet. "Dimensions of Power and IS Implementation." *Information and Management* 41, no. 5 (May, 2004): 635-644.

Guelich, Scott, Shishir Gundavaram, and Gunther Birznieks. *CGI Programming with Perl.* 2d ed. Cambridge, Mass.: O'Reilly, 2000.

Horstmann, Cay. *Big Java.* 4th ed. Hoboken, N.J.: John Wiley & Sons, 2010.

Lee, Kent D. *Foundations of Programming Languages.* Springer, 2014.

Scott, Michael L. *Programming Language Pragmatics.* 4th ed., Morgan Kaufmann, 2016.

Snow, Colin. "Embrace the Role and Value of Master Data Management," *Manufacturing Business Technology*, 26, no. 2 (February, 2008): 92-95.

Computer Memory

SUMMARY

Computer memory is the part of the device's hardware that stores information. There are different types of memory inside a computer, including permanent memory, temporary memory, read only memory (ROM), random access memory (RAM), and programmable read only memory (PROM). Different

DDR-SD-RAM, SD-RAM and two older sorts of memory. By Gyga (Own work)

memory storage devices in a computer are used for different purposes. Fast, temporary memory such as RAM is used to make quick calculations, while ROM and hard disk drives are used for long-term storage. Without memory, computers would not be able to function in any meaningful capacity.

HISTORY OF THE COMPUTER

In their earliest days, computers were strictly mechanical devices. They used punch cards for memory and output. These machines were developed for utility rather than for the multitude of tasks for which modern computers are designed. They were primarily used for complex calculations.

Alan Turing, a famous computer scientist, is credited with the idea for the first multipurpose computer. In the 1930s, J.V. Atanasoff created the first computer that contained no gears, cams, belts, or shafts. Atanasoff and his team then designed the first computer with functioning, nonmechanical memory devices. Primitive when compared to today's devices, Atanasoff's creation allowed the computer to solve up to twenty-nine equations simultaneously.

The next major leap in computing power was the usage of vacuum tubes. In 1994, professors John Mauchly and J. Presper Eckert built the first tube-powered electronic calculator. This is commonly considered the first digital computer. It was a massive machine, taking up the entirety of a large room. They soon began to market this computer to governments and businesses. However, tube computers became obsolete in 1947 with the invention of the transistor.

Ten years later, the transistor was used by Robert Noyce and Jack Kilby to create the first computer chip. This spurred the development of the first devices recognizable as modern computers. Computers took another leap forward with the graphical user interface (GUI), which projects options as images on a screen instead of requiring users to learn to code. Computers advanced further with the inventions of RAM in 1970 and floppy disks in 1971. Floppy disks were a form of permanent storage used in the early days of computers. They could easily be transferred from one computer to another, making them ideal for transporting information. Floppy disks were made obsolete by CD-ROMs, which are small plastic disks that store large amounts of information.

TYPES OF MEMORY

Computer memory is measured in binary digits, called bits. One bit is an extremely small amount of information. Eight bits is the equivalent of one byte. 1024 bytes is called a kilobyte (KB); 1024 KB makes a megabyte (MB); 1024 MB makes a gigabyte (GB); and 1024 GB makes a petabyte (PB). Over time, the cost of large amounts of computer storage has drastically fallen. However, the amount of memory required by modern computers has also drastically increased.

Computers contain several types of memory. The most common type is temporary memory, which is designed to hold information for only a short period. Permanent memory is designed to hold memory for a much longer time. Most of a computer's temporary memory is RAM. RAM is designed to quickly write and erase information. It performs calculations, runs scripts, and enacts most of the computer's functions. A computer with more RAM can perform more functions at once, making it more powerful.

Permanent storage may refer to several devices. In most cases, information is stored on the computer's hard disk drive. Most hard disk drives use a spinning disk and an actuator arm. The actuator arm writes to the spinning disk by rearranging electrons. In this scenario, the entire inside of the hard disk drive is located inside an airtight seal. These hard drives can be found in many sizes. However, modern hard disk drives are often found in capacities of hundreds of gigabytes to terabytes.

Many high-quality computer manufacturers have begun replacing hard disk drives with solid-state drives. Solid-state drives contain no moving parts. In

Solid-state drives are one example of non-volatile memory. By Intel Free Press (Flickr: Intel X25-M SATA Solid-State Drive (SSD))

most instances, these drives can read and write data much faster than hard disk drives. Because they have no moving parts, solid-state drives are also much quieter than hard disk drives. However, solid-state drives are also significantly more expensive than hard disk drives. For this reason, if a device needs large quantities of storage, it may be more cost-effective to use hard disk drives. However, if the device needs to be able to access data quickly, be durable, or be compact in size, manufacturers may use a solid-state drive.

Some computers utilize external memory sources. These are drives located outside the device. If it is connected to the device by a universal serial bus (USB) cable or other interface, it is called an external hard drive. External hard drives are easily transferable from one device to another, making them useful for quickly moving large media files. They may also be used to back up large amounts of important files, protecting them from computer malfunction or viruses.

If the external memory is accessed through the Internet, it is referred to as "the cloud." Many services offer large amounts of external memory for purchase. This may be used for server backups, media storage, or any number of other applications. Cloud storage allows users to expand the storage capacity of their machines without making any physical alterations to the computers. The cloud also features many of the same benefits as an external hard drive. Files can easily be relocated to a new machine in the event of a hardware or software failure. Additionally, renting server space from a cloud storage service may be cheaper than purchasing and installing additional physical storage devices.

—*Tyler Biscontini*

BIBLIOGRAPHY

Claerr, Jennifer. "What Are the Four Basic Functions of a Computer?" *Techwalla*, www.techwalla.com/articles/what-are-the-four-basic-functions-of-a-computer. Accessed 17 Nov. 2016.

"Computer – Introduction to Memory." *CCM Benchmark Group*, ccm.net/contents/396-computer-introduction-to-memory. Accessed 17 Nov. 2016.

"Data Measurement Chart." *University of Florida*, www.wu.ece.ufl.edu/links/dataRate/DataMeasurementChart.html. Accessed 17 Nov. 2016.

"Hard Drive." *Computer Hope*, www.computerhope.com/jargon/h/harddriv.htm. Accessed 17 Nov. 2016.

"How Computers Work: The CPU and Memory." *University of Rhode Island*, homepage.cs.uri.edu/faculty/wolfe/book/Readings/Reading04.htm. Accessed 17 Nov. 2016.

Subhash, D. "Types of Memory." *IT4NextGen*, 27 Aug. 2016, www.it4nextgen.com/computer-memory-types. Accessed 17 Nov. 2016.

"Timeline of Computer History." *Computer History Museum*, www.computerhistory.org/timeline/computers/. Accessed 17 Nov. 2016.

Zimmerman, Kim Ann. "History of Computers: A Brief Timeline." *Live Science*, 8 Sept. 2015, www.livescience.com/20718-computer-history.html. Accessed 17 Nov. 2016.

COMPUTER NETWORKS

SUMMARY

Computer networks consist of the hardware and software needed to support communications and the exchange of data between computing devices. The computing devices connected by computer networks include large servers, business workstations, home computers, and a wide array of smart mobile devices.

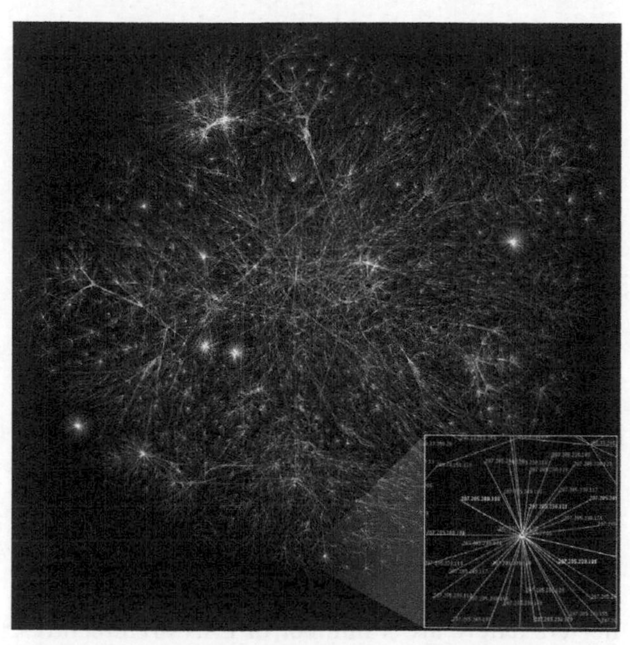

Internet map. The Internet is a global system of interconnected computer networks that use the standard Internet Protocol Suite (TCP/IP) to serve billions of users worldwide. By The Opte Project

The most popular computer network application is e-mail, followed by exchanging audio and video files. Computer networks provide an infrastructure for the Internet, which in turn provides support for the World Wide Web.

DEFINITION AND BASIC PRINCIPLES

A computer network is a collection of computer devices that are connected in order to facilitate the sharing of resources and information. Underlying the computer network is a communications network that establishes the basic connectivity of the computing devices. This communications network is often a wired system but can include radio and satellite paths as well. Devices used on the network include large computers, used to store files and execute applications; workstations, used to execute small applications and interact with the servers; home computers, connected to the network through an Internet service provider (ISP); and mobile devices, connected to the network by radio wave transmissions. Middleware is the software that operates on top of the communications network to provide a software layer for developers to add high-level applications, such as a search engine, to the basic network.

The high-level applications are what most people using the network see as the network. Some of the most important computer network applications provide communications for users. Older examples of this are e-mail, instant messaging, chat rooms, and videoconferencing. Newer examples of communications software are the multitude of social networks, such as Facebook. Other high-level applications allow users to share files. One of the oldest and still quite popular file-sharing programs is the file transfer protocol (FTP) program. Another way to use computer networks is to share computing power. The Telnet program (terminal emulation program for TCP/IP networks) allowed one to use an application on a remote mainframe in the early days of networking; and modern Web services allow one to run an application on a mobile device while getting most of the functionality from a remote server.

BACKGROUND AND HISTORY

The scientists who developed the first computers in the 1950s recognized the advantage of connecting computing devices. Teletype machines were in common use at that time, and many of the early computers were "networked" with these teletype machines over wired networks. By 1960, American Telephone and Telegraph (AT&T) had developed the first modem to allow terminal access to mainframes, and in 1964, IBM and American Airlines introduced the SABRE (Semi-Automated Business Research Environment) networked airline reservation system.

The Defense Department created ARPANET (Advanced Research Projects Agency Network) in 1966 to connect its research laboratories with college researchers. The early experience of ARPANET led the government to recognize the importance of being able to connect different networks so they could interoperate. One of the first efforts to promote interoperability was the addition of packet switching to ARPANET in 1969. In 1974, Robert Kahn and Vinton Cerf published a paper on packet-switching networks that defined the TCP/IP protocol, and in 1980, the U.S. government required all computers it purchased to support TCP/IP. When Microsoft added a TCP/IP stack to its Windows 95 operating system, TCP/IP became the standard wide area network in the world.

The development of the microcomputer led to the need to connect these devices to themselves and the wide area networks. In 1980, the Institute of Electrical and Electronics Engineers (IEEE) 802 standard was announced. It has provided most of the connectivity to networks since that time, although many wireless computing devices can connect with Bluetooth.

HOW IT WORKS

Computer networks consist of the hardware needed for networking, such as computers and routers; the software that provides the basic connectivity, such as the operating system; and middleware and applications, the programs that allow users to use the network. In understanding how these components work together, it is useful to look at the basic connectivity of the wide area network and contrast that to the way computers access the wide area network.

Wide Area Networks. A wide area network is one that generally covers a large geographic area and in which each pair of computers connect via a single line. The first wide area networks consisted of a number of connected mainframes with attached terminals. By connecting the mainframes, a user on one system could access applications on every networked computer. IBM developed the Systems Network Architecture (SNA), which was a popular early wide area network in the United States. The X.25 packet switching protocol standard provided a common architecture for the early wide area networks in Europe. Later, as all computers provided support for the TCP/IP protocol, it became possible for different computer networks to work together as a single network. In the twenty-first century, any device on a single network, whether attached as a terminal by X.25 or as a part of a local area network (LAN), can access applications and files on any other network on the Internet. This complete connectivity allows browsers on any type of mobile device to access content over the World Wide Web, because it runs transparently over the Internet.

Routing. A key to making the Internet operate efficiently is the large number of intermediate devices that route IP packets from the computer making a connection to the computer receiving the data. These devices are called routers, and they are the heart of the Internet. When a message is sent from a computer, it is decomposed into IP packets, each having the IP address of the receiver. Then each packet is forwarded to a border router of the sender. From the border router, the packet is routed through a set of intermediate routers, using an algorithm-like best path, and delivered to the border router of the receiver, and then the receiver. Once all the packets have arrived, the message is reassembled and used by the receiver.

Local Area Networks. A local area network (LAN) is a collection of computers, servers, printers, and the like that are logically connected over a shared media. A media access control protocol operates over the LAN to allow any two devices to communicate at the link level as if they were directly connected. There are a number of LAN architectures, but Ethernet is the most popular LAN protocol for connecting workplace computers on a LAN to other networks. Ethernet is also usually the final step in connecting home computers to other networks using regular, cable, and ADSL (Asymmetric Digital Subscriber Line) modems. The original coaxial cable Ethernet, developed by Robert Metcalfe in 1973, has been largely supplanted by the less expensive and easier to use twisted-pair networks. The IEEE 802.11 wireless LAN is a wireless protocol that is almost as popular as Ethernet.

Wireless Networks. A laptop computer's initial connection to a network is often through a wireless device. The most popular wireless network is the IEEE 802.11, which provides a reliable and secure connection to other networks by using an access point, a radio transmitter/receiver that usually includes border-routing capabilities. The Bluetooth network is another wireless network that is often used for peer-to-peer connection of cameras and cell phones to a computer and then through the computer to other networks. Cell phones also provide network connectivity of computing devices to other networks. They use a variety of techniques, but most of them include using a cell phone communications protocol, such as code division multiple access (CDMA) or Global System for Mobile Communications (GSM), and then a modem.

APPLICATIONS AND PRODUCTS

Computer networks have many applications. Some improve the operation of networks, while others provide services to people who use the network. The World Wide Web supplies many exciting new applications, and technologies such as cloud computing appear to be ready to revolutionize computing in the future.

Network Infrastructure. Computer networks are complex, and many applications have been developed to improve their operation and management. A typical example of the software developed to manage networks is Cisco's Internetwork Operating System (IOS) software, an operating system for routers that also provides full support for the routing functions. Another example of software developed for managing computer networks is the suite of software management programs provided by IBM's Tivoli division.

Communications and Social Networking. Communications programs are among the most popular computer applications in use. Early file-sharing applications, such as FTP, retain their popularity, and later protocols such as BitTorrent, which enables peer-to-peer sharing of very large files, and file hosting services like Dropbox, which provides cloud storage and online backup services, are used by all. Teleconferencing is used to create virtual workplaces; and Voice over Internet Protocol (VoIP) allows businesses and home users to use the Internet for digital telephone service with services provided by companies such as Vonage.

The earliest computer network communications application, e-mail, is still the largest and most successful network application. One accesses e-mail services through an e-mail client. The client can be a stand-alone program, such as Microsoft Outlook or Windows Live Mail, or it can be web-based (webmail), accessed using a browser such as Internet Explorer or Google Chrome. The e-mail server can be a proprietary system such as IBM Notes or one of the many webmail programs, such as Gmail, Outlook.com, or Yahoo! Mail.

One of the most important types of applications for computer networks is social networking sites. Facebook is by far the most widely used online social network, claiming nearly 1.4 billion users by the end of 2014, although sites such as Twitter and LinkedIn are also very popular. Facebook was conceived by Mark Zuckerberg in 2004 while he was still a student at Harvard University. Facebook members can set up a profile that they can share with other members. The site supports a message service, event scheduling service, and news feed. By setting up a "friend" relationship with other members, one can quickly rekindle old friendships and gain new acquaintances. Although Zuckerberg's initial goal was to create a friendly and interactive environment for individuals, Facebook has fast become a way to promote businesses, organizations, and even politicians. Facebook does not charge a subscription fee and has a reasonably successful business model selling customer information and advertising as well as working with partners.

Cloud Computing. Cloud computing refers to the hosting of data and applications on remote servers and networks ("the cloud") instead of on local computers. Among other things, this makes the data and software services accessible from multiple locations, such as both home and office computers, as well as from mobile devices. This has benefits for data accessibility as well as businesses' approach to their technology needs, as cloud computing essentially allows them to outsource some of their information technology (both hardware and software) for an ongoing service fee rather than up-front capital costs. For individuals, it allows ready access to one's files and programs anywhere one has access to the Internet. With cloud computing still very much an evolving phenomenon, existing platforms as of 2015 included Microsoft Azure, Amazon Web Services, and OpenStack, among others.

Next to e-mail, the most common application is word processing, and Microsoft Word is the dominant word processor. Microsoft offers online versions of Word and all its other Microsoft Office software, through a service called Office Online. Google offers similar cloud-based services such as Google Docs and Google Drive. As more and more applications are produced to run in the cloud, it is predicted that this will become the dominant form of computing, replacing the desktop computers of the early twenty-first century.

SOCIAL CONTEXT AND FUTURE PROSPECTS

The development of computer networks and network applications has often resulted in some sticky legal issues. File exchange programs, such as Flickr, are very popular but can have real copyright issues. Napster developed a very successful peer-to-peer file exchange program but was forced to close after only two years of operation by a court decision in 2001. Legislation has played an important role in the development of computer networks in the United States. The initial deregulation of the communications system and the breakup of AT&T in 1982 reduced the cost of the base communications system for computer networks and thus greatly increased the number of networks in operation. Opening up NSFNET for commercial use in 1991 made the Internet possible.

Networking and access to the Web have been increasingly important to business development and improved social networking. The emergence of cloud computing as an easy way to do computing promises to be just as transformative. Although there are still security and privacy issues to be solved, people are increasingly using mobile devices to access a wide variety of Web services through the cloud. Using just a smartphone, people are able to communicate with friends and associates, transact business, compose letters, pay bills, read books, and watch films.

—George M. Whitson III, BS, MS, PhD

Bibliography

Cerf, V. G., and R. E. Kahn. "A Protocol for Packet Network Interconnection." *IEEE Transactions on Communication Technology* 22 (May, 1974): 627–641.

Comer, Douglas. *Computer Networks and Internets.* Pearson Education, 2015.

Dordal, Peter L. "An Introduction to Computer Networks." Loyola University Chicago, 2015, intronetworks.cs.luc.edu/current/html/.

Dumas, M. Barry, and Morris Schwartz. *Principles of Computer Networks and Communications.* Upper Saddle River, N.J.: Pearson/Prentice Hall, 2009.

Forouzan, Behrouz. *Data Communications and Networking.* 5th ed. New York: McGraw-Hill, 2013.

Metcalfe, Robert, and David Boggs. "Ethernet: Distributed Packet Switching for Local Computer Networks." *Communications of the ACM* 19, no. 7 (July, 1976): 395-404.

O'Leary, Mike. *Cyber Operations: Building, Defending, and Attacking Modern Computer Networks.* Apress, 2015.

Stallings, William. *Data and Computer Communications.* 10th ed. Upper Saddle River, N.J.: Prentice Hall, 2014.

Tanenbaum, Andrew. *Computer Networks.* 5th ed. Boston: Pearson, 2011.

Computer Simulation

SUMMARY

Computer simulation is the use of computer technology to make digital models that represent real-world elements, behaviors, and systems. Programmers and engineers can then use the computerized model in experiments, changing conditions and observing the effects on the model. By observing and analyzing the results of these experiments, people can draw inferences about how the elements might behave in the real world. Testing a simulated model instead of a real-life object or scenario is generally much safer and less expensive, though simulated tests may prove to be less accurate. A type of applied mathematics, computer simulation is commonly used for a wide variety of purposes in fields such as science, politics, military studies, and entertainment.

USES AND TYPES

To perform a computer simulation, operators design a digital model of something to be tested. For instance, engineers may want to explore the feasibility of a new bridge linking two cities. In this case, operators program data about the proposed bridge—such as its dimensions, weight, construction style, and

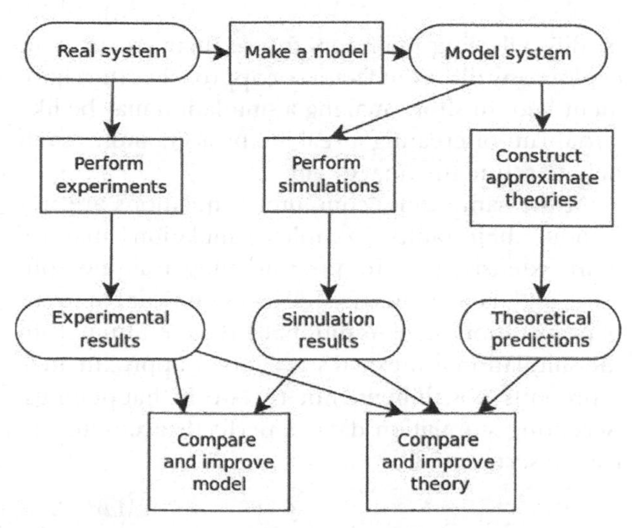

Process of building a computer model, and the interplay between experiment, simulation, and theory. By Danski14 (Own work)

materials—into an appropriate simulation program. Other factors, such as the location, climate, water flow, and typical wind speed, are programmed as well to complete the simulated scenario.

Once the simulation begins, the program shows how the proposed bridge would likely fare in the prevailing conditions. If engineers need specific information, such as how a tornado would affect the bridge, they can manipulate the data to reflect tornado-speed winds. They may also want to test different designs for the bridge, and may program different sizes or styles of bridge to test which works best in the given circumstances.

The simulation may run once or many times, and may be programmed to reflect a short time or a single event or potentially countless times or events. Operators study the data gathered by each simulation to help them gain better understanding of the dynamics of the system they have designed. Scientists with large quantities of data usually use statistics or other means of interpreting their findings, especially since simulations may not accurately reflect real-world possibilities.

People may use computer simulations to test proposed systems for many reasons. Sometimes the system being studied does not exist in the real world and cannot be easily or safely created, or may even be impossible to create. Sometimes a system does exist but operators want to simulate proposed changes to the system without actually altering the system in real life, which might involve serious expenses or dangers. In other cases, the simulation tests systems that have not occurred but may occur in the future—some examples of this are forecasting weather or predicting the spread of populations or diseases across a given area. Other simulations may test physical structures, economic principles, workplace practices, biological systems, or social trends.

There are three major kinds of computer simulations: continuous, Monte Carlo, and discrete. *Continuous simulations* show results of a system over time, with equations designed to show progressive changes. *Monte Carlo simulations* use random numbers and random events without regard for time. Finally, *discrete simulations* (sometimes classified as a subtype of Monte Carlo simulations) involve occasional events that break up otherwise uneventful blocks of time.

DEVELOPMENT AND MODERN IMPORTANCE

Although computer simulations are a relatively modern science, only arising after the dawn of the computer age in the twentieth century, the history of simulation reaches into ancient times. Before advanced science and technology were available to provide some accuracy to simulated results, people used mystical means to try to divine details of the future or hypothetical situations. Astrologers, oracles, prophets, and sorcerers were sought after in many lands for their purported abilities to gather information inaccessible to most.

As advances in science and technology replaced mysticism, people began creating computer programs that would use mathematical and scientific principles to create simulations. Some of the earliest computer-based simulations took place during World War II when scientists such as John von Neumann and Stanislaw Ulam used early computerized devices to simulate the effects of atomic bombs. As the primitive computers of the 1940s became faster, stronger, and more reliable, their ability to create simulations developed as well.

Modern computers are able to create comprehensive simulations for a wide range of different fields and applications. Simulation programs give operators more tools to customize the factors of their simulations, alter variables, and create animated

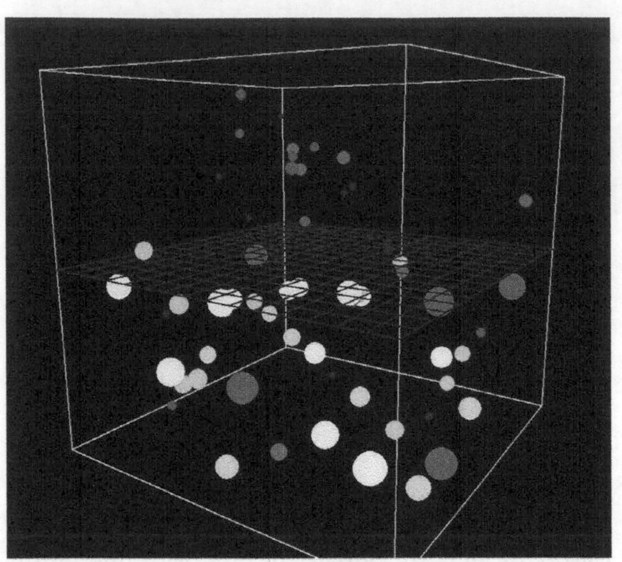

Computer simulation of the process of osmosis. By Lazarus666 at en.wikipedia (Transferred from en.wikipedia)

digital displays to represent the simulated scenarios. As the efficiency of simulations increases, so too do demands on simulation programmers to make their products more efficient and free of errors.

Computer simulations are generally less costly and difficult to prepare than real-life demonstrations, enabling them to be performed more quickly and frequently, thus creating more usable data. Even the process of creating a simulation could result in benefits for an individual or organization. To design the simulation, operators must painstakingly plan the model, which means analyzing all aspects of the proposed design and sometimes gathering new information from other sources. During this process, flaws in the concept may appear, or new potential approaches may come to light. When the simulation is complete, it is generally easy to customize and modify, allowing operators to use their creativity to explore any different factors or approaches they have identified. In short, making a simulation may be like a trial run of creating a real-life project, and as such help to refine the final design.

At the same time, computer simulations are not without their faults. Complex simulations may require extensive costs for programming, training, software, and data analysis as well as a significant amount of preparation time. Additionally, data resulting from the simulation is inexact, since it only approximately represents possibilities, not real-world happenings. Accepting simulation data as perfectly accurate can lead to serious risks.

—Mark Dziak

BIBLIOGRAPHY

"IBM SPSS Statistics–Monte Carlo Simulation." *IBM.* IBM. Web. 4 Aug. 2015. http://www-01.ibm.com/software/analytics/spss/products/statistics/monte-carlo-simulation.html

McHaney, Roger. *Computer Simulation: A Practical Perspective.* San Diego: Academic Press, Inc., 1991. 1–9. Print.

McHaney, Roger. *Understanding Computer Simulation.* StudyGuide.pk. Web. 4 Aug. 2015. http://studyguide.pk/Handouts/ACCA/acca_it/understanding.pdf

Winsberg, Eric. "Computer Simulations in Science." *Stanford Encyclopedia of Philosophy.* Center for the Study of Language and Information (CSLI), Stanford University. 23 Apr. 2015. Web. 4 Aug. 2015. http://plato.stanford.edu/entries/simulations-science/

Winsberg, Eric. *Science in the Age of Computer Simulation.* Chicago: University of Chicago Press, 2010. 1–6. Print.

COMPUTER SOFTWARE

SUMMARY

Computer software refers to one or more programs that determine how a computer operates. Software works along with a computer's *hardware,* or physical parts, to make the computer function efficiently for various tasks. There are two main kinds of computer software: *system software* and *application software.* System software is mainly used by the computer to help its parts communicate and cooperate. Application software is generally added to allow humans to perform specific tasks with the computer. Together, these kinds of software make computers valuable machines for work and entertainment alike.

COMPUTERS, HARDWARE, AND SOFTWARE

Computers are electronic machines that can accept and interpret different forms of data and perform various operations using the data. People use computers for a great variety of purposes ranging from complex scientific research to entertainment. After several generations of development, computers have taken on a wide range of forms in the modern world. People today may use large computers on desktops or smaller laptop computers. Notebook computers and relatively tiny mobile versions become more popular each year, and embedded computers are critical parts of most automobiles and many appliances.

Despite the great variety in modern computers, they share some important similarities. One of the main similar features of all computers is that they use hardware and software. Hardware is a term referring to the physical parts of a computer. These parts include easily identifiable external features such as the central processing unit (CPU), keyboard, and mouse or touchpad. They also include smaller, internal features such as *hard disks,* or devices that allow the computer to store and access information. Hardware is important because it creates a basis upon which people may add software that will make the computer perform specific tasks.

TYPES OF COMPUTER SOFTWARE

Computer software is a much broader category than computer hardware. Software refers to all the programs that manufacturers or users may install on a computer. Software may be used by the computer, by the user, or by both to organize and use files and other data for specific tasks.

The most important software on a computer, and the first program that should be installed, is called the *operating system.* The operating system serves as a wide-reaching interface, or a digital workstation through which users can install, organize, and operate other, smaller programs. It also helps the computer share its memory and other resources among its various tasks, handle input and output functions, and store new data.

For most computer users, the operating system is the first thing visible once the computer loads after it is turned on. Microsoft Windows was one of the most common operating systems in use, but some types of computers and computer hardware may work better with other operating systems. Some of the main competitors to Windows include Macintosh OS, iOS, Linux, and Android.

Although the operating system is the most important software, there are dozens of other kinds of software available to modern computer users for both personal and professional tasks. This software can generally be divided into two main categories based on their types, the work they do, and the fields in which they are most commonly used. These main categories are *system software* and *application software.*

SYSTEM SOFTWARE

System software includes any kind of program that controls other programs or the computer's systems. These programs are meant to function closely with the hardware and are useful for making the computer and all of its other programs run more efficiently. The operating system is one prime example of system software. Another important example is a *compiler,* a program that translates the codes, or computer "language," used by the programs and hardware.

Other types of system software include programming tools such as loaders, linkers, and assemblers, which also gather and translate data. There are also *drivers,* or programs that assist hardware in performing its necessary tasks. All of these kinds of system software help the various features of a computer communicate and cooperate. Some also allow users to diagnose and solve problems with, or maximize the efficiency of, the computer or its programs.

APPLICATION SOFTWARE

The other main category of software is application software. Whereas system software is mostly used to help the computer function properly, application software is generally more focused on the needs of the user. This category includes all programs that are added to a computer to perform tasks for people using the computer. These tasks can be for productivity, for fun, or both, and may be general in nature or meant for specific goals.

Hundreds of examples of application software are available to computer users, each designed to meet specific user needs. Most of this software falls into broad categories. For instance, multimedia software lets computer users view, listen to, or even create visual and audio files. Spreadsheet and database software helps users store, sort, and analyze large amounts of information. Word processing software lets users create various kinds of documents, and presentation software helps users display their creations for others. Internet browsers, gaming programs, and communication tools are also popular types of application software.

—*Mark Dziak*

Bibliography

Damien, Jose. *Introduction to Computers and Application Software.* Sudbury: Jones & Bartlett Learning, 2011. 4–7, 25. Print.

"Different Types of Software." *Introduction to IT English.* University of Victoria. Web. 9 July 2015. http://web2.uvcs.uvic.ca/elc/sample/ite/u01/u1_1_03.html

Minich, Curt. "Types of Software Applications." *INSYS 400 Home Page.* Penn State University, Berks Campus. Web. 9 July 2015. http://www.minich.com/education/psu/instructtech/softwareapps.htm

Patterson, David A. and John L. Hennessy. *Computer Organization and Design: The Hardware/Software Interface.* 5th ed. Waltham: Elsevier, Inc., 2014. 3–9, 13–17. Print.

Toal, Ray. "What Is Systems Programming?" *Computer Science Division.* College of Science and Engineering, Loyola Marymount University. Web. 9 July 2015. http://cs.lmu.edu/~ray/notes/sysprog/

Computer-Aided Design and Manufacturing

SUMMARY

Computer-aided design and manufacturing is a method by which graphic artists, architects, and engineers design models for new products or structures within a computer's "virtual space." It is variously referred to by a host of different terms related to specific fields: computer-aided drafting (CAD), computer-aided manufacturing (CAM), computer-aided design and drafting (CADD), computer-aided drafting/computer-aided manufacturing (CAD/CAM), and computer-aided drafting/numerical control (CAD/NC). The generic term "computer-aided technologies" is, perhaps, the most useful in describing the computer technology as a whole. Used by professionals in many different disciplines, computer-aided technology allows designers in traditional fields such as architecture to render and edit two-dimensional (2-D) and three-dimensional (3-D) structures. Computer-aided technology has also revolutionized once highly specialized fields, such as video-game design and digital animation, into broad-spectrum fields that have applications ranging from the film industry to ergonomics.

DEFINITION AND BASIC PRINCIPLES

Computer-aided design and manufacturing is the combination of two different computer-aided technologies: computer-aided design (CAD) and computer-aided manufacturing (CAM). Although the two are related by their mutual reliance on task-specific computer software and hardware, they differ in how involved an engineer or architect must be in the creative process. A CAD program is a three-dimensional modeling software package that enables a user to create and modify architectural or engineering diagrams on a computer. CAD software allows a designer to edit a project proposal or produce a model of the product in the virtual "space" of the computer

Example: 3-D CAD model. By Freeformer

screen. CAD simply adds a technological layer to what is still, essentially, a user-directed activity.

CAM, on the other hand, is a related computer-aided technology that connects a CAD system to laboratory or machine-shop tools. There are many industrial applications for computer-directed manufacturing. One example of CAD/CAM might be when an automotive designer creates a three-dimensional model of a proposed design using a CAD program, verifies the physical details of the design in a computer-aided engineering (CAE) program, and then constructs a physical model via CAD or CAE interface with CAM hardware (using industrial lasers to burn a particular design to exact specifications out of a block of paper). In this case, the automotive designer is highly involved with the CAD program but allows the CAD or CAE software to direct the activity of the CAM hardware.

BACKGROUND AND HISTORY

CAD/CAM is, essentially, a modernization of the age-old process of developing an idea for a structure or a product by drawing sketches and sculpting models. Intrigued by the idea that machines could assist designers in the production of mathematically exact diagrams, Ivan Sutherland, a doctoral student at the Massachusetts Institute of Technology, proposed the development of a graphical user interface (GUI) in his 1963 dissertation, "Sketchpad: A Man-Machine Graphical Communication System." GUIs allow users to direct computer actions via manipulation of images rather than by text-based commands. They sometimes require the use of an input device, such as a mouse or a light pen, to interact with the GUI images, also known as icons. Although it was never commercially developed, Sketchpad was innovative in that a user could create digital lines on a screen and determine the constraints and parameters of the lines via buttons located along the side of the screen. In subsequent decades, the CAD software package employed the use of a mouse, light pen, or touch screen to input line and texture data through a GUI, which was then manipulated by computer commands into a virtual, or "soft," model.

In the twenty-first century, CAM is more or less the mechanical-output aspect of CAD work, but in the early days, CAM was developed independently to speed up industrial production on assembly lines. Patrick J. Hanratty, regarded as the father of CAD/CAM, developed the first industrial CAM using numerical control (NC) in 1957. Numerical control—the automation of machining equipment by the use of numerical data stored on tape or punched on cards—had been in existence since the 1940s, as it was cheaper to program controls into an assembly machine than to employ an operator. Hanratty's PRONTO, a computer language intended to replace the punched-tape system (similar to punch cards in that the automaton motors reacted to the position of holes punched into a strip of tape) with analog and digital commands, was the first commercial attempt to modernize numerical control and improve tolerances (limiting the range of variation between the specified dimensions of multiple machined objects).

HOW IT WORKS

Traditionally, product and structural designers created their designs on a drawing board with pencils, compasses, and T squares. The entire design process was a labor-intensive art that demanded not just the ability to visualize a proposed product or structure in exact detail but also extensive artistic abilities to render what, up until that point, had been only imagined. Changes to a design, whether due to a client's specific needs or the limitations of the materials needed for construction, frequently required the designer to "go back to the drawing board" and create an entirely new series of sketches. CAD/CAM was intended to speed up the design and model-making processes dramatically.

COMPUTER-AIDED DESIGN (CAD). CAD programs function in the same way that many other types of computer software function: One loads the software into a computer containing an appropriate amount of memory, appropriate video card, and appropriate operating system to handle the mathematical calculations required by the three-dimensional modeling process. There are quite literally hundreds of different CAD programs available for different applications. Some programs are primarily two-dimensional drawing applications, while others allow three-dimensional drawing, shading, and rendering. Some are intended for use in construction management, incorporating project schedules, or for engineering, focused on structural design. There are a range of CAD programs of varying complexity for computer systems of different capabilities and operating systems.

COMPUTER-AIDED MANUFACTURING (CAM). CAM, on the other hand, tends to be a series of automated mechanical devices that interface with a CAD program installed in a computer system—typically, but not exclusively, a mainframe computer. The typical progress of a CAD/CAM system from design to soft model might involve an automotive designer using a light pen on a CAD system to sketch out the basic outlines of suggested improvements for an existing vehicle. Once the car's structure is defined, the designer can then command the CAD program to overlay a surface, or "skin," on top of the skeletal frame. Surface rendering is a handy tool within many CAD programs to play with the use of color or texture in a design. When the designer's conceptual drawings are complete, they are typically loaded into a computer-aided engineering program to ascertain the structural integrity and adherence to the requirements of the materials from which the product will ultimately be constructed. If the product design is successful, it is then transferred to a CAM program, where a product model, machined out of soft model board, can be constructed to demonstrate the product's features to the client more explicitly.

Advertisements for CAD programs tend to emphasize how the software will enhance a designer's imagination and creativity. They stress the many different artistic tools that a designer can easily manipulate to create a virtual product that is as close as possible to the idealized conceptual design that previously existed only in the designer's imagination. On the other hand, CAM advertisements stress a connection of the CAM hardware to real-world applications and a variety of alterable structural materials ranging from blocks of paper to machinable steel. The two processes have fused over time, partly because of the complementary nature of their intended applications and partly because of clients' need to have a model that engages their own imaginations, but certain aspects of CAD (those technological elements most appropriate to the creation of virtual, rather than actual, worlds) may eventually allow the software package to reduce the need for soft models in terms of product development.

CAD IN VIDEO GAMES AND FILM ANIMATION. CAD work intended for video games and film animation studios, unlike many other types of designs, may intentionally and necessarily violate one basic rule of engineering—that a design's form should never take precedence over its function. Industrial engineers also stress the consideration of ergonomics (the human body's requirements for ease, safety, and comfort when using a product) in the design process. In the case of the virtual engineer, however, the visual appeal of a video game or animation element may be a greater factor in how a design is packaged for a client. *Snow White and the Seven Dwarfs* was an animated film originally drawn between 1934 and 1937. Since animation was drawn by hand in those days—literally on drawing boards—with live models for the animated characters, the studio's founder, Walt Disney, looked for ways to increase the animated characters' visual appeal. One technique he developed was to draw human characters (in this case, Snow White herself) with a larger-than-standard size head, ostensibly to make her appear to have the proportions of a child and, thus, to be more appealing to film viewers. In order for the artists to draw their distorted heroine more easily, the actor playing Snow White, Marjorie Bell, was asked to wear a football helmet while acting out sections of the story's plot.

Unfortunately for Disney, Bell complained so much about the helmet that it had to be removed after only a few minutes. Ergonomics took precedence over design. Later animation styles, influenced by designers' use of CAD animation software, were able to be even more exaggerated without the need for a live model in an uncomfortable outfit. For example, the character of Jessica Rabbit in Amblin Entertainment's *Who Framed Roger Rabbit?* (1988) has a face and body that were designed exclusively for visual appeal, rather than to conform to external world reality. Form dominated function.

APPLICATIONS AND PRODUCTS

CAD and CAD/CAM have an extensive range and variety of applications.

MEDICINE. In the medical field, for example, software that employs finite element methods (FEM; also called finite element analysis, or FEA), such as Abaqus FEA, allows designers to envision new developments in established medical devices—improving artificial heart valves, for example—or the creation of new medical devices. CAD software ideally allows a user to simulate real-world performance.

AEROSPACE. Femap and NEi Nastran are software applications used for aerospace engineering. These programs can simulate the tensile strength and load capacity of a variety of structural materials as well as the effect of weather on performance and durability.

MECHANICAL ENGINEERING. Mechanical engineering firms sometimes use Siemens PLM software for CAM or CAE functions in order to create and revise structural models quickly.

Some forms of CAD have engineering applications that can anticipate a design's structural flaws to an even greater extent than an actual model might. One example of this application is SIMULIA, one of the simulation divisions of Dassault Systèmes that creates virtual versions of the traditional automobile safety tester—the crash-test dummy. The company's program of choice is Abaqus FEA, but it has altered the base software to use various specialized virtual models for a wider variety of weight and height combinations. In the real world, crash-test dummies are rather complex mechanical devices, but their virtual simulacra have been declared as effective in anticipating structural failure as the original models.

ANIMATION. CAD programs have caused the rapid expansion and development of the animation industry—both in computer games and in film animation. Models of living creatures, both animal and human, are generated through the use of three-dimensional CAD software such as Massive (which was used by the graphic artists creating the battle sequences in Peter Jackson's *Lord of the Rings* film trilogy) and Autodesk Maya (considered an industry standard for animated character designs). Autodesk 3ds Max is one of the forerunner programs in the video-game industry, since it allows a great deal of flexibility in rendering surfaces. Certain video games, such as Will Wright's *Sims* series, as well as any number of massively multiplayer online role-playing games (MMORPGS) such as *World of Warcraft*, *Second Life*, and *EverQuest*, not only have CAD functions buried in the design of the game itself but also accept custom content designed by players as a fundamental part of individualization of a character or role. A player must design a living character to represent him or her in the video game, including clothing, objects carried, and weapons used, and the process of creation is easily as enjoyable as the actual game play.

Avatar, released by Twentieth Century Fox in 2009, was envisioned by director James Cameron as a sweeping epic in much the same vein as George Lucas's original *Star Wars* trilogy (1977–83). *Avatar* was originally scheduled to start filming in 1997, after Cameron completed *Titanic*, but Cameron felt the computer-aided technology was not ready to create the detailed feature film he envisioned. Given that the film was also supposed to take advantage of newly developed three-dimensional stereoscopic film techniques, Cameron did not start work on the computer aided-technology aspects of his film until 2006, when he felt sure the technology was up to the task.

Even with an extensive three-dimensional software animation package at his command, the task of creating and populating an entire alien world was a daunting prospect. Most of the action of *Avatar* takes place on a distant planet, Pandora, which is intrinsically toxic to human beings. To get around this difficulty and successfully mine Pandora for a much-needed mineral, humans must use genetically created compatible bodies and interact with the native population, the Na'vi. Both the native creatures, who are ten feet tall with blue skin, and the hybrid human-Na'vi "avatar" bodies were a challenge for animators to render realistically even using computer-aided technology. The full IMAX edition was released on December 18, 2009, having ultimately cost its creators somewhere between $280 million and $310 million. Fortunately for Cameron, the film ended up rewarding its creators with gross revenues of more than $2 billion. Having created the highest-grossing film of all time, which also won Academy Awards for art direction, cinematography, and visual effects, Cameron was able to negotiate successfully with Fox for three planned sequels.

SOCIAL CONTEXT AND FUTURE PROSPECTS

Computer-aided technologies, in general, are expected to keep developing and evolving. The possibilities of virtual reality are continuing to evolve as the need for structural and materials modeling increases. One common description of the future of computer-aided design and manufacturing is the term "exponential productivity." In other words, so much productive time in the fields of design has been wasted by the constant need to create by hand what can be better (and more quickly) produced by a computer. An architect or engineer might spend hours carefully working out the exact proportions of a given structure, more hours using a straight edge and compass to determine the exact measurements needed by the design, and even more time to clean up mistakes. On the other hand, computers can help that same architect draw straight lines, evaluate numerical formulas for proportions, and edit mistakes in a matter of minutes. When one considers the added benefits of a compatible computer-aided manufacturing program, one can further see the time saved in the quick and efficient production of a soft model.

—Julia M. Meyers, MA, PhD

BIBLIOGRAPHY

Duggal, Vijay, Sella Rush, and Al Zoli, eds. *CADD Primer: A General Guide to Computer-Aided Design and Drafting*. Elmhurst: Mailmax, 2000. Print.

Kerlow, Issac. *The Art of 3D Computer Animation and Effects*. 4th ed. Hoboken: Wiley, 2009. Print.

Lee, Kunwoo. *Principles of CAD/CAM/CAE Systems*. Reading: Addison, 1999. Print.

Park, John Edgar. *Understanding 3D Animation Using Maya*. New York: Springer, 2005. Print.

Petty, Luke. "What is Computer Aided Manufacturing? And How Is It Leading the Digital Site Revolution?" *Chapman Taylor*, 19 May 2016, www.chapmantaylor.com/en/insights/article/what-is-computer-aided-manufacturing-and-how-is-it-leading-the-digital-site/en/. Accessed 24 Oct. 2016.

Sarkar, Jayanta. Computer Aided Design: A Conceptual Approach. CRC Press, 2015.

Schell, Jesse. *The Art of Game Design: A Book of Lenses*. 2nd ed. Boca Raton: CRC, 2015. Print.

Schodeck, Daniel, et al. *Digital Design and Manufacturing: CAD/CAM Applications in Architecture and Design*. Hoboken: Wiley, 2005. Print.

CONTINUOUS RANDOM VARIABLE

SUMMARY

A continuous random variable is one defined on an uncountably infinite sample space. A sample space is a set of possible outcomes of some procedure or trial where there is more than one possible outcome. An example would be the length of time it takes to read this article. The time could be measured to the nearest minute to yield a discrete random variable. Allowing the possibility of infinite reading time, there are many possible outcomes of any number greater than zero.

Some notion of probability must be associated with the sample space. A random variable is a function that maps the sample space onto a number. In this case, time elapsed is associated with a number. As we are mapping an event with an associated probability we induce some notion of probability onto the numbers we obtain. The probability that a random variable X takes on a specific value is expressed as x Prob($X = x$).

The key distinction between continuous and discrete random variables is that for continuous random variables the sample space is infinite and uncountable. The random variable can be turned into a discrete random variable by recording the time in rounded units, such as minutes or microseconds. However, even if the time available were restricted to a maximum of one hour, there are still uncountably many real numbers between 0 and 60.

The key implication of dealing with an uncountable sample space is that the probability associated with any specifically realised number is zero. A rather approximate analogy is the event of rolling a fair die, which has a finite countable sample space; you obtain dot, 2 dots,...,6 dots. Associated with this is the random variable X, which denotes the number of dots (1,2,...6) and has associated the probability 1/6 that any particular number is seen on any given die throw. If the number of sides of die is increased the probability of any particular event decreases with the number of sides. For example, with 12 sides the probability is 1/12. In rolling a die with an uncountably infinite number of sides, the probability of the die landing on any one side is zero.

Rather than Prob($X = x$), *Prob($a<X<b$)* is used to determine the probability that the random variable X lies in a subset of the sample space between real numbers a and b. This subset E can be defined formally as E: $\{a < x < b\}$. It is possible to find the probability that event E occurs. The experiment is conducted and the random variable X takes a number between a and b by using

$$P(a \leq X \leq b) = \int_{x \subset B} f(x)\,dx.$$

Here $f(x)$ denotes any function that acts as a probability density function. This function can model the notion of probability associated with the random variable X. If we integrate this function over the range of interest we obtain the probability that a random variable X lies in that range.

One common choice of probability model for the time it takes to read this article would be the so-called negative exponential distribution. For some parameter $\theta > 0$ and random variable $X > 0$ the probability density function is given by:

$f(x)\ \theta e = \theta e^{-\theta x}$

If, for example, the parameter $\theta > 0.2$ we could find

$$P(3 \leq X \leq 6) = \int_{3}^{6} 0.2e^{-0.2x}\,dx = 0.2476$$

—*Paul Hewson*

BIBLIOGRAPHY

Blitzstein, J. K., and J. Hwant. *Introduction to Probability.* Boca Raton, FL: CRC, 2015.

Mendenhall, William, Robert J. Beaver, Barbara M. Beaver. *Introduction to Probability and Statistics.* Boston: Cengage, 2013.

Pishro-Nik, Hossein. *Introduction to Probability, Statistics, and Random Processes.* Kappa Research, 2014.

CYBERNETICS

SUMMARY

Cybernetics is the science of control and communication in living organisms, machines, and other systems. The primary focus of cybernetics is how a system—biological or otherwise—processes information and initiates changes to improve its functioning in response to the information it receives. Accordingly, cybernetics is concerned with system models in which some sort of monitor sends information about what is happening within or around said system at a given time to a controller. This controller then initiates whatever changes are necessary to keep the system operating within normal parameters. Cybernetics is a broad scientific field that often intersects with and involves many other scientific studies, including mathematics, biology, logic, and sociology. Although it has roots that date back to antiquity, cybernetics as it is known today first emerged as a significant field of study in the 1940s and has continued to develop since that time.

HISTORY

The term *cybernetics* is derived from the Greek word *kybernetikos,* which means "good at steering." This derivation emphasizes the fact that cybernetics first emerged as a concept in the works of the ancient Greek philosopher Plato, who used it in reference to systems of government. This definition never enjoyed widespread popularity or use, but the idea of cybernetics as a science that investigates the control of governments persisted for centuries. Among the greatest modern advocates of this interpretation was French physicist André-Marie Ampère, who tried with little success to popularize cybernetics as a government science in the nineteenth century. Despite Ampère's best efforts, the concept of cybernetics as it was known up to that point fell from favor and virtually disappeared by the early twentieth century.

The modern interpretation of cybernetics first arose when American mathematician Norbert Wiener published a book called *Cybernetics: Or Control and Communication in the Animal and the Machine* in 1948. As its subtitle suggests, Wiener's landmark book laid the groundwork for cybernetics as a science that connects control with communication, implying a direct relationship between the action a system's control mechanism takes to achieve a certain goal and the connection and flow of information between the control mechanism and the surrounding environment. In short, Wiener's primary argument was that communication is necessary for effective action. In addition, Wiener's subtitle further indicates that the principles of cybernetics apply equally to biological and non-biological systems, meaning that machines can and do rely on communication and control as much as living organisms do. Two years after his original publication, Wiener subsequently suggested that societal systems could be considered a third realm of cybernetics. Thus, the contemporary definition of cybernetics was established.

BASICS OF CYBERNETICS

Although the study of cybernetics can be quite complex, the underlying principles of cybernetics as a concept are relatively simple. The basic premise of

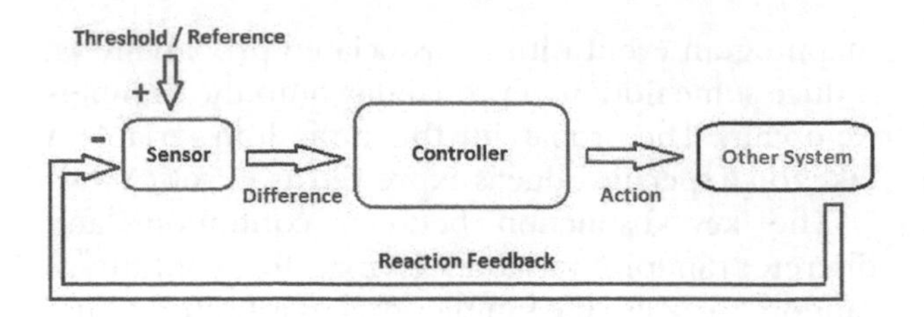

A Cybernetic Loop

Pathways and Components in a Cybernetic Loop. By Baango (Own work)

cybernetics is perhaps most easily understood when examined in a biological context. Seen through that particular lens, cybernetics is closely tied to the physiology of living things and the concept of automatic control. In humans and other animals, the brain frequently functions as a controller that receives information called *feedback* from monitors—such as the eyes, ears, or fingertips—and issues instructions meant to initiate a physical response to the feedback in question. For example, if a person is attempting to pick up a glass of water, the eyes, functioning as monitors, send critical feedback to the brain, which acts as a controller, about the distance between the person's hand and the glass of water. In response to this feedback, the brain may instruct the person's body to move the hand closer to the glass of water to help the person pick it up. In other words, it is cybernetics, or the combination of control and communication, that enables the person to pick up the glass of water.

The same principles apply to mechanical systems. In a computer, the microprocessor may act as a controller that receives feedback from a peripheral device such as a mouse or keyboard, which acts as a monitor. When input is received from the mouse or keyboard, the microprocessor responds by issuing instructions to other system components to allow the computer to complete a specific task.

Just as cybernetics plays an important role in the function of living things and machines, it also plays a role in the function of societal systems. On the societal level, one might consider the citizens of a town to be the town's monitors. If these citizens notice an increase in crime, they may choose to alert members of the town council to their concerns about the rising crime rate. The council members, who act as the controllers in this example, may respond to the citizens' concerns by enacting new ordinances or hiring more police officers—actions that may facilitate the ultimate goal of lowering the crime rate. In this way, the town citizens and council members have used communication and control—or cybernetics—to achieve a desired end.

COMMON MISCONCEPTIONS

In general, few people outside of the cybernetics field have an accurate understanding of cybernetics. Because of the widespread use of terms like *cyberspace* and *cyborgs* and the proliferation of books, films, and television shows that use the word *cybernetics* to reference computers and robots, many people who are unfamiliar with the actual principles of cybernetics believe that the term has something to do with robotics or artificial intelligence. Although cybernetics can certainly play an important role in robotic and artificial intelligence systems, it would be incorrect to use the words *cybernetics* or *cybernetic* to describe either of these systems as a whole.

—Jack Lasky

Bibliography

"'Getting Started' Guide to Cybernetics." *Pangaro.com.* Paul Pangaro. 2013. Web. 14 Mar. 2016. http://www.pangaro.com/definition-cybernetics.html

"History of Cybernetics." *American Society for Cybernetics.* American Society for Cybernetics. Web. 14 Mar. 2016. http://www.asc-cybernetics.org/foundations/history.htm

Kline, Ronald R. *The Cybernetics Movement, or Why We Call Our Age the Information Age.* Baltimore, MD: John Hopkins University Press, 2015. Print.

Novikov, D.A. *Cybernetics: From Past to Future.* Switzerland: Springer International Publishing, 2016. Print.

Cybersecurity

CYBERATTACKS

Any computer connected to the Internet is vulnerable (weakened) to a cyberattack. A cyberattack is harm done to a computer or network. There are many different kinds of cyberattacks.

When a computer is hacked, someone in another location is viewing or taking the computer's information without permission. A person who hacks into computers is a hacker. Some hackers try to find bank account or credit card information. They use this information to steal money or make purchases. Other hackers steal passwords. Passwords are special words or phrases that people use to protect their information. Hackers use passwords to access a person's e-mail or other accounts. Some hackers install viruses on computers just to cause damage.

Viruses are programs meant to harm computers. A program is a set of directions that tell a computer what to do. Viruses enter computers in several ways. Some viruses enter when people visit certain websites. Some look like advertisements or even security updates. When people click them, they allow the virus in. Other viruses spread through e-mail.

Spam is unwanted e-mail. Another name for "spam" is "junk mail." Many businesses send spam

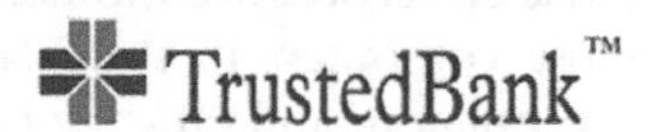

Dear valued customer of TrustedBank,

We have recieved notice that you have recently attempted to withdraw the following amount from your checking account while in another country: $135.25.

If this information is not correct, someone unknown may have access to your account. As a safety measure, please visit our website via the link below to verify your personal information:

http://www.trustedbank.com/general/custverifyinfo.asp

Once you have done this, our fraud department will work to resolve this discrepency. We are happy you have chosen us to do business with.

Thank you,
TrustedBank

Member FDIC © 2005 TrustedBank, Inc.

An example of a phishing email, disguised as an official email from a (fictional) bank. The sender is attempting to trick the recipient into revealing confidential information by "confirming" it at the phisher's website. Note the misspelling of the words received and discrepancy as recieved and discrepency respectively.

advertisements. They want people to buy the product in the ad. Most spam is harmless, but some spam has links to websites that contain viruses that can damage computers.

Cyberattacks are becoming more common. In June 2012 U.S. president Barack Obama stated that cyberattacks were a serious threat. Hackers have launched cyberattacks against several big companies. Stores such as T. J. Maxx, Marshall's, Target, and Barnes & Noble were all victims of cyberattacks between 2007 and 2013. A victim is anyone who has been hurt by a crime. Hackers stole credit card information and other data from millions of customers in these attacks. In 2014 hackers launched cyberattacks against online businesses including eBay and AOL. These attacks show the need for cybersecurity.

PROTECTING PERSONAL INFORMATION

Computers serve many purposes in today's world. Computers help people stay connected to friends and family. People talk or send messages to one another with the click of a button. They use the Internet to shop for anything from shirts to cars. Computers make it possible to send, receive, sort, and save all kinds of information quickly and easily. As a result, people are entering and saving more private information on computers.

Doctors and hospitals use computers to save patients' medical records. Banks and credit card companies use computers to store information about people's money. Businesses use computers to keep records of customers' purchases. Governments use computers to store tax records, driver's license information, and other documents. People use computers to store pictures and videos. They use computers to send and receive e-mail messages. They may use computers to connect to social networks. A social network is a group of people that a person connects to on the Internet.

The more that people save and share on computer networks, the more they are at risk for cyberattacks. Ways to avoid becoming a victim of cyberattacks include installing security software. Security software is a special program that checks computers for viruses and other threats. Security software should be added to anything that connects to the Internet, such as printers and game systems. Security software should be added to smartphones, which are cell phones that make calls, play music, and connect to the Internet. Tablets, which are thin, handheld computers, also should have security software. Security software should be updated often to protect against new viruses.

Another way to avoid becoming a victim of a cyberattack is to choose good passwords. A password should be long and use uppercase and lowercase letters. It should also include numbers and symbols. Do not use the same password for everything. Keep a list of passwords in a safe place.

Avoid sharing private information on the Internet. Never list phone numbers or addresses on social network websites.

Only shop and bank on trusted websites. Web addresses that begin with "https://" or "shttp://" and secure and safe. Addresses that begin "http://" are not secure and financial and other private information should not be shared on these websites.

People connect to the Internet now more than ever. As a result, personal information stored on computers has never been more at risk. Cybersecurity is the only way to keep this information safe.

—Lindsay Rock Rohland

BIBLIOGRAPHY

"Cyber Security Primer." Cybersecurity. University of Maryland University College, n.d. Web. 2 June 2014. http://www.umuc.edu/cybersecurity/about/cybersecurity-basics.cfm. This webpage tells what cybersecurity is and why it is important.

Finkle, Jim, Soham Chatterjee, and Lehar Maan. "eBay Asks 145 Million Users to Change Passwords after Cyber Attack." Reuters. Thomson Reuters, 21 May 2014. Web. 2 June 2014. http://www.reuters.com/article/2014/05/21/us-ebay-password-idU.S.BREA4K0B420140521. This article is about the May 2014 cyberattack on eBay, which resulted in hackers stealing the personal, but not financial, information of 145 million eBay users.

Jamieson, Alastair, and Eric McClam. "Millions of Target Customers' Credit, Debit Card Accounts May Be Hit by Data Breach." NBC News. NBCNews, 19 Dec. 2013. Web. 2 June 2014. http://www.nbcnews.com/business/consumer/millions-target-customers-credit-debit-card-accounts-may-be-hit-f2D11775203. This article is about the 2013 cyberattack at Target stores across the United States.

Obama, Barack. "Taking the Cyberattack Threat Seriously." Wall Street Journal. Dow Jones, 19 July 2012. Web. 2 June 2014. http://online.wsj.com/news/articles/SB10000872396390444330904577535492693044650. In this article, President Barack Obama tells how cyberattacks could threaten the United States.

"Tips & Advice." Stay Safe Online. National Cyber Security Alliance, n.d. Web. 2 June 2014. http://www.staysafeonline.org/stop-think-connect/tips-and-advice. This webpage tells how to protect computers from cyberattacks.

Cyberspace

SUMMARY

Cyberspace is the formless, nonphysical realm that theoretically exists as the result of the links between computers, computer networks, the Internet, and other devices and components involved in Internet use. Although it is commonly thought of as being synonymous with the Internet, the term *cyberspace* actually predates the Internet itself. When the concept was originally conceived in the 1980s, cyberspace was typically seen as a sort of alternative reality that arose from the early electronic networks of the pre-Internet era and was entirely separate from the physical world. With the later rise of the Internet, the term *cyberspace* came to be defined more specifically as the amorphous virtual world in which Internet users interact with one another. Since that time, the nature of cyberspace and even the question of its very existence have been the subject of considerable debate. In addition to arguments over whether or not real-world governments have the right to control cyberspace, many pundits ask whether cyberspace is truly a real place or just an outdated intellectual-political fad.

ORIGINS AND MEANING

The term *cyberspace* was first coined in 1982 by author William Gibson in a story that was initially published in *Omni* magazine and later in his science fiction novel *Neuromancer.* In this landmark work, Gibson described cyberspace as "a graphic representation of data abstracted from the banks of every computer in the human system." Rather than being a strictly real place, Gibson's cyberspace was a largely imaginary realm wherein information took on some of the properties of physical matter. Further, when hackers plugged themselves into Gibson's cyberspace, they literally exchanged their physical bodies for ones made of digital data. Although this construct was anchored far more in science fiction than reality, it firmly established the idea of cyberspace as a unique electronic realm that existed both within and outside the physical world.

The modern definition of cyberspace developed with the advent of the Internet as the dominant medium of electronic communication in the 1990s. Seen through the lens of the Internet, cyberspace was defined as the virtual space in which Internet users interact with one another when they are online. Although the term is often used interchangeably with "Internet" today, cyberspace is more accurately described as the virtual environment in which things like chat rooms, online games, and websites exist and to which individuals gain access through the Internet. As such, cyberspace is also a forum for political discourse, intellectual debate, and other culturally important social interactions.

CYBERSPACE AND GOVERNMENT

One of the biggest and most controversial concerns cyberspace has faced since the advent of the Internet is the question of what role, if any, real-world governments should play in its administration. From the outset of this debate, many activists maintained that real-world governments had no place in cyberspace at all and openly discouraged these bodies from trying to establish any sort of control over the online world. In 1996, author and lyricist John Perry Barlow famously penned "A Declaration of the

Independence of Cyberspace," a missive in which he wrote that "cyberspace does not lie within [government] borders ... it is an act of nature and it grows itself through our collective actions." More specifically, Barlow believed that cyberspace should and would operate by its own rules and conduct itself without being subject to the laws of any particular real-world government.

While Barlow's idealistic views on the relationship between cyberspace and real-world governments may still be viewed as the best case scenario by activists, things have played out quite differently in reality. Governments around the world have successfully established a wide variety of laws that regulate the use of cyberspace within their respective countries and formed numerous international agreements that do the same on a global basis. In the United States, the federal government has adopted a number of laws that restrict certain online activities, such as the illicit sharing of copyrighted digital data. Because the Internet has become such an integral part of the national infrastructure, the federal government has also taken steps to ensure the security of cyberspace as well as that of the nation itself by extension. In other countries, such as China and Iran, governments have taken a more aggressive approach to regulating cyberspace, even going so far as to restrict citizens' access to various websites and effectively censoring the Internet.

IS CYBERSPACE REAL?

The question of government regulation of the online world underlies a broader debate about whether the idea of cyberspace is a legitimate concept or simply an outdated term that should be abandoned. As the online world has increasingly intersected with and become a bigger part of the real world, many critics have begun to ask whether the term *cyberspace* is still an accurate, meaningful descriptor for the digital realm. Many of these critics argue that the concept of cyberspace as a digital utopia that was originally put forth by people like Barlow is no longer a useful or even reasonable way to think about the online world. Rather, they suggest that the traditional idea of cyberspace is little more than an outdated intellectual-political fad that no longer reflects the reality of the online world as it exists in the twenty-first century. To such critics, the online world is simply a digital layer of the real world that is shaped and molded by human desires and needs and is ultimately subject to the laws and regulations created and enforced by real-world governments. In the end, regardless of whether the use of cyberspace as a term continues or not, it is clear that the online world will retain its place as a vital component of real-world infrastructure.

—Jack Lasky

Bibliography

Lind, Michael. "Stop Pretending Cyberspace Exists." *Salon.* Salon Media Group, Inc. 12 Feb. 2013. Web. 2 Mar. 2016. http://www.salon.com/2013/02/12/the_end_of_cyberspace/

Meyer, David. "'Cyberspace' Must Die. Here's Why." *Gigaom.* Knowingly, Inc. 7 Feb. 2015. Web. 2 Mar. 2016. https://gigaom.com/2015/02/07/cyberspace-must-die-heres-why/

Rey, PJ. "The Myth of Cyberspace." *The New Inquiry.* The New Inquiry. 13 Apr. 2012. Web. 2 Mar. 2016. http://thenewinquiry.com/essays/the-myth-of-cyberspace/

Williams, Brett. "Cyberspace: What Is It, Where Is It, and Who Cares?" *Armed Forces Journal.* Sightline Media Group. 13 Mar. 2014. Web. 2 Mar. 2016. http://www.armedforcesjournal.com/cyberspace-what-is-it-where-is-it-and-who-cares/

D

Data Analytics (DA)

SUMMARY

Data analytics (often shortened to "analytics") is the examination, exploration, and evaluation of information (data) with the goal of discovering patterns. "Data analytics" is sometimes used interchangeably, or confused, with "data analysis." Both data analytics and data analysis rely heavily on mathematics and statistics, but data analysis is usually defined more broadly, to include any examination of bodies of data, using a large array of possible tools, for the purpose of discovering information in that data or supporting a conclusion. There is considerable overlap between the methodologies and concerns of the two disciplines, both of which might draw on data mining, modeling, and data visualization. The important difference is the overall focus on patterns in analytics. Analytics is used in numerous industries and academic disciplines, from finance to law enforcement and marketing.

BACKGROUND

Some of the key concepts involved in analytics include machine learning, neural networks, and data mining. Machine learning is the branch of computer science that grew out of computational statistics. It focuses on the development of algorithms that, through modeling, learn from the data sets on which they operate and can express outputs that are data-driven decisions or predictions. A familiar example of machine learning is e-mail spam filtering, which monitors the e-mails that users mark as spam, examines those e-mails for common characteristics, derives a profile from this examination, and predicts the probability of an e-mail being spam. Neural networks are used in machine learning to model functions that, like biological neural networks, have a large number of possible inputs. Neural networks are often used in special kinds of decisionmaking and pattern-recognition applications, such as software underlying handwriting or speech recognition, vehicle control, radar systems, and the artificial intelligence in computer games.

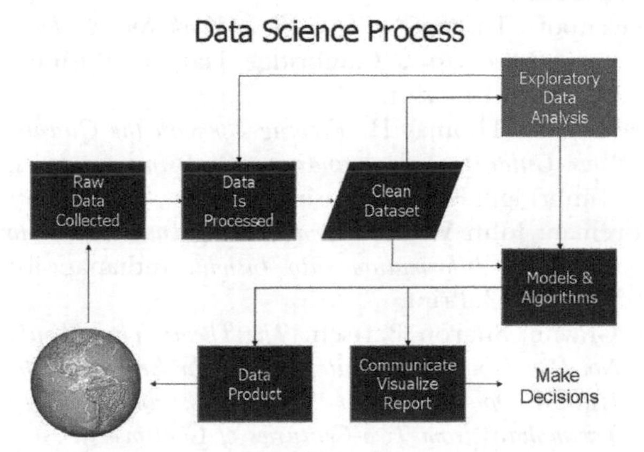

Flowchart showing the data visualization process. Farcaster at English Wikipedia

Data mining is a computer-science subfield devoted to using computers to discover patterns in data sets, especially data sets too large to be feasibly analyzed for such patterns by human operators. Computers can also search for patterns in ways that pose specific challenges to humans not because of the size of the data set but because of the nature of the data or the pattern; the way human vision operates, for instance, works against the ability to perceive certain patterns in visual data that become apparent when data is mined by a program. The methodology was first made possible by Bayes's theorem, named for the eighteenth-century English statistician Thomas Bayes, whose work was also one of the first significant contributions to predictive analytics (the use of data analytics to evaluate patterns in data in order to draw conclusions about future behavior). Bayes's work was used by Alan Turing and his team during World War II to break the German Enigma code.

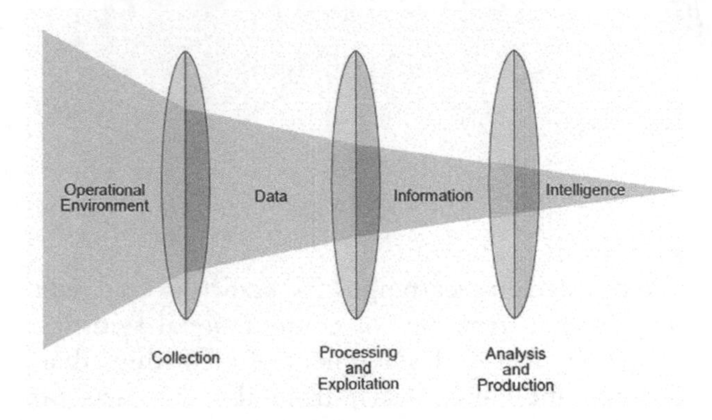

The phases of the intelligence cycle used to convert raw information into actionable intelligence or knowledge are conceptually similar to the phases in data analysis. By U.S. Joint Chiefs of Staff JP2-0

PRACTICAL APPLICATIONS

Analytics is used across numerous industries and in many kinds of applications. In the early 2000s, Oakland Athletics general manager Billy Beane and baseball statistician Bill James revolutionized baseball by bringing analytics to the sport, and to the ways the Athletics evaluated player performance and made draft decisions. James's "sabermetrics" downplayed traditional player performance metrics such as home runs in favor of analyzing the large body of data available on baseball games to determine the player performance statistics that had the most impact on team wins. Baseball happens to lend itself especially well to analytics because of the large number of games played per season (162 per team in the current era) combined with the more than century-long history of the sport and the large number of players per team. These three attributes create a large and rich data set.

Analytics has become fundamental to the way that the Internet has been monetized, especially after it became clear that e-commerce alone was not sufficient to do so. Web analytics is a specific area of marketing analytics, relying on data collected by websites about the Internet users who visit them. Modern web analytics typically includes not only search keywords and "referrer data" (information about how the user arrived at the website), data about their engagement with the website itself (the pages one has viewed, for example), and basic identity information (such as one's geographic location) but also extended information about one's Internet activity on other websites, which as a result of social media, can mean a wealth of demographic information. Tracking these activities allows marketers to tailor online advertisements for specific demographics. For instance, a user who watches an online superhero-movie trailer may be more likely to be shown a targeted ad for a new comic book; a woman who changes her Facebook relationship status to "engaged" may be shown ads for bridal magazines and wedding-planning services.

In the finance and banking industry, analytics is used to predict risk. Potential investments are analyzed according to their past performance when possible, or the performance of investments that fit a similar profile. Banks considering loans to individuals or businesses similarly rely on analytics to model the probable financial performance of the individual or business in the future. Naturally, there is some dispute about which factors are logical, practical, or acceptable to consider in these models.

Applied mathematicians (or even engineers or physicists) who specialize in developing the mathematical and statistical approaches underlying analytics are called quantitative analysts, or "quants." The increasing reliance on predictive analytics to guide greater numbers of business decisions is sometimes called big data, a term that can also refer to data sets too large to be processed by traditional means.

—Bill Kte'pi, MA

BIBLIOGRAPHY

Davenport, Thomas H. *Analytics at Work: Smarter Decisions, Better Results.* Cambridge: Harvard Business Review, 2010. Print.

Davenport, Thomas H. *Keeping Up with the Quants: Your Guide to Understanding and Using Analytics.* Cambridge: Harvard Business Review, 2013. Print.

Foreman, John W. *Data Smart: Using Data Science to Transform Information into Insight.* Indianapolis: Wiley, 2013. Print.

McGrayne, Sharon Bertsch. *The Theory That Would Not Die: How Bayes' Rule Cracked the Enigma Code, Hunted Down Russian Submarines, and Emerged Triumphant from Two Centuries of Controversy.* New Haven: Yale UP, 2012. Print.

Mlodinow, Leonard. *The Drunkard's Walk: How Randomness Rules Our Lives.* New York: Vintage, 2009. Print.

Provost, Foster, and Tom Fawcett. *Data Science for Business: What You Need to Know about Data Mining and Data-Analytic Thinking.* Sebastopol: O'Reilly, 2013. Print.

Silver, Nate. *The Signal and the Noise: Why So Many Predictions Fail—But Some Don't.* New York: Penguin, 2015. Print.

Wheelan, Charles. *Naked Statistics: Stripping the Dread from the Data.* New York: Norton, 2014. Print.

Date of original publication: December 31, 2016

Deep Learning

SUMMARY

Deep learning consists of artificial neurons in multilayered networks that through algorithms can teach software to train itself to recognize objects such as language and images. It is a subset of machine learning—the ability of a machine to "learn" with experience instead of programming—and is a part of artificial intelligence. Patterned after brain science, deep learning involves feeding a neural network with large amounts of data to train the machine for classification. The machine is given an object to identify and will process it through several network layers. As the process continues, the machine goes from simple layers to ones that are more complicated until an answer is reached. Algorithms instruct the neurons how to respond to improve the results.

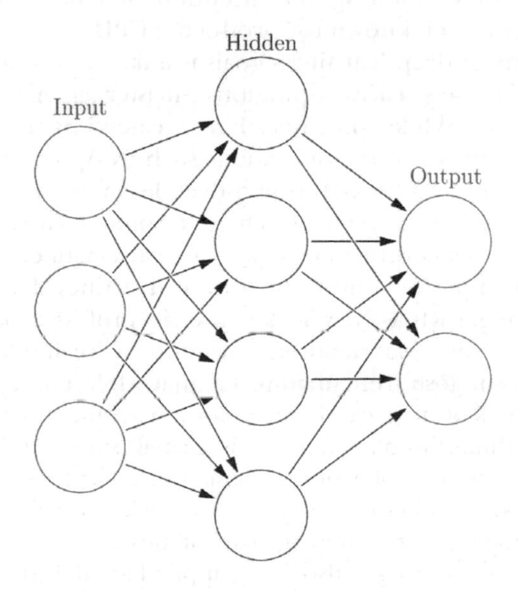

Artificial neural network with layer coloring. By Glosser.ca

BRIEF HISTORY

Neuron networks were first introduced in the 1950s as biologists were mapping out the workings of the human brain. Computer scientists were looking beyond logical applications to replicate thinking in machines. In 1958, research psychologist Frank Rosenblatt applied these theories to design the perceptron, a single-layered network of simulated neurons using a room-sized computer. Through their connections, the neurons would relay a value, or "weight," of either 1 or 0 to correspond with a shape. However, after several tries, the machine would not recognize the right shape. Rosenblatt applied supervised learning, training the perceptron to output the correct answer with the machine developing an algorithm to tweak the weights to get the correct answer.

Rosenblatt's algorithm, however, did not apply to multilayered networks, limiting the perceptron's ability to perform more complex tasks. In a 1969 book, artificial-intelligence scientists Marvin Minsky and Seymour Papert believed that making more layers would not make perceptrons more useful. Neural networks were abandoned for nearly two decades.

In the mid-1980s, researchers Geoffrey Hinton and Yann LeCun revived interest in neuron networks, with the belief that a brain-like structure was needed to fulfill the potential of artificial intelligence. Instead of only outputting an answer, the goal was to create a multilayered network that would allow the machine to learn from past mistakes. The duo and other researchers used a learning algorithm called backpropagation that would allow data to pass through multiple layers and the network to make adjustments to give the right answer. This spawned technology in the 1990s that could read handwritten text. However, like perceptrons, backpropagation had its limitations and required much data to be fed into a machine. Other researchers were developing alternative learning algorithms that did not require neurons. The networks fell out of fashion again.

Hinton returned to neuron research in the mid-2000s, and in 2006, he developed methods to teach larger networks with multiple layers of neurons. This required a progressive set of layers, as each layer recognized a certain feature and moved on to the next until the system identified the object. In 2012, Hinton and two students won an image recognition contest with their software, identifying one thousand objects. Focus returned to neuron networks and deep learning was pushed to the forefront.

PRACTICAL APPLICATIONS

Since 2011, deep learning has created advancements in artificial intelligence. Technology giants such as Google, Facebook, Microsoft, and Chinese web services company Baidu are using deep-learning applications to access massive amounts of data. Microsoft and Google employ it to power their speech- and image-recognition products, including their voice-activated searches, translation tools, and photo searches. Google Brain, with Hinton on its research team, has more than one thousand deep-learning projects under development, and Google-acquired DeepMind has combined deep learning with other technological approaches, such as deep-reinforcement learning, to solve issues such as energy efficiency and gaming. Facebook uses deep learning in its face-recognition software whenever a photo is uploaded to the service to identify or tag people. Through Facebook Artificial Intelligence Researchers (FAIR), started by LeCun, the company is exploring ways to improve neuron networks' ability to understand language.

Hardware has evolved to keep up with the demand for deep learning. With neural networks requiring large amounts of data, companies want computer

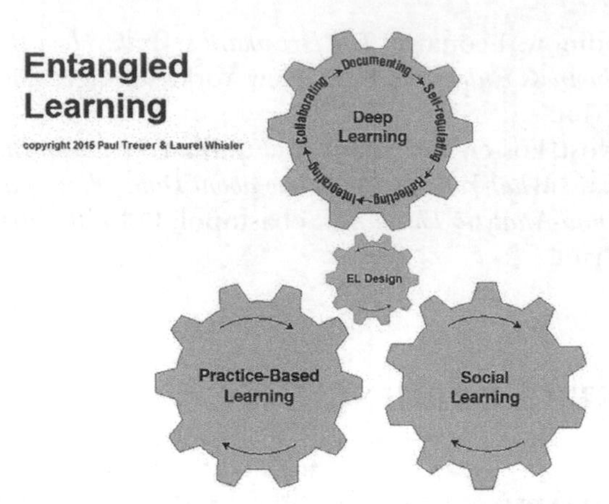

The three interlocking gears in the Entangled Learning model: practice-based learning, social learning, and design for deep learning. By Lwhisler (Own work)

chips that can deliver fast results. This has benefited firms such as Nvidia and Advanced Micro Devices that develop computer chips that render rich images for video game players. These chips, known as graphics processing units (GPUs), can compute hundreds more deep-learning processes compared to traditional computer processing units (CPUs).

Google has developed its own chips called tensor processing units, and IBM is making brain-like chips called TrueNorth. Start-up companies are designing chips specifically for deep learning, with some catching the attention of chip giants such as Intel, known for producing CPUs.

One of deep learning's goals is linked to artificial intelligence's early aspirations—mastering natural language. While companies have created interactive, speech-recognition applications such as Apple's Siri and Amazon's Alexa through deep learning, natural conversation has yet to be achieved. Some researchers believe that neuron networks operate in a manner that is too simple for conversations to occur. Critics of deep learning, such as New York University professor Gary Marcus and Massachusetts Institute of Technology professor Josh Tenenbaum, say that while immense amounts of data are required for a machine to learn something, humans need only a small sum of information and a collection of basic skills. Nevertheless, Facebook and Google continue to back research into developing better language capabilities.

Deep learning is also being applied to other fields. For example, it is instrumental in the development of self-driving cars. The vehicle's neuron networks process decisions based on a human driver's likely responses and making adjustments to improve its driving. Google and Baidu are backing research in this field. Deep-learning programs may be used in medicine. Start-ups are using improved image recognition to analyze medical imagery and foster drug development.

The push for deep learning has created a demand for computer scientists, with companies such as Microsoft, Amazon, and Google hiring the top researchers in the field. These firms have also made open source software available to developers interested in exploring the concept further.

—*Tamara Dunn*

BIBLIOGRAPHY

Bengio, Yoshua. "Machines Who Learn." *Scientific American*, vol. 314, no. 6, 2016, pp. 46–51.

Clark, Don. "Computer Chips Evolve to Keep Up with Deep Learning." *Wall Street Journal*, 11 Jan. 2017, www.wsj.com/articles/computer-chips-evolve-to-keep-up-with-deep-learning-1484161286. Accessed 16 Jan. 2017.

Hof, Robert D. "Deep Learning." *MIT Technology Review*, www.technologyreview.com/s/513696/deep-learning. Accessed 16 Jan. 2017.

Knight, Will. "AI's Unspoken Problem." *MIT Technology Review*, vol. 119, no .5, 2016, pp. 28–37.

______. "Kindergarten for Computers." *MIT Technology Review*, vol. 119, no. 1, 2016, pp. 52–58.

Metz, Cade. "2016: The Year That Deep Learning Took Over the Internet." *Wired*, 25 Dec. 2016, www.wired.com/2016/12/2016-year-deep-learning-took-internet. Accessed 16 Jan. 2017.

______. "Finally, Neural Networks That Actually Work." *Wired*, 21 Apr. 2015, www.wired.com/2015/04/jeff-dean. Accessed 16 Jan. 2017.

Parloff, Roger. "Why Deep Learning Is Suddenly Changing Your Life." *Fortune*, 28 Sept. 2016, fortune.com/ai-artificial-intelligence-deep-machine-learning. Accessed 16 Jan. 2017.

Simonite, Tom. "Teaching Machines to Understand Us." *MIT Technology Review*, 6 Aug. 2015, www.technologyreview.com/s/540001/teaching-machines-to-understand-us/. Accessed 16 Jan. 2017.

DIGITAL LOGIC

SUMMARY

Digital logic is electronic technology constructed using the discrete mathematical principles of Boolean algebra, which is based on binary calculation, or the "base 2" counting system. The underlying principle is the relationship between two opposite states, represented by the numerals 0 and 1. The various combinations of inputs utilizing these states in integrated circuits permit the construction and operation of many devices, from simple on-off switches to the most advanced computers.

DEFINITION AND BASIC PRINCIPLES

Digital logic is built upon the result of combining two signals that can have either the same or opposite states, according to the principles of Boolean algebra. The mathematical logic is based on binary calculation, or the base 2 counting system. The underlying principle is the relationship between two opposite states, represented by the numerals 0 and 1.

The states are defined in modern electronic devices as the presence or absence of an electrical signal, such as a voltage or a current. In modern computers and other devices, digital logic is used to control the flow of electrical current in an assembly of transistor structures called gates. These gates accept the input signals and transform them into an output signal. An inverter transforms the input signal into an output signal of exactly opposite value.

An AND gate transforms two or more input signals to produce a corresponding output signal only when all input signals are present. An OR gate transforms two or more input signals to produce an output signal if any of the input signals are present. Combinations of these three basic gate structures in integrated circuits are used to construct NAND (or not-AND) and NOR (or not-OR) gates, accumulators, flip-flops, and numerous other digital devices that make up the functioning structures of integrated circuits and computer chips.

BACKGROUND AND HISTORY

Boolean algebra is named for George Boole (1815–64), a self-taught English scientist. This form of algebra was developed from Boole's desire to express concrete logic in mathematical terms; it is based entirely on the concepts of true and false. The intrinsically opposite nature of these concepts allows the logic to be applied to any pair of conditions that are related as opposites.

The modern idea of computing engines began with the work of Charles Babbage (1791–1871), who envisioned a mechanical "difference engine" that would calculate results from starting values. Babbage did not see his idea materialize, though others using his ideas were able to construct mechanical difference engines.

The development of the semiconductor junction transistor in 1947, attributed to William Shockley, John Bardeen, and Walter Brattain, provided the means to produce extremely small on-off switches that could be used to build complex Boolean circuits. This permitted electrical signals to carry out Boolean algebraic calculations and marked the beginning of what has come to be known as the digital revolution. These circuits helped produce the many modern-day devices that employ digital technology.

HOW IT WORKS

BOOLEAN ALGEBRA. The principles of Boolean algebra apply to the combination of input signals rather than to the input signals themselves. If one associates one line of a conducting circuit with each digit in a binary number, it becomes easy to see how the presence or absence of a signal in that line can be combined to produce cumulative results. The series of signals in a set of lines provides ever larger numerical representations, according to the number of lines in the series. Because the representation is binary, each additional line in the series doubles the amount of information that can be carried in the series.

Assorted discrete transistors. Packages in order from top to bottom: TO-3, TO-126, TO-92, SOT-23. By Transisto at en.wikipedia (Own work)

BITS AND BYTES. Digital logic circuits are controlled by a clock signal that turns on and off at a specific frequency. A computer operating with a CPU (central processing unit) speed of 1 gigahertz (10_9 cycles per second) is using a clock control that turns on and off 1 billion times each second. Each clock cycle transmits a new set of signals to the CPU in accord with the digital logic circuitry. Each individual signal is called a bit (plural byte) of data, and a series of 8 bits is termed one byte of data. CPUs operating on a 16-bit system pass two bytes of data with each cycle, 32-bit systems pass four bytes, 64-bit systems pass eight bytes, and 128-bit systems pass sixteen bytes with each clock cycle.

Because the system is binary, each bit represents two different states (system high or system low). Thus, two bits can represent four (or 2_2) different states, three bits represents eight (or 2_3) different states, four bits represents sixteen (or 2_4) different states, and so on. A 128-bit system can therefore represent 2_{128} or more than $3.40 \times 10_{38}$ different system states.

DIGITAL DEVICES. All digital devices are constructed from semiconductor junction transistor circuits. This technology has progressed from individual transistors to the present technology in which millions of transistors can be etched onto a small silicon chip.

Digital electronic circuits are produced in "packages" called integrated circuit, or IC, chips. The simplest digital logic device is the inverter, which converts an input signal to an output signal of the opposite value. The AND gate accepts two or more input signals such that the output signal will be high only if all of the input signals are high. The OR gate produces an output high signal if any one or the other of the input signals is high.

All other digital logic devices are constructed from these basic components. They include NAND gates, NOR gates, X-OR gates, flip-flops that produce two simultaneous outputs of opposite value, and counters and shift registers, which are constructed from series of flip-flops. Combinations of these devices are used to assemble accumulators, adders, and other components of digital logic circuits.

One other important set of digital devices is the converters that convert a digital or analog input signal to an analog or digital output signal, respectively. These find extensive use in equipment that relies on analog, electromagnetic, signal processing.

KARNAUGH MAPS. Karnaugh maps are essential tools in designing and constructing digital logic circuits. A Karnaugh map is a tabular representation of all possible states of the system according to Boolean algebra, given the desired operating characteristics of the system. By using a Karnaugh map to identify the allowed system states, the circuit designer can select the proper combination of logic gates that will then produce those desired output states.

APPLICATIONS AND PRODUCTS

Digital logic has become the standard structural operating principle of most modern electronic devices, from the cheapest wristwatch to the most advanced supercomputer. Applications can be identified as programmable and nonprogrammable.

Nonprogrammable applications are those in which the device is designed to carry out a specific set of operations as automated processes. Common examples include timepieces, CD and DVD players, cellular telephones, and various household appliances. Programmable applications are those in which an operator can alter existing instruction sets or provide new ones for the particular device to carry out. Typical examples include programmable logic controllers, computers, and other hybrid devices into which some degree of programmability has been incorporated, such as gaming consoles, GPS (global positioning system) devices, and even some modern automobiles.

Digital logic is utilized for several reasons. First, the technology provides precise control over the processes to which it is applied. Digital logic circuits function on a very precise clock frequency and with a rigorously defined data set in which each individual bit of information represents a different system state that can be precisely defined millions of times per second, depending on the clock speed of the system. Second, compared with their analog counterparts, which are constructed of physical switches and relays, digital circuits require a much lower amount of energy to function. A third reason is the reduced costs of materials and components in digital technology. In a typical household appliance, all of the individual switches and additional wiring that would be required of an analog device are replaced by a single small printed circuit containing a small number of IC chips, typically connected to a touchpad and LCD (liquid crystal display) screen.

PROGRAMMABLE LOGIC CONTROLLER. One of the most essential devices associated with modern production methods is the programmable logic controller, or PLC. These devices contain the instruction set for the automated operation of various machines, such as CNC (computer numerical control) lathes and milling machines, and all industrial robotics. In operation, the PLC replaces a human machinist or operator, eliminating the effects of human error and fatigue that result in undesirable output variability. As industrial technology has developed, the precision with which automated machinery can meet demand has far exceeded the ability of human operators.

PLCs, first specified by General Motors Corporation (now Company) in 1968, are small computer systems programmed using a reduced-instruction-set programming language. The languages often use a ladder-like structure in which specific modules of instructions are stacked into memory. Each module consists of the instructions for the performance of a specific machine function. More recent developments of PLC systems utilize the same processors and digital logic peripherals as personal computers, and they can be programmed using advanced computer programming languages.

DIGITAL COMMUNICATIONS. Digital signal processing is essential to the function of digital communications. As telecommunication devices work through the use of various wavelengths of the electromagnetic spectrum, the process is analog in nature. Transmission of an analog signal requires the continuous, uninterrupted occupation of the specific carrier frequency by that signal, for which analog radio and television transmission frequencies are strictly regulated.

A digital transmission, however, is not continuous, being transmitted as discrete bits or packets of bits of data rather than as a continuous signal. When the signal is received, the bits are reassembled for audio or visual display. Encryption codes can be included in the data structure so that multiple signals can utilize the same frequency simultaneously without interfering with each other. Data reconstruction occurs at a rate that exceeds human perception so that the displayed signal is perceived as a continuous image or sound. The ability to interleaf signals in this way increases both the amount of data that can be transmitted in a limited frequency range and the efficiency with which the data is transmitted.

A longstanding application of digital logic in telecommunications is the conversion of analog source signals into digital signals for transmission, and then the conversion of the digital signal back into an analog signal. This is the function of digital-to-analog converters (DACs), and analog-to-digital converters (ACDs). A DAC uses sampling to measure the magnitude of the analog signal, perhaps many millions of times per second depending upon the clock speed of the system. The greater the sampling rate, the closer its digital representation will be to the real nature of the analog signal. The ACD accepts the digital representation and uses it to reconstruct the analog signal as its output.

One important problem that exists with this method, however, is what is called aliasing, in which the DAC analog output correctly matches the digital representation, but at the wrong frequencies. Present-day telecommunications technology is eliminating these steps by switching to an all-digital format that does not use analog signals.

SERVOMECHANISMS. Automated processes controlled by digital logic require other devices by which the function of the machine can be measured. Typically, a measurement of some output property is automatically fed back into the controlling system and used to adjust functions so that desired output parameters are maintained. The adjustment is carried out through the action of a servomechanism, a mechanical device that performs a specific action in the operation of the machine. Positions and rotational speeds are the principal properties used to gauge machine function.

In both cases, it is common to link the output property to a digital representation such as Gray code, which is then interpreted by the logic controller of the machine. The specific code value read precisely describes either the position or the rotational speed of the component, and is compared by the controller to the parameters specified in its operating program. Any variance can then be corrected and proper operation maintained.

SOCIAL CONTEXT AND FUTURE PROSPECTS

At the heart of every consumer electronic device is an embedded digital logic device. The transistor quickly became the single most important feature of electronic technology, and it has facilitated the rapid development of everything from the transistor radio to space travel. Digital logic, as embedded devices, is becoming an ever more pervasive feature of modern technology; an entire generation has now grown up not knowing anything but digital computers and technology. Even the accoutrements of this generation are rapidly being displaced by newer versions, as tablet computers and smartphones displace more traditional desktop personal computers, laptops, and cell phones. The telecommunications industry in North America is in the process of making a government-regulated switch-over to digital format, making analog transmissions a relic of the past.

Research to produce new materials for digital logic circuits and practical quantum computers is ongoing. The eventual successful result of these efforts, especially in conjunction with the development of nanotechnology, will represent an unparalleled advance in technology that may usher in an entirely new age for human society.

—*Richard M. Renneboog, MSc*

BIBLIOGRAPHY

Brown, Julian. *Minds, Machines, and the Multiverse: The Quest for the Quantum Computer.* Simon & Schuster, 2000.

Bryan, L. A., and E. A. Bryan. *Programmable Controllers: Theory and Application.* Industrial Text, 1988.

Holdsworth, Brian, and R. Clive Woods. *Digital Logic Design.* 4th ed., Newnes, 2002.

Jonscher, Charles. *Wired Life: Who Are We in the Digital Age?* Bantam Press, 1999.

Marks, Myles H. *Basic Integrated Circuits.* TAB Books, 1986.

Miczo, Alexander. *Digital Logic Testing and Simulation.* 2nd ed., John Wiley & Sons, 2003.

DNA COMPUTING

SUMMARY

DNA computing makes use of the structure of genetic material to solve problems, process data, and make calculations. The concept was first discovered in 1994. While typical computer hardware consists of silicon-based circuitry, DNA computers use DNA as hardware, making use of its natural functions of storing and processing information. Its advantage over more traditional computer designs is that due to DNA's structure, it can compute many different factors simultaneously. DNA computers are extremely small in comparison to their silicon counterparts. They have also shown the capacity to process information about organisms and administer treatments on a molecular level. They have displayed great potential to detect and eliminate cancers and other diseases with great precision.

However, DNA computers have significant drawbacks that prevent them from replacing traditional computers. They process data at a rate that is relatively slow. They also rely on very specific biochemical reactions to obtain data. Particular molecules can only be used once because they are forever changed by these reactions, meaning DNA computers require a great deal of maintenance and manipulation between computations.

BACKGROUND

DNA, or deoxyribonucleic acid, is the material that contains life-forms' genetic code. It is made up of molecules called nucleotides, which consist of phosphate, sugar, and one of four different nitrogen bases. Different traits of an organism are formed by different arrangements of these bases. Scientists have learned to identify many of these sequences, which are often represented by writing out the letters that the four bases start with. The bases are adenine, cytosine, guanine, and thymine. Short sections of DNA are transcribed onto another molecule called RNA (ribonucleic acid), and many are then translated into proteins. These allow an organism's various cells to function properly, and each particular genetic code serves a different purpose, making up the wide variety of organisms and the unique individuals present on Earth. Proteins called enzymes are responsible for singling out the sections of DNA that are transcribed.

Leonard Adleman was a University of Southern California professor of molecular biology and computer science who was considering ways to make computers more efficient. In 1994, he realized that DNA naturally served the purpose he was looking for: storing and efficiently communicating large amounts of information. He used DNA to help solve a common logic problem, known as the Traveling Salesman Problem. It asks for the shortest distance a salesman can take between several different cities connected by a network of roads without entering the same city more than once. With DNA computing, he calculated the optimal route through seven cities in less than a second. Traditional computers could take months to arrive at the same conclusion.

His demonstration showcased the format's advantages: It processed numerous calculations simultaneously, while traditional computers followed a path that must solve one calculation at a time. His discovery generated a great deal of excitement in the scientific community, particularly chemists, biologists, and computer scientists.

However, the limitations of DNA computers became apparent in the following years. They could only perform a massive amount of calculations when each individual problem was relatively simple. While there was great potential for data storage, every possible solution to a problem needed to be represented with DNA strands, which made determining results an extremely cumbersome process. By the end of the decade, scientists regarded DNA computing as more of a novelty than a competitor to traditional computing.

IMPACT

In the early 2000s, researchers at Israel's Weizmann Institute of Science made great strides in the field of DNA computing. They combined enzymes and DNA to search for precise genetic codes. Dr. Ehud Shapiro helmed a 2004 project that showed the potential to seek out signs of genetic diseases. Shapiro's team produced a computer designed to solve an extremely simple problem: whether or not a DNA sample contained a particular genetic code. This could theoretically be used to identify genes that code for cancers. Shapiro's model successfully identified prostate cancer in a small, controlled test tube environment. The next step would be to do so in a living organism.

This was the beginning of a major shift in the perception of DNA computing. While it was determined to be extremely impractical for performing most functions that traditional computers were used for, this line of research emphasized its advantages. DNA computing could process numerous, very simple tasks with great speed and efficiency compared to traditional computers. It could also interact with and gain information from organic matter in ways that traditional computers never could. This led to increased interest and research in DNA computing due to its potential in the medical field.

In 2006, researchers demonstrated the ability to "fold" DNA, manipulating its structure and suppressing any reactions until a desired time. Harvard and Israel's Bar-Ilan University collaborated on a project to combine this development with DNA computing, which they demonstrated in 2014. They created microscopic nanobots with folded DNA and programmed them similarly to the computers Shapiro's team had designed. The devices would "unfold" if they encountered particular molecules and remain folded if they did not. Upon unfolding, the DNA would be able to react to outside stimuli.

The team programmed the nanobots to react to certain genetic code within cockroaches. The DNA would unfold and begin a reaction, which would administer a drug to the insects. The tests worked, proving that DNA computers could function within organisms. It also showed how cancers could not only be detected but also treated at a molecular level. The great degree of precision in treatment meant that side effects would be greatly reduced. Cancer treatments such as chemotherapy can be harmful because, in addition to the cancerous cells, they affect parts of the individual that were previously undamaged. If there was a reliable method of pinpointing cancer at a molecular level, it would not only leave healthy cells unscathed, but it would also greatly improve the chances of detecting all traces of the disease.

The tests performed as intended in cockroaches, but human immune systems proved too robust for the nanobots. The team continued to conduct tests to improve the technology. They have expressed confidence that with further research and development, DNA computing can become an essential medical tool.

—Reed Kalso

Bibliography

Amir, Yaniv, et al. "Universal Computing by DNA Origami Robots in a Living Animal." *Nature Nanotechnology*, vol. 9, 2014, pp. 353–357. *PubMed Central*, doi: 10.1038/nnano.2014.58. Accessed 20 Nov. 2016.

Cannon, Leah. "What Can DNA-Based Computers Do?" *MIT Technology Review*, 4 Feb. 2015, www.technologyreview.com/s/534721/what-can-dna-based-computers-do/. Accessed 20 Nov. 2016.

De Chant, Tim. "A Drug-Delivering DNA Nanobot Computer, Built Inside a Cockroach." *Nova Next*, 10 Apr. 2014, www.pbs.org/wgbh/nova/next/body/a-drug-delivering-dna-nanobot-computer-built-inside-a-cockroach/. Accessed 20 Nov. 2016.

"New Research Reveals How DNA Could Power Computers." *Live Science*, 17 May 2010, www.livescience.com/6445-research-reveals-dna-power-computers.html. Accessed 20 Nov. 2016.

Parker, Jack. "Computing with DNA." *EMBO Reports*, vol. 4, no.1, Jan 2003, pp. 7–10. *PubMed Central*, doi:10.1038/sj.embor.embor719. Accessed 20 Nov. 2016.

Pollack, Andrew. "A Glance at the Future of DNA: MD's Inside the Body." *New York Times*, 29 Apr. 2004, www.nytimes.com/2004/04/29/us/a-glimpse-at-the-future-of-dna-md-s-inside-the-body.html. Accessed 20 Nov. 2016.

Rettner, Rachael. "DNA: Definition, Structure & Discovery." *Live Science*, 6 June 2013, www.livescience.com/37247-dna.html. Accessed 20 Nov. 2016.

Shapiro, Ehud, and Yaakov Benenson. "Bringing DNA Computers to Life." *Scientific American*, vol. 294, no.5, 2006, pp. 44–51, www.scientificamerican.com/article/bringing-dna-computers-to/. Accessed 20 Nov. 2016.

Domain-Specific Language (DSL)

SUMMARY

In computer science, a domain-specific language (DSL) is a computer language designed for a specific application domain. Application domains consist of the set of functionality, requirements, and terminology for a particular application. In contrast with general-purpose languages (GPLs) like the programming languages of C or Python or the markup language XML, DSLs are small programming languages because they cover only those features and functionalities required for their specific purpose. The line between the two can blur, however. PostScript fits the requirements of a GPL but in actual practice has found use almost exclusively in page description. Perl was designed as a DSL for text processing but quickly attracted interest as a GPL.

Domain-specific language is a major part of domain engineering, the science (in software development) of reusing domain knowledge—knowledge about a particular environment in which software operates, or an end to which the software will be used. DSLs are used not only to avoid reinventing the wheel but also to preserve the perfected wheel and continue to use it in new applications. DSLs have the advantage of being more streamlined than GPLs. Also, DSLs, when well-designed, are easier to use and quicker to learn.

PRACTICAL APPLICATIONS

DSLs are often written for business software, and they are designed for specific areas of business operations—such as billing, accounting, bookkeeping, inventory, and the calculation of salaries and benefits. DSLs also may be designed for operations specific to a particular kind of business, such as the tailoring of insurance policies to customers based on specific actuarial data, or the tracking of the development and costs of the disparate elements of a publisher's in-progress projects. The more ubiquitous that software in the running of a business, the more specific DSLs have been developed for narrow areas of application. Early, common business software, which had a much smaller customer base, focused on areas such as accounting and inventory.

An early example of DSL is Logo, an educational programming language designed in 1967 and taught in many schools through the twentieth century. Logo taught LISP-like programming concepts by letting students write programs that would move a cursor called a turtle around the screen and draw lines. (Applications with a robot turtle were also developed.) Later, languages like Mathematica, Maple, and Maxima were developed for spreadsheet mathematics, while GraphViz was developed for graphing. Other languages combine features or commands from GPLs in a simple to use format for a specific purpose, such as the Game Maker Language developed for the Game

Maker software package, designed to let users more easily write their own computer games with a combination of C++ and Delphi commands.

One of the most familiar DSLs, taught in some schools as Logo's popularity waned, is HTML, the Hypertext Markup Language still used to encode web pages (though hypertext applications preceded the web). Similar markup languages had been used in word processing applications, and were important in the early days of the Internet.

—Bill Kte'pi, MA

Bibliography

Fowler, Martin. *Domain-Specific Languages.* Upper Saddle River: Addison, 2011.

Ghosh, Debasish. *DSLs in Action.* New York: Manning, 2010.

Parr, Terence. *The Definitive ANTLR Reference: Building Domain-Specific Languages.* Raleigh: Pragmatic, 2013.

Sadalage, Pramod, and Martin Fowler. *NoSQL Distilled: A Brief Guide to the Emerging World of Polyglot Persistence.* Upper Saddle River: Addison, 2012.

Vernon, Vaughn. *Implementing Domain-Driven Design.* Upper Saddle River: Addison, 2013.

Empirical Formula

SUMMARY

An empirical formula is an equation that expresses an empirical relationship. An empirical relationship is based on real-world observations rather than on theoretical conjecture. The most common context in which empirical formulas are used is chemistry, where mathematical calculations are used to discover the amounts of various chemical substances required to produce the desired results.

The empirical formula does not give a precise indication of how much of a chemical ingredient is present in a particular sample. Instead, an empirical formula describes the proportional relationships that must exist between the various ingredients for them to be combined in a reaction that creates a given compound. These proportional relationships are usually expressed as either percentages or in ratio form. For example, one could say that the chemical composition of water is two hydrogen atoms for every one oxygen atom (hence, H2O) or that a given quantity of water will be approximately 66% hydrogen and 30% oxygen. These are both empirical formulas that tell an observer of a sample of material how much of one substance there is in relation to another substance. But they do not tell how much of the substance is present in the sample.

PRACTICAL APPLICATIONS

Knowing the empirical formula of a compound is useful to scientists because (while it does not directly tell exactly how much of a substance is present in a sample) it can be used to calculate this value with the addition of more information. If one knows the empirical formula of the compound C4H6O3, for example, and if there are 100 grams of the compound present—and then if one also knows the molecular weight of each element in the compound (for example, Cs molecular weight is 12.011)—it is possible to calculate an approximate value for the amount of each element present in the compound.

It is important to remember that an empirical formula does not tell one anything about the molecular structure of a compound; that is, the empirical formula does not indicate how the atoms of each element are arranged within a molecule of the compound. The empirical formula simply shows how much of each element will be present in a given molecule or sample. Thus, it is possible for a single empirical formula to describe more than one compound.

A well known example of this can be seen in the empirical formulas of acetylene and benzene. Each of these compounds contains equal amounts of carbon and hydrogen. Therefore, the empirical formula for both benzene and acetylene is CH. What accounts for the differences between the two substances is the way the atoms are arranged in each molecule. In acetylene, the molecular formula is C2H2, meaning that each acetylene molecule has two C and two H atoms. Benzene, though, has a molecular formula of C6H6.

A helpful analogy for remembering the difference between empirical and molecular formulas is that of children's toy blocks. We can give a child five red blocks and seven blue blocks, which would have an empirical formula of R5B7. The child could then use these blocks to create two very different structures: one a truck and the other an airplane. The properties of these structures are as distinct as benzene and acetylene are from one another.

—Scott Zimmer, MS, MLS

BIBLIOGRAPHY

Berg, Hugo. *Mathematical Models of Biological Systems.* New York: Oxford UP, 2011.

Dahm, Donald J., and Eric A. Nelson. *Calculations in Chemistry: An Introduction.* New York: Norton, 2013.

Leszczynski, Jerzy. *Handbook of Computational Chemistry.* Dordrecht: Springer, 2012.

Monk, Paul, and Lindsey J. Munro. *Maths for Chemistry: A Chemist's Toolkit of Calculations.* New York: Oxford UP, 2010.

Nassiff, Peter, and Wendy A. Czerwinski. "Using Paperclips to Explain Empirical Formulas to Students." *Journal of Chemical Education* 91.11 (2014): 1934-38.

Evaluating Expressions

SUMMARY

An expression is a mathematical sentence made up of variables and terms, such as $3x$ and $2y$, as well as constants, such as 3 and 7. Terms are often separated by operations, such as + and −, and each variable stands for a number that begins as unknown. In $3x + 6$, for example, x is the variable, and the more complicated $2x(3 + y)$ has both x and y as variables.

HISTORY AND BACKGROUND

Throughout the history of algebra, symbolism went through three different stages. Rhetorical algebra, in which equations were written as full sentences ("One thing plus two things equals three things"). This was how the ancient Babylonians expressed their mathematical ideas, and it stayed this way until the sixteenth century. Then came syncopated algebra, in which the symbolism became more complex but still didn't contain all the qualities of today's symbolic algebra. Restrictions, such as using subtraction only on one side of an equation was lifted when symbolic algebra came along and Diophantus' book *Arithmetica* was eventually replaced by Brahmagupta's *Brahma Sphuta Siddhanta*. Eventually, Ibn al-Banna, François Viète and René Descartes completed the modern notation used today, which is known as Cartesian geometry.

Most expressions have at least one unknown variable that is a symbol or letter, frequently x or y. These variables are typically used as a way of solving a problem and determining the value of something that isn't known yet. Unknown variables can be defined according to what they are added to, subtracted from, multiplied by, or divided by. An example of an unknown variable that has been added to something is $x + 89 = 137$ where 89 had been added to x and it is known that the sum equals 137. The variable x is unknown, but it can be found by performing the subtraction of 89 from 137. Similarly in multiplication an unknown variable can be multiplied by 5 so that $5y = 75$. The variable y is still unknown, but 75 can be divided by 5 to find the value of y is 15. The process of solving for the unknown variable is called evaluating the expression, and it can be done with one or two variables depending on the complexity of the expression.

EVALUATING EXPRESSIONS WITH ONE VARIABLE

When evaluating an expression with one variable, a number is simply plugged into the only variable in the expression, like turning x into 3, or replacing x with 3, as in the evaluation of $3x + 6$. When $x = 3$ the expression becomes $3(3) + 6$, which is also means $9 + 6$, which is also the same as 15. This means that the expression $3x + 6$ has been evaluated as being 15 when $x = 3$. Whenever a specific value is substituted in for a single variable, and the operations are performed, this is called evaluating an expression with one variable.

The steps for evaluating an expression always begin with substituting each letter with an assigned value. Then the numbers are enclosed with parenthesis when they are being multiplied to other numbers, and the order of operations is used to determine which step must take place first, second, and third. Understanding the order of operations is very important. If there aren't any exponents, then multiplication and division are done next. Finally, addition and subtraction are done last in the order of operations and the expression is evaluated. In mathematics, the order of operations is the basic rule that determines which procedures are done first: exponents and roots, multiplication and division; addition and subtraction.

Often, it is easier to treat division as multiplication by creating the reciprocal or inverse of a number that is being used in division. For example, it may be easier to view $2 \div 4$ as being the same as $2 \times 1/4$. The same is often done with subtraction being changed into addition so that $2 - 4$ can be viewed as "$2 + (-4)$. Whether these tricks are used or not, the order of operations always remains the same, and the evaluation proceeds according to the three steps above.

An example using exponents might look like $5x^2 - 12x + 12$ where $x = 4$. First, the variable is substituted with 4 to get "$5(4)^2 - 12(4) + 12$." Then the order of operations dictates that exponents come before multiplication or addition. Squaring the 4 then gives 16 so the expression becomes $5(16) - 12(4) + 12$. Next in the order of operations is multiplication and division, so the expression becomes $(5 \times 16) - (12 \times 4) + 12$, which means $80 - 48 + 12$. The final order of operations can then be done with addition and subtraction to get 44. This means that the evaluation results in $5x^2 - 12x + 12 = 44$ when $x = 4$.

EVALUATING EXPRESSIONS WITH TWO VARIABLES

Evaluating expressions with two variables is exactly the same as evaluating only one variable but the process is done twice. An example is the evaluation of $3x + 4y$ when $x = 2$ and $y = 6$. First, there is a replacement of x with 2 so that the result is $3(2) + 4y$. Then y is replaced with 6 to get $3(2) + 4(6)$. There are no exponents so the first order of operations is multiplication, giving $6 + 24$ so that the final order of operation can be done in addition to get 30. Therefore $3x + 4y = 30$ when $x = 2$ and $y = 6$.

Equivalent Forms of Expression

$$\int_0^T f(t)e^{-i\frac{2\pi mt}{T}}\,\mathrm{dt} = \int_0^T \sum_{n=-\infty}^{\infty} c_n e^{i\frac{2\pi(n-m)t}{T}}\,dt$$

Complex Equation

Figure 1. © EBSCO

EQUIVALENT FORMS OF EXPRESSIONS

Expressions are equivalent if their evaluations are equal. When two expressions are placed on each side of the = sign, and the same number results for every substitution and of all variables, then an equivalent form of expression is the result. For example, $x + 3$ and $y - 7$. If the two expressions are placed on each side of an = sign, the two expressions are said to be equivalent. Equivalent forms of expression can be extremely complex, as in Figure 1. No matter how complex these expressions get, they are still referred to as equivalent if they stand on both sides of an = sign. This form of expression is also known as an equation.

UNDERSTANDING FUNCTION NOTATION

Function notation is another way of expressing equations as functions. Instead of the previous form already discussed as $y = x + 2$, a new notation $f(x)$, which is pronounced as "eff-of-eks," was introduced by Leonhard Euler in 1734. The parentheses in this notation do not indicate multiplication as might first be suspected. Instead, the $f(x)$ symbol is really just a substitution for y and written in a different way. It expresses the variable y in terms of x and is therefore written as $f(x)$. With this new notation, it can be shown how substitutions are being made in the expressions and, where the previous variable y didn't suggest anything about its partner on the other side of the equation, it now suggests a relationship to x that can be visualized from both sides.

For example, when $x = 3$ it may also be said that $y = 3 + 2$, but this makes x vanish completely from the equation. Now, with function notation, the equation is written as $f(x) = x + 2$ and the substitution becomes $f(3) = 3 + 5$. Function notation shows another way of writing out the same expressions by showing a relationship that was previously hidden from view. It is simple and easy, and there is no need to complicate the new notation beyond its simple purpose.

EVALUATING EXPRESSIONS WITH FUNCTION NOTATION

When evaluating functional notation, the term $f(x)$ is first shown instead of the old term y but, just as an expression would be evaluated in y, it is also evaluated in $f(x)$. For example, $f(x) = x^2 + 2x - 1$. This entire statement is actually an equation, but the limited expression $x^2 + 2x - 1$ can be evaluated by substituting in a value for x once $f(x)$ is given a more clear value. For example, it might be suggested that an evaluation be done for $f(2)$ instead of the previous manner of saying that $y = 2$. In this case, $f(2) = (2)2 + 2(2) - 1$, which is the same as saying that $f(2) = 4 + 4 - 1$ and that $f(x) = 7$.

—Paul Ambrose, MA

BIBLIOGRAPHY

Angel, Allen R., and Dennis C. Runde. *Elementary Algebra for College Students.* Boston: Pearson, 2011.

Larson, Ron, and David C. Falvo. *Precalculus.* Boston: Cengage, 2014.

Lial, Margaret L., E. J. Hornsby, and Terry McGinnis. *Intermediate Algebra.* Boston: Pearson, 2012.

McKellar, Danica. *Hot X: Algebra Exposed!* New York: Plume, 2011.

Miller, Julie. *College Algebra Essentials.* New York: McGraw, 2013.

Wheater, Carolyn C. *Practice Makes Perfect Algebra.* New York: McGraw, 2010.

Young, Cynthia Y. *College Algebra.* Hoboken, NJ: Wiley, 2012.

EXPERT SYSTEM

SUMMARY

An *expert system* is a computer program that uses reasoning and knowledge to solve problems. Expert systems are usually *interactive* in that users input information and receive feedback or a solution based on this input. Well-designed expert systems are said to simulate human intelligence so closely that the results are similar to those that would come from a highly learned human being—an expert.

Expert systems are an important area in the field of *artificial intelligence* (AI). AI researchers aim to use technology to develop intelligent machines—that is, machines that can use deduction, reasoning, and knowledge to solve problems or produce answers. Expert systems are highly complex and utilize advanced technology as well as scientific ideas about thought and rationality.

Researchers are interested in creating intelligent machines in part because they can be helpful to humans. Expert systems technology can produce machines that are capable of "learning" through experience—that is, they learn as they are exposed to different kinds of input.

Most expert systems include at least the following elements:

- Knowledge base: An expert system's knowledge base is all of the facts the system has about a subject. Expert systems are usually supplied with a huge amount of information on a particular topic. This information comes from experts, databases, electronic encyclopedias, etc.
- Rules set: The knowledge base also includes sets of rules, or *algorithms.* These rules tell the system how to evaluate and work with the information and how to answer or approach queries from users.
- Inference engine: An inference engine uses the knowledge base and set of rules to interpret and evaluate the facts to provide the user with a response or solution.
- User interface: The user interface is the part of the expert system that users interact with. It allows users to enter their input and shows them the results. User interfaces are designed to be *intuitive,* or easy to use.

Different types of expert systems exist and are used for a variety of purposes. For example, some can diagnose disease in humans while others can identify malfunctions in machinery. Still other expert systems can classify objects based on their characteristics or monitor and control processes and schedules.

A Symbolics Lisp Machine: An Early Platform for Expert Systems. Note the unusual "space cadet keyboard". (wikipedia). Michael L. Umbricht and Carl R. Friend (Retro-Computing Society of RI)

BRIEF HISTORY

Simple expert systems have existed for decades. In the 1970s, researchers at Stanford University created an expert system that could diagnose health problems. The system was more effective than some junior doctors and performed almost as well as some medical experts. Throughout the 1970s and 1980s, different researchers continued to work on expert systems that diagnosed medical conditions. Eventually, the technology began to be applied in other areas. For example, new expert systems helped geologists to identify the best locations to drill for natural resources, while others helped financial advisers to invest funds wisely.

In modern times, expert systems continue to advance. For example, automotive companies are working toward driverless vehicles; most of these vehicles include an expert system. The expert system must make decisions about accelerating, turning, and stopping, just as a human driver would. These tasks are much more complicated than the tasks of early expert systems, but they are based on some of the same principles.

APPLICATIONS

Expert systems continue to affect many different aspects of society. Businesses can benefit from expert systems because they can save money by relying on a system rather than a human. Current technologies allow expert systems to handle large amounts of data, which can be beneficial for companies that crunch numbers, such as financial companies. For example, companies such as Morgan Stanley already benefit from the use of an expert system to make decisions about investments. Similarly, transportation companies can use expert systems to operate complicated vehicles such as trains or airplanes. The auto-pilot that is installed on modern airplanes is an example of an expert system; it can make decisions about navigation more quickly than human pilots.

ADVANTAGES AND DISADVANTAGES

Using expert systems rather than human experts can have some advantages. For example, an expert system's knowledge is permanent. The system does not forget key details, as a human might. Another advantage is that expert systems are consistent; they make similar recommendations for similar situations without the burden of human bias.

Additionally, expert systems can sometimes solve problems in less time than humans, allowing them to react more quickly than people, which can be especially useful in situations where time is of the essence. They can also be replicated with relative ease, allowing for the availability and sharing of information in multiple places.

Although expert systems have many advantages, some experts have pointed out some disadvantages, too. Today's expert systems do not have the same "common sense" as humans. That is, the system might produce answers that cannot or should not be applied in the real world. Additionally, expert systems may not recognize that some situations have no solution.

Finally, expert systems are only as good as the people who designed them, the accuracy of their data, and the precision of the rules. Thus, an expert system might make a bad choice because it is working from incorrect or incomplete information or because its rules are illogical.

—*Elizabeth Mohn*

Bibliography

"Definition: Expert System." *PCMag.* PCMag Digital Group. Web. 9 Mar 2016. http://www.pcmag.com/encyclopedia/term/42865/expert-system

"Expert System." *TechTarget.* SearchHealthIT. 2014 Nov. Web. 9 Mar 2016. http://searchhealthit.techtarget.com/definition/expert-system

Joshi, Kailash. "Chapter 11: Expert Systems and Applied Artificial Intelligence." Management Information Systems, College of Business Administration. University of Missouri, St. Louis. Web. 9 Mar 2016. http://www.umsl.edu/~joshik/msis480/chapt11.htm

Reingold, Eyal and Jonathan Nightingale. "Expert Systems." *Artificial Intelligence.* Department of Psychology, University of Toronto. Web. 9 Mar 2016. http://psych.utoronto.ca/users/reingold/courses/ai/expert.html

Russell, Stuart and Peter Norvig. *Artificial Intelligence: A Modern Approach.* New York: Prentice-Hall (1995): 3-27. Web. 9 Mar 2016. http://www.cs.berkeley.edu/~russell/aima1e/chapter01.pdf

Extreme Value Theorem

SUMMARY

The extreme value theorem comes from the field of calculus, and it concerns functions that can take on real number values. The extreme value theorem states that for a function f that is continuous over an interval that is closed and bounded by values a and b, there must exist values c and d such that for all $f(x)$, $c \geq x \geq d$. Stated another way, for a function that is continuous between values a and b (meaning that there are no points between a and b for which the function is undefined), the function will always have a maximum value c and a minimum value d. This will hold true for all values that x can take between a and b. At first blush, the extreme value theorem can seem to be almost too obvious to require the development of a formal theorem to describe it. Nevertheless, the theorem does contain several components that are worth further analysis, because understanding why they are necessary parts of the theorem provides one with a more comprehensive understanding of both the theorem and of mathematics in general. One of these concepts is that of continuity, and another is the notion of boundedness.

PRACTICAL APPLICATIONS

Continuity is a central concept underlying the extreme value theorem, because the theorem seeks to make a general statement that will apply to all values a function is capable of taking on within a specified range. In order for it to be possible to make such a statement about the properties of all values in the range, there must be the possibility of the function taking on any of those values and producing a defined, knowable result. If the function were undefined at any point in the range, then there would be the potential for the theorem to be a false statement rather than one that is true in all possible cases. As a result, the extreme value theorem only applies in cases where $f(x)$ is defined for all values between a and b. It is even acceptable for the value of $f(x)$ to be the same throughout the entire interval running from a to b. The graph of this function would look like a horizontal line, and the value of x would be the same at all points in the interval, making x its own minimum and maximum value—nothing in the extreme value theorem prevents the minimum and maximum values represented by c and d from being the same number. In a similar vein, the idea of boundedness is also a necessary limitation on the situations in which the extreme value theorem applies. Without such a requirement, the function values could approach infinitely close to a or b without ever becoming equal to a or b, in the manner of an asymptote. This would violate the extreme value theorem because the function value would continually increase but it would never reach a maximum value because it would never meet the boundary of a or b.

—Scott Zimmer, MLS, MS

BIBLIOGRAPHY

Banner, Adrian D. *The Calculus Lifesaver: All the Tools You Need to Excel at Calculus.* Princeton UP, 2007.

Courant, Richard. *Differential and Integral Calculus.* Hoboken: Wiley, 2011.

Lax, Peter D., and Maria S. Terrell. *Calculus with Applications.* New York: Springer, 2013.

Liu, Baoding. "Extreme Value Theorems of Uncertain Process with Application to Insurance Risk Model." *Soft Computing* 17.4 (2013): 549–556.

Makarov, Mikhail. "Applications of Exact Extreme Value Theorem." *Journal of Operational Risk* 2.1 (2007): 115-120.

Richardson, Leonard F. *Advanced Calculus: An Introduction to Linear Analysis.* Chicester: Wiley, 2011.

F

Fiber Technologies

SUMMARY

Fibers have been used for thousands of years, but not until the nineteenth and twentieth centuries did chemically modified natural fibers (cellulose) and synthetic plastic or polymer fibers become extremely important, opening new fields of application. Advanced composite materials rely exclusively on synthetic fibers. Research has also produced new applications of natural materials such as glass and basalt in the form of fibers. The current "king" among fibers is carbon, and new forms of carbon, such as carbon nanotubes, promise to advance fiber technology even further.

DEFINITION AND BASIC PRINCIPLES

A fiber is a long, thin filament of a material. Fiber technologies are used to produce fibers from different materials that are either obtained from natural sources or produced synthetically. Natural fibers are either cellulose-based or protein-based, depending on their source. All cellulosic fibers come from plant sources, while protein-based fibers such as silk and wool are exclusively from animal sources; both fiber types are referred to as biopolymers. Synthetic fibers are manufactured from synthetic polymers, such as nylon, rayon, polyaramides, and polyesters. An infinite variety of synthetic materials can be used for the production of synthetic fibers.

A TOSLINK fiber optic audio cable being illuminated at one end. By Hustvedt (Template:One)

Production typically consists of drawing a melted material through an orifice in such a way that it solidifies as it leaves the orifice, producing a single long strand or fiber. Any material that can be made to melt can be used in this way to produce fibers. There are also other ways in which specialty fibers also can be produced through chemical vapor deposition. Fibers are subsequently used in different ways, according to the characteristics of the material.

BACKGROUND AND HISTORY

Some of the earliest known applications of fibers date back to the ancient Egyptian and Babylonian civilizations. Papyrus was formed from the fibers of the papyrus reed. Linen fabrics were woven from flax fibers. Cotton fibers were used to make sail fabric. Ancient China produced the first paper from cellulose fiber and perfected the use of silk fiber.

Until the nineteenth century, all fibers came from natural sources. In the late nineteenth century, nitrocellulose was first used to develop smokeless gunpowder; it also became the first commercially successful plastic: celluloid.

As polymer science developed in the twentieth century, new and entirely synthetic materials were discovered that could be formed into fine fibers. Nylon-66 was invented in 1935 and Teflon in 1938. Following World War II, the plastics industry grew rapidly as new materials and uses were invented. The immense variety of polymer formulations provides an almost limitless array of materials, each with its own unique characteristics. The principal fibers used today are varieties of nylons, polyesters, polyamides, and epoxies that are capable of being produced in fiber form. In addition, large quantities of carbon and glass fibers are used in an ever-growing variety of functions.

HOW IT WORKS

The formation of fibers from natural or synthetic materials depends on some specific factors. A material must have the correct plastic characteristics that allow it to be formed into fibers. Without exception, all natural plant fibers are cellulose-based, and all fibers from animal sources are protein-based. In some cases, the fibers can be used just as they are taken from their source, but the vast majority of natural fibers must be subjected to chemical and physical treatment processes to improve their properties.

Cellulose Fibers. Cellulose fibers provide the greatest natural variety of fiber forms and types. Cellulose is a biopolymer; its individual molecules are constructed of thousands of molecules of glucose chemically bonded in a head-to-tail manner. Polymers in general are mixtures of many similar compounds that differ only in the number of monomer units from which they are constructed. The processes used to make natural and synthetic polymers produce similar molecules having a range of molecular weights. Physical and chemical manipulation of the bulk cellulose material, as in the production of rayon, is designed to provide a consistent form of the material that can then be formed into long filaments, or fibers.

Synthetic Polymers. Synthetic polymers have greatly expanded the range of fiber materials that are available, and the range of uses to which they can be applied. Synthetic polymers come in two varieties: thermoplastic and thermosetting. Thermoplastic polymers are those whose material becomes softer and eventually melts when heated. Thermosetting polymers are those whose the material sets and becomes hard or brittle through heating. It is possible to use both types of polymers to produce fibers, although thermoplastics are most commonly used for fiber production.

The process for both synthetic fibers is essentially the same, but with reversed logic. Fibers from thermoplastic polymers are produced by drawing the liquefied material through dies with orifices of the desired size. The material enters the die as a viscous liquid that is cooled and solidifies as it exits the die. The now-solid filament is then pulled from the die, drawing more molten material along as a continuous fiber. This is a simpler and more easily controlled method than forcing the liquid material through the die using pressure, and it produces highly consistent fibers with predictable properties.

Fibers from thermosetting polymers are formed in a similar manner, as the unpolymerized material is forced through the die. Rather than cooling, however, the material is heated as it exits the die to drive the polymerization to completion and to set the polymer.

Other materials are used to produce fibers in the manner used to produce fibers from thermoplastic polymers. Metal fibers were the first of these materials. The processes used for their production provided the basic technology for the production of fibers from polymers and other nonmetals. The best-known of these fibers is glass fiber, which is used with polymer resins to form composite materials. A somewhat more high-tech variety of glass fiber is used in fiber optics for high-speed communications networks. Basalt fiber has also been developed for use in composite materials. Both are available commercially in a variety of dimensions and forms.

Production of carbon fiber begins with fibers already formed from a carbon-based material, referred to as either pitch or PAN. Pitch is a blend of polymeric substances from tars, while PAN indicates that the carbon-based starting material is polyacrylonitrile. These starting fibers are then heat-treated in such a way that essentially all other atoms in the material are driven off, leaving the carbon skeletons of the original polymeric material as the end-product fiber.

Boron fiber is produced by passing a very thin filament of tungsten through a sealed chamber, during which the element boron is deposited onto the tungsten fiber by the process of chemical vapor deposition.

APPLICATIONS AND PRODUCTS

All fiber applications derive from the intrinsic nature of the material from which the fibers are formed. Each material, and each molecular variation of a material, produces fibers with unique characteristics and properties, even though the basic molecular formulas of different materials are very similar. As well, the physical structure of the fibers and the manner in which they were processed work to determine the properties of those fibers. The diameter of the fibers is a very important consideration. Other considerations are the temperature of the melt from which fibers of a material were drawn; whether the fibers were stretched or not, and the degree by which they were stretched; whether the fibers are hollow, filled, or solid; and the resistance of the fiber material to such environmental influences as exposure to light and other materials.

Structural Fibers. Loosely defined, all fibers are structural fibers in that they are used to form various structures, from plain weave cloth for clothing to advanced composite materials for high-tech applications. That they must resist physical loading is the common feature identifying them as structural fibers. In a stricter sense, structural fibers are fibers (materials such as glass, carbon, aramid, basalt, and boron) that are ordinarily used for construction purposes. They are used in normal and advanced composite materials to provide the fundamental load-bearing strength of the structure.

A typical application involves "laying-up" a structure of several layers of the fiber material, each with its own orientation, and encasing it within a rigid matrix of polymeric resin or other solidifying material. The solid matrix maintains the proper orientation of the encased fibers to maintain the intrinsic strength of the structure.

Materials so formed have many structural applications. Glass fiber, for example, is commonly used to construct different fiberglass shapes, from flower pots to boat hulls, and is the most familiar of composite fiber materials. Glass fiber is also used in the construction of modern aircraft, such as the Airbus A-380, whose fuselage panels are composite structures of glass fibers embedded in a matrix of aluminum metal.

Carbon and aramid fibers such as Kevlar are used for high-strength structures. Their strength is such that the application of a layer of carbon fiber composite is frequently used to prolong the usable lifetime of weakened concrete structures, such as bridge pillars and structural joists, by several years. While very light, Kevlar is so strong that high-performance automotive drive trains can be constructed from it. It is the material of choice for the construction of modern high-performance military and civilian aircraft, and for the remote manipulators that were used aboard the space shuttles of the National Aeronautics and Space Administration. Kevlar is recognizable as the high stretch-resistance cord used to reinforce vehicle tires of all kinds and as the material that provides the impact-resistance of bulletproof vests.

In fiber structural applications, as with all material applications, it is important to understand the manner in which one material can interact with another. Allowing carbon fiber to form a galvanic connection to another structural component such as aluminum, for example, can result in damage to the overall structure caused by the electrical current that naturally results.

Fabrics and Textiles. The single most recognized application of fiber technologies is in the manufacture of textiles and fabrics. Textiles and fabrics are produced by interweaving strands of fibers consisting of single long fibers or of a number of fibers that have been spun together to form a single strand. There is no limit to the number of types of fibers that can be combined to form strands, or on the number of types of strands that can be combined in a weave.

The fiber manufacturing processes used with any individual material can be adjusted or altered to produce a range of fiber textures, including those that are soft and spongy or hard and resilient. The range of chemical compositions for any individual polymeric material, natural or synthetic, and the range of available processing options, provides a variety of properties that affect the application of fabrics and textiles produced.

Clothing and clothing design consume great quantities of fabrics and textiles. Also, clothing designers seek to find and utilize basic differences in fabric and textile properties that derive from variations in chemical composition and fiber processing methods.

Fibers for fabrics and textiles are quantified in units of deniers. Because the diameter of the fiber can be produced on a continuous diameter scale, it is therefore possible to have an essentially infinite range of denier weights. The effective weight of a fiber may also be adjusted by the use of sizing materials added to fibers during processing to augment or improve their stiffness, strength, smoothness, or weight. The gradual loss of sizing from the fibers accounts for cotton denim jeans and other clothing items becoming suppler, less weighty, and more comfortable over time.

The high resistance of woven fabrics and textiles to physical loading makes them extremely valuable in many applications that do not relate to clothing. Sailcloth, whether from heavy cotton canvas or light nylon fabric, is more than sufficiently strong to move the entire mass of a large ship through water by resisting the force of wind pressing against the sails. Utility covers made from woven polypropylene strands are also a common consumer item, though used more for their water-repellent properties than for their strength. Sacks made from woven materials are used worldwide to carry quantities of goods ranging from coffee beans to gold coins and bullion. One reason for this latter use is that the fiber fabric can at some point be completely burned away to permit recovery of miniscule flakes of gold that chip off during handling.

CORDAGE. Ropes, cords, and strings in many weights and winds traditionally have been made from natural fibers such as cotton, hemp, sisal, and manila. These require little processing for rough cordage, but the suppleness of the cordage product increases with additional processing. Typically, many small fibers are combined to produce strands of the desired size, and these larger strands can then be entwined or plaited to produce cordage of larger sizes. The accumulated strength of the small fibers produces cordage that is stronger than cordage of the same size consisting of a single strand. The same concept is applied to cordage made from synthetic fibers.

Ropes and cords made from polypropylene can be produced as a single strand. However, the properties of such cordage would reflect the properties of the bulk material rather than the properties of combined small fibers. It would become brittle when cold, overly stretchy when warm, and subject to failure by impact shock. Combined fibers, although still subject to the effects of heat, cold, and impact shock, overcome many of these properties as the individual fibers act to support each other and provide superior resistance.

SOCIAL CONTEXT AND FUTURE PROSPECTS

One could argue that the fiber industry is the principal industry of modern society, solely on the basis that everyone wears clothes of some kind that have been made from natural or synthetic fibers. As this is unlikely ever to change, given the climatic conditions that prevail on this planet and given the need for protective outerwear in any environment, there is every likelihood that there will always be a need for specialists who are proficient in both fiber manufacturing and fiber utilization.

—Richard M. Renneboog, MSc

BIBLIOGRAPHY

Fenichell, Stephen. *Plastic: The Making of a Synthetic Century*. New York: HarperCollins, 1996. A well-researched account of the plastics industry, focusing on the social and historical contexts of plastics, their technical development, and the many uses for synthetic fibers.

Morrison, Robert Thornton, and Robert Nielson Boyd. *Organic Chemistry*. 5th ed. Newton, Mass.: Allyn & Bacon, 1987. Provides one of the best and most readable introductions to organic chemistry and polymerization.

Selinger, Ben. *Chemistry in the Marketplace*. 5th ed. Sydney: Allen & Unwin, 2002. The seventh chapter of this book provides a concise overview of many fiber materials and their common uses and properties.

Weinberger, Charles B. "*Instructional Module on Synthetic Fiber Manufacturing.*" Gateway Engineering Education Coalition: 30 Aug. 1996. This article presents an introduction to the chemical engineering of synthetic fiber production, giving an idea of the sort of training and specialization required for careers in this field.

Fullerene

SUMMARY

Fullerenes are molecules made completely of carbon. They were discovered in the 1980s by scientists investigating particles found in space. Resembling very small soccer balls in shape, fullerenes are hollow. They occur naturally and can be manufactured. Their unique properties mean that fullerenes have applications in nanotechnology, where tiny particles of matter are manipulated to achieve a specific purpose. Fullerenes have many potential uses in medicine, electrical engineering, and military technology.

BACKGROUND

Carbon is a very prevalent and versatile element. About one-fifth of the human body includes carbon atoms. It is present in everyday items such as the graphite in pencils and the gasoline in cars. These different forms have different structural compositions that help determine each one's strength and purpose. For example, each carbon atom in a diamond bonds with four additional carbon atoms, making three-dimensional bonds that are hard to break. Each carbon atom in graphite is only bonded in two-dimensional layers that allow them to separate easily.

In 1985 scientists Richard Smalley, Harry Kroto, and Robert Curl began investigating the structure of long snake-like carbon molecules that Kroto thought might have originated in the atmospheres of red giant stars. Smalley had a special laser-supersonic cluster beam device known as the AP_2, or the "app-two," in his laboratory at Rice University. Kroto wanted to use this to test the carbon molecules. The three men, along with three graduate students, soon began tests. Over the course of ten days, they were able to replicate the molecules Kroto was looking for. The experiments also revealed an additional never-before-seen carbon molecule.

The newly discovered C_{60} molecule had sixty carbon atoms and was very stable. It did not react with other elements, which puzzled the researchers. Carbon molecules usually have leftover atoms that "dangle" and react with other molecules. The fact

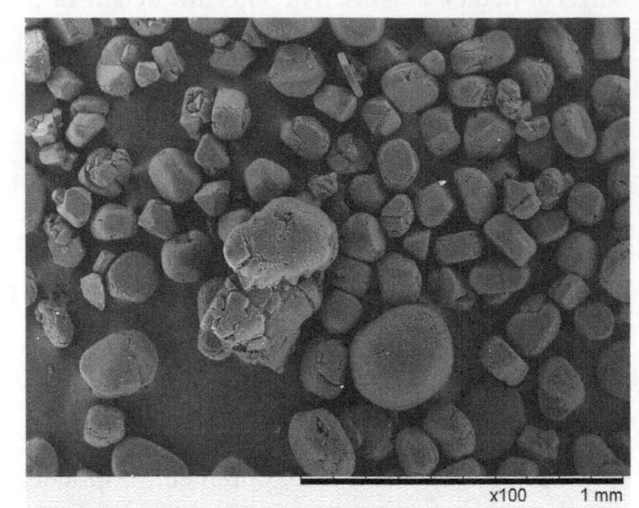

Fullerite C60 Scanning Electron Mycroscope image. © NASA

that these new molecules did not do this left the scientists wondering about its shape. None of the shapes they considered would give the stability the molecule displayed with no leftover molecules.

One of them—even the scientists differ on who came up with the idea first—suggested that the molecule might be hollow, with its atoms arranged to form an outer skin like as a hollow sphere. As they considered the idea of a sphere, they tried forming replicas of the image in computer programs and with paper and tape using hexagons, but they discovered that the sphere would not close. Eventually, Kroto thought about geodesic domes, architectural structures that use pentagons and hexagons in combination to form a closed sphere. Smalley then constructed a paper model with sixty flat surfaces, or vertices, that formed a perfect sphere. It also closely resembled a soccer ball.

This, the scientists decided, was the way the new carbon molecule had to be shaped. They named the C_{60} molecule the Buckminsterfullerene after Richard Buckminster "Bucky" Fuller. Fuller was an American architect who designed a number of geodesic domes. The new molecule was also sometimes referred to as a "buckyball."

The team continued to gather information to support their idea about the shape of the C_{60} molecule.

Other scientists duplicated the process of synthesizing C_{60} in multiple labs and confirmed the original hypothesis about its shape. In 1996, Smalley, Kroto, and Curl were awarded the Nobel Prize in Chemistry for their work.

PRACTICAL APPLICATIONS

Since their discovery, fullerenes have also been found in soot, especially soot created by flames from burning a combination of acetylene and benzene. They can also be reliably replicated by the Krätschmer-Huffman method. This method, named after Wolfgang Krätschmer and Donald Huffman, two of the scientists who corroborated the shape of the fullerene, involves heating carbon rods in a helium-rich atmosphere to generate fullerene-rich soot. Further testing has revealed that fullerenes are extremely strong and resilient, able to withstand the pressure of as many as three thousand atmospheres and the force of a fifteen thousand mile-per-hour collision and return to their original shape.

As a result of these properties, fullerenes show great promise for use in industrial settings as part of molecular wires for very small computers, in certain kinds of sensors, and as part of a hydrogen gas storage system. Researchers believe that fullerenes could be used to make an organic energy storage film with many possible applications, including creating such things as an outer coating that would make cell phones self-charging and advertising signs that would light without external power. The strength and resilience of the molecule could result in uses for military armor. It is also thought that fullerenes could give companies that manufacture diamonds a head start on recreating the natural process of diamond formation.

Fullerenes are already at use in some cosmetics that deliver antioxidants. It is thought that methods could someday be devised to use hollow fullerenes as drug delivery devices. Some researchers believe they can be used to help deliver medications such as antibiotics and AIDS medications in ways that could improve treatment outcomes. They may also provide a way to coat contrast agents used in hi-tech scans, such as magnetic resonance imagery (MRI), to help them remain in the body longer and provide better images.

Continued research into fullerenes has led scientists to uncover more than one thousand other new compounds. These include nanotubes, another form of carbon similar to fullerenes that resembles a tube made of rolled-up wire. The applications for fullerenes and the impact of their discovery extend to the fields of physics, chemistry, and geology.

—Janine Ungvarsky

Bibliography

Bakry, Rania, et al. "Medicinal Applications of Fullerenes." *International Journal of Nanomedicine,* vol. 2, no. 4, pp. 639–49, Dec. 2007, www.ncbi.nlm.nih.gov/pmc/articles/PMC2676811/. Accessed 14 Dec. 2016.

Berne, Olivier, et al. "30 Years of Cosmic Fullerenes." *Cornell University Library,* 27 Oct. 2015, arxiv.org/abs/1510.01642. Accessed 14 Dec. 2016.

"Carbon Nanotubes and Buckyballs." *University of Wisconsin-Madison,* education.mrsec.wisc.edu/nanoquest/carbon/. Accessed 14 Dec. 2016.

"The Discovery of Fullerenes," *American Chemical Society,* www.acs.org/content/dam/acsorg/education/whatischemistry/landmarks/lesson-plans/discovery-of-fullerenes.pdf. Accessed 14 Dec. 2016.

"Discovery of Fullerenes National Historic Chemical Landmark." *American Chemical Society,* www.acs.org/content/acs/en/education/whatischemistry/

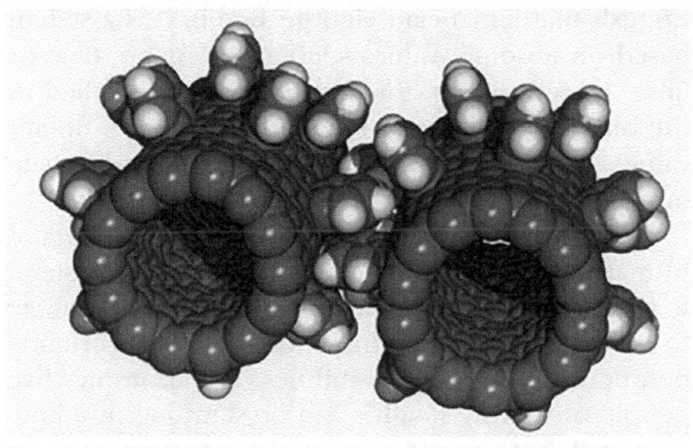

This photograph depicts two molecule-sized "Fullerene Nano-gears" with multiple teeth. The hope is that products can be constructed made of thousands of tiny machines that could self-repair and adapt to the environment in which they exist. By Ludvig14 (Own work)

landmarks/fullerenes.html. Accessed 14 Dec. 2016.
"Fullerenes." *BBC Bitesize,* www.bbc.co.uk/schools/gcsebitesize/science/add_ocr_gateway/chemical_economics/nanochemistryrev3.shtml. Accessed 14 Dec. 2016.
"Fullerenes." *Stanford University,* web.stanford.edu/group/mota/education/Physics%2087N%20Final%20Projects/Group%20Delta/fullerenes.html. Accessed 14 Dec. 2016.
"Fullerenes and Their Applications in Science and Technology." *Florida International University College of Engineering and Computing,* Spring 2013, web.eng.fiu.edu/~vlassov/EEE-5425/Ulloa-Fullerenes.pdf. Accessed 14 Dec. 2016.

Fuzzy Logic

SUMMARY

Fuzzy logic is a system of logic in which statements can vary in degrees of truthfulness rather than be confined to true or false answers. The human brain operates on fuzzy logic as it breaks down information into areas of probability. The idea was first developed in the 1960s by a computer scientist trying to discover a way to make computers mimic the human decision-making process. Fuzzy logic is widely used in the field of artificial intelligence (AI), with its more Practical Applications in consumer products such as air conditioners, cameras, and washing machines.

BACKGROUND

The standard form of logic is often referred to as Boolean logic, named after nineteenth-century English mathematician George Boole. It is a system based on absolute values such as yes or no, true or false. In computing, this form of logic is applied to the binary values used in machine language. Binary values consist of 1s and 0s, which correspond to true or false, or on or off within a computer's circuitry.

The task of processing information within the human brain does not operate on an absolute system. It uses a method that closely resembles fuzzy logic to arrive at a result within a degree of uncertainty between a range of possibilities. For example, five people of varying heights can be standing in a line. Standard logic would answer the question of "Is a person tall?" with a yes or no response. Six feet may be considered tall, while five feet, eleven inches would not be. Fuzzy logic allows for a number of responses between yes and no. A person many be considered tall, or somewhat tall, short or somewhat short. It also allows for personal interpretation, as six feet may be considered tall if a person is standing alone but not if he or she is standing next to an individual who is seven feet.

PRACTICAL APPLICATIONS

In 1965, professor Lotfi Zadeh at the University of California, Berkeley, developed the concept of fuzzy logic while attempting to find a way for computers to understand human language. Because humans do not think or communicate in 0s and 1s, Zadeh created a data set that assigned objects within the set values between 0 and 1. The values of 0 and 1 were included in the set but marked its outlying borders. For example, instead of a computer categorizing a person as old or young, it was assigned values that allowed it to classify a person as a percentage of young. Age five may be considered 100 percent young, while age twenty may be 50 percent young. Instead of determining data in absolutes, computers that used fuzzy logic measured the degree of probability that a statement was correct.

A concept important in fuzzy logic is the idea of a fuzzy set. A fuzzy set is a data set without a crisp, easily defined boundary. It is in contrast to a classical set in which the elements can clearly be placed within defined parameters. For example, it is easy to find a classical data set for months of the year from a list that includes June, Monday, July, Tuesday, August, Wednesday, and September. The answer is obviously June, July, August, and September. However, change the data set to summer months and it becomes a fuzzy set. While June, July, and August are often associated with summer, only about ten June days occur after the summer solstice. Conversely, the majority of September, which is often considered a fall month, actually corresponds with the end of summer. As a result, fuzzy logic holds that July and August are 100 percent summer months, while June is roughly 33 percent summer and September is 66 percent summer.

Fuzzy logic also involves an input of information that is run through a series of "if-then" statements called rules to produce an output of information. In order to achieve a valid result, all the variables must be defined prior to the input of information. A program designed to detect temperature changes in a room and adjust the air-conditioning accordingly must first be told what values constitute hot, warm, and cold and the proper air-conditioning output corresponding to those values. Then the program will run through a series of rules such as "if the temperature value is a certain percentage of hot and rising, then increase air conditioner output," or "if temperature value is a percentage of cold and falling, then decrease output."

While Zadeh developed fuzzy logic in the 1960s, it took almost a decade of advances in computer technology to allow it to be used in Practical Applications. Fuzzy logic applications became more common in Japan than in the United States, where the computing concept was slow to catch on. Because it tries to replicate the human thought process, fuzzy logic is often used in the field of artificial intelligence and robotics. It has found a more practical application, however, on lower-level AI systems such as those found in smart consumer and industrial products. Fuzzy logic is the process that allows vacuum cleaners to detect the amount of dirt on a surface and adjust its suction power to compensate. It allows cameras to adjust to the proper amounts of light or darkness in an environment, microwaves to coordinate cooking time with the amount of food, or washing machines to compensate for added volume of laundry. Fuzzy logic programs are also very flexible and continue to function if they encounter an unexpected value. They are also easily fine-tuned, often needing only an input of a new set of values to change the system's production.

—Richard Sheposh

Bibliography

Cintula, Petr, et al. "Fuzzy Logic." *Stanford Encyclopedia of Philosophy*, 15 Nov. 2016, plato.stanford.edu/entries/logic-fuzzy/. Accessed 24 Jan. 2017.

Dingle, Norm. "Artificial Intelligence: Fuzzy Logic Explained." *Control Engineering*, 4 Nov. 2011, www.controleng.com/single-article/artificial-intelligence-fuzzy-logic-explained/8f3478c1338 4a2771ddb7e93a2b6243d.html. Accessed 24 Jan. 2017.

"Foundations of Fuzzy Logic." *MathWorks*, www.mathworks.com/help/fuzzy/foundations-of-fuzzy-logic.html. Accessed 24 Jan. 2017.

"Fuzzy Logic Introduction." *Imperial College London*, www.doc.ic.ac.uk/~nd/surprise_96/journal/vol2/jp6/article2.html. Accessed 24 Jan. 2017.

Kaehler, Steven D. "Fuzzy Logic – An Introduction." *Seattle Robotics Society*, www.seattlerobotics.org/encoder/mar98/fuz/fl_part1.html#INTRODUCTION. Accessed 24 Jan. 2017.

McNeill, Daniel, and Paul Freiberger. *Fuzzy Logic: The Revolutionary Computer Technology That Is Changing Our World.* Touchstone, 1993.

Ross, Timothy J. *Fuzzy Logic with Engineering Applications.* 3rd ed., Wiley, 2010.

"What Is 'Fuzzy Logic'? Are There Computers That Are Inherently Fuzzy and Do Not Apply the Usual Binary Logic?" *Scientific American*, www.scientificamerican.com/article/what-is-fuzzy-logic-are-t/. Accessed 24 Jan. 2017.

G

Game Theory

SUMMARY

Game theory is a tool that has come to be used to explain and predict decisionmaking in a variety of fields. It is used to explain and predict both human and nonhuman behavior as well as the behavior of larger entities such as nation states. Game theory is often quite complex mathematically, especially in new fields such as evolutionary game theory.

DEFINITION AND BASIC PRINCIPLES

Game theory is a means of modeling individual decisionmaking when the decisions are interdependent and when the actors are aware of this interdependence. Individual actors (which may be defined as people, firms, and nations) are assumed to maximize their own utility—that is, to act in ways that will provide the greatest benefits to them. Games take a variety of forms such as cooperative or noncooperative; they may be played one time only or repeated either indefinitely or with a finite ending point, and they may involve two or more players. Game theory is used by strategists in a variety of settings.

The underlying mathematical assumptions found in much of game theory are often difficult to understand. However, the explanations drawn from game theory are often intuitive, and people can engage in decisionmaking based on game theory without understanding the underlying mathematics. The formal approach to decisionmaking that is part of game theory enables decision makers to clarify their options and enhance their ability to maximize their utility.

Classical game theory has quite rigid assumptions that govern the decision process. Some of these assumptions are so rigid that critics have argued that game theory has few applications beyond controlled settings.

Later game theorists developed approaches to strategy that have made game theory applicable in many decisionmaking applications. Advances such as evolutionary game theory take human learning into account, and behavioral game theory factors emotion into decisionmaking in such a way that it can be modeled and predictions can be made.

BACKGROUND AND HISTORY

The mathematical theory of games was developed by the mathematician John von Neumann and the economist Oskar Morgenstern, who published *The Theory of Games and Economic Behavior* in 1944. Various scholars have contributed to the development of game theory, and in the process, it has become useful for scholars and practitioners in a variety of fields, although economics and finance are the disciplines most commonly associated with game theory.

Some of the seminal work in game theory was done by the mathematician John Nash in the early 1950's, with later scholars building on his work. Nash along with John Harsanyi and Reinhard Selten received a Nobel Prize in Economic Sciences in 1994 for their work in game theory. The political scientist Thomas Schelling received a Nobel Prize in Economic Sciences in 2005 for his work in predicting the outcomes of international conflicts.

HOW IT WORKS

Game theory is the application of mathematical reasoning to decisionmaking to provide quantitative estimates of the utilities of game players. Implicit in the decisionmaking process is the assumption that all actors act in a self-interested fashion so as to maximize their own utilities. Rationality, rather than altruism or cooperation, is a governing principle in much of

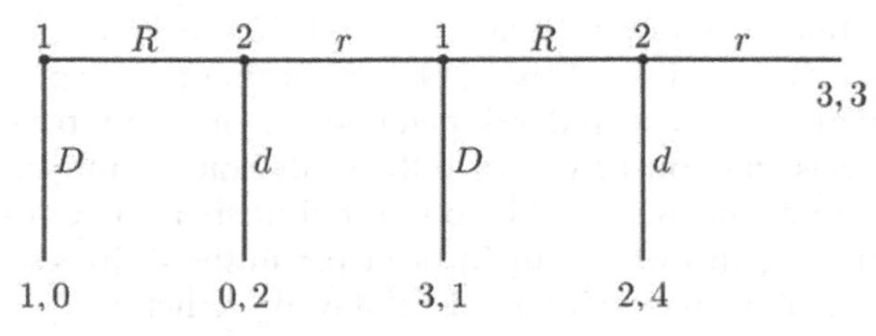

A four-stage centipede game.

game theory. Actors are expected to cooperate only when doing so benefits their own self-interest.

NONCOOPERATIVE GAMES. Most games are viewed as noncooperative in that the players act only in their own self-interest and will not cooperate, even when doing so might lead to a superior outcome. A well-known noncooperative game is the prisoners' dilemma. In this game, the police have arrested Moe and Joe, two small-time criminals, for burglary. The police have enough information to convict the two men for possession of burglar's tools, a crime that carries a sentence of five years in prison. They want to convict the two men for burglary, which carries a ten-year sentence, but this requires at least one man to confess and testify against the other. It would be in each man's best interest to remain silent, as this will result in only a five-year sentence for each. The police separate Moe and Joe. Officers tell Moe that if he confesses, he will receive a reduced sentence (three years), but if Joe confesses first, Moe will receive a harsher sentence (twelve years). At the same time, other officers give Joe the same options. The police rely on the self-interest of each criminal to lead him to confess, which will result in ten-year sentences for both men. If the various payoffs are examined, cooperation (mutual silence), which means two five-year sentences, is the most rewarding overall. However, if the two criminals act rationally, each will assume that the other will confess and therefore each will confess. Because each man cannot make sure that the other will cooperate, each will act in a self-interested fashion and end up worse off than if he had cooperated with the other by remaining silent.

As some game theorists such as Robert Axelrod have demonstrated, it is often advantageous for players to cooperate so that both can achieve higher payoffs. In reality, as the prisoners' dilemma demonstrates, players who act rationally often achieve an undesirable outcome.

SEQUENTIAL GAMES. Many games follow a sequence in which one actor takes an action and the other reacts. The first actor then reacts, and so the game proceeds. A good way to think of this process is to consider a chess game in which each move is countered by the other player, but each player is trying to think ahead so as to anticipate his or her opponent's future moves. Sequential games are often used to describe the decision process of nations, which may lead to war if wrong reactions occur.

Sequential games can be diagramed using a tree that lists the payoffs at each step, or a computer program can be used to describe the moves. Some sequential games are multiplayer games that can become quite complicated to sort out. Work in game theory suggests that equilibrium points change at each stage of the game, creating what are called sequential equilibria.

SIMULTANEOUS GAMES. In some games, players make their decisions at the same time instead of reacting to the actions of the other players. In this case, they may be trying to anticipate the actions of the other player so as to achieve an advantage. The prisoners' dilemma is an example of a simultaneous game. Although rational players may sometimes cooperate in a sequential game, they will not cooperate in a simultaneous game.

APPLICATIONS AND PRODUCTS

Game theory is most commonly used in economics, finance, business, and politics, although its applications have spread to biology and other fields.

ECONOMICS, FINANCE, AND BUSINESS. Game theory was first developed to explain economic decision-making, and it is widely used by economists, financial analysts, and individuals. For example, a firm may want to analyze the impact of various options for responding to the introduction of a new product by a competitor. Its strategists might devise a payoff matrix that encompasses market responses to the competitor's product and to the product that the company introduces to counter its competitor. Alternatively, a firm might prepare a payoff matrix as part of decision process concerning entry into a new market.

Strategy for the Prisoner's Dilemma

	Prisoner B stays silent (cooperates)	Prisoner B confesses (defects)
Prisoner A stays silent (cooperates)	Each serves 1 year	Prisoner A: 3 years Prisoner B: goes free
Prisoner A confesses (defects)	Prisoner A: goes free Prisoner B: 3 years	Each serves 2 years

Game theory cannot be used to predict the stock market. However, some game theorists have devised models to explain investor response to such events as an increase in the interest rate by the Federal Reserve Board. At the international level, there are various games that can be used to explain and predict the responses of governments to the imposition of various regulations such as tariffs on imports. Because of the large number of variables involved, this sort of modeling is quite complex and still in its infancy.

Some businesses might be tempted to develop a game theoretic response (perhaps using the prisoners' dilemma) to the actions of workers. For example, a company can develop a game theoretic response to worker demands for increased wages that will enable the company to maximize its utility in the negotiating process. Even if a game does not play out exactly as modeled, a company gains by clarifying its objectives in the development of a formal payoff matrix. Companies can also develop an approach to hiring new employees or dealing with suppliers that draws on game theory to specify goals and strategies to be adopted.

Politics. Some of the most interesting work in game theory has occurred in explaining developments in international affairs, enabling countries to make better decisions in the future. Game theory is often used to explain the decision process of the administration of President John F. Kennedy during the Cuban Missile Crisis. Other game theoretic explanations have been developed to examine a country's decision to start a war, as in the Arab-Israeli conflicts. Game theorists are able to test their formal game results against what actually occurred in these cases so as to enhance their models' ability to predict.

Other game theorists have devised game theoretic models that explain legislative decisionmaking. Much of the legislative process can be captured by game theoretic models that take into account the step-by-step process of legislation. In this process, members from one party propose legislation and the other party responds, then the first part often responds, and the responses continue until the legislation is passed or defeated. Most of these models are academic and do not seem to govern legislative decisionmaking, at least not explicitly.

Biology. Some evolutionary biologists have used game theory to describe the evolutionary pattern of some species. One relationship that is often described in game theoretic terms is the coevolution of predators and prey in a particular area. In this case, biologists do not describe conscious responses but rather situations in which a decline in a prey species affects predators or an increase in predators leads to a decline in prey and a subsequent decline in predators. Biologists use this relationship (called the hawk-dove game by the biologist John Maynard Smith) to show how species will evolve so as to better fit an evolutionary niche, such as the development of coloration that enables prey to better conceal itself from predators.

SOCIAL CONTEXT AND FUTURE PROSPECTS

Game theory is an evolving field that can become esoteric and divorced from the realities of practical decisionmaking. It can also be an intuitive, essentially nonmathematical approach to enhancing decisionmaking. Both formal and informal game theoretic approaches are likely to be used in business and government in the future. However, most observers agree that sophisticated actors who are aware of the principles of game theory are likely to prevail over those who follow an ad hoc approach to decisionmaking.

Game theory is not a perfect guide to decisionmaking. At times, it has led to overly simplistic approaches that are derived from a narrow view of utility. With the introduction of newer conceptual frameworks, game theory has become less rigid and better able to model human decisionmaking. In the future, sophisticated computer simulations based on game theory are likely to be used for corporate and governmental decisionmaking.

—John M. Theilmann, PhD

Bibliography

Binmore, K. G. *Game Theory: A Very Short Introduction.* New York: Oxford University Press, 2007.

Bueno de Mesquita, Bruce. *The Predictioneer's Game: Using the Logic of Brazen Self-Interest to See and Shape the Future.* New York: Random House, 2009.

Chakravarty, Satya R., Manipushpak Mitra, and Palash Sarkar. *A Course on Cooperative Game Theory.* Cambridge UP, 2015.

Fisher, Len. *Rock, Paper, Scissors: Game Theory in Everyday Life.* New York: Basic Books, 2008.

Gintis, Herbert. *Game Theory Evolving: A Problem-Centered Introduction to Modeling Strategic Behavior.* 2d ed. Princeton, N.J.: Princeton University Press, 2009.

Mazalov, Vladimir. *Mathematical Game Theory and Applications.* John Wiley & Sons, 2014.

Miller, James. *Game Theory at Work: How to Use Game Theory to Outthink and Outmaneuver YourCompetition.* New York: McGraw-Hill, 2003.

Young, Peyton, and Shmuel Zamir. *Handbook of Game Theory.* Elsevier, 2015.

Geoinformatics

SUMMARY

The term "geoinformatics" refers to a collection of information systems and technologies used to create, collect, organize, analyze, display, and store geographic information for specific end-user applications. The field represents a paradigm shift from traditional discipline-based systems such as cartography, geodesy, surveying, photogrammetry, and remote sensing to a data systems management protocol that includes all earlier technologies and combines them to create new models of spatial information. Computation is an essential foundation of all geoinformatics systems.

DEFINITION AND BASIC PRINCIPLES

Geoinformatics is a complex, multidisciplinary field of knowledge specializing in the creation, collection, storage, classification, manipulation, comparison, and evaluation of spatially referenced information for use in a variety of public and private practices. Its technologies are rooted in mapping, land surveying, and communication technologies that are thousands of years old. The exponential growth of science-based knowledge and mathematical expertise during the nineteenth and twentieth centuries greatly assisted the accumulation of verifiable geographic information describing the Earth and its position among the myriad celestial entities occupying the known universe. Detailed geometric descriptions and photographic materials make it possible to translate, measure, and order the surface of the Earth into multidimensional coordinate systems that provide a rich and detailed visual language for understanding the relationships and features of locations too vast to be easily comprehended by the human senses. Geographic communication systems are greatly enhanced by the proliferation of computation technologies including database management systems, laser-based surveys, digital satellite photo technologies, and computer-aided design (CAD).

BACKGROUND AND HISTORY

The components and design structures of maps form the fundamental language of geographic information systems (GIS). Maps illustrate a variety of environments, both real and imagined, and the structures and life forms that fill them. The geographies of the Earth's land masses and the relationships of land and water to the Moon and stars are the foundations of advanced mathematics and the physical sciences, subjects that are continually modified by new measurements of spatial-temporal coordinates. Aerial and nautical photography introduced a new and vital reality to cartographic representation in the twentieth century. Advances in sonar and radar technologies, the telephone and telegraph, radio broadcast, mass transportation, manned space flight, spectrometry, the telescope, nuclear physics, biomedical engineering, and cosmology all rely on the power of the map to convey important information about the position, structure, and movement of key variables.

Cartography. Mapping is an innate cognitive ability; it is common in a variety of animate species. In human practice, mapping represents an evolutionary process of symbolic human communication. Its grammar, syntax, and elements of style are composed of highly refined systems of notation, projections, grids, symbols, aesthetics, and scales. These time-honored features and practices of cartographic representation demand careful study and practice. Maps are the most ancient documents of human culture and civilization and, as such, many are carefully preserved by private and public institutions all over the world. The World Wide Web has made it possible to share, via the Internet, facsimiles of these precious cultural artifacts; the originals are protected for posterity. Scholars study the intellectual and technical processes used to create and document maps. These provide valuable clues about the beliefs and assumptions of the cultures, groups, and individuals that contract and prepare them.

Geodesy. The cartographer artfully translates the spatial features of conceptual landscapes into two- and three-dimensional documents using data and symbols selected to illustrate specific relationships or physical features of a particular place for the benefit or education of a group or individual enterprise. Geodesy is an earth science the practitioners of which provide timely geographical measurements used by cartographers to create accurate maps. Since earliest recorded times, geodesists have utilized the most current astronomical and geographical knowledge to measure the surface of the Earth and its geometric relationship to the Sun and Moon. The roots of geodetic measurement systems are buried in the ancient cultures and civilizations of Egypt, China, Mesopotamia, India, and the Mediterranean. Enlightened scholars and astute merchants of land and sea traveled the known world and shared manuscripts, instruments, personal observation, and practical know-how for comprehending the natural world. Hellenic scholars brilliantly advanced the study of astronomy and the Earth's geography. Their works formed the canon of cartographical and astronomical theory in the Western world from the time of the great voyages of discovery in the fifteenth and sixteenth centuries.

Alexandria was the intellectual center of the lives of the Greek polymath Eratosthenes of Cyrene and the Roman astronomer Ptolemy. Both wrote geographical and astronomical treatises that were honored and studied for centuries. Eratosthenes wrote *Peri tes avametreoeos tes ges* (On the Measurement of the Earth) and the three-volume *Geographika* (Geography), establishing mathematics and precise linear measurements as necessary prerequisites for the accurate geographical modeling of the known world. His works are considered singular among the achievements of Greek civilization. He is particularly noted for his calculations of the earth's circumference, measurements that were not disputed until the seventeenth century.

Ptolemy, following Eratosthenes, served as the librarian at Alexandria. He made astronomical observations from Alexandria between the years 127 and 141, and he was firm in his belief that accurate geographical maps were derived from the teachings of astronomy and mathematics. The *Almagest* is a treatise devoted to a scientific, methodical description of the movement of the stars and planets. The beauty and integrity of his geocentric model set a standard for inquiry that dominated astronomy for centuries. He also wrote the eight-volume *Geographia* (Geography), in which he established latitude and longitude coordinates for the major landmasses and mapped the world using a conic projection of prime and linear meridians and curved parallels. His instructions for creating a coordinate system of mapping, including sectional and regional maps to highlight individual countries, are still in practice.

At the end of the nineteenth century, geophysics became a distinct science. Its intellectual foundations provided rising petroleum corporations new technologies for identifying and classifying important land features and resources. In the United States, extensive surveys were conducted for administrative and military purposes. The U.S. Coast Survey was established by President Thomas Jefferson in 1807. The Army Corps of Engineers, the Army Corps of Topographical Engineers, and the United States Naval Observatory supported intensive geophysical research including studies of harbors and rivers, oceans and land topographies. In 1878, the U.S. Coast Survey became the U.S. Coast and Geodetic Survey. In 1965, the U.S. Coast and Geodetic Survey was reincorporated under the Environmental Sciences Services Administration, and in 1970 it was reorganized as the National Oceanic and Atmospheric Administration (NOAA).

Underwater acoustics and geomagnetic topographies were critical to the success of naval engagements during World War I and World War II. The International Union of Geodesy and Geophysics (IUGG) was founded in 1919. As of 2018, it is one of thirty-one scientific unions participating in the International Social Science Council (ISSC). In 2011, the American National Committee of the International Union of Geodesy and Geophysics combined with the Committee on Geophysics of the National Research Council; in 1972 the committee was independently organized as the American Geophysical Union. Graduate geophysics programs became prominent in major universities after World War II.

Geodesy is a multidisciplinary effort to calculate and document precisely the measurements of the Earth's shape and gravitational field in order to define accurate spatial-temporal locations for points on its surface. Land surveys and geomensuration, or the measure of the Earth as a whole, are essential practices. Geodetic data is used GIS—these include materials from field surveys, satellites, and digital maps. The Earth's shape is represented as an ellipsoid in current mathematical models. Three-dimensional descriptors are applied to one quadrant of the whole in a series of calculations. All cartographic grid systems and subsequent measurements of the Earth begin with a starting point called a datum. The first datum was established in North America in 1866 and its calculations were used for the 1927 North American Datum (NAD 27). In 1983, a new datum was established (NAD 83) and it is the basis of the standard geodetic reference system (GRS 80). It was again modified and became the World Geodetic System in 1984 (WGS 84). WGS 84 utilizes constant parameters for the definition of the Earth's major and minor axes, semimajor and semiminor axes, and various ratios for calculating the flattening at the poles. The geoid is another geodetic representation of the Earth, as is the sphere.

Photogrammetry. The word photogrammetry refers to the use of photographic techniques to produce accurate three-dimensional information about the topology of a given area. Accurate measurements of spaces and structures are obtained through various applications including aerial photography, aerotriangulation, digital mapping, topographic surveys, and database and GIS management. Precise, detailed photographs make it possible to compare and analyze particular features of a given environment. Aerial photographs are particularly useful to engineers, designers, and planners who need visual information about the site of a project or habitat not easily accessible by other means. The International Society for Photogrammetry and Remote Sensing (ISPRS) advances the knowledge of these technologies in more than 100 countries worldwide.

Remote Sensing. Like photogrammetry, remote sensing is an art, science, and technology specializing in the noncontact representation of the Earth and the environment by measuring the wavelengths of different types of radiation. These include passive and active technologies, both of which collect data from natural or emitted sources of electromagnetic energy. These include cosmic rays, X rays, ultraviolet light, visible light, infrared and thermal radiation, microwaves, and radio waves. Aerial photography and digital imaging are traditional passive remote sensing technologies based on photographic techniques and applications. Later applications include manned space and space shuttle photography and Landsat satellite imagery. Radar interferometry and laser scanning are common examples of active remote sensing technologies. These products are used for the documentation of inaccessible or dangerous spaces. Examples include studies of particular environments at risk or in danger.

Global Navigation Satellite Systems (GNSS). Satellites and rapidly advancing computer technologies have transformed GIS products. Satellite systems of known distance from the Earth receive radio transmissions, which are translated and sent back to Earth as a signal giving coordinates for the position and elevation of a location. Navstar is an American Global Positioning System (GPS) originating from a World War II radio transmission system known as loran, which is an acronym for long-range navigation. It is the only GNSS system in operation. On April 27, 2008, the European Space Agency announced the successful launch of the second Galileo In-Orbit Validation Element (GIOVE-B), one of thirty identical satellites planned to complete a European satellite navigation system similar to the Global Positioning System in use in the United States and the Global Navigation Satellite System (GLONASS) in Russia. Students can monitor the activities of remote sensing satellite systems such as the Landsat, Seasat, and Terra satellites used to create new map profiles of terrestrial and extraterrestrial landscapes, many of which are available online for study. These satellite systems are used to collect data including measurements of the Earth's ozone shield, cloud mappings, rainfall distributions, wind patterns, studies of ocean phenomena, vegetation, and land-use patterns.

HOW IT WORKS

Geographic information systems are built on geometric coordinate systems representing particular locations and terrestrial characteristics of the Earth or other celestial bodies. The place and position of particular land features form the elementary and most regular forms of geographic data. Maps provide visual information about key places of interest and structural features that assist or impede their access. Land-survey technologies and instrumentation are ancient and are found in human artifacts and public records that are thousands of years old. They are essential documents of trade, land development, and warfare. Spatial coordinates describing the Moon and stars and essential information about particular human communities are chinked into rocks, painted on the walls of caves, and hand-printed on graphic media all over the world. Digital photographs, satellite data, old maps, field data, and measurements provide new contexts for sharing geographic information and knowledge about the natural world and human networks of exchange.

GIS devices include high-resolution Landsat satellite imagery, light detecting and ranging (lidar) profiles, computer-aided design (CAD) data, and database management systems. Cross-referenced materials and intricately detailed maps and overlays create opportunities for custom-designed geographic materials with specific applications. Data that is created and stored in a GIS device provide a usable base for building complex multi-relational visualizations of landscapes, regions, and environments.

How a GIS component is used depends on the particular applications required. Service providers will first conduct a needs assessment to understand what information will be collected, by whom, and how it will be used by an individual or organization. The careful design of relationships connecting data sets to one another is an essential process of building an effective GIS system. Many data sets use different sets of coordinates to describe a geographical location and so algorithms need to be developed to adjust values that can have significant effects on the results of a study.

Depending on the needs of an organization, the rights to use some already-established data sets can be acquired. Other data can be collected by the user and stored in appropriate data files. This includes records already accumulated by an individual or organization. Converting and storing records in GIS format is particularly useful for creating the documents needed for a time-series analysis of particular land features and regional infrastructures. Digital records protect against the loss of valuable information and create flexible mapping models for communicating with internal and external parties.

APPLICATIONS AND PRODUCTS

Satellite and computer technologies have transformed the way spatial information is collected and communicated worldwide. With exponential increases in speed and detail, satellites provide continuous streams of information about Earth's life systems. More than fifty satellites provide free access to GPS coordinates. These are used on a daily basis by people from all walks of life, providing exceptional mapping applications for determining the location and elevation of roadways and waterways.

The Environmental Systems Research Institute (ESRI) is a pioneer in the development of GIS landscape-analysis systems. This includes the development of automated software products with Web applications. Users can choose from a menu of services including community Web-mapping tools; ArcGIS Desktop with ArcView, ArcEditor, and ArcInfo applications; and two applications for use in educational settings, ArcExplorer Java Edition for Education (AEJEE) and Digital Worlds (DW). ESRI software is an industry standard used worldwide by governments and industries. The ESRI International User Conference is attended by thousands of users and is one of the largest events of its kind.

The Global Geodetic Observing System (GGOS) is another application developed by the International Association of Geodesy. It provides continuing observations of the Earth's shape, gravity field, and rotational motion. These measurements are integrated into the Global Earth Observing System of Systems (GEOSS), an application that provides high-quality reference materials for use by groups such as the Group on Earth Observations and the United Nations.

SOCIAL CONTEXT AND FUTURE PROSPECTS

As a result of the World Wide Web, computer and telecommunications technologies continue to provide novel platforms for connecting individuals, groups, and communities worldwide. Location is an essential feature of such networks, and geographic information systems are needed to provide timely spatial information for individual and cooperative ventures. The safety and integrity of global communications systems and data systems continue to challenge political values in free and democratic societies. Nevertheless, geographic information systems will continue to be integrated into a variety of applications, contributing to the safety of individuals, the integrity of the world's natural resources, and the profitability of enterprises around the globe.

—Victoria M. Breting-García, BA, MA

Bibliography

Bender, Oliver, et al., eds. *Geoinformation Technologies for Geocultural Landscapes: European Perspectives.* Leiden, the Netherlands: CRC Press, 2009. These essays explain the many ways geoinformation technologies are used to model and simulate the changing reality of landscapes and environments.

DeMers, Michael N. Fundamentals of Geographic Information Systems. 4th ed. Hoboken, N.J.: John Wiley & Sons, 2009. Provides useful historical Background for the technologies and systems presented. Each chapter ends with a review of key terms, review questions, and a helpful list of references for further study.

Galati, Steven R. *Geographic Information Systems Demystified.* Boston: Artech House, 2006. Introductory textbook provides graphic examples of the concepts presented and includes a very useful glossary of terms.

Harvey, Francis. *A Primer of GIS: Fundamental Geographic and Cartographic Concepts.* New York: Guildford Press. 2008. This well-organized and clearly written textbook presents a thorough review of GIS with questions for analysis and vocabulary lists at the end of every chapter.

Kolata, Gina Bari. "Geodesy: Dealing with an Enormous Computer Task." *Science* vol. 200, no. 4340 (April 28, 1978): pp. 421-466. This article explains the enormous task of computing the 1983 North American Datum.

Konecny, Gottfried. Geoinformation: Remote Sensing, Photogrammetry and Geographic Information Systems. London: Taylor & Francis, 2003. This detailed introduction to geoinformation includes hundreds of illustrations and tables, with a chapter-by-chapter Bibliography.

Rana, Sanjay, and Jayant Sharma, eds. *Frontiers of Geographic Information Technology.* New York: Springer, 2006. This collection of essays covers rising geographic technologies. Chapter 1: "Geographic Information Technologies—An Overview" and Chapter 4: "Agent-Based Technologies and GIS: Simulating Crowding, Panic, and Disaster Management" are of note.

Scholten, Henk J., Rob van de Velde, and Niels van Manen, eds. *Geospatial Technology and the Role of Location in Science.* New York: Springer, 2009. The essays collected in this volume represent a series of presentations made at the September, 2007, conference held by the Spatial Information Laboratory of the VU University Amsterdam. They address the use of geospatial technologies at the junction of historical research and the sciences.

Thrower, Norman J. W. *Maps and Civilization: Cartography in Culture and Society*. 3d ed. Chicago: University of Chicago Press. 2007. This book is an excellent introduction to the field of cartography.

Go

SUMMARY

Go, also known as golang or Google Go, is an open source programming language created by engineers of the technology company Google. Go is a systems programming language, the class of programming languages used for writing operating systems, game engines, and other systems, as opposed to languages writing application software run on those systems. Google's development of Go addressed some of the issues engineers had with leading systems programming languages such as C, while attempting to retain the elements that make such languages valuable.

PRACTICAL APPLICATIONS

Go was developed by Robert Griesemer, Rob Pike, and Ken Thompson in 2007 and later announced by Google in 2009 once it was decided to implement the language. It is one of several languages influenced by communicating sequential processes (CSP), a language first described in 1978 for handling patterns of interaction in concurrent systems, and C, a programming language developed in the early 1970s for the Unix operating system. Like C, Go was designed for efficiency. It improves on C (from the perspective of the implementations the design team had in mind) by improving memory safety in most areas, facilitating a functional programming style, and being overall smaller and more lightweight by eschewing those features in C that are the most cumbersome and complex (including pointer arithmetic and generic programming).

The Go gopher: mascot of the Go programming language (sometimes dubbed Gordon). By Renee French. reneefrench.blogspot.com

Go was intended to have a lot of the features of dynamic languages such as JavaScript and Python, specifically avoiding the need for repetition and aiming for readability. It also needed to support networking and multiprocessing. At the same time, it was a language clearly conceived in the tradition of C and C++ since, if the experiment went well, it was intended as a potential C replacement.

Go is used in similar circumstances as the Erlang and occam programming languages, but the three languages handle the semantics of interprocess communication in different ways. For instance, Go and occam cannot read messages from channel out of order, while Erlang can use matching to read events from the message queue out of order. While Erlang and occam closely resemble CSP in semantics and syntax, Go differs in syntax and is only somewhat similar in semantics.

The year it was introduced, Go was named programming language of the year by coding standards organization TIOBE for its rapid increase in popularity. Since its introduction, Go has been implemented not only by Google but by a number of companies including cloud storage provider Dropbox, video-streaming service Netflix, online audio platform SoundCloud, and ridesharing service Uber. Some specific programs written in Go include the CoreOS open-source operating system; Docker, a Linux system for automating the deployment of applications inside software containers; Kubernetes, a cross-platform container cluster manager Google developed for the Cloud Native Computing Foundation; and Juju, an orchestration management tool for Ubuntu Linux. Juju (then called Ensemble) was originally written in Python, marking one of several notable occasions in which something written in a dynamic programming language was migrated to Go.

An earlier programming language, Go!, is unrelated.

—Bill Kte'pi, MA

Bibliography

Chang, Sau Sheong. *Go Web Programming.* Shelter Island: Manning, 2016. Print.

Donovan, Alan A. A., and Brian W. Kernighan. *The Go Programming Language.* New York: Addison-Wesley, 2015. Print.

Doxsey, Caleb. *Introducing Go: Build Reliable, Scalable Programs.* Sebastopol: O'Reilly, 2016. Print.

Finnegan, Matthew. "Google Go: Why Google's Programming Language Can Rival Java in the Enterprise." *TechWorld.* IDG UK, 25 Sept.2015. Web. 19 Aug. 2016.

Golang.org. Go Programming Language, 2016. Web. 19 Aug. 2016.

Goriunova, Olga. *Fun and Software: Exploring Pleasure, Paradox, and Pain in Computing.* New York: Bloomsbury Academic, 2014. Print.

Kennedy, William, Brian Ketelsen, and Erik St. Martin. *Go in Action.* Shelter Island: Manning, 2016. Print.

Kincaid, Jason. "Google's Go: A New Programming Language That's Python Meets C++." *TechCrunch.* AOL, 10 Nov. 2009. Web. 19 Aug. 2016.

Grammatology

SUMMARY

Grammatology is the science of writing. It is closely related to the field of linguistics but concentrates on written expression whereas linguistics is primarily oriented toward oral expression. Grammatology is an interdisciplinary field that is applied to research in a variety of fields studying human interaction and communication and the various cultures and civilizations. Anthropologists, psychologists, literary critics, and language scholars are the main users of grammatology. Electracy, digital literacy, is a concept that applies the principles of grammatology to electronic media.

DEFINITION AND BASIC PRINCIPLES

Grammatology is the scientific examination of writing. Oral communication is the natural means of communication for human beings. Oral language uses neither instruments nor objects that are not part of the human body. Writing, by contrast, is not natural language; it requires an object (paper, stone, wood, or even sand) on which marks can be made and some kind of tool with which to make the marks. Although it is possible to use one's fingers to make marks in a soft medium such as sand, such "writing" lacks an essential feature of writing as defined by grammatology: Writing lasts beyond the moment of its creation. It preserves thoughts, ideas, stories, and accounts of various human activities. In addition, writing, in contrast to pictography (drawn pictures), can be understood only with knowledge of the language that it represents, because writing uses symbols to create a communication system. The discipline defines what writing is by using a system of typologies, or types. An example of this is the classification of languages by script (the kind of basic symbols used). Another way of classifying languages is by the method used to write them; linear writing systems use lines to compose the characters.

Grammatology also investigates the importance of writing in the development of human thought and reasoning. Grammatologists address the role of writing in the development of cultures and its importance in the changes that occur in cultures. Grammatology proposes the basic principle that changes in ways of communication have profound effects on both the individual and the societal life of people. It addresses the technology of writing and its influence on the use and development of language. It deals extensively with the impact of the change from oral to handwritten communication, from handwritten texts to printed texts, and from printed texts to electronic media texts.

BACKGROUND AND HISTORY

Polish American historian Ignace Gelb, the founder of grammatology, was the first linguist to propose a science of writing independent of linguistics. In 1952, he published *A Study of Writing: The Foundations of Grammatology*, in which he classified writing systems according to their typologies and set forth general principles for the classification of writing systems. As part of his science of writing systems, Gelb traced the evolution of writing from pictures that he referred to as "no writing" through semasiography to full writing or scripts that designate sounds by characters. Gelb divided languages into three typologies: logo-syllabic systems, syllabic systems, and alphabetic systems. Peter T. Daniels, one of Gelb's students, added two additional types to this typology: abugida and abjad. Gelb also included as part of the science of writing an investigation of the interactions of writing and culture and of writing and religion as well as an analysis of the relationship of writing to speech or oral language. Although Gelb did not elaborate extensively on the concept of writing as a technology, he did address the importance of the invention of printing and the changes that it brought about in dissemination of information. This aspect of writing (by hand or device) as technology, as an apparatus of communication, has become one of the most significant areas of the application of grammatology to other disciplines.

Gelb's work laid the foundation for significant research and development in several fields involved in the study of language. Eric Havelock, a classicist working in Greek literature, elaborated on the idea of how the means of communication affects thought. He developed the theory that Greek thought underwent a major change or shift when the oral tradition was replaced by written texts. In Havelock's *Preface to Plato* (1963), he presented his argument that the transition to writing at the time of Plato altered Greek thought and consequently all Western thought by its effect on the kind of thinking available through written expression. Havelock stated that what could be expressed and how it could be expressed in written Greek was significantly different from what could be expressed orally. He devoted the rest of his scholarship to elaboration of this premise and investigation of the difference between the oral tradition and writing. Havelock's theory met with criticism from many of his colleagues in classics, but it received a favorable reception in many fields, including literature and anthropology.

Starting from Havelock's theory, Walter J. Ong, a literary scholar and philosopher, further investigated the relationship between written and spoken language and how each affects culture and thought. He proposed that writing is a technology and therefore has the potential to affect virtually all aspects of an oral culture. Addressing the issue of writing's effect on ways of thinking, Ong stated that training in writing shifted the mental focus of an individual. Ong reasoned that oral cultures are group directed, and those using the technology of writing are directed toward the individual. His most influential work is *Orality and Literacy: The Technologizing of the Word* (1982). The British anthropologist Sir John "Jack" Rankine Goody applied the basic premise of grammatology (that the technology of writing influences and changes society) in his work in comparative anthropology. For Goody, writing was a major cause of change in a society, affecting both its social interaction and its psychological orientation.

In *De la grammatologie* (1967; *Of Grammatology*, 1976), French philosopher and literary critic Jacques Derrida examined the relationship of spoken language and writing. By using Gelb's term "grammatology" as a term of literary criticism, Derrida narrowed the significance of the term. Grammatology became associated with the literary theory of deconstruction. However, Derrida's work had more far-reaching implications, and for Derrida, as for Gelb, grammatology meant the science of writing. Derrida was attempting to free writing from being merely a representation of speech.

In *Applied Grammatology: Post(e)-pedagogy from Jacques Derrida to Joseph Beuys* (1985), Gregory Ulmer moved from deconstructing writing to a grammatology that he defined as inventive writing. Although Ulmer's work moves away from the areas of systems and classifications of writing as established by Gelb, it does elaborate on Gelb's notion of how literacy or the technology of writing changes human thought and interaction. Ulmer's work brings Gelb's theories and work into the electronic age. Gelb and his followers investigated the development of writing systems and the technologies of handwritten and printed texts and their influences on human beings, both in their social interactions and in their thought processes. Just as print influenced the use, role, and significance of writing, electronic media is significantly affecting writing, both in what it is and what it does.

HOW IT WORKS

Classification by Typologies. Written languages are classified based on particular features of the language. Grammatologists analyze written languages to determine whether the basic element of the language is a syllable, a letter, a word, or some other element. They also look for similarities among languages that they compare. From these analyses, they classify languages in existing typologies or create new ones. The established typologies are logographic, syllabic, alphabetic, abjad, and abugida.

Mechanisms and Apparatuses. Grammatology identifies the mechanisms or the apparatuses of writing. It examines how writing is produced, what type of tools are used to produce it, and the kind of texts that result from writing. It also considers the speed and breadth of dissemination of information through writing. Grammatology looks at writing systems from a practical viewpoint, considering questions of accuracy and ease of usage.

Relationship of Oral and Written Language. Grammatologists investigate the relationship of oral and written language by comparing a language in its oral and written form. They look for differences in sentence structure, vocabulary, and complexity of usage.

Writing and Social Structure. Grammatologists note the time that writing appears in a culture and analyze the social, political, economic, and family structures to ascertain any and all changes that appear after the introduction of writing. They also attempt to identify the type of thinking, ideas, and beliefs prevalent in the society before and after the introduction of writing and determine if writing brought about changes in these areas. Once they have established evidence of change, they address the ways in which writing causes these modifications. In addition, they investigate these same areas when the mechanism of writing in a culture changes, such as when print replaces handwritten texts and when electronic media replaces print.

APPLICATIONS AND PRODUCTS

Pedagogy. The science of grammatology is important in the development of pedagogies that adequately address the needs and orientations of students. The way information is presented and taught varies with the medium by which students are accustomed to receiving knowledge. In an oral culture, stories, poems, and proverbs that use repetitive signifiers are employed to enable students to memorize information. Handwritten texts are suited to cultures in which a small quantity of information is disseminated among a limited number of students. Print text cultures rely heavily on the dissemination of information through books and other printed sources.

Instruction is classroom based with face-to-face contact between the teacher and student. Assignments are written out and turned in to the professor to be corrected and returned. Cultures that rely on electronic media call for different pedagogical structures. Grammatology provides a means of analyzing and addressing this need.

Anthropology. Anthropologists use grammatology as an integral part of their research techniques to ascertain the development of civilizations and cultures. The time when writing appeared in a culture (or the lack of writing in a culture), the type of writing, and the classes of society that use writing are all important to the study and understanding of a culture or civilization.

Literary Criticism. Grammatology enables literary critics to dissect or deconstruct texts and perform close analysis of both the structure and the meaning of a work. It has also been the basis for the creation of new types of literary criticism, including deconstruction and postdeconstruction.

Linguistics. Grammatology, which deals with the written form of language, plays an important role in linguistics, the study of languages. It classifies languages, addresses how graphemes (units of a writing system) and allographs (letters) are used, compares written to oral forms of languages, and identifies archaic forms and vocabularies. The study of writing forms is necessary to linguists tracing the development of a particular language. Texts written in the language during different periods permit linguists to compare vocabulary, syntax, and prosaic and poetic forms of the language throughout its development. Written texts also provide evidence of borrowings from other languages.

Archaeology. The ability to analyze and classify written language is an essential part of the work of the archaeologist. Knowledge regarding when particular written forms were used and what cultures used them plays an important role in the dating of artifacts.

Electronic Media. Grammatology is highly applicable to cyberlanguage and electronic media as these phenomena affect writing. Texting and e-mail have created what may be viewed as a subtext or alternative to the traditional print culture. The principles and techniques of analysis of grammatology provide a means for classifying the language of texting and e-mail and for evaluating the effects of these new forms of language on thought and on societal interaction.

SOCIAL CONTEXT AND FUTURE PROSPECTS

With the changes in the mechanisms and apparatus of writing that have taken place in the second half of the twentieth century and continue to take place in the twenty-first, grammatology continues to be an essential science. Research in electronic media and its relation to human thought processes will be an area in which grammatology will play an important role. Because grammatology is a science used in conjunction with a considerable number of areas of scientific research, it should continue to play a significant role in human intellectual life.

—Shawncey Jay Webb, PhD

Bibliography

Daniels, Peter T., and William Bright, eds. *The World's Writing Systems.* New York: Oxford University Press, 1996. These essays continue and refine the work done by Gelb.

Gelb, Ignace. *A Study of Writing: The Foundations of Grammatology.* 1952. Reprint. Chicago: University of Chicago Press, 1989. The foundational text for grammatology.

Olson, David R., and Michael Cole, eds. *Technology, Literacy and the Evolution of Society: Implications of the Work of Jack Goody.* Mahwah, N.J.: Lawrence Erlbaum Associates, 2006. An interdisciplinary look at Goody's work on modes of communication in societies and their influence.

Ong, Walter J. *Orality and Literacy: The Technologizing of the Word.* 2d ed. Reprint. New York: Routledge, 2009. Ong elaborates on the work of Havelock concerning the impact of writing on oral culture. Identifies writing as a learned technology.

Rogers, Henry. *Writing Systems: A Linguistic Approach.* Malden, Mass.: Wiley-Blackwell, 2005. A very detailed presentation of world languages and their writing systems that describes the classification of writing systems.

Ulmer, Gregory. *Applied Grammatology: Post(e)-pedagogy from Jacques Derrida to Joseph Beuys.* 1985. Reprint. Baltimore: The Johns Hopkins University Press, 1992. Presents Ulmer's early thoughts on how electronic media was changing the technology of writing. Also provides an excellent explanation of Derrida's views.

_______. *Electronic Monuments.* Minneapolis: University of Minnesota Press, 2005. Ulmer's view of the effects of electracy, or digital literacy, on society and human interaction.

GRAPHENE

SUMMARY

Graphene is a property of carbon. Carbon is the chemical basis for all life on earth. Harnessing graphene has led to major advances in electronics and biotechnology. The value of graphene is in its transparency, elasticity, density, flexibility, hardness, resistance, and electric and thermal conductivity. Graphene is a chemical element existing in different forms in the same physical two-dimensional state of the atomic scale. When tightly packed, graphene is a layer of carbon atoms bonding together in a honeycomb lattice. Graphene is the thinnest compound scientists have uncovered. It is one atom thick, one hundred to three hundred times stronger than steel with massive tensile stiffness. It conducts heat and electricity more efficiently than other chemical elements. It conducts light and is a major factor in spintronics, i.e., affecting the spin of electrons, magnetic movements, and electronic charge, in solid-state devices. Graphene generates chemical reactions with other substances, and scientists believe graphene has potential for advancing the technology revolution.

BACKGROUND

Graphite, a mineral occurring naturally on earth, is the most stable form of carbon. Southeastern Europeans were using graphite 3000 years BCE in decorative ceramic paints for pottery. People discovered wider uses for graphite such as making lead pencils, leading scientists to speculate there must be another undiscovered element to graphite. In 1859, Benjamin Brodie studied the structure of graphite. His work was followed up with scientific progress between 1918 and 1925 by other physicists. P.R. Wallace's study of the theory of graphene in 1947 opened a new field of inquiry that theoretical physicists and chemists continued pursuing for another half-century, setting aside the awareness that graphene was discovered in lab analyses on nickel and silicon carbide.

Andre Geim, a Soviet Union-born University of Manchester professor, discovered graphene from graphite in lab experiments by examining under an electron microscope the tiniest atomic-size particles from graphite grindings. Geim isolated the first two-dimensional material ever discovered. It was a layer of carbon an atom thick with structure and properties about which the physicists had been theorizing. Geim has dedicated most of the rest of his professional career to studying graphene. He and his students uncovered graphene's field effect, allowing scientists to control the conductivity of graphene. This is one of the defining characteristics of silicon that advanced the entire world of computer chips and computers. Graphene is the thinnest material known in the universe. In the 1970s, chemists figured a way to place carbon monolayers onto other materials. One of the first U.S. patents for graphene production was granted in 2006. Small amounts Geim produced in his lab were not going to satisfy demand. New processes were discovered to produce graphene in large quantities. Geim and his associate were awarded the Nobel Prize in Physics in 2010 for their ground-breaking experiments on graphene, not for discovering it.

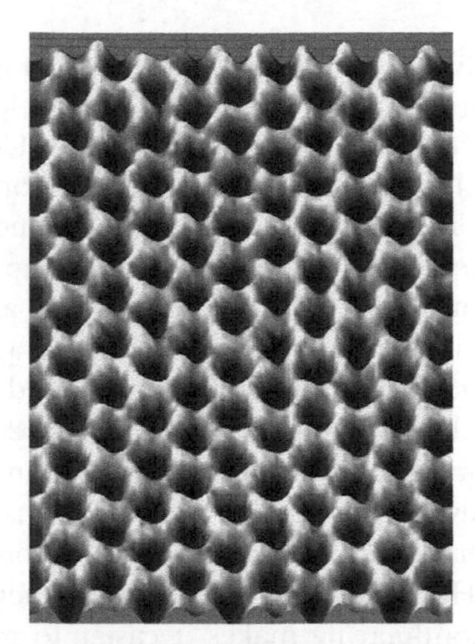

Graphene. By U.S. Army Materiel Command

Further study found ways graphene is used in electrical engineering, medicine, and chemistry. Between 2011 and 2013, graphene-related patents issued by the U.K. Intellectual Property Office rose from 3,018 to 8,416 for products like long-life batteries, computer screens, and desalinization of water.

GRAPHENE TODAY

New research into graphene is making great strides into finding ways to increase the power density of batteries. Scientists have hopes for graphene to produce ultra-long-life batteries that have less weight, are quicker to charge, and are thinner and less expensive to produce than lithium batteries. Korean-based company Samsung has been awarded the most patents in graphene. Samsung funds research on graphene at a Korean university. Chinese universities are second and third in the number of patents for graphene discoveries, and Rice University in the United States has filed thirty-three patent applications since 2014.

Professor James Tour, a synthetic organic chemist at Rice University, is among the leaders, researching graphene and looking into its possible commercial applications. His lab sold its patents for graphene-infused paint; its conductivity makes it easier to remove ice from helicopter blades; mixed with fluids, graphene increases oil drill efficiency; and graphene is used in materials to make airplane emergency slides and life rafts lighter and safer for passengers. Thus it is expected to save millions of dollars in fuel costs for airlines.

Tour's associates are experimenting with graphene for help for people with spinal cord injuries. Graphene oxide bonds with radioactive elements, thus the importance of early discoveries in bonding graphene to other substrates. Experiments are attempting to turn the mix into sludge to be scooped away for effective environmental cleanup following radioactive disasters. Improving the mobility of electronic information to flow over graphene surfaces from one point to another will mean increasing the speed of communication a hundred fold or more. In addition to work on graphene by physicists and chemists, biologists are looking to use the graphene nanomaterial. Biologists are working on bonding graphene with chemical groups that might improve therapies and have an effect on cancer and neuronal cells and immune systems. Graphene is being studied for its relation to and interactions with boron nitride, molybdenum sulphate, tungsten, and silicene, all two-dimensional materials as small as atoms. If means of bonding with graphene are discovered, scientists might be able to create new properties.

A molecular model of the structure of graphene. By Mohammad Javad Kiani, Fauzan Khairi Che Harun, Mohammad Taghi Ahmadi, Meisam Rahmani, Mahdi Saeidmanesh, Moslem Zare

In June 2016, the University of Exeter announced that their research engineers and physicists discovered a lightweight graphene-adapted material for conducting electricity that substantially improves the effectiveness of large flat flexible lighting. Brightness is increased 50 percent, and GraphExeter as it is called, greatly extends shelf life before needing replacement. Researchers are looking into application for health-light therapies as well. MIT researchers cannot yet fully explain their lab findings showing a relationship between hyper-thin carbon structures and light. Light moves slowly in graphene, and graphene is too thin for light to remain inside, but electrons move quickly inside the graphene honeycomb matrix. Shockwaves are somehow created. Researchers believe their work might lead to new materials making optical computer cores of the future.

Engineers are studying uses for graphene in electronics to make smaller transistors, consume less energy, and scatter heat faster. Many scientists and engineers consider commercial and health applications of graphene are still in a stage of immature or novice technology. There are experiments into bendable smartphones using graphene to create the screens. The primary drawbacks are the cost of high-price equipment and lack of better knowledge of mass production. Corporate funding of university laboratories and leeway given them in other fields of research are today's model for graphene research and applications.

—Harold Goldmeier, EdD

BIBLIOGRAPHY

Bradley, David. "A Chemical History of Graphene." *MaterialsToday*. Elsevier Ltd. 10 June 2014. Web. 22 June 2016.

Colapinto, John. "Graphene May Be the Most Remarkable Substance Ever Discovered. But What's It for?" *The New Yorker*. Conde Nast, 22 and 29 Dec., 2014. Web. 22 June 2016.

Emmio, Nicolette. "Why Do Engineers Care So Much About Graphene?" *Electronics 360*. Electronics 360, 22 June 2016. Web. 23 June 2016.

Geim, A.K., and K.S. Novoselov. "The Rise of Graphene." *Nature Materials.* Macmillan Publishers Limited 6 (2007): 183-91. Web. 22 June 2016.
"Graphene-Based Material Illuminates Bright New Future for Flexible Lighting Devices." *Nanowerk News*: Nanowerk, June 2016. Web. 23 June 2016.
Mertens, Ron. *The Graphene Handbook.* Graphene-Info, 2016. Print and online.
Poulter, Sean. "Bendable Smartphones Are Coming! Devices With Screens Made From Graphene Are So Flexible They Can Be Worn Like a Bracelet." *Mail Online.* The Daily Mail. Associated Newspapers Ltd, 24 May 2016. Web. 22 June 2016.
Sabin, Dyani. "Graphene-Based Computers Could Turn Electricity Into Light, Speeding Processing." *Inverse*, n.p., 23 June 2016. Web. 23 June 2016.

Graphics Technologies

SUMMARY

Graphics technology, which includes computer-generated imagery, has become an essential technology of the motion picture and video-gaming industries, of television, and of virtual reality. The production of such images, and especially of animated images, is a complex process that demands a sound understanding of not only physics and mathematics but also anatomy and physiology.

DEFINITION AND BASIC PRINCIPLES

While graphics technologies include all of the theoretical principles and physical methods used to produce images, it more specifically refers to the principles and methods associated with digital or computer-generated images. Digital graphics are displayed as a limited array of colored picture elements (pixels). The greater the number of pixels that are used for an image, the greater the resolution of the image and the finer the detail that can be portrayed. The data that specifies the attributes of each individual pixel are stored in an electronic file using one of several specific formats. Each file format has its own characteristics with regard to how the image data can be manipulated and utilized.

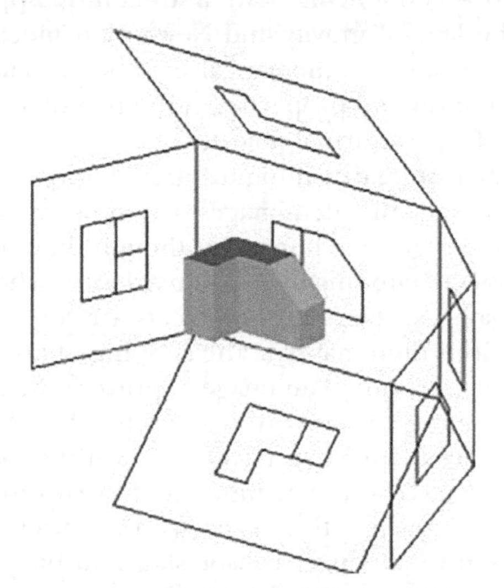

Image of a part represented in First Angle Projection. NASA

Because the content of images is intended to portray real-world objects, the data for each image must be mathematically manipulated to reflect real-world structures and physics. The rendering of images, especially for photo-realistic animation, is thus a calculation-intensive process. For images that are not produced photographically, special techniques and applications are continually being developed to produce image content that looks and moves as though it were real.

BACKGROUND AND HISTORY

Imaging is as old as the human race. Static graphics have historically been the norm up to the invention of the devices that could make a series of still pictures appear to move. The invention of celluloid in the late nineteenth century provided the material for photographic film, with the invention of motion picture cameras and projectors to follow. Animated films, commonly known as cartoons, have been produced since the early twentieth century by repeatedly photographing a series of hand-drawn cels. With the development of the digital computer and color displays in the last half of the century, it became possible to generate images without the need for hand-drawn intermediaries.

Computer graphics in the twenty-first century can produce images that are indistinguishable from traditional photographs of real objects. The methodology continues to develop in step with the development of new computer technology and new programming methods that make use of the computing abilities of the technology.

HOW IT WORKS

Images are produced initially as still or static images. Human perception requires about one-thirtieth of a second to process the visual information obtained through the seeing of a still image. If a sequential series of static images is displayed at a rate that exceeds the frequency of thirty images per second, the images are perceived as continuous motion. This is the basic principle of motion pictures, which are nothing more than displays of a sequential series of still pictures. Computer-generated still images (now indistinguishable from still photographs since the advent of digital cameras) have the same relationship to computer animation.

Images are presented on a computer screen as an array of colored dots called pixels (an abbreviation of "picture elements"). The clarity, or resolution, of the image depends on the number of pixels that it contains within a defined area. The more pixels within a defined area, the smaller each pixel must be and the finer the detail that can be displayed. Modern digital cameras typically capture image data in an array of between 5 and 20 megapixels. The electronic data file of the image contains the specific color, hue, saturation, and brightness designations for each pixel in the associated image, as well as other information about the image itself.

To obtain photorealistic representation in computer-generated images, effects must be applied that correspond to the mathematical laws of physics. In still images, computational techniques such as ray tracing and reflection must be used to imitate the effect of light sources and reflective surfaces. For the virtual reality of the image to be effective, all of the actual physical characteristics that the subject would have if it were real must be clearly defined as well so

Computer animation of Galileo probe separating from orbiter.

that when the particular graphics application being used renders the image to the screen, all of the various parts of the image are displayed in their proper positions.

To achieve photorealistic effects in animation, the corresponding motion of each pixel must be coordinated with the defined surfaces of the virtual object, and their positions must be calculated for each frame of the animation. Because the motions of the objects would be strictly governed by the mathematics of physics in the real world, so must the motions of the virtual objects. For example, an animated image of a round object bouncing down a street must appear to obey the laws of gravity and Newtonian mechanics. Thus, the same mathematical equations that apply to the motion and properties of the real object must also apply to the virtual object.

Other essential techniques are required to produce realistic animated images. When two virtual objects are designed to interact as though they are real, solid objects, clipping instructions identify where the virtual solid surfaces of the objects are located; the instructions then mandate the clipping of any corresponding portions of an image to prevent the objects from seeming to pass through each other. Surface textures are mapped and associated with underlying data in such a way that movement corresponds to real body movements and surface responses. Image animation to produce realistic skin and hair effects is based on a sound understanding of anatomy and physiology and represents a specialized field of graphics technology.

APPLICATIONS AND PRODUCTS

Software. The vast majority of products and applications related to graphics technology are software applications created specifically to manipulate electronic data so that it produces realistic images and animations. The software ranges from basic paint programs installed on most personal computers (PCs) to full-featured programs that produce wireframe structures, map surface textures, coordinate behaviors, movements of surfaces to underlying structures, and 360-degree, three-dimensional animated views of the resulting images.

Other types of software applications are used to design objects and processes that are to be produced as real objects. Computer-assisted design is commonly used to generate construction-specification drawings and to design printed-circuit boards, electronic circuits and integrated circuits, complex machines, and many other real-world constructs. The features and capabilities of individual applications vary.

The simplest applications produce only a static image of a schematic layout, while the most advanced are capable of modeling the behavior of the system being designed in real time. The latter are increasingly useful in designing and virtual-testing such dynamic systems as advanced jet engines and industrial processes. One significant benefit that has accrued from the use of such applications has been the ability to refine the efficiency of systems such as production lines in manufacturing facilities.

Hardware. The computational requirements of graphics can quickly exceed the capabilities of any particular computer system. This is especially true of PCs. Modern graphics technology in this area makes use of separate graphics processing units (GPUs) to handle the computational load of graphics display. This allows the PC's central processing unit (CPU) to carry out the other computational requirements of the application without having to switch back and forth between graphic and nongraphic tasks.

Many graphics boards also include dedicated memory for exclusive use in graphics processing. This eliminates the need for large sectors of a PC's random access memory (RAM) to be used for storing graphics data, a requirement that can render a computer practically unusable.

Another requirement of hardware is an instruction system to operate the various components so that they function together. For graphics applications, with the long periods of time they require to carry out the calculations needed to render a detailed image, it is essential that the computer's operating system be functionally stable. The main operating systems of PCs are Microsoft Windows, its open-source competitor Linux, and the Apple Mac OS. Some versions of other operating systems, such as Sun Microsystems' Solaris, have been made available but do not account for a significant share of the PC market.

The huge amounts of graphics and rendering required for large-scale projects such as motion pictures demand the services of mainframe computers. The operating systems for these units have a longer history than do PC operating systems. Mainframe computers function primarily with the UNIX operating system, although many now run under some variant of the Linux operating system. UNIX and Linux are similar operating systems, the main difference being that UNIX is a proprietary system whereas Linux is open-source.

Motion Pictures and Television. Graphics technology is a hardware- and software-intensive field. The modern motion picture industry would not be possible without the digital technology that has been developed since 1980. While live-action images are still recorded on standard photographic film in the traditional way, motion picture special effects and animation have become the exclusive realm of digital graphics technologies. The use of computer generated imagery (CGI) in motion pictures has driven the development of new technology and continually raised the standards of image quality. Amalgamating live action with CGI through digital processing and manipulation enables film-makers to produce motion pictures in which live characters interact seamlessly with virtual characters, sometimes in entirely fantastic environments. Examples of such motion pictures are numerous in the science-fiction and fantasy film genre, but the technique is finding application in all areas, especially in educational programming.

Video Gaming and Virtual Training. The most graphics-intensive application is video gaming. All video games, in all genres, exist only as the graphic representation of complex program code. The variety of video game types ranges from straightforward computer versions of simple card games to complex three-dimensional virtual worlds.

Many graphics software applications are developed for the use of game designers, but they have also made their way into many other imaging uses. The same software that is used to create a fictional virtual world can also be used to create virtual copies of the real world. This technology has been adapted for use in pilot- and driver-training programs in all aspects of transportation. Military, police, and security personnel are given extensive practical and scenario training through the use of virtual simulators. A simulator uses video and graphic displays of actual terrain to give the person being trained hands-on experience without endangering either personnel or actual machinery.

SOCIAL CONTEXT AND FUTURE PROSPECTS

Graphics technology is inextricably linked to the computer and digital electronics industries. Accordingly, graphics technology changes at a rate that at minimum equals the rate of change in those industries. Since 1980, graphics technology using computers has developed from the display of just sixteen colors on color television screens, yielding blocky image components and very slow animation effects, to photorealistic full-motion video, with the capacity to display more colors than the human eye can perceive and to display real-time animation in intricate detail. The rate of change in graphics technology exceeds that of computer technology because it also depends on the development of newer algorithms and coding strategies. Each of these changes produces a corresponding new set of applications and upgrades to graphic technology systems, in addition to the changes introduced to the technology itself.

Each successive generation of computer processors has introduced new architectures and capabilities that exceed those of the preceding generation, requiring that applications update both their capabilities and the manner in which those capabilities are performed. At the same time, advances in the technology of display devices require that graphics applications keep pace to display the best renderings possible. All of these factors combine to produce the unparalleled value of graphics technologies in modern society and into the future.

—Richard M. Renneboog, MSc

Bibliography

Abrash, Michael. *Michael Abrash's Graphics Programming Black Book.* Albany, N.Y.: Coriolis Group, 1997. Consisting of seventy short chapters, the book covers graphics programming up to the time of the Pentium processor.

Brown, Eric. "True Physics." *Game Developer* 17, no. 5 (May, 2010): 13-18. This article provides a clear explanation of the relationship between real-world physics and the motion of objects in animation.

Jimenez, Jorge, et al. "Destroy All Jaggies." *Game Developer* 18, no. 6 (June/July, 2011): 13-20. This article describes the method of anti-aliasing in animation graphics.

Oliver, Dick, et al. *Tricks of the Graphics Gurus.* Carmel, Ind.: Sams, 1993. Provides detailed explanations of the mathematics of several computer graphics processes.

Ryan, Dan. *History of Computer Graphics: DLR Associates Series.* Bloomington, Ind.: AuthorHouse, 2011. A comprehensive history of computer graphics and graphics technologies.

Holographic Technology

SUMMARY

Holographic technology employs beams of light to record information and then rebuilds that information so that the reconstruction appears three-dimensional. Unlike photography, which traditionally produces fixed two-dimensional images, holography re-creates the lighting from the original scene and results in a hologram that can be viewed from different angles and perspectives as if the observer were seeing the original scene. The technology, which was greatly improved with the invention of the laser, is used in various fields such as product packaging, consumer electronics, medical imaging, security, architecture, geology, and cosmology.

DEFINITION AND BASIC PRINCIPLES

Holography is a technique that uses interference and diffraction of light to record a likeness and then rebuild and illuminate that likeness. Holograms use coherent light, which consists of waves that are aligned with one another. Beams of coherent light interfere with one another as the image is recorded and stored, thus producing interference patterns. When the image is re-illuminated, diffracted light allows the resulting hologram to appear three dimensional. Unlike photography, which produces a fixed image, holography re-creates the light of the original scene and yields a hologram, which can be viewed from different angles and different perspectives just as if the original subject were still present.

Several basic types of holograms can be produced. A transmission hologram requires an observer to see the image through light as it passes through the hologram. A rainbow hologram is a special kind of transmission hologram, in which colors change as the observer moves his or her head. This type of transmission hologram can also be viewed in white light, such as that produced by an incandescent lightbulb. A reflection hologram can also be viewed in white light. This type allows the observer to see the image with light reflected off the surface of the hologram. The holographic stereogram uses attributes of both holography and photography. Industry and art utilize the basic types of holograms as well as create new and advanced technologies and applications.

BACKGROUND AND HISTORY

Around 1947, Hungarian-born physicist Dennis Gabor developed the basics of holography while attempting to improve the electron microscope. Early efforts by scientists to develop the technique were restricted by the use of the mercury arc lamp as a monochromatic light source. This inferior light source contributed to the poor quality of holograms, and the field advanced little throughout the next decade. Laser light was introduced in the 1960s and was considered stable and coherent. Coherent light contains waves that are aligned with one another and is well suited for high-quality holograms. Subsequently, discoveries and innovations in the field began to increase and accelerate.

In 1960, American physicist Theodore Harold Maiman of Hughes Research Laboratories developed the pulsed ruby laser. This laser used rubies to operate and generated powerful bursts of light lasting only nanoseconds. The pulsed ruby laser, which acted much like a camera's flashbulb, became ideal for capturing images of moving objects or people.

In 1962, while working at the University of Michigan, scientists Emmett Leith and Juris Upatnieks decided to improve on Gabor's technique. They produced images of three-dimensional (3-D) objects—a toy bird and train. These were the first transmission holograms and required an observer to see the image through light as it passed through the holograms.

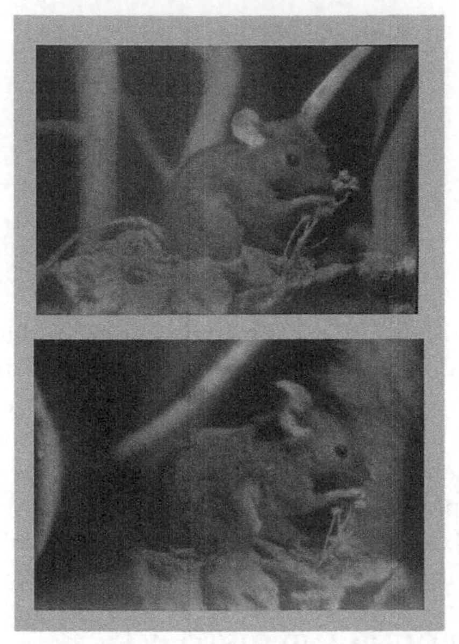

Two photographs of a single hologram taken from different viewpoints. By Holo-Mouse.jpg: Georg-Johann Lay derivative work: Epzcaw (Holo-Mouse.jpg)

Also in 1962, Russian scientist Yuri Nikolaevich Denisyuk combined his own work with the color photography work of French physicist Gabriel Lippmann. This resulted in a reflection hologram that could be viewed with white light reflecting off the surface of a hologram. Reflection holograms do not need laser light to be viewed.

In 1968, electrical engineer Stephen Benton developed the rainbow hologram. When an observer moves his or her head, he or she sees the spectrum of color, as in a rainbow. This type of hologram can also be viewed in white light.

Holographic art appeared in exhibits beginning in the late 1960s and early 1970s, and holographic portraits made with pulsed ruby lasers found some favor beginning in the 1980s. Advances in the field have continued, and many and varied types of holograms are used in many different areas of science and technology, while artistic applications have lagged in comparison.

HOW IT WORKS

A 3-D subject captured by conventional photography becomes stored on a medium, such as photographic film, as a two-dimensional (2-D) scene. Information about the intensity of the light from a static scene is acquired, but information about the path of the light is not recorded. Holographic creation captures information about the light, including the path, and the whole field of light is recorded.

A beam of light first reaches the object from a light source. Wavelengths of coherent light, such as laser light, leave the light source "in phase" (in sync) and are known collectively as an object beam. These waves reach the object, are scattered, and then are interfered with when a reference beam from the same light source is introduced. A pattern occurs from the reference beam interfering with the object waves. This interference pattern is recorded on the emulsion. Re-illumination of the hologram with the reference beam results in the reconstruction of the object light wave, and a 3-D image appears.

Light Sources. An incandescent bulb generates light in a host of different wavelengths, whereas a laser produces monochromatic wavelengths of the same frequency. Laser light, also referred to as coherent light, is used most often to create holograms. The helium-neon laser is the most commonly recognized type.

Types of lasers include all gas-phase iodine, argon, carbon dioxide, carbon monoxide, chemical oxygen-iodine, helium-neon, and many others.

To produce wavelengths in color, the most frequently used lasers are the helium-neon (for red) and the argon-ion (for blue and green). Lasers at one time were expensive and sometimes difficult to obtain, but modern-day lasers can be relatively inexpensive and are easier to use for recording holograms.

Recording Materials. Light-sensitive materials such as photographic films and plates, the first resources used for recording holograms, still prove useful. Since the color of light is determined by its wavelength, varying emulsions on the film that are sensitive to different wavelengths can be used to record information about scene colors. However, many different types of materials have proven valuable in various applications.

Other recording materials include dichromated gelatin, elastomers, photoreactive polymers, photochromics, photorefractive crystals, photoresists, photothermoplastics, and silver-halide sensitized gelatin.

APPLICATIONS AND PRODUCTS

Art. Holographic art, prevalent in the 1960s through the 1980s, still exists. Although fewer artists practice holography exclusively, many artistic creations contain holographic components. A small number of schools and universities teach holographic art.

Digital Holography. Digital holography is one of the fastest-growing realms and has applications in the artistic, scientific, and technological communities. Computer processing of digital holograms lends an advantage, as a separate light source is not needed for re-illumination.

Digital holography first began to appear in the late 1970s. The process initially involved two steps: first writing a string of digital images onto film, then converting the images into a hologram. Around 1988, holographer Ken Haines invented a process for creating digital holograms in one step.

Digital holographic microscopy (DHM) can be used noninvasively to study changes in the cells of living tissue subjected to simulated microgravity. Information is captured by a digital camera and processed by software.

Display. Different types of holograms can be displayed in store windows; as visual aids to accompany lectures or presentations; in museums; at art, science, or technology exhibits; in schools and libraries; or at home as simple decorations hung on a wall and lit by spotlights.

Embossed Holograms. Embossed holograms, which are special kinds of rainbow holograms, can be duplicated and mass produced. These holograms can be used as means of authentication on credit cards and driver's licenses as well as for decorative use on wrapping paper, book covers, magazine covers, bumper stickers, greeting cards, stickers, and product packaging.

Holograms in Medicine and Biology. The field of dentistry provided a setting for an early application in medical holography. Creating holograms of dental casts markedly reduced the space needed to store dental records for Britain's National Health Service (NHS). Holograms have also proved useful in regular dental practice and dentistry training.

The use of various types of holograms has proved beneficial for viewing sections of living and nonliving tissue, preparing joint replacement devices, noninvasive scrutiny of tumors or suspected tumors, and viewing the human eye. A volume-multiplexed hologram can be used in medical-scanning applications.

Moving Holograms. Holographic movies created for entertaining audiences in a cinema are in development. While moving holograms can be made, limitations exist for the production of motion pictures. Somewhat more promising is the field of holographic video and possibly television.

Security. A recurring issue in world trade is that of counterfeit goods. Vendors increasingly rely on special holograms embedded in product packaging to combat the problem. The creation of complex brand images using holographic technology can offer a degree of brand protection for almost any product, including pharmaceuticals.

Security holography garners a large segment of the market. However, makers of security holograms, whose designs are used for authentication of bank notes, credit cards, and driver's licenses, face the perplexing challenge of counterfeit security images. As time progresses, these images become increasingly easier to fake; therefore, this area of industry must continually create newer and more complex holographic techniques to stay ahead of deceptive practices.

Stereograms. Holographic stereograms, unique and divergent, use attributes of both holography and photography. Makers of stereograms have the potential of creating both very large and moving images. Stereograms can be produced in color and also processed by a computer.

Nonoptical Holography. Types of holography exist that use waves other than light. Some examples include acoustical holography, which operates with sound waves; atomic holography, which is used in applications with atomic beams; and electron holography, which utilizes electron waves.

SOCIAL CONTEXT AND FUTURE PROSPECTS

Holography in one form or another is prevalent in modern society, whether as a security feature on a credit card, a component of a medical technique, or a colorful wrapping paper. Holograms have been interwoven into daily life and will likely continue to increase their impact in the future.

Next-generation holographic storage devices have been developed, setting the stage for companies to compete for future markets. Data is stored on the surface of DVDs; however, devices have been invented to store holographic data within a disk. The significantly enlarged storage capacity is appealing for customers with large storage needs who can afford the expensive disks and drives, but some companies are also interested in targeting an even larger market by revising existing technology. Possible modification of current technology, such as DVD players, could potentially result in less expensive methods of playing 3-D data.

—Glenda Griffin, BA, MLS

Bibliography

Ackermann, Gerhard K., and Jürgen Eichler. *Holography: A Practical Approach.* Wiley, 2007.

Hariharan, P. *Basics of Holography.* Cambridge UP, 2002.

Harper, Gavin D. J. *Holography Projects for the Evil Genius.* McGraw-Hill, 2010.

Johnston, Sean F. "Absorbing New Subjects: Holography as an Analog of Photography." *Physics in Perspective,* vol. 8, no. 2, 2006, pp. 164–88.

Saxby, Graham, and Stanislovas Zacharovas. *Practical Holography.* 4rd ed., CRC Press, 2016.

Yaroslavsky, Leonid. *Digital Holography and Digital Image Processing: Principles, Methods, Algorithms.* Kluwer Academic, 2004.

Human-computer Interaction

SUMMARY

Human-computer interaction (HCI) is a field concerned with the study, design, implementation, evaluation, and improvement of the ways in which human beings use or interact with computer systems. The importance of human-computer interaction within the field of computer science has grown in tandem with technology's potential to help people accomplish an increasing number and variety of personal, professional, and social goals. For example, the development of user-friendly interactive computer interfaces, websites, games, home appliances, office equipment, art installations, and information distribution systems such as advertising and public awareness campaigns are all applications that fall within the realm of HCI.

DEFINITION AND BASIC PRINCIPLES

Human-computer interaction is an interdisciplinary science with the primary goal of harnessing the full potential of computer and communication systems for the benefit of individuals and groups. HCI researchers design and implement innovative interactive technologies that are not only useful but also easy and pleasurable to use and anticipate and satisfy the specific needs of the user. The study of HCI has applications throughout every realm of modern life, including work, education, communications, health care, and recreation.

The fundamental philosophy that guides HCI is the principle of user-centered design. This philosophy proposes that the development of any product or interface should be driven by the needs of the person or people who will ultimately use it, rather than by any design considerations that center around the object itself. A key element of usability is affordance, the notion that the appearance of any interactive element should suggest the ways in which it can be manipulated. For example, the use of shadowing around a button on a website might help make it look three-dimensional, thus suggesting that it can be pushed or clicked. Visibility is closely related to affordance; it is the notion that the function of all the controls with which a user interacts should be clearly mapped to their effects. For example, a label such as "Volume Up" beneath a button might indicate exactly what it does. Various protocols facilitate the creation of highly usable applications. A cornerstone of HCI is iterative design, a method of development that uses repeated cycles of feedback and analysis to improve each prototype version of a product, instead of simply creating a single design and launching it immediately. To learn more about the people who will eventually use a product and how they will use it, designers also make use of ethnographic field studies and usability tests.

BACKGROUND AND HISTORY

Before the advent of the personal computer, those who interacted with computers were largely technology specialists. In the 1980s, however, more and more individual users began making use of software such as word-processing programs, computer games, and spreadsheets. HCI as a field emerged from the growing need to redesign such tools to make them practical and useful to ordinary people with no technical training. The first HCI researchers came from a variety of related fields: cognitive science, psychology, computer graphics, human factors (the study of how human capabilities affect the design of mechanical systems), and technology. Among the thinkers and researchers whose ideas have shaped the formation of HCI as a science are John M. Carroll, best known for his theory of minimalism (an approach to instruction that emphasizes real-life applications and the chunking of new material into logical parts), and Adele Goldberg, whose work on early software interfaces at the Palo Alto Research Center (PARC) was instrumental in the development of the modern graphical user interface.

In the early days of HCI, the notion of usability was simply defined as the degree to which a computer system was easy and effective to use. However, usability has come to encompass a number of other qualities, including whether an interface is enjoyable, encourages creativity, relieves tension, anticipates points of confusion, and facilitates the combined efforts of multiple users. In addition, there has been a shift in HCI away from a reliance on theoretical findings from cognitive science and toward a more hands-on approach that prioritizes field studies and usability testing by real participants.

HOW IT WORKS

Input and Output Devices. The essential goal of HCI is to improve the ways in which information is transferred between a user and the machine he or she is using. Input and output devices are the basic tools HCI researchers and professionals use for this purpose. The more sophisticated the interaction between input and output devices—the more complex the feedback loop between the two directions of information flow—the more the human user will be able to accomplish with the machine.

An input device is any tool that delivers data of some kind from a human to a machine. The most familiar input devices are the ones associated with personal computers: keyboards and mice. Other commonly used devices include joysticks, trackballs, pen styluses, and tablets. Still more unconventional or elaborate input devices might take the shape of head gear designed to track the movements of a user's head and neck, video cameras that track the movements of a user's eyes, skin sensors that detect changes in body temperature or heart rate, wearable gloves that precisely track hand gestures, or automatic speech recognition devices that translate spoken commands into instructions that a machine can understand. Some input devices, such as the sensors that open

Human use of computers is a major focus of the field of HCI. By Todd Huffman

automatic doors at the fronts of banks or supermarkets, are designed to record information passively, without the user having to take any action.

An output device is any tool that delivers information from a machine to a human. Again, the most familiar output devices are those associated with personal computers: monitors, flat-panel displays, and audio speakers. Other output devices include wearable head-mounted displays or goggles that provide visual feedback directly in front of the user's field of vision and full-body suits that provide tactile feedback to the user in the form of pressure.

PERCEPTUAL-MOTOR INTERACTION. When HCI theorists speak about perceptual-motor interaction, what they are referring to is the notion that users' perceptions—the information they gather from the machine—are inextricably linked to their physical actions, or how they relate to the machine. Computer systems can take advantage of this by using both input and output devices to provide feedback about the user's actions that will help him or her make the next move. For example, a word on a website may change in color when a user hovers the mouse over it, indicating that it is a functional link. A joystick being used in a racing game may exert what feels like muscular tension or pressure against the user's hand in response to the device being steered to the left or right. Ideally, any feedback a system gives a user should be aligned to the physical direction in which he or she is moving an input device. For example, the direction in which a cursor moves on screen should be the same as the direction in which the user is moving the mouse. This is known as kinesthetic correspondence.

Another technique HCI researchers have devised to facilitate the feedback loop between a user's perceptions and actions is known as augmented reality. With this approach, rather than providing the user with data from a single source, the output device projects digital information, such as labels, descriptions, charts, and outlines, on the physical world. When an engineer is looking at a complex mechanical system, for example, the display might show what each part in the system is called and enable him or her to call up additional troubleshooting or repair information.

APPLICATIONS AND PRODUCTS

COMPUTERS. At one time, interacting with a personal computer required knowing how to use a command-line interface in which the user typed in instructions—often worded in abstract technical language—for a computer to execute. A graphical user interface, based on HCI principles, supplements or replaces text-based commands with visual elements such as icons, labels, windows, widgets, menus, and control buttons. These elements are controlled using a physical pointing device such as a mouse. For instance, a user may use a mouse to open, close, or resize a window or to pull down a list of options in a menu in order to select one. The major advantage graphical user interfaces have over text-based interfaces is that they make completing tasks far simpler and more intuitive. Using graphic images rather than text reduces the amount of time it takes to interpret and use a control, even for a novice user. This enables users to focus on the task at hand rather than to spend time figuring out how to manipulate the technology itself. For instance, rather than having to recall and then correctly type in a complicated command, a user can print a particular file by selecting its name in a window, opening it, and clicking on an icon designed to look like a printer. Similarly, rather than choosing options from a menu in order to open a certain file within an application, a user might drag and drop the icon for the file onto the icon for the application.

Besides helping individuals navigate through and execute commands in operating systems, software engineers also use HCI principles to increase the usability of specific computer programs. One example is the way pop-up windows appear in the word-processing program Microsoft Word when a user types in the salutation in a letter or the beginning item in a list. The program is designed to recognize the user's task, anticipate the needs of that task, and offer assistance with formatting customized to that particular kind of writing.

CONSUMER APPLIANCES. Besides computers, a host of consumer appliances use aspects of HCI design to improve usability. Graphic icons are ubiquitous parts of the interfaces commonly found on cameras, stereos, microwave ovens, refrigerators, and televisions. Smartphones such as Apple's iPhone rely on the same graphic displays and direct manipulation techniques as used in full-sized computers. Many also add extra tactile, or haptic, dimensions of usability such as touchscreen keyboards and the ability to rotate windows on the device by physically rotating the device itself in space. Entertainment products such as video game consoles have moved away from keyboard and joystick interfaces, which may not have kinesthetic correspondence, toward far more sophisticated controls. The hand-held device that accompanies the Nintendo Wii, for instance, allows players to control the motions of avatars within a game through the natural movements of their own bodies. Finally, HCI research influences the physical design of many household devices. For example, a plug for an appliance designed with the user in mind might be deliberately shaped so that it can be inserted into an outlet in any orientation, based on the understanding that a user may have to fit several plugs into a limited amount of space, and many appliances have bulky plugs that take up a lot of room.

Increasingly, HCI research is helping appliance designers move toward multimodal user interfaces. These are systems that engage the whole array of human senses and physical capabilities, match particular tasks to the modalities that are the easiest and most effective for people to use, and respond in tangible ways to the actions and behaviors of users. Multimodal interfaces combine input devices for collecting data from the human user (such as video cameras, sound recording devices, and pressure sensors) with software tools that use statistical analysis or artificial intelligence to interpret these data (such as natural language processing programs and computer vision applications). For example, a multimodal interface for a GPS system installed in an automobile might allow the user to simply speak the name of a destination aloud rather than having to type it in while driving. The system might use auditory processing of the user's voice as well as visual processing of his or her lip movements to more accurately interpret speech. It might also use a camera to closely follow the movements of the user's eyes, tracking his or her gaze from one part of the screen to another and using this information to helpfully zoom in on particular parts of the map or automatically select a particular item in a menu.

Similarly, in 2015, Amazon took the technology of the Bluetooth speaker one step further with its release of the Echo device. This speaker has a built-in program that allows the user to give voice commands to instruct the device to play certain music or to sync up with other applications and devices.

Workplace Information Systems. HCI research plays an important role in many products that enable people to perform workplace tasks more effectively. For example, experimental computer systems are being designed for air traffic control that will increase safety and efficiency. Such systems work by collecting data about the operator's pupil size, facial expression, heart rate, and the forward momentum and intensity of his or her mouse movements and clicks. This information helps the computer interpret the operator's behavior and state of mind and respond accordingly. When an airplane drifts slightly off its course, the system analyzes the operator's physical modalities. If his or her gaze travels quickly over the relevant area of the screen, with no change in pupil size or mouse click intensity, the computer might conclude that the operator has missed the anomaly and attempt to draw attention to it by using a flashing light or an alarm.

Other common workplace applications of HCI include products that are designed to facilitate communication and collaboration between team members, such as instant messaging programs, wikis (collaboratively edited Web sites), and videoconferencing tools. In addition, HCI principles have contributed to many project management tools that enable groups to schedule and track the progress they are making on a shared task or to make changes to common documents without overriding someone else's work.

Education and Training. Schools, museums, and businesses all make use of HCI principles when designing educational and training curricula for students, visitors, and staff. For example, many school districts are moving away from printed textbooks and toward interactive electronic programs that target a variety of information-processing modalities through multimedia. Unlike paper and pencil worksheets, such programs also provide instant feedback, making it easier for students to learn and understand new concepts. Businesses use similar programs to train employees in such areas as the use of new software and the company's policies on issues of workplace ethics. Many art and science museums have installed electronic kiosks with touchscreens that visitors can use to learn more about a particular exhibit. HCI principles underlie the design of such kiosks. For example, rather than using a text-heavy interface, the screen on an interactive kiosk at a science museum might display video of a museum staff member talking to the visitor about each available option.

SOCIAL CONTEXT AND FUTURE PROSPECTS

As HCI moves forward with research into multimodal interfaces and ubiquitous computing, notion of the computer as an object separate from the user may eventually be relegated to the archives of technological history, to be replaced by wearable machine interfaces that can be worn like clothing on the user's head, arm, or torso. Apple released its second version of a "smartwatch" in 2016, which is designed to have all of the features of smartphones in a wearable, theoretically more convenient format. Much like other wearable gadgets such as the Fitbit, playing into society's increased concern with exercise and overall health, the watch has the ability to track human components such as heart rate and serve as a GPS that can map running, walking, and biking routes. Virtual reality interfaces have been developed that are capable of immersing the user in a 360-degree space that looks, sounds, feels, and perhaps even smells like a real environment—and with which they can interact naturally and intuitively, using their whole bodies. As the capacity to measure the physical properties of human beings becomes ever more sophisticated, input devices may grow more and more sensitive; it is possible to envision a future, for instance, in which a machine might "listen in" to the synaptic firings of the neurons in a user's brain and respond accordingly. Indeed, it is not beyond the realm of possibility that a means could be found of stimulating a user's neurons to produce direct visual or auditory sensations. The future of HCI research may be wide open, but its essential place in the workplace, home, recreational spaces, and the broader human culture is assured. As further evidence of the significance of human-computer interactions and its place in modern technology, the organization Advancing Technology for Humanity held its second annual International Conference on Human-Computer Interactions in 2016.

—M. Lee, BA, MA

Bibliography

Bainbridge, William Sims, editors. *Berkshire Encyclopedia of Human-Computer Interaction: When Science Fiction Becomes Fact.* Berkshire, 2004. 2 vols.

Gibbs, Samuel, and Alex Hern. "Apple Watch 2 Brings GPS, Waterproofing and Faster Processing." *The Guardian,* 8 Sept. 2016, www.theguardian.com. Accessed 28 Oct. 2016.

Helander, Martin. *A Guide to Human Factors and Ergonomics.* 2nd ed., CRC Press, 2006.

Hughes, Brian. "Technology Becomes Us: The Age of Human-Computer Interaction." *The Huffington Post,* 20 Apr. 2016, www.huffingtonpost.com/brian-hughes/technology-becomes-us-the_b_9732166.html. Accessed 28 Oct. 2016.

Jokinen, Jussi P. P. "Emotional User Experience: Traits, Events, and States." *International Journal of Human-Computer Studies,* vol. 76, 2015, pp. 67–77.

Purchase, Helen C. *Experimental Human-Computer Interaction: A Practical Guide with Visual Examples.* Cambridge UP, 2012.

Sears, Andrew, and Julie A. Jacko, editors. *The Human-Computer Interaction Handbook: Fundamentals, Evolving Technologies, and Emerging Applications.* 2nd ed., Erlbaum, 2008.

Sharp, Heken, et al. *Interaction Design: Beyond Human-Computer Interaction.* 2nd ed., John Wiley & Sons, 2007.

Soegaard, Mads, and Rikke Friis Dam, editors. *The Encyclopedia of Human-Computer Interaction.* 2nd ed., Interaction Design Foundation, 2014.

Thatcher, Jim, et al. *Web Accessibility: Web Standards and Regulatory Compliance.* Springer, 2006.

Tufte, Edward R. *The Visual Display of Quantitative Information.* 2nd ed., Graphics, 2007.

Hydraulic Engineering

SUMMARY

Hydraulic engineering is a branch of civil engineering concerned with the properties, flow, control, and uses of water. Its applications are in the fields of water supply, sewerage evacuation, water recycling, flood management, irrigation, and the generation of electricity. Hydraulic engineering is an essential element in the design of many civil and environmental engineering projects and structures, such as water distribution systems, wastewater management systems, drainage systems, dams, hydraulic turbines, channels, canals, bridges, dikes, levees, weirs, tanks, pumps, and valves.

DEFINITION AND BASIC PRINCIPLES

Hydraulic engineering is a branch of civil engineering that focuses on the flow of water and its role in civil engineering projects. The principles of hydraulic engineering are rooted in fluid mechanics. The conservation of mass principle (or the continuity principle) is the cornerstone of hydraulic analysis and design. It states that the mass going into a control volume within fixed boundaries is equal to the rate of increase of mass within the same control volume. For an incompressible fluid 0with fixed boundaries, such as water flowing through a pipe, the continuity equation is simplified to state that the inflow rate is equal to the outflow rate. For unsteady flow in a channel or a reservoir, the continuity principle states that the flow rate into a control volume minus the outflow rate is equal to the time rate of change of storage within the control volume.

Hydraulic Flood Retention Basin (HFRB). By Qinli Yang/ Will McMinn

Energy is always conserved, according to the first law of thermodynamics, which states that energy can neither be created nor be destroyed. Also, all forms of energy are equivalent. In fluid mechanics, there are mainly three forms of head (energy expressed in unit of length). First, the potential head is equal to the elevation of the water particle above an arbitrary datum. Second, the pressure head is proportional to the water pressure. Third, the kinetic head is proportional to the square of the velocity. Therefore, the conservation of energy principle states that the potential, pressure, and kinetic heads of water entering a control volume, plus the head gained from any pumps in the control volume, are equal to the potential, pressure, and kinetic heads of water exiting the control volume, plus the friction loss head and any head lost in the system, such as the head lost in a turbine to generate electricity.

Hydraulic engineering deals with water quantity (flow, velocity, and volume) and not water quality, which falls under sanitary and environmental engineering. However, hydraulic engineering is an essential element in designing sanitary engineering facilities such as wastewater-treatment plants.

Hydraulic engineering is often mistakenly thought to be petroleum engineering, which deals with the flow of natural gas and oil in pipelines, or the branch of mechanical engineering that deals with a vehicle's engine, gas pump, and hydraulic breaking system. The only machines that are of concern to hydraulic engineers are hydraulic turbines and water pumps.

BACKGROUND AND HISTORY

Irrigation and water supply projects were built by ancient civilizations long before mathematicians defined the governing principles of fluid mechanics. In the Andes Mountains in Peru, remains of irrigation canals were found, radiocarbon dating from the fourth millennium BCE. The first dam for which there are reliable records was built before 4000 BCE on the Nile River in Memphis in ancient Egypt. Egyptians built dams and dikes to divert the Nile's floodwaters into irrigation canals. Mesopotamia (now Iraq and western Iran) has low rainfall and is supplied with surface water by two major rivers, the Tigris and the Euphrates, which are much smaller than the Nile but have more dramatic floods in the spring. Mesopotamian engineers, concerned about water storage and flood control as well as irrigation, built diversion dams and large weirs to create reservoirs and to supply canals that carried water for long distances. In the Indus Valley civilization (now Pakistan and northwestern India), sophisticated irrigation and storage systems were developed.

One of the most impressive dams of ancient times is near Marib, an ancient city in Yemen. The 1,600-foot-long dam was built of masonry strengthened by copper around 600 BCE. It holds back some of the annual floodwaters coming down the valley and diverts the rest of that water out of sluice gates and into a canal system. The same sort of diversion dam system was independently built in Arizona by the Hohokam civilization around the second or third century CE.

In the Szechwan region of ancient China, the Dujiangyan irrigation system was built around 250 BCE and still supplies water in modern times. By the second century CE, the Chinese used chain pumps, which lifted water from lower to higher elevations, powered by hydraulic waterwheels, manual foot pedals, or rotating mechanical wheels pulled by oxen.

The Minoan civilization developed an aqueduct system in 1500 BCE to convey water in tubular conduits in the city of Knossos in Crete. Roman aqueducts were built to carry water from large distances to Rome and other cities in the empire. Of the 800 miles of aqueducts in Rome, only 29 miles were above ground. The Romans kept most of their aqueducts underground to protect their water from enemies and diseases spread by animals.

The Muslim agricultural revolution flourished during the Islamic golden age in various parts of Asia and Africa, as well as in Europe. Islamic hydraulic engineers built water management technological complexes, consisting of dams, canals, screw pumps, and *norias*, which are wheels that lift water from a river into an aqueduct.

The Swiss mathematician Daniel Bernoulli published *Hydrodynamica* (1738; *Hydrodynamics by Daniel Bernoulli*, 1968), applying the discoveries of Sir Isaac Newton and Gottfried Wilhelm Leibniz in mathematics and physics to fluid systems. In 1752, Leonhard Euler, Bernoulli's colleague, developed the more generalized form of the energy equation.

In 1843, Adhémar-Jean-Claude Barré de Saint Venant developed the most general form of the differential equations describing the motion of fluids, known as the Saint Venant equations. They are sometimes called Navier-Stokes equations after Claude-Louis Navier and Sir George Gabriel Stokes, who were working on them around the same time.

The German scientist Ludwig Prandtl and his students studied the interactions between fluids and solids between 1900 and 1930, thus developing the boundary layer theory, which theoretically explains the drag or friction between pipe walls and a fluid.

HOW IT WORKS

Properties of Water. Density and viscosity are important properties in fluid mechanics. The density of a fluid is its mass per unit volume. When the temperature or pressure of water changes significantly, its density variation remains negligible. Therefore, water is assumed to be incompressible. Viscosity, on the other hand, is the measure of a fluid's resistance to shear or deformation. Heavy oil is more viscous than water, whereas air is less viscous than water. The viscosity of water increases with reduced temperatures. For instance, the viscosity of water at its freezing point is six times its viscosity at its boiling temperature. Therefore, a flow of colder water assumes higher friction.

Hydrostatics. Hydrostatics is a subdiscipline of fluid mechanics that examines the pressures in water at rest and the forces on floating bodies or bodies submerged in water. When water is at rest, as in a tank or a large reservoir, it does not experience shear stresses; therefore, only normal pressure is present. When the pressure is uniform over the surface of a body in water, the total force applied on the body is a product of its surface area times the pressure. The direction of the force is perpendicular (normal) to the surface. Hydrostatic pressure forces can be mathematically determined on any shape. Buoyancy, for instance, is the upward vertical force applied on floating bodies (such as boats) or submerged ones (such as submarines). Hydraulic engineers use hydrostatics to compute the forces on submerged gates in reservoirs and detention basins.

Fluid Kinematics. Water flowing at a steady rate in a constant-diameter pipe has a constant average velocity. The viscosity of water introduces shear stresses between particles that move at different velocities. The velocity of the particle adjacent to the wall of the pipe is zero. The velocity increases for particles away from the wall, and it reaches its maximum at the center of the pipe for a particular flow rate or pipe discharge. The velocity profile in a pipe has a parabolic shape. Hydraulic engineers use the average velocity of the velocity profile distribution, which is the flow rate over the cross-sectional area of the pipe.

Bernoulli's Theorem. When friction is negligible and there are no hydraulic machines, the conservation of energy principle is reduced to Bernoulli's equation, which has many applications in pressurized flow and open-channel flow when it is safe to neglect the losses.

APPLICATIONS AND PRODUCTS

Water Distribution Systems. A water distribution network consists of pipes and several of the following components: reservoirs, pumps, elevated storage tanks, valves, and other appurtenances such as surge tanks or standpipes. Regardless of its size and complexity, a water distribution system serves the purpose of transferring water from one or more sources to customers. There are raw and treated water systems. A raw water network transmits water from a storage reservoir to treatment plants via large pipes, also called transmission mains. The purpose of a treated water network is to move water from a water-treatment plant and distribute it to water retailers through transmission mains or directly to municipal and industrial customers through smaller distribution mains.

Some water distribution systems are branched, whereas others are looped. The latter type offers more reliability in case of a pipe failure. The hydraulic engineering problem is to compute the steady velocity or flow rate in each pipe and the pressure at each junction node by solving a large set of continuity equations and nonlinear energy equations that characterize the network. The steady solution of a branched network is easily obtained mathematically; however, the looped network initially offered challenges to engineers. In 1936, American structural engineer Hardy Cross developed a simplified method that tackled networks formed of only pipes. In the 1970's and 1980's, three other categories of numerical methods were developed to provide solutions for complex networks with pumps and valves. In 1996, engineer Habib A. Basha and his colleagues offered a perturbation solution to the nonlinear set of equations in a direct, mathematical fashion, thus eliminating the risk of divergent numerical solutions.

Hydraulic Transients in Pipes. Unsteady flow in pipe networks can be gradual; therefore, it can be modeled as a series of steady solutions in an extended period simulation, mostly useful for water-quality analysis. However, abrupt changes in a valve position, a sudden shutoff of a pump because of power failure, or a rapid change in demand could cause a hydraulic transient or a water hammer that travels back and forth in the system at high speed, causing large pressure fluctuations that could cause pipe rupture or collapse.

The solution of the quasi-linear partial differential equations that govern the hydraulic transient problem is more challenging than the steady network solution. The Russian scientist Nikolai Zhukovsky offered a simplified arithmetic solution in 1904. Many other methods–graphical, algebraic, wave-plane analysis, implicit, and linear methods, as well as the method of characteristics–were introduced between the 1950's and 1990's. In 1996, Basha and his colleagues published another paper solving the hydraulic transient problem in a direct, noniterative fashion, using the mathematical concept of perturbation.

Open-Channel Flow. Unlike pressure flow in full pipes, which is typical for water distribution systems, flow in channels, rivers, and partially full pipes is called gravity flow. Pipes in wastewater evacuation and drainage systems usually flow partially full with a free water surface that is subject to atmospheric pressure. This is the case for human-built canals and channels (earth or concrete lined) and natural creeks and rivers.

The velocity in an open channel depends on the area of the cross section, the length of the wetted perimeter, the bed slope, and the roughness of the channel bed and sides. A roughness factor is estimated empirically and usually accounts for the material, the vegetation, and the meandering in the channel.

Open-channel flow can be characterized as steady or unsteady. It also can be uniform or varied flow, which could be gradually or rapidly varied flow. A famous example of rapidly varied flow is the hydraulic jump.

When high-energy water, gushing at a high velocity and a shallow depth, encounters a hump, an obstruction, or a channel with a milder slope, it cannot sustain its supercritical flow (characterized by a Froude number larger than 1). It dissipates most of its energy through a hydraulic jump, which is a highly turbulent transition to a calmer flow (subcritical flow with a Froude number less than 1) at a higher depth and a much lower velocity. One way to solve for the depths and velocities upstream and downstream of the hydraulic jump is by applying the conservation of momentum principle, the third principle of fluid mechanics and hydraulic engineering. The hydraulic jump is a very effective energy dissipater that is used in the designs of spillways.

Hydraulic Structures. Many types of hydraulic structures are built in small or large civil engineering projects. The most notable by its size and cost is the dam. A dam is built over a creek or a river, forming a reservoir in a canyon. Water is released through an outlet structure into a pipeline for water supply or into the river or creek for groundwater recharge and environmental reasons (sustainability of the biological life in the river downstream). During a large flood, the reservoir fills up and water can flow into a side overflow spillway–which protects the integrity of the face of the dam from overtopping–and into the river.

The four major types of dams are gravity, arch, buttress, and earth. Dams are designed to hold the immense water pressure applied on their upstream face. The pressure increases as the water elevation in the reservoir rises.

Hydraulic Machinery. Hydraulic turbines transform the drop in pressure (head) into electric power. Also, pumps take electric power and transform it into water head, thereby moving the flow in a pipe to a higher elevation.

There are two types of turbines, impulse and reaction. The reaction turbine is based on the steam-powered device that was developed in Egypt in the first century CE by Hero of Alexandria. A simple example of a reaction turbine is the rotating lawn sprinkler.

Pumps are classified into two main categories, centrifugal and axial flow. Pumps have many industrial, municipal, and household uses, such as boosting the flow in a water distribution system or pumping water from a groundwater well.

SOCIAL CONTEXT AND FUTURE PROSPECTS

In the twenty-first century, hydraulic engineering has become closely tied to environmental engineering. Reservoir operators plan and vary water releases to keep downstream creeks wet, thus protecting the biological life in the ecosystem.

Clean energy is the way to ensure sustainability of the planet's resources. Hydroelectric power generation is a form of clean energy. Energy generated by ocean waves is a developing and promising field, although wave power technologies still face technical challenges.

—Bassam Kassab, BEng., MEng., MSc

Bibliography

Basha, Habib A. "Nonlinear Reservoir Routing: Particular Analytical Solution." *Journal of Hydraulic Engineering* 120, no. 5 (May, 1994): 624-632.

Basha, Habib A., and W. El-Asmar. "The Fracture Flow Equation and Its Perturbation Solution." *Water Resources Research* 39, no. 12 (December, 2003): 1365.

Boulos, Paul F. "H2ONET Hydraulic Modeling." *Journal of Water Supply Quarterly, Water Works Association of the Republic of China (Taiwan)* 16, no. 1 (February, 1997): 17-29.

Boulos, Paul F., Kevin E. Lansey, and Bryan W. Karney. *Comprehensive Water Distribution Systems Analysis Handbook for Engineers and Planners.* 2d ed. Pasadena, Calif.: MWH Soft Press, 2006.

Chow, Ven Te. *Open-Channel Hydraulics.* 1959. Reprint. Caldwell, N.J.: Blackburn, 2008. Contains chapters on uniform flow, varied flow, and unsteady flow.

Finnemore, E. John, and Joseph B. Franzini. *Fluid Mechanics With Engineering Applications.* 10th international ed. Boston: McGraw-Hill, 2009.

Walski, Thomas M. *Advanced Water Distribution Modeling and Management.* Waterbury, Conn.: Haestad Methods, 2003.

HYPERTEXT MARKUP LANGUAGE (HTML)

SUMMARY

HTML, which stands for hypertext markup language, is a code used by computer programmers to control the functionality of Web pages. It was invented in the late twentieth century and has become an indispensable part of online technology.

HISTORY

Timothy Berners-Lee developed HTML over several years during the 1980s. Berners-Lee was a computer consultant at a nuclear physics lab in Switzerland. There he developed an early program that allowed someone reading one document to connect, or "link," to another document through a network. His invention was called *hypertext.* Later in the decade, he invented systems that enabled computers within the same network to communicate with each other. And, in 1989, he proposed developing a network that would allow computers all over the world to connect. Rather than communicating by email, files could be located on a page on the Web, where they could be accessed by anyone with the right technology and credentials_____the code that would make this possible was HTML. Although the specific name for this language has changed significantly since its invention, its purpose has remained the same: to facilitate the smooth operation of web pages of all types.

Tim Berners-Lee. By Enrique Dans from Madrid, Spain (Con Tim Berners-Lee)

FUNCTION

Although the "language" of HTML is not necessarily second-nature to all users, it is extremely systematic, with a few basic, important components. The file in which HTML code is written is called a *document,* which usually has an .html extension. The word "html" is in brackets that look like this –< >–at the beginning and end of a document. The words inside the brackets are called *tags.* Tags are a crucial component of an HTML document.

Tags are generally in pairs—the first tag, called the *start tag,* looks like this: <tag>, and the second tag, called the *end tag,* looks like this: </tag>. *Keywords,* the text between tags, depend on the content of a page. Keywords may be of many different types, with many different degrees of significance. Many tags will simply designate the exact words and phrases that will

Logo for HTML code. By New MooonHabuhiah at fr.wikipedia

appear on a page. For example, the words <title>My Favorite Job Ever</title> would probably appear at the top of the screen on the web version of an html document. This code might appear in the first paragraph under the title: <p> I have had many jobs. But one stood out as a favorite.</p>

It is important to remember that HTML is a unique language. The text within tags is read and interpreted by a computer's software, so the rules of the language must be followed. For example, the tag <!DOCTYPE html> must be used at the beginning of an HTML document to indicate the document type. When an HTML document is opened within a web browser, the language of the document guides the appearance and function of the page. The tags <title>text</title> indicate the title of a document, <body> and </body> indicate the start and end of the main text of a document, and <p>text</p> indicate the starting and ending points of a paragraph within a document. To place headers within a document, these tags must be used: <h1> and </h1>. To create headers of different sizes, these tags are used: <h2> and </h2>, <h3> and </h3>. Images to be inserted in a document are preceded and followed with the tags <img> and </img>.

The characteristics of an HTML document, such as the font, the color of the font, or the size of an image, are called *attributes*. They are usually placed between the brackets of a start tag and are followed by an equals (=) sign; the exact attribute itself is in quotes. For example, to display an image that is 600 pixels wide, this text would be used: width="600". Attributes such as color or font are referred to as *styles* and they must be designed in a specific way. A font is indicated with the phrase "font-family." For example, to create a title in Times font, these tags would be used: <title font-family="Times">The Best Job I Ever Had</title>. Every element of a Web page and every characteristic of every element has its own designation in HTML. HTML has codes for tables, animations, different tabs within a page, and the size and shape of a page. Once an HTML document is complete, it is opened in a browser, which is when the person writing the HTML code can see whether or not the code has been properly written.

As time has passed, HTML has been modified and used in different ways, such as in mobile devices. A notable development is XHTML. The "X" stands for "Extensible," indicating that the language can be used in a wider variety of situations than web development. As programmers continue to write code and as the Internet continues to become more pervasive in daily life, the true significance of Berner-Lee's invention becomes more and more evident.

—*Max Winter*

BIBLIOGRAPHY

Boudreaux, Ryan. "HTML 5 Trends and What They Mean for Developers. Web Designer. 9 Feb. 2012. Web. 28 Aug, 2015. http://www.techrepublic.com/blog/web-designer/html-5-trends-and-what-they-mean-for-developers/

"The History of HTML." Ironspider. Iron Spider. Web. 20 Aug. 2015. http://www.ironspider.ca/webdesign101/htmlhistory.htm

"The History of HTML." Landofcode.com. Web. 20 Aug. 2015. http://landofcode.com/html-tutorials/html-history.php

"HTML Basics." MediaCollege.com. Web 20 Aug. 2015. http://www.mediacollege.com/internet/html/html-basics.html

"HTML Introduction." W3Schools.com. W3 Schools. Web. 28 Aug. 2015. http://www.w3schools.com/html/html_intro.asp

"HTML Tutorial." Refsnes Data. Web. 20 Aug. 2015. http://www.w3schools.com/html/default.asp

Mobbs, Richard. "HTML Developments." University of Leicester. Oct. 2009. Web. 28 Aug. 2015. http://www.le.ac.uk/oerresources/bdra/html/page_04.htm

"A Short History of HTML." W3C-HTML.com. W3C Foundation. Web. 28 Aug. 2015. http://w3c-html.com/html-history.html

I

Integral

SUMMARY

The development of integral calculus was motivated by two seemingly disparate mathematical problems: finding a function whose derivative is known; and finding the area between the graph of a function and the x-axis over an interval $[a,b]$. The first is called the antiderivative problem, and the second is called the area problem.

Integral calculus, or integration, has been studied since the third century BCE when Archimedes presented his method of exhaustion to calculate the area of a circle. Integral calculus, as it is known today, came into sharper focus in the seventeenth century AD with the works of Sir Isaac Newton and Gottfried Leibniz, who each developed modern calculus independently.

Leibniz' notation is the most widely used one today, and that is what will be used in this article. The indefinite integral, or antiderivative, of a function $f(x)$ is denoted by:

$\int f(x)\,dx,$

where $\int$ is an integral sign, and dx is the differential of x. This integral arises in the first of the two problems mentioned above. For the definite integral of a function on an interval $[a,b]$, the notation is:

$$\int_a^b f(x)\,dx.$$

This integral arises in the second of the two aforementioned problems.

PRACTICAL APPLICATIONS

From differential calculus it is known that $2x$ is the unique derivative of x^2. Antidifferentiation can be loosely thought of as an inverse operation to differentiation. This means that x^2 is an antiderivative of $2x$ because $\frac{d}{dx}(x^2) = 2x$. However, x^2 is not the only antiderivative of $2x$, because also, so x^{2+1} is also an antiderivative of $2x$. In fact, for any constant C, x^2+C is an antiderivative of $2x$. This implies that antiderivatives are not unique, as derivatives are. In general terms the antiderivative problem can be stated as follows: Given $f(x)$, find a family of functions $y = F(x) + C$ such that $f(x) = \frac{dy}{dx}$. The problem is approached as follows.

$$f(x) = \frac{dy}{dx}$$

$$f(x)dx = dy$$

$$\int f(x)\,dx = \int dy$$

The solution is on the right, as $\int dy = y + C = F(x) + C$. The final result is:

$\int f(x)\,dx = F(x) + C.$ (1)

Using (1), the earlier example can be rewritten as follows.

$\int 2x dx = x^2 + C$

Now consider the problem of finding the area A between the graph of $f(x) = x^2 + 1$ and the x-axis on the interval $[0,2]$. The problem is attacked by filling the region with rectangles. See Figure 1.

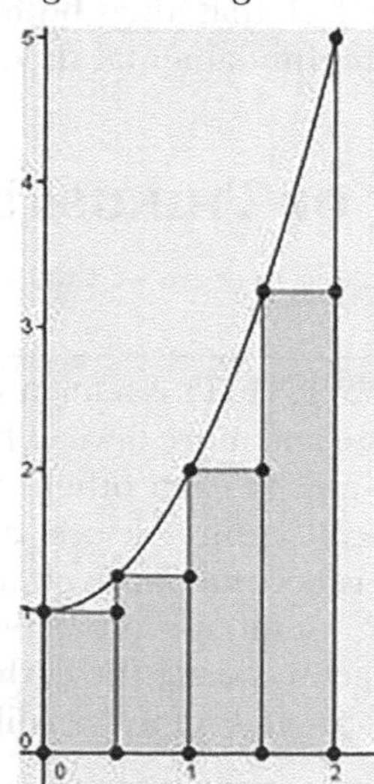

The base of each rectangle is

$$\Delta x = \frac{2-0}{4} = \frac{1}{2}$$

and the heights of the rectangles are

$$f(0), f\left(\frac{1}{2}\right), f(1), \text{ and } f\left(\frac{3}{2}\right).$$

The area A is approximated by summing the areas of the rectangles.

$$A \approx f(0)\Delta x + f\left(\frac{1}{2}\right)\Delta x + f(1)\Delta x + f\left(\frac{3}{2}\right)\Delta x = \frac{15}{4}$$

The exact area A is found by taking the limit as the number of rectangles $n \to \infty$, which gives precisely the definite integral of $f(x) = x^2 + 1$ over [0,2]:

$$\int_0^2 \left(x^2+1\right)dx = A.$$

In general, let $f(x)$ be a function on $[a,b]$, and let $[a,b]$ be divided into n subintervals by a partition Δ, given as follows:

$$\Delta : a = x_0 < x_1 < \ldots x_{n-1} < x_n = b.$$

Let c_1 be any point in the ith subinterval, $\Delta x_i = x_i - x_{i-1}$ be the width of the ith subinterval, and $\|\Delta\|$ be the width of the widest subinterval. Then the definite integral of f on $[a,b]$ is defined as:

$$\int_a^b f(x)\,dx = \lim_{\|\Delta\| \to 0} \sum_{i=1}^{n} f(c_i)\Delta x_i,$$

provided the limit exists.

On the surface it appears that the indefinite and definite integrals have nothing to do with each other apart from the fact that they both have the integral sign. But the fundamental theorem of calculus teaches us otherwise. It states that if f is continuous on $[a,b]$ and if F is an antiderivative of f on $[a,b]$, then:

$$\int_a^b f(x)\,dx = F(b) - F(a).$$

The significance of this theorem is that it unites the concepts of definite and indefinite integration, and its importance cannot be overstated. As an example of an application of (2), the exact area of the region in the last paragraph is found as follows.

$$A = \int_0^2 (x^2+1)\,dx$$

$$A = \left(\frac{1}{3}x^3 + x\right)_{x=2} - \left(\frac{1}{3}x^3 + x\right)_{x=0}$$

$$A = \frac{14}{3}$$

—*Thomas M. Mattson, MS*

BIBLIOGRAPHY

Anton, Howard, Irl Bivens, and Stephen Davis. *Calculus.* Hoboken, NJ: Wiley, 2012.

Archimedes, and Thomas L. Heath. *The Works of Archimedes.* Mineola, NY: Dover, 2002.

LaTorre, D. R. *Calculus Concepts: An Informal Approach to the Mathematics of Change.* Boston: Cengage, 2012.

Larson, Ron, and Bruce H. Edwards. *Calculus.* Boston: Cengage, 2014.

Ross, Kenneth A. *Elementary Analysis: The Theory of Calculus.* New York: Springer, 2013.

Stewart, James. *Calculus: Early Transcendentals.* Belmont, CA: Cengage, 2012.

INTERNET OF THINGS (IOT)

SUMMARY

The Internet of Things (IoT) is a term used to describe the trend of more and more devices being connected to the Internet and to each other. As smartphones, wearable data collection devices, and other "connected" products become more common, the IoT is becoming more useful and more popular. The IoT became a real factor during the early 2010s, as high-speed Internet became more readily available and hardware and software became less expensive to produce. It has been said that IoT could potentially affect all devices with an on/off switch. The impact of the IoT will most likely be far-reaching, as it will affect people's personal lives as well as businesses.

WHAT IS THE IOT?

Objects that can be connected using the IoT include almost anything that can be turned on and off. These devices include home security systems, vehicles, medical devices, appliances, smartphones, and much more. Some examples of the IoT at work would be a thermostat that a homeowner can change using a smartphone or a wearable device that tells a person how far she ran during a workout.

The IoT depends a great deal on data. Many modern devices collect and track data. When devices are connected through the IoT, they can transmit the data they collect. The IoT will help devices connect with people and other devices. In addition, it will help people connect with other people.

THE EMERGENCE OF THE IOT

The IoT has become possible for a number of reasons. In recent years, broadband Internet became widely available, allowing more people to connect to the Web. As more people become connected, the demand for data and connectivity increases. At the same time, the cost of hardware and software production has decreased. Another reason the IoT has become more prevalent is that more people are using smartphones, wearable data collection devices, and other "smart" products. Computing and technology now gives people access to the Internet and data at all times and in all places.

People are also becoming more aware of the IoT's importance. The IoT will have far-reaching effects, and many people believe that it will also have great economic value. Connected devices that send information can help reduce the amount of time that humans spend collecting and analyzing data. In other words, the IoT can help businesses and people be more efficient and save them money.

EFFECTS OF THE IOT

The IoT could possibly affect every business sector in the world. The fields that will most likely see important effects include manufacturing and production, health and medicine, and transportation.

MANUFACTURING AND PRODUCTION

The IoT will affect many business sectors, but it has the potential to change manufacturing and production greatly. A manufacturing facility could use the IoT to reorder supplies when they are running low without a human having to check inventory.

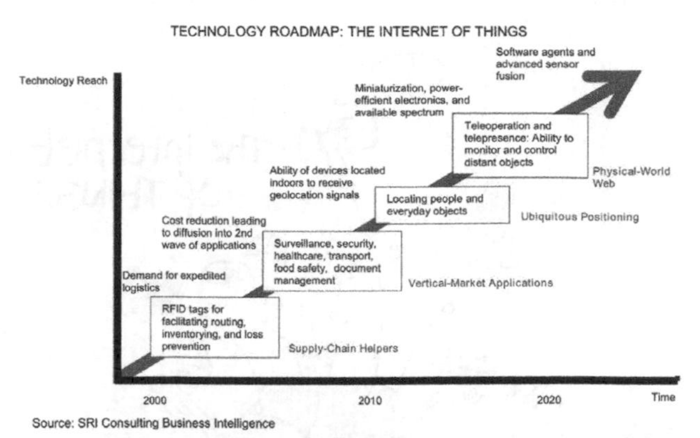

A technology roadmap of the Internet of Things. By SRI Consulting Business Intelligence/National Intelligence Council

Company machines connected to the IoT could inform an employee when they need to be repaired.

HEALTH AND MEDICINE

Wearable devices that track exercise and fitness are already a significant part of the IoT. Other medical devices connected to the IoT have the potential to be even more important for health and medicine. It is possible that someday scanning devices could analyze results and help doctors prescribe treatments for patients. Medical devices connected to the IoT could also connect doctors located in different parts of the world. These types of collaborations could help patients receive improved care.

TRANSPORTATION

The IoT could help automobiles locate empty parking spaces. In addition, cars could be fitted with technology that would automatically send a message to friends, family members, or colleagues when drivers are stuck in heavy traffic. IoT technology might also help airplanes avoid deadly collisions or enable the planes to report when they require maintenance.

EVERYDAY EFFECTS

Most people can see the effects that the IoT will have on business. However, the IoT also has the potential to impact everyday life. Homeowners are already using security systems and smoke detectors that can be accessed from smartphones. People could also see their lives change in other small ways thanks to the IoT. For example, a refrigerator could tell people when they are running low on milk or eggs. Or a person could

An artistic representation of the Internet of Things. By Wilgengebroed on Flickr

locate a wallet, purse, phone, remote control, or lost pet using an app on a computer or smartphone.

The IoT could also affect towns and cities. Smart cities could use the IoT to control traffic lights to make roads less congested. Devices connected with the IoT could also track water and air pollution in an attempt to identify the causes. Cities could easily provide visitors with information about local attractions, including real-time data. For instance, an amusement park could provide guests with data about the current wait times at particular attractions.

SECURITY CONCERNS AND THE IOT

The IoT relies a great deal on the sharing of data. Because of that, some excerpts are concerned about data security on the IoT. Some people estimate that more than 100 billion devices will be connected with the IoT by the year 2020. This gives hackers access to potentially sensitive information, including medical records, city utility details, and more. Experts have pointed out that many devices are unencrypted (which means the data is easy for hackers to read and understand) and many devices and software platforms do not require complex passwords. To ensure data is secure in the future, people, businesses, and governments will have to make Internet security a priority and develop new ways to protect data on the IoT.

—Elizabeth Mohn

BIBLIOGRAPHY

"Create the Internet of Your Things." *Microsoft.* Microsoft. Web. 3 Feb. 2015. http://www.microsoft.com/en-us/server-cloud/internet-of-things.aspx

"The Internet of Things Research Study." *HP.* Hewlett-Packard Development Company, L.P. Sept. 2014. Web. 3 Feb. 2015. http://www8.hp.com/h20195/V2/GetPDF.aspx/4AA5-4759ENW.pdf

Miller, Zach. "Move Over Internet of Things, Here Is Pixie's Location of Things." *Forbes.* Forbes.com LLC. 3 Feb. 2015. Web. 3 Feb. 2015. http://www.forbes.com/sites/zackmiller/2015/02/03/move-over-internet-of-things-here-is-pixies-location-of-things/

Morgan, Jacob. "Everything You Need to Know about the Internet of Things." *Forbes.* Forbes.com LLC. 30 Oct. 2015. Web. 3 Feb. 2015. http://www.microsoft.com/en-us/server-cloud/internet-of-things.aspx

Morgan, Jacob. "A Simple Explanation of 'The Internet of Things.'" *Forbes.* Forbes.com LLC. 13 May 2015. Web. 3 Feb. 2015. http://www.forbes.com/sites/jacobmorgan/2014/05/13/simple-explanation-internet-things-that-anyone-can-understand/

INTEROPERABILITY

SUMMARY

Interoperability describes the capacity different systems, products, or vendors have to interact with each other. Perfectly interoperable systems can exchange and interpret information with complete freedom. The term most commonly applies to information technology, computer systems, and computer programs, but it can also refer to different organizations, such as law enforcement. The medical industry, government programs, and software development are sectors that place emphasis on interoperability.

Achieving interoperability requires a few different factors. Sufficient technology needs to be in place to make it feasible. There also need to be standards that are easily recognizable and usable to different system developers. Economic considerations are also important. Often, particular systems are tied up by copyright or licensing agreements, preventing developers from settling on a single standard. Business process interoperability (BPI) is a specific type of interoperability designed to improve workloads and projects. It focuses on automating work as much as possible and only involving live decisions when necessary.

BACKGROUND

The spread of interoperability is tied closely with the development of the Internet, which greatly improved the capacity for people to communicate and exchange data over long distances. At first, the Internet was primarily used by governments and academics. In 1989, the creation of the World Wide Web made the Internet far more accessible to average civilians. With a wide range of people with different needs and interests online, and personal computers becoming much more common, a wider variety of both computers and software emerged. Whether these different systems could share data with each other became a concern for businesses, academics, the government, and consumers alike.

As computers became a more common resource for storing and sending information in the late twentieth century and into the twenty-first, the health industry took a particular interest in interoperability. Proponents argued that interoperability would help medical research advance much more efficiently while also improving day-to-day operations and drastically cutting costs at hospitals. However, with different developers eager for their products to stand out from others, licensing and trademark issues often interfered with free information exchange.

The Human Genome Project demonstrated some of the benefits of interoperability. This project began in 1990 as an effort to map the sequences of genetic code that make up human deoxyribonucleic acid (DNA) and involved collaboration on a global scale. In 1996, conflict over the patent rights of findings arose, and the project leaders met to discuss how

The Office for Interoperability and Compatibility of the Department of Homeland Security aims to integrate public safety systems. By DHS, as noted below. (http://www.uscg.mil/)

to avoid these issues in the future. They established that all project participants must agree to share all findings and data with all other laboratories within twenty-four hours of discovery. This decision led to accelerated progress, and the project was considered completed in 2003.

The success of the Human Genome Project helped bring about Creative Commons, an organization dedicated to licensing works so that they are readily accessible while protecting the creators' interests. It was a high-profile case of information exchange being prioritized over licensing restrictions, but it largely applied to published materials. Demand for interoperable systems for the exchange of information significantly increased into the twenty-first century.

LEVELS OF INTEROPERABILITY

Interoperability can be categorized into three different levels: foundational, structural, and semantic. Foundational refers to systems that can transfer data from one system to another. However, the data may arrive in a different format and need to be reinterpreted on the receiving end. The structural level describes a system that can transfer data without need for reinterpretation. At the semantic level, data can be transferred in a state where it is instantly usable afterward.

The George W. Bush administration gave significant support to the health industry's cause in 2004. It established the Office of the National Coordinator for Health Information Technology. The ONC quickly went to work, creating the Nationwide Health Information Network. The NwHIN was designed to encourage the exchange of medical information

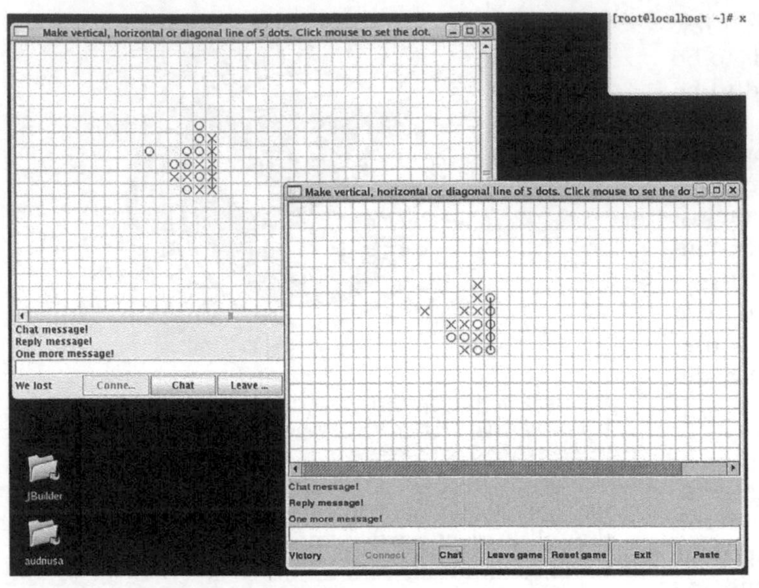

Software interoperability allows for communication between different host systems. By Audriusa at en.wikipedia (Transferred from en.wikipedia)

among a large assortment of resources, including different government agencies and private entities. The ONC went on to establish two committees: one to provide standards to improve data exchange, and the other to reach out to businesses and help persuade them to support the unrestricted exchange of information. The committees gained some support, but both soon folded. Under the Barack Obama administration, the ONC was rejuvenated with funding and new legal power. It formed the Standards and Interoperability Framework, a more advanced version of the NwHIN. It led to the development of programs that allowed patients to access their medical records with ease.

In 2015, the U.S. government revealed an initiative to invest in what it called smart cities. This program was designed to emphasize technology and cooperation between local governments and businesses to help improve quality of life. According to the initiative, smart cities should have technology that can collect data efficiently and transfer it immediately to the appropriate sources to help run different city services. This includes traffic-monitoring systems, crime alerts, public transportation, and delivery services. It also calls for city governments to be in direct communication with universities, businesses, and fellow cities. This involved significant financial investments into improving infrastructure in chosen cities and investments into research. The European Telecommunications Standards Institute (ETSI) has helped make strides toward the vision of smart cities. It developed technology that could interface with different city software programs and provide a standard that would allow them to communicate easier.

The development of mobile phones in the mid-2000s had a significant impact on interoperability. In a short time, mobile phones with numerous built-in features, the ability to download vast amounts of software applications, and the ability to interact with other devices became common. The sheer volume of different products and applications has resulted in many interoperable systems, but it has also led to many legal restrictions.

By the 2010s, several governments offered incentives or requirements to embrace interoperability. Its benefits to businesses and consumers alike were well known, especially in the health industry. Economists generally considered businesses that ran on interoperable technology more efficient. Even technology itself was not seen as the obstacle to interoperability that it was in the twentieth century. However, many technology developers did not stand to benefit from giving extra time, effort, and money to ensure that their product could work with a competitor's offerings. Proponents often advocate for more incentives to be aimed at developers to adopt interoperability rather than at businesses that use their products.

—Reed Kalso

BIBLIOGRAPHY

"Business Process Interoperability Framework." *Australian Government Department of Finance*, www.finance.gov.au/archive/policy-guides-procurement/interoperability-frameworks/bpif/. Accessed 23 Jan. 2017.

Gur-Arie, Margalit. "The History of Healthcare Interoperability." *HIT*, 11 Apr. 2013, hitconsultant.net/2013/04/11/history-of-healthcare-interoperability/. Accessed 23 Jan. 2017.

"How Did the Human Genome Project Make Science More Accessible?" *YourGenome.org*, 13 June 2016, www.yourgenome.org/stories/how-did-the-human-genome-project-make-science-more-accessible. Accessed 23 Jan. 2017.

"Impact of eHealth Exchange on Nationwide Interoperability." *HealthIT Interoperability*, 10 Jan. 2017, healthitinteroperability.com/news/impact-of-ehealth-exchange-on-nationwide-interoperability. Accessed 23 Jan. 2017.

"Making Technology Talk: How Interoperability Can Improve Care, Drive Efficiency, and Reduce Waste." *Center for Medical Interoperability*, 27 Apr. 2016, medicalinteroperability.org/making-technology-talk-how-interoperability-can-improve-care-drive-efficiency-and-reduce-waste/.Accessed 23 Jan. 2017.

Mostashari, Farzad. "Mostashari: Value-Based Care Demands Free-Flowing Data." *Healthcare IT News*, 20 Jan. 2017, www.healthcareitnews.com/news/mostashari-value-based-care-demands-free-flowing-data. Accessed 23 Jan. 2017.

"NEC Pushes Improved Interoperability for Smart City Solutions." *Enterprise Innovation*, 18 Jan. 2017, www.enterpriseinnovation.net/article/nec-pushes-improved-interoperability-smart-city-solutions-1519206727. Accessed 23 Jan. 2017.

Sutner, Shaun. "Marc Probst Calls for Government Healthcare Interoperability Standards." *TechTarget*, 29 Dec. 2016, searchhealthit.techtarget.com/video/Marc-Probst-calls-for-government-healthcare-interoperability-standards. Accessed 23 Jan. 2017.

"Technology Not Enough for Interoperability." *IT-Online*, 17 Jan. 2017, it-online.co.za/2017/01/17/technology-not-enough-for-interoperability/. Accessed 23 Jan. 2017.

"What Is Interoperability?" *HIMSS*, www.himss.org/library/interoperability-standards/what-is-interoperability. Accessed 23 Jan. 2017.

INTERVAL

SUMMARY

An interval is a set of real numbers, defined as those numbers between (and by default, including) two designated endpoints—though one of the endpoints may be infinite, which means that the real numbers themselves, taken as a set, are an interval, as is the set of all positive real numbers. The empty set is also an interval. Typically, though, an interval will be given as the numbers between *a* and *b*.

The conventional representation of an interval is [*a*, *b*], where *a* and *b* are the endpoints and *a* is usually the smaller endpoint. It is usually assumed that the interval includes the endpoints and the numbers between them: that [*a*, *b*] is the set of numbers greater than or equal to *a* and less than or equal to *b*. To exclude an endpoint, the bracket is usually reversed: [*a*, *b*]; [*a*, *b*[; or]*a*, *b*[, for instance. In some contexts, parentheses are used instead of brackets to indicate intervals excluding their endpoints. In some publications—because some countries use a decimal comma in numeric notation—a semicolon is used instead of a comma to separate the endpoints of the interval. In some cases when an interval consists only of the integers in the set (especially common in computer programming), the notation [*a* . *b*] is used instead of a comma.

PRACTICAL APPLICATIONS

Intervals are frequently encountered in everyday life. Children entering first grade are often limited to those who turn 6 between November 1 (for instance) of one year and October 31 of the next, essentially an interval consisting of a set of birthdays. Public streets permit driving within an interval bound by the speed limit at one endpoint and an unofficial minimum speed at the other endpoint. The possible price of a particular product in a particular area is an interval, the endpoints of which are determined by various economic factors: For instance, the maximum price (one endpoint) of a book is its cover price, but it may be put on sale by the bookstore for as little as 50% of that price (the other endpoint) or may be found used for even less. And gasoline varies by price even on the same city block, as gas stations compete with one another for customers.

Intervals that exclude the endpoints are often called open intervals, while endpoint-inclusive intervals are closed. If its smallest endpoint is not negative infinity, it is left-bounded; if its largest endpoint is not infinity, it is right-bounded; by extension, if it is both left- and right-bounded, both of its endpoints are finite real numbers—and it is called a bounded interval or finite interval. An interval consisting of only a single real number is a degenerate interval, which some mathematicians consider to include the empty set. A real interval that is neither degenerate nor the empty set is called a proper interval, and with the exception of the computer programming cases mentioned above, consists of an infinite number of elements.

—Bill Kte'pi, MA

BIBLIOGRAPHY

Aigner, Martin, and Gunter M. Ziegler. *Proofs from the Book.* New York: Springer, 2014.

Dshalalow, Eugene. *Foundations of Abstract Analysis.* New York: Springer, 2014.

Grunbaum, Branko, and G. C. Shephard. *Tilings and Patterns.* New York: Dover, 2015.

Hanna, Gila. *Explanation and Proof in Mathematics.* New York: Springer, 2014.

Larson, Ron. *Algebra and Trigonometry.* Boston: Cengage, 2015.

Millman, Richard, Peter Shiue, and Eric Brendan Kahn. *Problems and Proofs in Numbers and Algebra.* New York: Springer, 2015.

K

Kinematics

SUMMARY

Kinematics is the branch of classical mechanics concerned with the motion of particles or objects without explicit consideration of the masses and forces involved. Kinematics focuses on the geometry of motion and the relationship between position, velocity, and acceleration vectors of particles as they move. Kinematics is a subfield of classical mechanics, along with statics (the study of physical systems for which the forces are in equilibrium) and dynamics (the study of objects in motion under the influence of unequilibrated forces). In practice, kinematic equations appear in fields as diverse as astrophysics (e.g., to describe planetary orbits and the motion of other celestial bodies) and robotics (e.g., to describe the motion of an articulated arm on an assembly line).

MOTION OF A PARTICLE IN ONE DIMENSION

Kinematics focuses on the geometry of the motion of a particle, or point-like object, by investigating the relationship between position, velocity, and acceleration vectors of a particle without involving mass or force.

To describe mathematically the motion of a particle, it is first necessary to define a reference frame, or coordinate system, relative to which the motion of the particle is measured. For a particle moving in one dimension, a coordinate frame would be a number line.

The position, at time t, of a particle moving on a straight line can be described by its distance, $x(t)$, from a fixed point on the coordinate number line identified as the origin ($x = 0$).

The velocity $v(t)$ of the particle is the rate of change of its position with respect to time. Velocity is a vector quantity, which includes both the speed of the particle and its direction of motion. In one dimension, the speed of the particle is given by absolute value of the velocity, $|v(t)|$. The direction of motion is given by the sign of $v(t)$, with negative and positive values corresponding to motion to the left and right, respectively.

In general, the velocity is the first derivative, or rate of change, of the particle's position with respect to time: $v = dx/dt$.

In the special case of a particle moving at constant velocity v in one dimension, the position of the particle is given by $x(t) = vt + x_0$, where x_0 is the initial position of the particle at time $t = 0$.

The acceleration $a(t)$ of a particle is the rate of change of its velocity vector with respect to time. Acceleration is also vector quantity and includes both magnitude and direction. In one dimension, the absolute value $|a(t)|$ indicates the strength of the acceleration, or how quickly the velocity is changing. Note that the sign of $a(t)$ is the direction of the acceleration, not of the particle itself. When the sign of the acceleration and velocity are opposite, the speed of the particle will decrease.

In general, the acceleration is the time rate of change of the particle's velocity: $a = dv/dt$. Since velocity $v = dx/dt$, the acceleration is, equivalently, the second derivative of the particle's position with respect to time: $a = d^2 x/dt^2$.

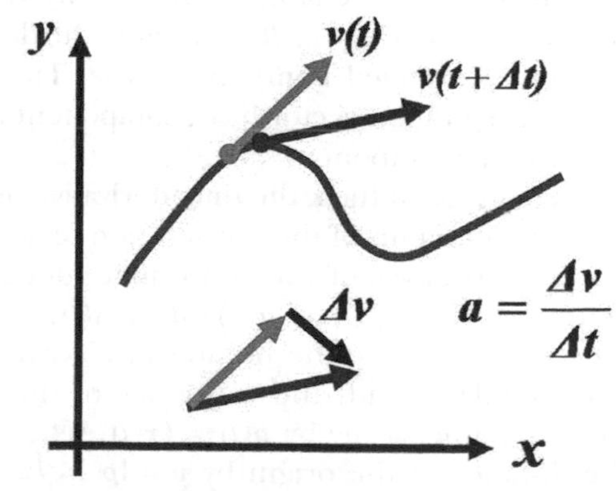

Acceleration as derivative of velocity along trajectory. By Brews ohare (Own work)

In the special case of a particle moving with constant acceleration a in one dimension, the velocity of the particle is given by $v(t) = at + v_0$, where v_0 is the initial velocity of the particle at time $t = 0$, and the position of the particle is given by $x(t) = ½\, at^2 + v_0 t + x_0$, where x_0 is the initial position of the particle.

MOTION OF A PARTICLE IN TWO OR THREE DIMENSIONS

The position of a particle moving in two dimensions is a vector quantity. The position vector $\boldsymbol{p}$ is given by its coordinates (x, y) with respect to a coordinate reference frame. Together, the coordinate functions, $x(t)$ and $y(t)$, give the position of the particle at time t.

The *trajectory*, or path, $\boldsymbol{p}(t) = (x(t), y(t))$ of the particle is the curve in the plane defined parametrically by the coordinate functions.

Note that magnitude of the position vector, $\boldsymbol{p} = |\boldsymbol{p}| = \sqrt{x^2 + y^2}$, measures the distance from the particle to the origin.

As in one dimension, the velocity of a particle moving in two dimensions is the rate of change of its position with respect to time: $v(t) = d/dt\, \boldsymbol{p}(t)$, where the derivative is performed component wise. Thus, at time t, the velocity vector has x component $v(t) = d/dt\, x(t)$ and y component $v(t) = d/dt\, y(t)$.

Of key importance is the fact that the velocity vector $v(t) = d/dt\, \boldsymbol{p}(t)$ always points in the direction tangent to the trajectory $\boldsymbol{p}(t)$ of the particle, as the instantaneous velocity of the particle points in the direction the particle is moving at that instant.

The acceleration of a particle moving in two dimensions is the rate of change of its velocity vector with respect to time: $a(t) = d/dt\, v(t)$, where the derivative is again performed component wise. Thus, at time t, the acceleration vector has x component $a_x(t) = d/dt\, v_x(t)$ and y component $a_y(t) = d/dt\, v_y(t)$.

Since velocity is, in turn, the time-derivative of position, the components of the acceleration vector are the second derivatives of the coordinate functions. That is, $a_x(t) = d^2/dt^2 x(t)$ and $a_y(t) = d^2/dt^2 y(t)$.

In three dimensions, the motion of a particle is described similarly, with the trajectory of the particle given parametrically by $\boldsymbol{p}(t) = (x(t), y(t), zy(t))$ and the distance to the origin by $\boldsymbol{p} = |\boldsymbol{p}| = \sqrt{x^2 + y^2 + z^2}$. The velocity, $v(t)$, and acceleration, $a(t)$, vectors gain z components as well; namely, $vz(t) = d/dt\, z(t)$ and $a_z(t) = d^2/dt^2 z(t)$.

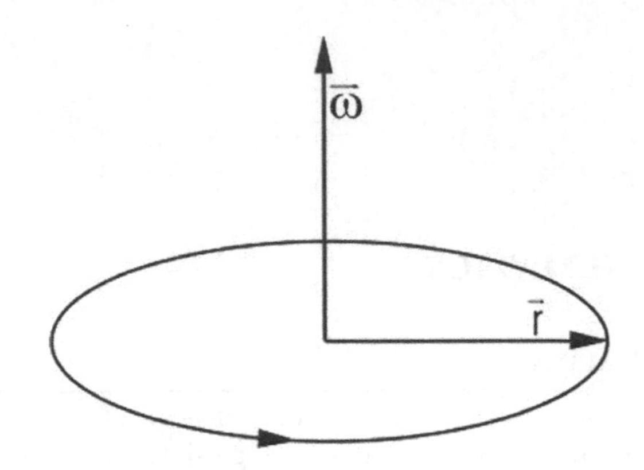

Angular velocity. Public domain, via Wikicommons

OTHER TYPES OF KINEMATIC MOTION

Kinematic considerations can be extended to particles that are rotating about an axis and the motion of a particle with respect to another particle.

A point on a rotating circle of fixed radius is also constrained to move in one dimension. The kinematic equations of a rotating particle have the same form as above, with the linear quantities x, v, and a replaced by their rotational counterparts, angular position, angular velocity, and angular acceleration.

The position of a particle may be defined with respect to a point other than the origin. The relative position of a particle with position vector $\boldsymbol{p}$ with respect to the point $\boldsymbol{q}$ is simply the vector difference, $\boldsymbol{p} - \boldsymbol{q}$. For speeds much less than the speed of light c, the relative velocity and relative acceleration are the first and second time-derivatives of the relative position vector. However, at speeds approaching c, the relative motion of two particles is dictated by the laws of special relativity.

—*Anne Collins*

BIBLIOGRAPHY

Cohen, Michael. "Classical Mechanics: A Critical Introduction." *Physics & Astronomy*. The Trustees of the University of Pennsylvania. PDF. Web. 14 Mar. 2016. https://www.physics.upenn.edu/resources/online-textbook-mechanics

"The Kinematic Equations." *The Physics Classroom.* The Physics Classroom. Web. 14 Mar. 2016. http://www.physicsclassroom.com/class/1DKin/Lesson-6/Kinematic-Equations

"What Is a Projectile?" *The Physics Classroom.* The Physics Classroom. Web. 14 Mar. 2016. http://www.physicsclassroom.com/class/vectors/Lesson-2/What-is-a-Projectile

L

LIMIT OF A FUNCTION

SUMMARY

Ancient thinkers had major conceptual challenges with the concepts of zero and infinity, tied up with abstract notions of nothingness, nonexistence, and unattainable perfection. As a classic example by Zeno of Elea (c. 490–420 BCE), consider an ant moving a certain distance. First, the ant must cover half the distance, then half the remaining distance, then half the next remaining distance, and so forth. The ant must cover an infinite number of half-distances, each of which takes some finite amount of time to cross. An infinite number of finite times seemed as if it would be infinite, so it will take the ant an infinite amount of time to cover the full distance.

Even the proto-scientific thinking available in the ancient Greek world made it clear that this contradicted the clear evidence that ants could cross distances, but mathematicians were hard pressed how to address Zeno's paradox because of the philosophical complications caused by zero and infinity.

Solving problems such as this, and developing the mathematics of calculus, would require talking about the unattainable concept of infinity through the method of approaching a limit.

DEFINING THE LIMIT OF A FUNCTION

The limit of a function is a simple concept, though the language to define it precisely in mathematical terms can seem intimidating. For a continuous function with no gaps or breaks, the concept is relatively straightforward: the limit of a function represents, generally speaking, the idea that it is possible to approach a given point on the function by taking ever more tiny steps to approach that point.

Consider the number sequence of halves from Zeno's paradox.

$$\frac{1}{2}, \frac{1}{4}, \frac{1}{8}, \frac{1}{16}, \frac{1}{32}, \ldots, \frac{1}{2^x}$$

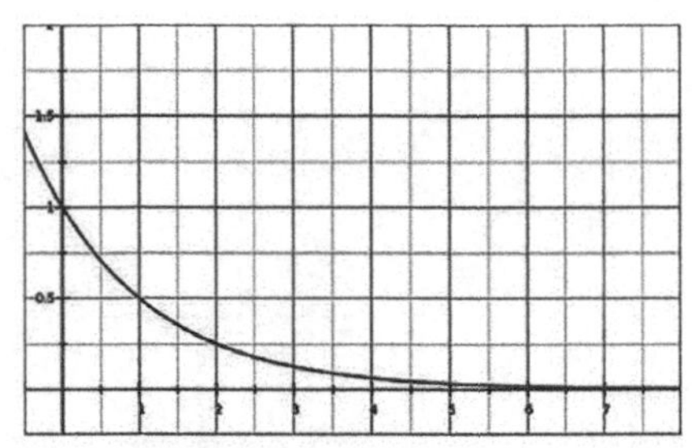

Figure 1. The graph of 1/2n. © EBSCO

As the number of half-distances, n, gets larger, the remaining distance gets smaller. It is helpful to look at the graph (Figure 1) representing the distance left for the nth term in the series.

As the number n gets larger, as it approaches infinity, the term gets smaller and smaller, approaching ever closer to the value of 0. Mathematicians would then say that "the limit of one over 2^n as n approaches infinity equals 0," expressed in mathematical notation as:

$$\lim_{x \to \infty} \frac{1}{2^x} = 0$$

This one-sided limit approaches the limit from only one direction as n increases. There is no other way for the value of n to approach infinity.

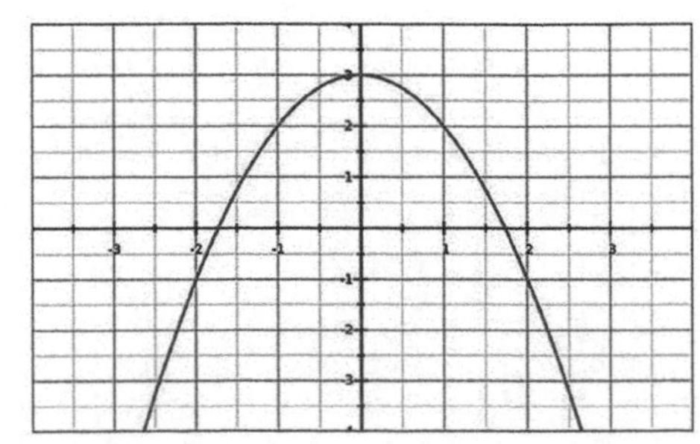

Figure 2. The graph of f(x) = −x2 + 3 approaches a value of 3 as x approaches 0.

A two-sided limit can also exist, such as for $f(x) = -x^2 + 3$ ($\lim_{x \to 0} f(x)$>). This limit can be approached from either the left or right directions. Because the function is continuous, this limit from each direction yields the same value, though this isn't the case for non-continuous functions.

Finding the limit of $f(x)$ at 0 can also be done by algebraically evaluating $f(0) = 3$ or looking at the graph. Some functions cannot be defined at a point, such as the following function $g(x)$, and these cases make the use of limits more clear.

$$g(x) = \frac{-x^3 + 3x}{x}$$

Since the point $g(0)$ is undefined, this function doesn't exist at that point, but the limit of the function does exist there.

Mathematicians such as Guillaume François l'Hopital (1661–1715 CE) and Augustin-Louis Cauchy (1789–1857 CE) contributed heavily to the understanding of rules about how to work with limits, largely due to the central role they play in differential and integral calculus.

—*Andrew Zimmerman Jones, MS*

BIBLIOGRAPHY

Borovik, Alexandre, and Mikhail Katz. "Who Gave You the Cauchy-Weierstrass Tale? The Dual History of Rigorous Calculus." *Foundations of Science* 17.3 (2012): 245-76.

Hornsby, John E., Margaret L. Lial, Gary K. Rockswold. A *Graphical Approach to Precalculus with Limits.* 6th ed. London: Pearson, 2014.

Larson, Ron. *Precalculus with Limits.* Boston: Cengage, 2014.

Ouellette, Jennifer. *The Calculus Diaries: How Math Can Help You Lose Weight, Win in Vegas, and Survive a Zombie Apocalypse.* New York: Penguin, 2010.

Pourciau, Bruce. "Newton and the Notion of Limit." *Historia Mathematica* 28.1 (2001): 18-30.

Schuette, Paul. "A Question of Limits." *Mathematics Magazine* 77.1 (2004): 61-68.

LINEAR PROGRAMMING

SUMMARY

Linear programming is a mathematical method for determining how to maximize or minimize a specific process when one or more restrictions on the process outcome exist. It is sometimes called "linear maximization." Linear programming is frequently used by businesses that want to determine important factors such as how to maximize output or profit or how to minimize labor and materials costs. Linear programming has been in use since the mid-twentieth century. In the decades since, it has been enhanced with computer programs that allow for more complicated processes to be examined with relative ease.

BACKGROUND

The process of linear programming began as a largely manual operation during World War II (1939–1945). George B. Dantzig was an American mathematician working in the Pentagon during the latter part of the war. He used his mathematical skills to design programs and plans for operations that made the most sense in terms of the time, manpower, and other available resources.

After the war ended, Dantzig continued to serve at the Pentagon as the mathematical adviser to the comptroller for the U.S. Air Force. He was asked to

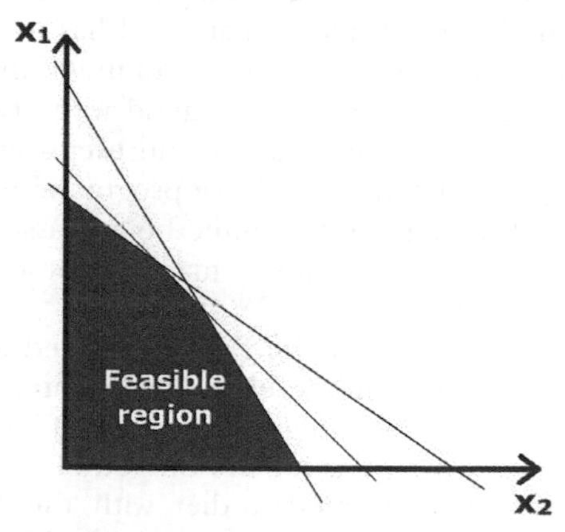

A graph showing how a series of linear constraints on two variables produce a feasible region in a linear programming problem. By en:User:Jacj (Own work by en:User:Jacj

Nobel laureate Leonid Kantorovich. By Андрей Богданов (Andrei-bogdanoff-yandex.ru)

plan ways to maximize training, deployment, and supply logistics. Dantzig's first efforts were accomplished using analog calculators and other manual devices, as computers capable of doing such calculations had not yet been invented.

Dantzig determined that his model for accomplishing the tasks he was assigned needed to be general enough to be used for multiple purposes and easy enough to adapt and modify to meet changing needs. It needed to accommodate several variables and had to be able to handle the large-scale variables inherent to moving and supplying military units. The process had to be capable of not only determining values for the variables but also determining the best way to maximize or minimize certain specific aspects of it.

Consider the following example. The army receives one hundred new recruits and has to assign them to new positions among four units. Of the one hundred recruits, fifty are very good with artillery, twenty-five are good mechanics, and twenty-five are of average skill in both. The linear programming process provides a way to determine how to assign the recruits to one of the units to make the best use of each one's skills.

Between 1947 and 1949, Dantzig worked on his programming until he developed what came to be known as the simplex algorithm. One of the first problems he solved was for the Bureau of National Standards, which wanted a diet with the lowest possible food cost. The programming, which had nine equations with seventy-seven variables, took nine workers four months to solve on calculators. A similar algorithm is now built into the standard Excel spreadsheet and can be solved almost instantly.

By 1946, computers were beginning to be available, and Dantzig's office in the Pentagon was equipped with one by the early 1950s. At that point, linear programming switched from a manual operation to a computerized function. The oil industry was among the first to use linear programming; industry leaders wanted to know the best way to blend petroleum products to provide a product that met specific performance requirements at the lowest possible cost.

PRACTICAL APPLICATIONS

While nearly all linear programming problems are handled by computers today, the steps are similar to the process Dantzig used in the 1940s and 1950s. The first step to manually solving the problem is to define the variables. For example, if a coat manufacturer wants to sell a coat for no more than $100 and make a 40 percent profit, he or she may use linear programming to determine the maximum cost to make the coat. The manufacturer would need to know the labor cost to make each coat and how much material the coat pattern requires. The variable would be the amount of material. Next, it's necessary to write an equation defining the linear problem, or the specific function to be maximized or minimized and the conditions that are related to this. These are called linear inequalities. Some of these linear inequalities place limits on the final outcome. In this case, the limits would be maximum price of the coat minus the labor cost, materials, and profit margin. These are called problem constraints. One of the problem constraints in this example would be the maximum price of the coat, and it would be written ≤100 (less than or equal to 100).

When computed manually, linear programming uses a graph to chart the equation and its solution. The points that define the answers on this chart define an area known as the feasible region. Once this is known, the fundamental theorem of linear programming can be applied. The theorem says that when a linear programming problem can be solved, the solution will be at a corner point of the feasible region or on a line between these two corner points. When used to find a maximum value, the larger corner point value is the answer. When used to find a minimum value, the lower corner point value provides the solution. This provides a geometric solution to the problem. Linear programming problems can also be solved using algebraic equations.

In the twenty-first century, real-world linear programming questions are generally resolved with the assistance of computer programs designed to perform specific functions. For instance, the airline industry uses linear programming to determine how many tickets to sell in advance, how many to hold for last-minute purchases, and how much each ticket should cost. The multimillion-dollar diet industry uses linear programming to determine how much of each type of food a person should eat within the other parameters of a diet (high protein, low gluten, low sugar, low fat, etc.) to meet a certain calorie requirement and lose a certain amount of weight each week. The military and other government agencies use linear programming for budgeting, supply acquisition, and other purposes. Other industries that make use of linear programming include mass transportation, agriculture, and amusement parks.

—*Janine Ungvarsky*

Bibliography

Dantzig, George B. "Linear Programming." *Operations Research,* vol. 50, no. 1, 2002, pp. 42–47, pubsonline.informs.org/doi/pdf/10.1287/opre.50.1.42.17798. Accessed 27 Jan. 2017.

Ferguson, Thomas S. "Linear Programming: A Concise Introduction." *University of California Los Angeles,* www.math.ucla.edu/~tom/LP.pdf. Accessed 27 Jan. 2017.

"George Dantzig, 90; Created Linear Programming." *Los Angeles Times,* 22 May 2005, articles.latimes.com/2005/may/22/local/me-dantzig22. Accessed 27 Jan. 2017.

Glydon, Natasha. "Linear Programming." *Regina University,* mathcentral.uregina.ca/beyond/articles/LinearProgramming/linearprogram.html. Accessed 27 Jan. 2017.

Kulkarni-Thaker, Shefali. "The Diet Problem." *The OR Café, University of Toronto,* 14 Jan. 2013, org.mie.utoronto.ca/blog/?p=45. Accessed 27 Jan. 2017.

"Linear Programming." *High School Operations Research,* www.hsor.org/what_is_or.cfm?name=linear_programming. Accessed 27 Jan. 2017.

"Linear Programming." *IBM,* www-01.ibm.com/software/commerce/optimization/linear-programming/. Accessed 27 Jan. 2017.

"Linear Programming." *Richland University,* people.richland.edu/james/lecture/m116/systems/linear.html. Accessed 27 Jan. 2017.

Local Area Network (LAN)

SUMMARY

A *local area network* is a group of computers that communicate through a network cable or other wireless device. Unlike the similar *wide area network,* local area networks, commonly referred to as LANs, cover only a small area. In most cases, this area is confined to a room or building. Local area networks are commonly used in offices, schools, and government buildings.

INTRODUCTION TO LOCAL AREA NETWORKS

In order for computers to communicate, they need a way to transfer data from one machine to another. In some cases, this is accomplished via an Internet connection. In others, computers share data through a local area network. This data sharing may take place through a variety of means, including Ethernet cables, token ring protocols, Wi-Fi, Fiber Distributed Data Interfaces (FDDI), and ARCNET. The most common interfaces are Ethernet cables and Wi-Fi.

To create a local area network with Ethernet cables, each computer must have a properly installed *Ethernet port.* An Ethernet port is a physical receiver for an Ethernet cable attached to the computer's motherboard. A computer lacking an Ethernet port can have one installed. Alternatively, a temporary external Ethernet port may be installed through a USB slot.

While computers can be connected directly from Ethernet port to Ethernet port, it is impractical to connect a large number of computers in this fashion.

Pacific Ocean (July 21, 2004) - Information Systems Technician 1st Class Jamie J. Andrews of Sacramento, Calif., checks connections on Local Area Network (LAN) cables aboard USS Kitty Hawk (CV 63). By U.S. Navy photo by Photographer's Mate Airman Jason D. Landon

It would require each computer to have a massive number of Ethernet ports to support any type of large network. Instead, Ethernet-based networks require a *hub*. A network hub is a piece of hardware to which all the machines on the network are connected. This hub manages and directs all communication between the machines. Hubs often have large numbers of ports to allow multiple machines to exist simultaneously on the network. In modern networks, a router often acts as a network hub.

Wireless local area networks are possible using Wi-Fi. Initially, these networks were so expensive to set up that they were used only in rare situations where wiring was impossible. However, with the increased prevalence of Wi-Fi and wireless routers, this networking option has become much more affordable. Wireless LANs allow any device within the network's range to connect and disconnect simply by manipulating the settings on the device. They also allow users of mobile devices—such as smartphones, tablets, and laptop computers—to remain active on the LAN while moving throughout a facility.

A *token ring* network is a type of local area network designed to avoid conflicts during complex operations. It may utilize Ethernet cables or Wi-Fi, but it is different from a traditional LAN. Token ring networks use a number of hubs or routers to connect a large number of computers. Several machines will be connected to each hub. However, information only travels among the hubs in a set loop. In a traditional LAN, computers all send information to one hub, which sorts the information and sends it back out. In a token ring, each computer sends information to its hub. That hub then sends the information to another hub and receives information from yet another hub. Because the information is always traveling in a set pattern, it is easier for the system to manage coherently. Some token rings include a second backup ring of hubs in case one of the hubs in the first ring fails.

An *FDDI* network is a specific type of token ring. It uses ANSI and ISO standards for transmission on specialized fiber-optic cables. Unlike normal LANs, these networks can stretch over city blocks, and they can handle download speeds of more than 100 megabytes per second. Many FDDI networks are made up of two token rings.

WHY USE A LOCAL AREA NETWORK?

Local area networks provide people, businesses, and governments with many advantages. They allow networks of computers to share expensive machines—such as copiers, scanners, and printers—so that businesses do not have to buy multiples of each machine. They also allow Internet services to be shared across the network, so that one powerful connection can supply large numbers of machines with online access.

Local area networks may be used to share critical information among machines without the need to copy the information from one machine to the others. Users may edit and access information kept in a database simultaneously. Local area networks can also be used to copy applications and programs to each machine and to set up periodic backups of every machine on the network. Lastly, local area networks may be used to set up internal messaging services and Internet pages. These pages and messaging services would appear similar to other Internet websites and messaging systems. However, they would only be available to machines directly connected to the LAN. This provides companies with an added layer of protection as it would be difficult for an outsider to hack into these services because that person would need to get access to the LAN.

Of course, local area networks do have disadvantages. A hacker who gains access to a private network could easily view or download many important company files. Additionally, if a virus, Trojan, or other form of malware infects one computer on the network, the malware will quickly spread to every other device connected to the network. For this reason, security experts recommend setting up LANs with built-in security protocols, such as firewalls.

—Tyler Biscontini

BIBLIOGRAPHY

"Fiber Optic Network Topologies for ITS and Other Systems." *Fiber-Optics Info.* Fiber-Optics Info. Web. 11 Feb. 2016. http://www.fiber-optics.info/articles/fiber_optic_network_topologies_for_its_and_other_systems

"LANs – Local Area Networks." *Revision World.* Revision World Networks, Ltd. Web. 11 Feb. 2016. http://revisionworld.com/gcse-revision/ict/networks-internet/computer-computer-communication/lans-%E2%80%93-local-area-networks

Rouse, Margaret. "Definition: FDDI (Fiber Distributed Data Interface)." *Search Networking.* TechTarget. Web. 11 Feb. 2016. http://searchnetworking.techtarget.com/definition/FDDI

Rouse, Margaret. "Definition: Local Area Network (LAN)." *Search Networking.* TechTarget. Web. 11 Feb. 2016. http://searchnetworking.techtarget.com/definition/local-area-network-LAN

Rouse, Margaret. "Definition: Token Ring." *Search Networking.* TechTarget. Web. 11 Feb. 2016. http://searchnetworking.techtarget.com/definition/Token-Ring

M

Machine Code

SUMMARY

Machine code, also known as machine language, is a type of language understood by computers. It is referred to as first-generation programming language (1GL), or the lowest level programming language of a computer. *Low-level programming language* is a type of vocabulary and set of grammatical rules used to instruct a computer to complete certain tasks.

Machine code is made up of a long series of numbers: zeros and ones—known as *binary digits,* or bits. Groups of binary digits called *instructions* are translated into a command that could be understood by the central processing unit, or CPU, of a computer. The *CPU,* also called the processor, is the most important part of a computer. It can be thought of as the brain of the computer and is responsible for translating machine code. A CPU can understand and perform millions of instructions per second. Computers from different manufacturers, such as IBM's PC (personal computer) or Apple's Mac, have different CPUs and use different types of machine code, meaning separate machine code programs must be written to run on the different brands of computers.

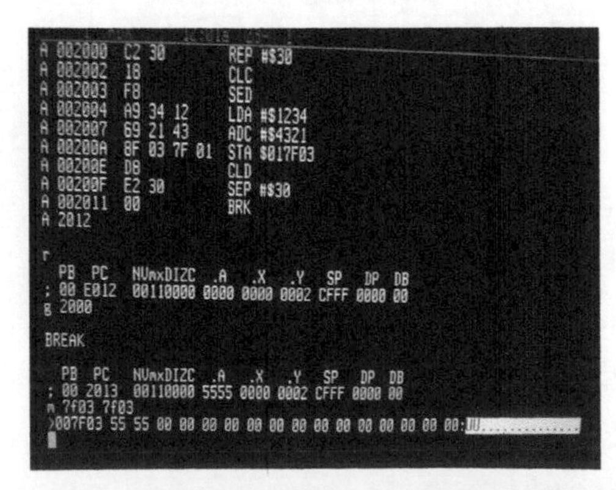

Screen shot of the output of the machine code monitor of a W65C816S single-board computer following execution of a short program (also illustrated). By BigDumbDinosaur, representing BCS Technology Limited (Own work)

LOW-LEVEL PROGRAMMING LANGUAGE

Machine code is written in binary digits (zeros and ones), which represent electric impulses; for example, off (zero) and on (one) states. The instructions are difficult for a human to understand because they are written in patterns of varying lengths that can be sixteen, twenty-four, thirty-two, or sixty-four digits long. Each CPU has its own machine code. The CPU translates instructions into commands that it can understand such as transferring files or storing data in RAM (random access memory).

The CPU is programmed to know which binary digits in a set of instructions tell it what to do. Each instruction consists of an opcode, or operator, and operand. The *opcode* refers to the first few digits of an instruction, and it signifies the operation to be performed by the computer. The rest of the digits in an instruction, or the *operand,* tell the computer the location where the task is to be performed. For example, in the instruction 000001000111000000000 100000010, the opcode 000001 tells the CPU to store the contents of the accumulator in a memory address. The operand 0001110000000000100000010 tells the CPU where to go to complete the task.

Assembly language, also known as assembler language, is a type of low-level programming language. It is known as second-generation programming language (2GL), which means it is a step up from machine code. It was written to help simplify machine code and speed up programming. It still contains long sequences of numbers and is very difficult for humans to read and write, but it does contain some characters. Just like with machine code, CPUs from different manufacturers have different assembly languages, meaning assembly language programs must be written to the specific type of computer.

Machine code runs very quickly and efficiently because the CPU executes the instructions. There are several downfalls, however. Instructions written in machine code can be very lengthy. Some simple tasks require more than ten separate instructions. One tiny change in machine code such as switching binary digits or leaving off one digit can affect the whole instruction sequence and cause errors. It also can cause the CPU to perform a different task altogether.

Computer programmers today typically do not write or read machine code or assembly language. Low-level programming language is difficult for humans to understand because of all the sequences of binary digits and characters required to direct a CPU to complete a task. High-level programming languages typically are used to write programs. However, low-level programming language is still used to program drivers, interfaces, and hardware devices. It also is used in a dump in some debugging programs used to find and correct errors. A *dump* shows the machine code from the program to help the user pinpoint errors.

HIGH-LEVEL PROGRAMMING LANGUAGE

High-level programming language is more like human language and easier to understand than machine code because it requires fewer characters to direct a CPU. It also uses characters and words rather than just binary digits. Some commands are very close to regular language. This makes high-level programming language either third-, fourth-, or fifth-generation programming language (3GL, 4GL, 5GL). The closer the language is to regular human language, the higher generation it is.

Regardless of the particular type of computer, any CPU can understand high-level programming language. However, it must be translated into machine code by a compiler or an interpreter for a CPU to understand it. Both compilers and interpreters are available for most high-level programming languages. Certain languages, however, are designed in a way that they must use either one or the other, but this is rare. Both types also are available for specific brands of computers.

The *compiler* looks at all of the code from the high-level programming language and then collects and reorganizes it into instructions that could be understood by the CPU. Compilers are more commonly used than interpreters.

An *interpreter* analyzes and executes high-level programming language without looking at all of the code. An interpreter is faster at translating the language than a compiler, but the program produced by the compiler runs faster than the one produced by an interpreter.

—*Angela Harmon*

Bibliography

"Assembly Language." *Techopedia*. Techopedia Inc. Web. 10 Mar. 2016. https://www.techopedia.com/definition/3903/assembly-language

Beal, Vangie. "Machine Language." *Webopedia*. QuinStreet Inc. Web. 10 Mar. 2016. http://www.webopedia.com/TERM/M/machine_language.html

"Machine Code." *Techopedia*. Techopedia Inc. Web. 10 Mar. 2016. https://www.techopedia.com/definition/8179/machine-code-mc

"Machine Code (Machine Language)." *TechTarget*. TechTarget. Web. 10 Mar. 2016. http://whatis.techtarget.com/definition/machine-code-machine-language

"Machine Language vs High-Level Languages." *Bright Hub Engineering*. Bright Hub Inc. Web. 10 Mar. 2016. http://www.brighthubengineering.com/consumer-appliances-electronics/115635-machine-language-vs-high-level-languages/

"Programming Language Generations." *TechTarget*. TechTarget. Web. 10 Mar. 2016. http://whatis.techtarget.com/definition/programming-language-generations

Magnetic Storage

SUMMARY

Magnetic storage is a durable and nonvolatile way of recording analog, digital, and alphanumerical data. In most applications, an electric current is used to generate a variable magnetic field over a specially prepared tape or disk that imprints the tape or disk with patterns that, when "read" by an electromagnetic

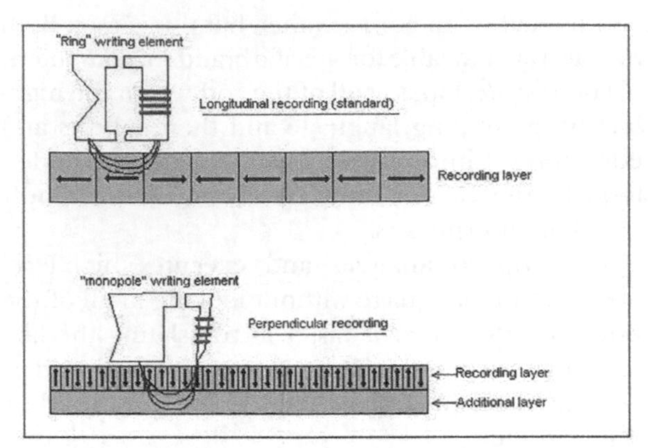

Two types of writing heads on a hard disk.

drive "head," duplicates the wavelengths of the original signal. Magnetic storage has been a particularly enduring technology, as the original conceptual designs were published well in the late nineteenth century.

DEFINITION AND BASIC PRINCIPLES

Magnetic storage is a term describing one method in which recorded information is stored for later access. A magnetized medium can be one 0or a combination of several different substances: iron wire, steel bands, strips of paper or cotton string coated with powdered iron filings, cellulose or polyester tape coated with iron oxide or chromium oxide particles, or aluminum or ceramic disks coated with multiple layers of nonmetallic alloys overlaid with a thin layer of a magnetic (typically a ferrite) alloy. The varying magnetic structures are encoded with alphanumerical data and become a temporary or permanent nonvolatile repository of that data. Typical uses of magnetic storage media range from magnetic recording tape and hard and floppy computer disks to the striping material on the backs of credit, debit, and identification cards as well as certain kinds of bank checks.

BACKGROUND AND HISTORY

American engineer Oberlin Smith's 1878 trip to Thomas Alva Edison's laboratory in Menlo Park, New Jersey, was the source for Smith's earliest prototypes of a form of magnetic storage. Disappointed by the poor recording quality of Edison's wax cylinder phonograph, Smith imagined a different method for recording and replaying sound. In the early 1820s, electrical pioneers such as Hans Ørsted had demonstrated basic electromagnetic principles: Electrical current, when run through a iron wire, could generate a magnetic field, and electrically charged wires affected each other magnetically. Smith toyed with the idea but did not file a patent—possibly because he never found the time to construct a complete, working model. On September 8, 1888, he finally published a description of his conceptual design, involving a cotton cord woven with iron filings passing through a coil of electrically charged wire, in *Electrical World* magazine. The concept in the article, "Some Possible Forms of Phonograph," though theoretically possible, was never tested.

The first actual magnetic audio recording was Danish inventor Valdemar Poulsen's telegraphone, developed in 1896 and demonstrated at the Exposition Universelle in Paris in 1900. The telegraphone was composed of a cylinder, cut with grooves along its surface, wrapped in steel wire. The electromagnetic head, as it passed over the tightly wrapped iron wire, operated both in recording sound and in playing back the recorded audio. Poulsen, trying to reduce distortion in his recordings, had also made early attempts at biasing (increasing the fidelity of a recording by including a DC current in his phonograph model) but, like Oberlin Smith's earlier model, his recorders, based on wire, steel tape, and steel disks, could not easily be heard and lacked a method of amplification.

Austrian inventor Fritz Pfleumer was the originator of magnetic tape recording. Since Pfleumer was accustomed to working with paper (his business was cigarette-paper manufacturing), he created the original magnetic tape by gluing pulverized iron particles (ferrous oxide) onto strips of paper that could be wound into rolls. Pfleumer also constructed a tape recorder to use his tape. On January 31, 1928, Pfleumer received German patent DE 500900 for his sound record carrier (*lautschriftträger*), unaware that an American inventor, Joseph O'Neill, had filed a patent—the first—for a device that magnetically recorded sound in December 1927.

HOW IT WORKS

The theory underlying magnetic storage and magnetic recording is simple: An electrical or magnetic current imprints patterns on the magnetic-storage medium. Magnetic tape, magnetic hard and floppy disks, and other forms of magnetic media operate in a very similar way: An electric current is generated and applied to a demagnetized surface to vary the substratum and form a pattern based on variations in the electrical current. The biggest differences between the three dominant types of magnetic-storage media (tape, rigid or hard disks, and flexible or floppy disks) are the varying speeds at which stored data can be recovered.

Magnetic Tape. Magnetic tape used to be employed extensively for archival computer data storage as well as analog sound or video recording. The ferrous- or chromium-impregnated plastic tape, initially demagnetized, passes at a constant rate over a recording head, which generates a weak magnetic field proportional to the audio or video impulses being recorded and selectively magnetizes the surface of the tape. Although fairly durable, given the correct storage conditions, magnetic tape has the significant disadvantage of being consecutively ordered—the recovery of stored information depends on how quickly the spooling mechanism within the recorder can operate. Sometimes the demand for high-density, cheap data storage outweighs the slower rate of data access. Large computer systems commonly archived information on magnetic tape cassettes or cartridges. Despite the archaic form, advances in tape density allowed magnetic tape cassettes to store up to five terabytes (TB) of data in uncompressed formats.

For audio or video applications, sequential retrieval of information (watching a movie or listening to a piece of music) is the most common method, so a delay in locating a particular part is regarded with greater tolerance. Analog tape was an industry standard for recording music, film, and television until the advent of optical storage, which uses a laser to encode data streams into a recordable media disk and is less affected by temperature and humidity.

Magnetic Disks. Two other types of recordable magnetic media are the hard and floppy diskettes—both of which involve the imprinting of data onto a circular disk or platter. The ease and speed of access to recorded information encouraged the development of a new magnetic storage media form for the computer industry. The initial push to develop a nonlinear system resulted, in 1956, with the unveiling of IBM's 350 Disk Storage Unit—an early example of what became known as a hard drive. Circular, ferrous-impregnated aluminum disks were designed to spin at a high rate of speed and were written upon or read by magnetic heads moving radially over the disk's surface.

Hard and floppy disks differ only in the range of components available within a standard unit. Hard disks, composed of a spindle of disks and a magnetic read-write apparatus, are typically located inside a metal case. Floppy disks, on the other hand, were packaged as a single or dual-density magnetic disk (separate from the read-write apparatus that encodes them) inside a plastic cover. Floppy disks, because they do not contain recording hardware, were intended to be more portable (and less fragile) than hard disks—a trait that made them extremely popular for home-computer users.

Another variant of the disk-based magnetic storage technology is magneto-optical recording. Like an optical drive, magneto-optical recording operates by burning encoded information with a laser and accesses the stored information through optical means. Unlike an optical-storage medium, a magneto-optical drive directs its laser at the layer of magnetic material. In 1992, Sony released the MiniDisc, an unsuccessful magneto-optical storage medium.

APPLICATIONS AND PRODUCTS

Applications for magnetic storage range from industrial or institutional uses to private-sector applications, but the technology underlying each of these formats is functionally the same. The technology that created so many different inventions based on electrical current and magnetic imprinting also had a big impact on children's toys. The Magna Doodle, a toy developed in 1974, demonstrates a simple application of the concept behind magnetic storage that can shed light on how the more complex applications of the technology also work. In this toy, a dense, opaque fluid encapsulates fine iron filings between two thin sheets of plastic. The upper layer of plastic is transparent and thin enough that the weak magnetic current generated by a small magnet encased in a cylinder of plastic (a magnetic pen) can make the iron filings float to the surface of the opaque fluid and form a visible dark line. Any images produced by the pen are, like analog audio signals encoded into magnetic tape, nonvolatile and remain visible until manually erased by a strip of magnets passing over the plastic and drawing the filings back under the opaque fluid.

Magnetic Tape Drives. It is this basic principle of nonvolatile storage that underlies the usage of the three basic types of magnetic storage media—magnetic tape, hard disks, and floppy disks. All three have been used for a wide variety of purposes. Magnetic tape, whether in the form of steel bands, paper, or any one of a number of plastic formulations, was the original magnetic media and was extensively used by the technologically developed nations to capture audio and, eventually, video signals. It long remained the medium of choice for archival mainframe data storage because large computer systems intended for mass data archival require a system of data storage that was both capable of recording vast amounts of information in a minimum of space (high-density) and was extremely inexpensive—two qualities inherent to magnetic tape. Internet-based (or "cloud") storage began to replace physical archival tapes in the 2010s.

Early versions of home computers also had magnetic tape drives as a secondary method of data storage. In the 1970s, IBM offered their own version of a magnetic cassette tape recorder (compatible with its desktop computer) that used the widely available cassette tape. By 1985, however, hard disks and floppy disks had dominated the market for computer systems designed to access smaller amounts of data frequently and quickly, and cassette tapes became obsolete for home-computer data storage.

Hard Disk Drives. In 1955, IBM's 350 Disk Storage Unit, one of the computer industry's earliest hard drives, had only a five megabyte (MB) capacity despite its massive size (it contained a spindle of fifty twenty-four-inch disks in a casing the size of a large refrigerator). However, the 350 was just the first of a long series of hard drives with ever-increasing storage capacity. Between the years of 1950 and 2010, the average area density of a hard drive has doubled every few years, starting from about three megabytes to the high-end availability of three terabytes by 2010. Ever higher-capacity drives are in development, as computer companies, such as Microsoft, Seagate, and Western Digital, are redefining the basic unit of storage capacity on a hard drive from 512 bytes (IBM's standard unit, established during the 1980s) to the immensely larger 4 to 10 terabytes, as of 2017. Initially, the size of the typical hard drive made it difficult to transport and caused the development, in 1971, of another, similar form of magnetic media—the floppy disk.

Early hard drives such as IBM's 350 were huge (88 cubic feet) and prohibitively expensive (costing about $15,000 per megabyte of data capacity). Given that commercial sales of IBM computers to nongovernmental customers were increasing rapidly, IBM wanted some way to be able to deliver software updates to clients cheaply and efficiently. Consequently, engineers conceived of a way of separating a hard disk's components into two units—the recording mechanism (the drive) and the recording medium (the floppy disk).

Floppy Disk Drives. The floppy disk itself, even in its initial eight-inch diameter, was a fraction of the weight and size needed for a contemporary hard disk. Because of the rapidly increasing storage needs of the most popular computer programs, smaller disk size and higher disk density became the end goal of the major producers of magnetic media—Memorex, Shugart, and Mitsumi, among others. As with the hard drive, floppy disk size and storage capacity, respectively, decreased and increased over time until the 3.5-inch floppy disk became the industry standard. Similarly, the physical dimensions of hard drives also shrank (from 88 cubic feet to 2.5 cubic inches), allowing the introduction of portable external hard drives into the market. Floppy disks were made functionally obsolete when another small, cheap, portable recording device came on the market—the thumb, or flash, drive. Sony, the last major manufacturer of floppy disks, announced that, as of March 2011, it would stop producing floppy disks.

Magnetic Striping. Magnetic storage, apart from tape and disks, is also widely used for the frequent transmission of small amounts of exclusively personal data—namely, the strip of magnetic tape located on the back of credit, debit, and identification cards as well as the ferrous-impregnated inks that are used to print numbers along the bottom of paper checks.

Since durability over time is a key factor, encasing the magnetic stripe into a durable, plastic card became the industry standard for banks and other lending institutions. In 2015 credit and debit cards started to feature embedded microchips in addition to traditional magnetic stripes in order to enhance data security.

SOCIAL CONTEXT AND FUTURE PROSPECTS

The ongoing social trend, both locally and globally, is to seek to collect and store vast amounts of data that, nevertheless, must be readily accessible. An example of this is in meteorology, which has been attempting for decades to formulate complex computer models to predict weather trends based on previous trends observed and recorded globally over a century. Weather and temperature records are increasingly detailed, so computer storage for the ideal weather-predicting computer will need to be able to keep up with the storage requirements of a weather-system archive and meet reasonable deadlines for the access and evaluation of a century's worth of atmospheric data.

Other storage developments will probably center on making more biometric data immediately available in identification-card striping. Another archival-related goal might be the increasing need for quick comparisons of DNA evidence in various sectors of law enforcement.

—Julia M. Meyers, MA, PhD

BIBLIOGRAPHY

Bertram, H. Neal. *Theory of Magnetic Recording.* Cambridge, England: Cambridge University Press, 1994.

Hadjipanayis, George C, ed. *Magnetic Storage Systems Beyond 2000.* Dordrecht, The Netherlands: Kluwer Academic, 2001.

Kessler, Sarah. "For a Glimpse of the Future, Try Reading a 3.5-Inch Floppy Disk." *Fast Company,* January 7, 2013.

Mee, C. Denis, and Eric D. Daniel. *Magnetic Recording Technology.* 2d ed. New York: McGraw-Hill, 1996.

National Research Council of the National Academies. *Innovation in Information Technology.* Washington, D.C.: National Academies Press, 2003.

Prince, Betty. *Emerging Memories: Technologies and Trends.* Norwell, Mass.: Kluwer Academic, 2002.

Santo Domingo, Joel. "SSD vs. HDD: What's the Difference?" *PC,* 9 June 2017, www.pcmag.com/article2/0,2817,2404258,00.asp. Accessed 20 Mar. 2018.

Wang, Shan X., and Alexander Markovich Taratorin. *Magnetic Information Storage Technology.* San Diego: Academic Press, 1999.

MECHATRONICS

SUMMARY

Mechatronics is a field of science that incorporates electrical and mechanical engineering, computer control, and information technology. The field is used to design functional and adaptable products and is concerned with traditional mechanical products that are controlled by electronic and computerized mechanisms. A robot is an example of an item made using mechatronics. Its physical parts are produced using mechanical engineering, but its parts that power it and allow it to function are produced using electronic and computerized means. A robot itself is also an example of a mechatronic product because it contains both mechanical and electronic parts controlled by a computer. Mechatronics can be applied to a variety of fields.

BACKGROUND

Mechatronics can be used to describe the process in which machines are automated by way of computers and other electronic equipment. Prior to the 1970s, many everyday items such as machine tools, manufacturing equipment, and home appliances were made using mechanical principles; they had few—if any—electronic elements.

For example, cars of this era had mostly mechanical parts such as the engine, gearbox, and drive shaft. The lights and windshield wipers were the only electrical components of the car. As time passed, cars were made with both mechanical and electrical parts. Cars with electrical sensors warned drivers of potential problems that a car may have such as low tire pressure or low oil levels. Modern cars have features such as traction control, which is controlled by both mechanical elements such as the tires and electrical elements such as sensors. Antilock brakes, climate control, and memory-adjust seats also are examples of features made possible by mechatronics.

Near the early 1970s, Japanese engineer Tetsuro Mori coined the term *mechatronics* from the words *mechanical* and *electronic* to describe the functions that the Yaskawa Electric Corp. used to manufacture mechanical factory equipment. The factory used both mechanical and electronic elements to produce electric motors that were controlled by computers. Yaskawa later registered the term, but it was not used regularly until electrical, electronic, and computerized systems were integrated into mechanical products.

The use of new technology, such as computer hardware and software, made controlling and operating machines much easier and less costly. This technology allowed factories to manufacture many new products that were both mechanical and electrical faster, more accurately, and less expensively since computers could be programmed to instruct machines.

Throughout the 1970s, mechatronic technology was used in the robotics field, such as to help robotic arms coordinate movements, and the automotive industry, such as to equip cars with electronic features. It was also used to advance servotechnology, which controlled automatic door openers, vending machines, and more. In the following decade, the introduction of information technology allowed engineers to improve the performance of mechanical systems by installing microprocessors in the machinery. By the 1990s, advanced computer technology further influenced the field of mechatronics and expanded its reach into numerous fields.

TOPIC TODAY

As of the twenty-first century, mechatronics is used to make numerous items such as automobiles, home appliances, computer hard drives, medical devices, and automotive parts as well as the machinery used to produce these items, such as computerized assembly systems. Any machinery that requires the use of a sensor is made using mechatronic technology.

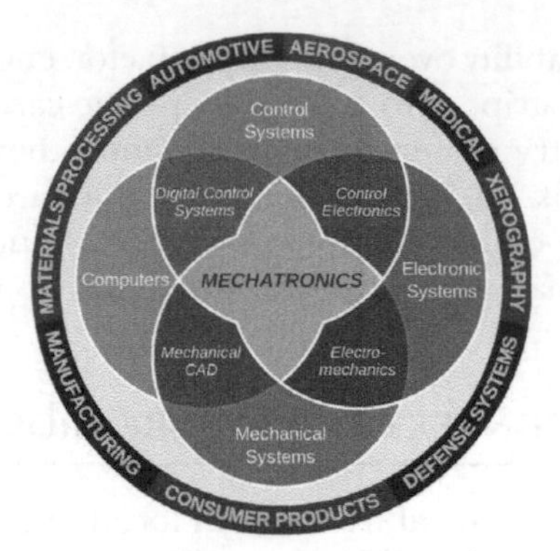

Mechatronics diagram. By original image: Ahm2307 vectorization: Own work (Own work, based on File:Mecha.gif)

For example, a clothes dryer is programmed to stop when clothes are dry or windshield wipers slow down for drizzle or speed up for harder rain.

Mechatronics is used in an array of applications, including biomedical systems; computer-aided design; computer numerical control (CNC) and direct numerical control (DNC) of machinery; energy and power systems; data communication systems; industrial goods; machine automation; and vehicular systems. Mechatronics can be classified into ten technical areas: actuators and sensors; automotive systems; intelligent control; manufacturing; modeling and design; micro devices and optoelectronics (technology that uses electronics and light); motion control; robotics; system integration; and vibration and noise control. These areas depend on a blend of mechanical, computerized, and electronic systems for product development and production technology.

While the subject of mechatronics has been studied in Japan and Europe for several decades, the field has been slow to gain both industrial and academic acceptance in the United States. Before engineers were skilled in mechatronics, mechanical engineers focused on designing machines and products, while software and computer engineers worked on the control and programming elements of these machines. The introduction of mechatronics streamlined this process since engineers skilled in mechatronics can both construct machinery and write the control systems to operate it. They understand both the mechanical design and all of the electronic and computerized elements that allow the devices to operate.

Students interested in mechatronics can pursue degrees in mechanical engineering with a focus on electrical engineering and computer control. Some

Robotics is an application of mechatronics. By Richard Greenhill and Hugo Elias (myself) of the Shadow Robot Company (http://www.shadowrobot.com/media/pictures.shtml)

universities offer specialized mechatronics programs. Mechatronics is used to create an array of products in many different industries. Many manufacturing companies are requiring engineers to be trained in electronics, mechanics, computer control, and more. Individuals trained in mechatronics can also work in a variety of fields outside the manufacturing industry. They can oversee robots in factories where the machines assemble products such as vehicles. They can run greenhouses using controls that manage lighting, irrigation, and temperature to more effectively produce plants or construct wind farms. Engineers skilled in mechatronics can help design systems that allow cars to drive themselves or drones to fly themselves. They can design high-tech security measures such as fingerprint sensors, voice-recognition programs, and retinal scans. They can design virtual reality interfaces for the gaming industry. The increased need for technological advances in the industrial and manufacturing sectors will continue to have an influence on the growing mechatronics field.

—*Angela Harmon*

BIBLIOGRAPHY

Brown, Alan S. "Mechatronics and the Role of Engineers." *American Society of Mechanical Engineers*, Aug. 2011, www.asme.org/engineering-topics/articles/mechatronics/mechatronics-and-the-role-of-engineers. Accessed 19 Jan. 2017.

Doehring, James. "What Is Mechatronics Engineering?" *Wisegeek*, 20 Dec. 2016, www.wisegeek.com/what-is-mechatronics-engineering.htm. Accessed 19 Jan. 2017.

Jeffress, D. "What Does a Mechatronics Engineer Do?" *Wisegeek*, 12 Jan. 2017, www.wisegeek.com/what-does-a-mechatronics-engineer-do.htm#comments. Accessed 19 Jan. 2017.

"Mechatronics History." *Bright Hub Engineering*, 13 May 2010, www.brighthubengineering.com/manufacturing-technology/71180-the-history-of-mechatronics. Accessed 19 Jan. 2017.

Spiegel, Rob. "Mechatronics: Blended Engineering for the Robotic Future." *DesignNews*, 23 Nov. 2016, www.designnews.com/automation-motion-control/mechatronics-blended-engineering-robotic-future/199011163046152. Accessed 19 Jan. 2017.

"10 Surprising Things You Might Do with a Mechatronics Degree." *East Coast Polytechnic Institute Blog*, www.ecpi.edu/blog/10-surprising-things-you-might-do-with-mechatronics-degree. Accessed 19 Jan. 2017.

"What Is Mechatronic Engineering?" *Brightside*, www.brightknowledge.org/knowledge-bank/engineering/features-and-resources/copy_of_what-is-mechatronic-engineering. Accessed 19 Jan. 2017.

"What Is Mechatronics?" *Institution of Mechanical Engineers*, www.imeche.org/get-involved/special-interest-groups/mechatronics-informatics-and-control-group/mechatronics-informatics-and-control-group-information-and-resources/what-is-mechatronics. Accessed 19 Jan. 2017.

"What Is Mechatronics?" *North Carolina State University*, www.engr.ncsu.edu/mechatronics/what-mech.php. Accessed 19 Jan. 2017.

MICROCOMPUTER

SUMMARY

The term *microcomputer* is an umbrella term used to describe all computers whose operations depend on microchips rather than large and cumbersome processors. This group includes laptops, smartphones, and automated teller machines (ATMs). The microcomputer has been in use since the 1970s, and computer scientists continue to work to find ways to make this type of computer more useful.

HISTORY

The path toward making microcomputers widely available began in 1971. That year, the Intel Corporation presented the first microprocessor to the public. This microprocessor, known as the MCS-4, included a set of four microchips. A *microchip* is a very small computer component that aids in specific computer functions. The MCS-4 had a *central processing unit* (CPU) microchip that was called the 4004. Both the microprocessor and its CPU chip were fairly basic but highly programmable, which allowed computer engineers to utilize these components in different machines. In 1974, Intel introduced the 8080, which was another microprocessor.

The 8080 was soon used in a computer called the Altair 8800 produced by Micro Instrumentation and Telemetry Systems (MITS). The Altair 8800 was sold as a kit that could be assembled at a customer's home. Because the computer was cumbersome, it presented obvious problems. It was also limited in its functionality.

Around this time, a young computer programmer named Bill Gates and his business partner Paul Allen developed a version of the computer language BASIC that they believed would make the Altair much more functional. After changes to the computer's software and hardware had been made, it was only a few short steps to creating a radically different experience for the home computer user. The key was to merge microchip technology with computer hardware and the BASIC programming language. Gates and Allen used this trifecta as the basis for their company Microsoft, which would go on to develop the popular Windows operating systems and become one of the most successful technology companies in the world.

The development of microcomputer technology continued into the mid-1970s, when Steve Jobs and Steve Wozniak started the Apple Corporation. The company made a huge impact on the microcomputer industry with the launch of the Apple II microcomputer in 1977. Given its small size and affordable price, the Apple II became one of the company's most successful products. At the start of the 1980s, the International Business Machines Corporation (IBM) introduced its own personal microcomputer, the Commodore 64. This machine eventually became one of the best-selling computers in history.

The Commodore 64 was one of the most popular microcomputers of its era, and is the best-selling model of home computer of all time. By Evan-Amos (Own work)

Microcomputers became popular with consumers because they provided people with useful applications. One such early application was VisiCalc. This program was an early example of an electronic spreadsheet. VisiCalc performed calculations based on the values users entered into its cells. Other useful applications included Apple Writer and WordStar, two early word processing programs.

The first Windows software for IBM computers was introduced in 1985. Since that point, developments in both IBM and Apple microcomputers have continued to make them more useful. Today, the popularity of a practically microcomputer brand depends on the needs of the consumer. Windows microcomputers tend to be less expensive than Apple microcomputers, though some users consider the ease of use and increased functionality of Apple products to make the extra expense worthwhile.

FUNCTION

A microcomputer, despite the suggestion of smallness in its name, is best thought of as a team of computer parts, working together to assist the user in completing a particular task. All of these parts function smoothly because they are linked by an integrated circuit (IC). This circuitry is typically contained within a circuit board inside the computer.

One of the most important and obvious parts of a microcomputer is the keyboard. The keyboard, along with the mouse or touchpad, is an input device. An input device allows the user to give the computer specific commands.

Once the commands are given, they run through the computer's CPU. When information is entered into a microcomputer, it is stored in the computer's memory. Computers often have two types of memory: *random access memory* (RAM) and *read-only memory* (ROM). RAM

A collection of early microcomputers, including a Processor Technology SOL-20 (top shelf, right), an MITS Altair 8800 (second shelf, left), a TV Typewriter (third shelf, center), and an Apple I in the case at far right. By Swtpc6800 en:User:Swtpc6800 Michael Holley (Own work)

has to do with the documents and other files the user saves on a computer, while ROM is primarily linked to the programs and software that help the computer operate. RAM can be altered, but ROM cannot.

Finally, users respond to computer processes and input additional commands based on what they see on the computer monitor. This output device links the user to the internal workings of the machine.

Another important element of the microcomputer's functionality is its ability to interact and communicate with other computers through the Internet. Of course, the hardware of the microcomputer should not be overlooked either. A microcomputer's hardware included the power cord, the USB slot, the CD drive, the Ethernet port, the speakers, and the modem.

—*Max Winter*

BIBLIOGRAPHY

Alfred, Randy. "April 4, 1975: Bill Gates, Paul Allen Form a Little Partnership." *Wired.* Conde Nast. 4 Apr. 2008. Web. 27 Aug. 2015. http://archive.wired.com/science/discoveries/news/2008/04/dayintech_0404

"Altair 8800 Microcomputer." *The National Museum of American History.* Smithsonian. Web. 27 Aug. 2015. http://americanhistory.si.edu/collections/search/object/nmah_334396

Delaney, Frank. "History of the Microcomputer Revolution." *Micro Technology Associates.* Micro Technology Associates. Web. 27 Aug. 2015. http://mtamicro.com/microhis.htm

"Intel's First Microprocessor." *Intel Corporation.* Intel Corporation. Web. 27 Aug. 2015. http://www.intel.com/content/www/us/en/history/museum-story-of-intel-4004.html

Kwon, Jaerock. "Lecture 1: Introduction to Microcomputers." *Mobile: Intelligent Robotics Lab.* Kettering University. 1 Jan. 2010. Web. 27 Aug. 2015. http://paws.kettering.edu/~jkwon/teaching/10-t1/ce-320/lecture/01-Introduction%20to%20Microcomputers.pdf

Mack, Pamela E. "The Microcomputer Revolution." *Clemson University.* Clemson University. 30 Nov. 2005. Web. 27 Aug. 2015. http://www.clemson.edu/caah/history/FacultyPages/PamMack/lec122/micro.htm

"RAM, ROM, and Flash Memory." *For Dummies.* John Wiley & Sons, Inc. Web. 27 Aug. 2015. http://www.dummies.com/how-to/content/ram-rom-and-flash-memory.html

Rockwell, Geoffrey. "Types of Computers: The Microcomputer." *McMaster University.* McMaster University. 15 June 1995. Web. 27 Aug. 2015. http://www.humanities.mcmaster.ca/~hccrs/MICRO.HTM

Veit, Stan. "Pre-IBM PC Computers." *PC-History.org.* PC-History.org. Web. 27 Aug. 2015. http://www.pc-history.org/

Wood, Lamont. "The 8080 Chip at 40: What's Next for the Mighty Microprocessor?" *ComputerWorld.* ComputerWorld, Inc. 8 Jan. 2015. Web. 27 Aug. 2015. http://www.computerworld.com/article/2865938/the-8080-chip-at-40-whats-next-for-the-mighty-microprocessor.html

MICROPROCESSOR

SUMMARY

Microprocessors are one of the critical components of the personal computer. They process all of the information entered into the computer through input devices, including mice, keyboards, cameras, and microphones. A more powerful microprocessor allows a computer to process more information at once.

Primo microprocessore Intel, l'it:Intel 4004. By LucaDetomi at it.wikipedia (Transfered from it.wikipedia)

Microprocessors have become increasingly powerful over the years. Smaller, more powerful microprocessors allow devices such as smartphones to exist.

HISTORY OF MICROPROCESSORS

Computers were not always a big business. In the era before personal computers, building and maintaining computers was done by rare hobbyists. At this time, computers required specialized knowledge to build. They were usually ordered in kits, which had to be assembled by a skilled technician. Computer owners would have to solder their own computer components into place and write their own code. Because of these limitations, computers appealed to a very limited market. For this reason, they were not mass produced.

Employees at major technology firms such as Intel and Hewlett-Packard urged their employers to enter the personal computer market. Both companies began manufacturing various computer chips, including different types of memory and processors. At the time, computer processors were large, inefficient, and composed of several separate devices.

In 1969, a small technology firm approached Intel with a simplified computer design. Their computer would have only four chips: a chip for input and output devices, a read-only memory (ROM) chip, a random access memory (RAM) chip, and a general processor chip. The general processor chip would handle sorting information between the other parts. Intel bought the rights to one of these chips, the 4004, in 1970 and began commercial production the following year. While the 4004 was designed to serve as a tiny central processing unit for digital calculators, engineers at Intel believed it could be applied to many of their devices. Under Intel, the 4004 became the first commercially marketed microprocessor.

Intel continued to improve their chips, dominating the fledgling microprocessor market. Their 8008 processor chip helped grow the computer-building hobbyist movement and was used in the famous Altair 8800 computer build. The Altair 8800 was the first machine to use the BASIC programming language. BASIC was the first programming language written specifically for use on personal computers. This led to the development of the Intel 8080 chip, which was used on some of the first popular personal computers. Eventually, chip manufacturers developed even more powerful chips. This began in 1978 with Intel's 16-bit microprocessors. These were followed by 32-bit microprocessors in the 1980s, which slowly gained market share until the 1990s. That decade, 32-bit microprocessors were replaced by 64-bit microprocessors, which first became available in 1992. Over the following decades, increasingly powerful 64-bit microprocessors became the industry standard.

MODERN MICROPROCESSORS

Modern microprocessors are used in a variety of devices. They are found in personal computers, cell phones, microwaves, dishwashers, refrigerators, DVD players, televisions, alarm systems, electric toothbrushes, and washing machines. They also may be found in any other electronic device that uses a small computer for basic operations.

The most recent generations of microprocessors have a variety of advantages over older models. Despite being exponentially more powerful, modern microprocessors are small, low-profile, and easy for computer engineers to utilize in a variety of hardware setups. They are cheap to manufacture, even though the process of doing so is extremely complex and delicate. Finished, tested products are very reliable and should not fail for many years. Lastly, microprocessors require very little electrical power to operate, making them energy efficient.

The microprocessor manages all data that enters a device. It then sorts the data through the appropriate parts. For example, any information typed on a keyboard first would enter the microprocessor. The

Intel 4004. By Photo by John Pilge. (en.wikipedia.org)

microprocessor then would know to send that information to the video card, showing it on the computer's screen, and to one of the computer's memory units. The microprocessor usually has to communicate with every chip in the device, and it is one of the core parts of the modern personal computer. Other central parts include RAM, hard drives or flash drives, the motherboard, and the power supply.

PARTS OF A MICROPROCESSOR

All microprocessors can be divided into six parts: the arithmetic and logic unit (ALU), the registers, the control unit, the instruction register, the program counter, and the bus. The ALU is the part of the microprocessor that performs all of the mathematical calculations necessary for the device to function. Higher-bit processors can perform more calculations at once. Most computer functions are broken down into mathematical functions and are processed here. The registers are the internal storage device of the microprocessor. Any data that the microprocessor needs to temporarily store is stored in this part. The control unit sorts all data entering and leaving the microprocessor, directing it to the appropriate parts. The instruction register is a separate temporary memory storage device used only for storing task instructions. It makes sure the microprocessor remembers how to perform all of its necessary operations. The program counter temporarily stores the address of the next instruction to be executed. Lastly, the bus is a set of connective cables that allows all of the microprocessor parts to communicate with one another.

—*Tyler Biscontini*

Bibliography

"Computer Bus—What Is It?" *CCM.* CCM Benchmark Group. Feb. 2016. Web. 23 Feb. 2016. http://ccm.net/contents/375-computer-bus-what-is-it

Gordon, Whitson. "How to Build a Computer, Lesson 1: Hardware Basics." *Lifehacker.* Gawker Media Group. Web. 23 Feb. 2016. http://lifehacker.com/5826509/how-to-build-a-computer-from-scratch-lesson-1-hardware-basics

"History of the Microprocessor." *Meeting Tomorrow.* Meeting Tomorrow, Inc. Web. 23 Feb. 2016. http://meetingtomorrow.com/content-library/history-of-the-microprocessor

"Intel's First Microprocessor: Its Invention, Introduction, and Lasting Influence." *Intel.* Intel Corporation. Web. 23 Feb. 2016. http://www.intel.com/content/www/us/en/history/museum-story-of-intel-4004.html

"Microprocessors in the Home." *Teach-ICT.com.* Teach-ICT.com. Web. 23 Feb. 2016. http://www.teach-ict.com/gcse_new/computer%20systems/microprocessors/miniweb/pg2.htm

Shankar. "Evolution of Microprocessors." *Bright Hub Engineering.* Bright Hub Inc. Web. 23 Feb. 2016. http://www.brighthubengineering.com/diy-electronics-devices/50149-evolution-of-microprocessors/

Singer, Graham. "The History of the Microprocessor and the Personal Computer." *TechSpot.* TechSpot, Inc. 17 Sep. 2014. Web. 23 Feb. 2016. http://www.techspot.com/article/874-history-of-the-personal-computer/

Singh, Ankit Kumar. "Evolution of Microprocessor." *Scanftree.com.* Scanftree. Web. 23 Feb. 2016. http://scanftree.com/microprocessor/Evolution-and-Classification-of-Microprocessors

Motion

SUMMARY

Motion in physics is a change in the position of an object over time relative to a frame of reference. Motion is one of the fundamental building blocks of physics and can be observed in terms of distance, speed, velocity, and acceleration. For motion to occur or change, an object must be acted upon by a force, an interaction with another object. The study of motion by itself is called kinematics, the study of motion and force is referred to as dynamics, while the study of motion, force, and energy is called mechanics. In general, motion follows laws discovered by Isaac Newton in the seventeenth century. Other laws of motion take effect when examining subatomic particles or when traveling close to the speed of light.

There are generally considered to be four types of motion. Linear motion occurs when an object moves in a straight line. Rotary motion is when an object moves in a circular path, such as a wheel or a merry-go-round. Oscillating motion is when an object swings or moves side to side in a repetitive manner. Examples of this are a pendulum or the movement of a rotating fan. Reciprocating motion is a back-and-forth motion in a straight line, such as the cutting motion of a saw. Some scientists add a fifth category called random motion, which is a type of motion that does not adhere to a fixed pattern.

BRIEF HISTORY

In the fourth century BCE, Greek philosopher Aristotle attempted to understand the concept of motion in his scientific work *The Physics*, a name taken from the Greek term for "lessons on nature." To Aristotle, motion was an eternal force and involved a change from potentiality to actuality. Natural motion, he argued, was the movement of an object without being forced, such as the rising of the Sun or the downward movement of a rock when it is dropped. Violent motion occurred when a "mover," or outside influence, pushed an object. While Aristotle's thinking was advanced for his time, his ideas on motion and physics were fundamentally flawed and incomplete. Nevertheless, they remained the scientific standard for almost two thousand years.

Motion involves a change in position, such as in this perspective of rapidly leaving Yongsan Station. © EBSCO

During the Renaissance, scientists began reexamining the accepted scientific methods and developed new ways to view the physical world. In the late sixteenth century, Italian astronomer Galileo Galilei recorded his own theories on motion, determining that objects dropped from a height will fall at the same speed no matter what their weight. In the early seventeenth century, German astronomer Johannes Kepler applied the concept of motion to heavenly bodies, developing his three laws of planetary motion. He found that the orbit of planets around the Sun is elliptical, or oval, in shape. Kepler also said that planets move more slowly in their orbits the further they are from the Sun. Despite this, he noted that a planet covers the same distance in equal intervals of time as it orbits the Sun.

LAWS OF MOTION

In 1687, English physicist Isaac Newton revolutionized science when he published the *Mathematical Principles of Natural Philosophy*, which included his three laws of motion. The first law states that a body at rest will remain at rest, and a body in motion will remain in motion unless it is acted upon by an external force. This means that objects will not move, change direction, or stop moving unless they are influenced by another force. For example, a baseball on the ground will remain motionless unless it is picked up and thrown by a pitcher. If that same baseball is hit by a batter, the ball would continue to move unless it encounters another object or is slowed by the force of friction caused by the air.

Newton's second law states that the force acting on an object is equal to the mass of that object times its acceleration. The equation is written out as $F=ma$, with F standing for force, m for mass, and a for acceleration. Mass is the amount of matter in an object, and acceleration is the rate at which an object changes its velocity. While speed is the distance an object travels over a certain period, velocity refers to the distance an object travels over time in a given direction. According to Newton's second law, if a constant force is applied to an object, that object will accelerate at a constant rate. It also states that the more mass an object has, the more force is needed to move

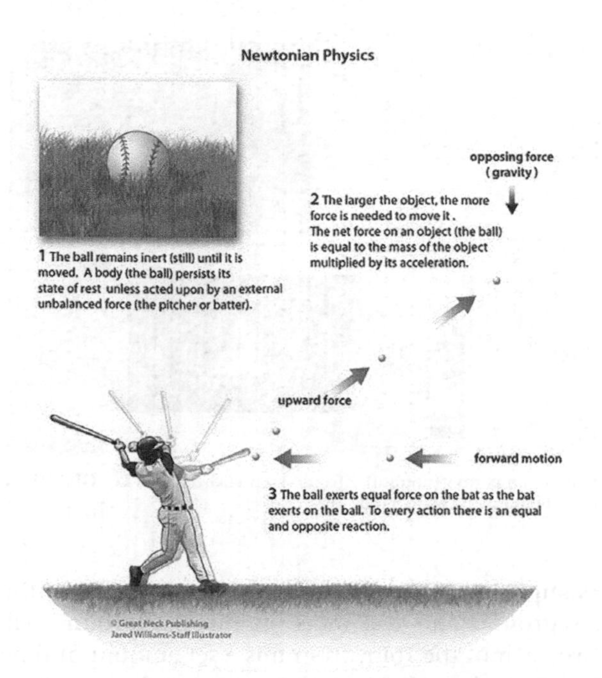

Newtonian Physics
opposing force (gravity)
2 The larger the object, the more force is needed to move it. The net force on an object (the ball) is equal to the mass of the object multiplied by its acceleration.
1 The ball remains inert (still) until it is moved. A body (the ball) persists its state of rest unless acted upon by an external unbalanced force (the pitcher or batter).
upward force
forward motion
3 The ball exerts equal force on the bat as the bat exerts on the ball. To every action there is an equal and opposite reaction.
© Great Neck Publishing
Jared Williams-Staff Illustrator

Newtonian Physics: Newtonian Physics explain the Three Laws of Motion as forces act on different objects. Danleo~commonswiki assumed (based on copyright claims).

it. Using the baseball analogy, if a pitcher throws a baseball, it will travel toward the batter at a certain speed. Assuming the batter makes solid contact, when the ball encounters the force of the bat, it will move in a different direction at a faster speed. If a pitcher throws a softball, which has a larger mass, and the batter makes the same contact, the softball would travel at a slower speed than the baseball due to its larger mass. To move both the baseball and softball at the same speed, the batter would have to use more force on the softball.

Newton's third law states that for every action, there is an equal and opposite reaction. This means that one force is always accompanied by another, acting in the opposite direction. For example, when a rocket lifts off, its fuel ignites behind it and pushes downward at the ground, propelling it upwards.

Newton's laws of motion work very well when referring to conditions on Earth or with planetary bodies. However, German physicist Albert Einstein showed in his theories on relativity that different rules apply when traveling close to the speed of light—186,000 miles per second. Einstein's famous equation, $E=mc^2$, means that energy (E) is equal to mass (m) times the speed of light squared (c^2). At speeds approaching that of light, an object's energy will increase, as will its mass. Very small subatomic particles such as electrons, protons, neutrons, and quarks do not adhere to Newton's laws at all. They are governed by a different set of physicals laws called quantum mechanics.

—Richard Sheposh

BIBLIOGRAPHY

"Description of Motion in One Dimension." *HyperPhysics*, hyperphysics.phy-astr.gsu.edu/hbase/mot.html. Accessed 13 Dec. 2016.

Kubitz, Alan A. *The Elusive Notion of Motion: The Genius of Kepler, Galileo, Newton, and Einstein.* Dog Ear Publishing, 2010.

Lang, Helen S. *Aristotle's Physics and Its Medieval Varieties.* State U of New York P, 1992.

"Learning to Live with the Laws of Motion." *European Space Agency*, 12 Nov. 2012, www.esa.int/Our_Activities/Human_Spaceflight/Astronauts/Learning_to_live_with_the_laws_of_motion. Accessed 13 Dec. 2016.

Lucas, Jim. "Newton's Laws of Motion." *Live Science*, 26 June 2014, www.livescience.com/46558-laws-of-motion.html. Accessed 13 Dec. 2016.

Myers, Rusty L. *The Basics of Physics.* Greenwood Press, 2006.

Van Helden, Al. "On Motion." *The Galileo Project*, 1995, galileo.rice.edu/sci/theories/on_motion.html. Accessed 13 Dec. 2016.

Zimba, Jason. *Force and Motion: An Illustrated Guide to Newton's Laws.* Johns Hopkins UP, 2009.

MULTITASKING

SUMMARY

Multitasking is a term that describes when a computer operating system (OS) is performing more than once process at a time. OSs have been designed to multitask for many years, but they have not always had that ability. Some older OSs were single user/single task systems, which meant that only one user and only one program could be active at a time. A multiuser OS allows more than one user to access the computer at a time. A multitasking OS allows multiple programs to be executed at the same time.

To perform multitasking, an OS will complete tasks (each task lasting only a fraction of a second) from various programs in a certain order. Although the computer is not actually performing more than one task at a time, it is performing each task quickly enough that it makes it seem that multiple programs are doing work at the exact same time. Even though the computer is not actually completing more than one task at a time, more than one program is open because the programs are saved on the central processing unit (CPU) and tasks from all the different programs are split up and executed at different times.

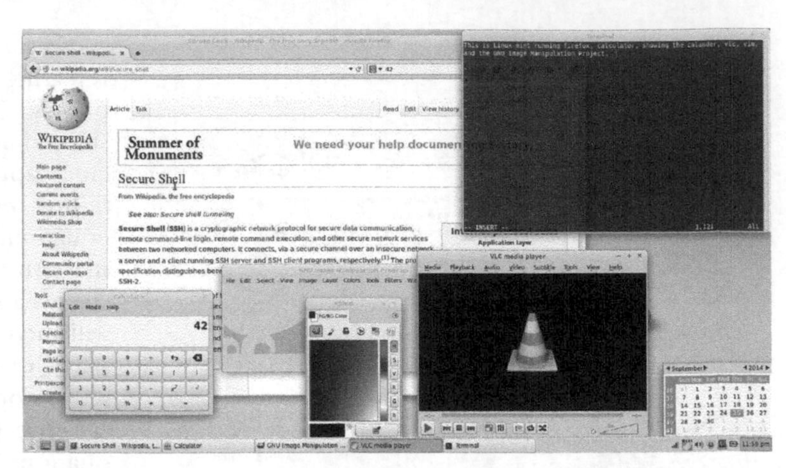

This computer is running linux mint(ubuntu) Xfce desktop enviroment VLC, GIMP, VIM, Calculator, Calander and Firefox. By Benjamintfl (Own work)

When multitasking, the computer will also choose to perform the tasks in the most efficient order possible. Performing these tasks in the most efficient order, rather than in a particular sequential order, allows the computer to seem to complete two different processes at the same time. The computer also tracks the tasks ordered by the system so it does not "forget" tasks or lose its place while processing tasks from different programs. The number of programs that can be run on a computer with a multitasking OS will vary depending on the sophistication of the OS and the type of hardware the computer contains.

Multitasking on personal computers started in the 1990s with the release of Microsoft's Windows 95. Other companies then released operating systems with this functionality. Today, multitasking OSs are the norm.

PREEMPTIVE AND COOPERATIVE MULTITASKING

Multitasking is a function of the operating system. To multitask, the OS has to choose to complete one task for a certain program. The way the OS makes that choice depends on the type of multitasking taking place.

Preemptive multitasking occurs when the programs running on a computer take turns receiving a set amount of processing time for their tasks. For example, a computer running two programs in the foreground and two more in the Background will give each of the four programs a set amount of time for processing. However, the computer does not have to give each of the programs the same amount of processing time. The computer may give programs running in the Background more time for processing because they are more important to the function of the computer. Other times, the computer gives equal processing time to all programs—no matter what the programs are.

Cooperative multitasking is different from preemptive in that not all the programs that are running are given time for processing. If one program is dominating the processor, the other programs may not be able to accomplish any processing. This form of multitasking is not as efficient as preemptive multitasking, and it is not popular in modern computing.

MULTITASKING, MULTIPROCESSING, AND MULTITHREADING

Multitasking is related to a number of other processes that help computer programs run efficiently and quickly. As previously stated, multitasking is using an OS to perform tasks from multiple programs, seemingly at the same time. The word *multiprocessing* is sometimes used in place of multitasking, but the two processes do have differences. Both processes allow users to complete multiple tasks, but multitasking is performing multiple tasks with one CPU.

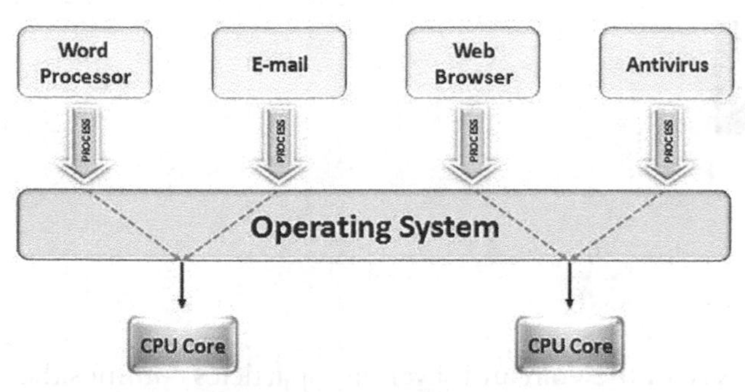

Application-level multitasking is a form of task parallelism, where the operating system auto load balances the tasks across the available CPUs in a system. By Jeff Meisel

Multiprocessing indicates that more than one CPU is completing the tasks.

Multithreading is another term that is often confused with multitasking. Multithreading is much like multitasking, but it breaks down tasks inside individual programs into separate threads. In multithreading, the OS and computer have to determine how to manage not only the tasks from different programs but also the different "threads" of tasks from each running program. Multithreading takes multitasking one step farther. Computers and users can see many benefits from multithreading. For example, multithreading makes CPU usage much more efficient, it improves the performance of the processor, and it makes the system more reliable.

ADVANTAGES OF MULTITASKING AND MULTITHREADING

Multitasking benefits computer users because they can use two different applications (e.g., word processing software and database software) at the same time, even while both are completing operations (though the specific tasks are not being executed at the exact same time). This allows computer users to work in multiple programs at once. Users have more flexibility, which makes working on complex projects or problems simpler.

Multitasking and multithreading also have advantages for computer programmers. When an operating system is capable of completing more than one process at a time, a computer programmer can create software that is more flexible. A programmer creating software for a multitasking and multithreading OS can design a program that has two different threads of execution. One thread could be dealing with the interactions from the user, and the other thread could be working on another operation in the Background. A program with this ability is able to do more work while maintaining a positive experience for the user.

—Elizabeth Mohn

Bibliography

"Advantages of Multitasking." *Microsoft.* Microsoft. Web. 13 Aug. 2015. https://msdn.microsoft.com/en-us/library/windows/desktop/ms681940(v=vs.85).aspx

"Definition of: Non-Preemptive Multitasking." *PC Mag.* PC Mag. Web. 14 Aug. 2015. http://www.pcmag.com/encyclopedia/term/48051/non-preemptive-multitasking

"Definition of: Preemptive Multitasking." *PC Mag.* PC Mag. Web. 14 Aug. 2015. http://www.pcmag.com/encyclopedia/term/49632/preemptive-multitasking

"Differences between Multithreading and Multitasking for Programmers." *National Instruments.* National Instruments Corporation. 20 Jan. 2014. Web. 14 Aug. 2015. http://www.ni.com/white-paper/6424/en/

Lemley, Linda. "Chapter 8: Operating Systems and Utility Programs." *University of West Florida.* University of West Florida. Web. 13 Aug. 2015. http://uwf.edu/clemley/cgs1570w/notes/Concepts-8.htm

"Multitasking." *WhatIs.Com.* TechTarget. Web 15 Aug. 2015. http://whatis.techtarget.com/definition/multitasking

Stokes, Jon. "Introduction to Multithreading, Superthreading and Hyperthreading." *ARS Technica.* Conde Nast. 3 Oct. 2002. Web. 14 Aug. 2015. http://arstechnica.com/features/2002/10/hyperthreading/

"When to Use Multitasking." *Microsoft.* Microsoft. Web. 13 Aug. 2015. https://msdn.microsoft.com/en-us/library/windows/desktop/ms687084(v=vs.85).aspx

N

NANOPARTICLE

SUMMARY

A *nanoparticle* is a microscopic particle of matter that is generally defined as being between one and one hundred nanometers in size. A nanometer is a unit of measurement corresponding to one billionth of a meter. For reference, a nanometer is about 25.4 millionth of an inch or one hundred thousand times smaller than the thickness of a sheet of paper. The size of nanoparticles allows them to exhibit unique properties. They are not governed by the same physical laws that effect larger objects, yet they are too big to be subject to the laws found at the atomic or molecular level. Nanoparticles can occur naturally or be manufactured. Their distinct properties were first noticed more than a thousand years ago and used to make stained glass windows and exceptionally strong sword blades. In the modern era, the particles are used in everyday items such as electronics and sunscreen, and aid in medical science to detect and treat disease.

EARLY HISTORY

While they did not understand the concept of nanoparticles, ancient cultures in China and Egypt had discovered a substance known as soluble gold as early as the second millennium BCE. Soluble gold is a material made up of nanoparticles of ionized gold atoms—atoms that have lost an electron—suspended in a solution. Around the beginning of the fourth century CE, Roman artisans were using gold and silver nanoparticle substances to create a form of glass that changed properties depending on the direction of a light source. The only intact surviving example of this glass is the Lycurgus Cup, a drinking vessel made around 325 CE. The cup appears green and opaque when light shines on it from the outside, but turns red and transparent when the light is placed on the inside. This occurs because the small size of the gold and silver nanoparticles confines the movement of the electrons in their atoms. As a result, the nanoparticles react differently to light than do larger particles.

Nanoparticle substances were also used to create the vivid colors of medieval stained-glass windows. Glassmakers discovered that by adding specific amounts of gold and silver substances to the glass they could change its physical characteristics. For example, adding a gold substance to the glass produced a distinct red color, while adding silver made a yellow color. Adjusting the type and amount of the substances produced different shades. From the thirteenth to the eighteenth centuries, bladesmiths developed a process using carbon nanoparticle tubes to create a durable substance known as Damascus steel. Swords made with this steel became legendary for their strength.

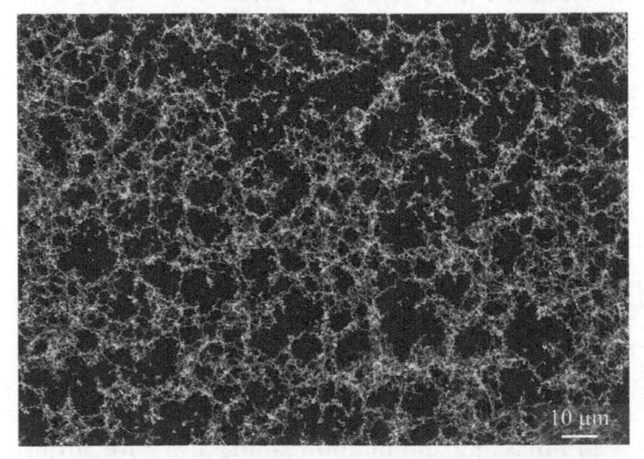

A series of SEM images obtained to study the effect of charged spots which have formed during a motion of charged particles on a solid surface. Silicon dioxide nanoparticles form a complex network of agglomerates under influence of electromagnetic field on the surface of a diode structure formed from a single crystal silicon substrate. The particles were formed from tetraethoxysilane by remote plasma-enhanced chemical vapor deposition under atmospheric pressure AP-RPECVD. The images have been obtained with a scanning electron microscope Zeiss Supra 55VP. By Sergey Langin (Own work)

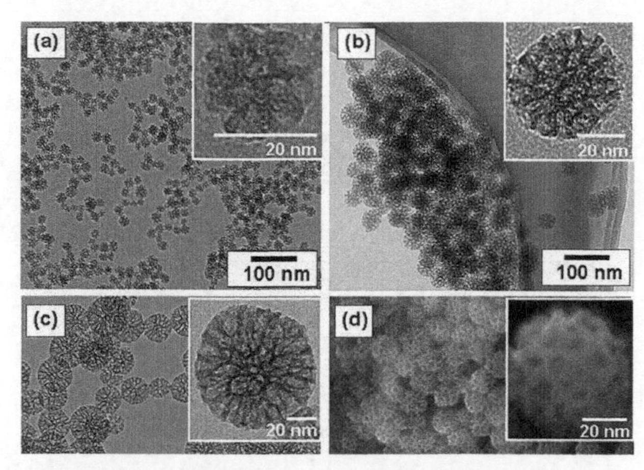

TEM (a, b, and c) images of prepared mesoporous silica nanoparticles with mean outer diameter: (a) 20nm, (b) 45nm; and (c) 80nm. SEM (d) image corresponding to (b). The insets are a high magnification of mesoporous silica particle. By Nandiyanto (Own work)

TOPIC TODAY

It was not until the twentieth century that scientists were able to better understand nanoparticles. The earliest microscope that allowed scientists to see at the level of nanoparticles was invented in the 1930s. The term *nanotechnology* was coined in 1974 by Japanese scientist Norio Taniguchi to describe working with materials at the atomic and nanoparticle scale. Further scientific advancements allowed researchers to see individual atoms and manipulate nanoparticles in the 1980s. Around the start of the twenty-first century, scientists were able to create substances utilizing nanoparticles to make stronger, more lightweight materials. These are used in numerous consumer products such as automobile bumpers, tennis rackets, and golf balls. Silver nanoparticles are added to socks to protect the material from absorbing perspiration, reducing the smell of sweaty feet. Nanoparticles are also infused into sunscreen, making it clear upon application while keeping its protective capabilities. In electronics, they are used to make improved digital displays in televisions and cell phones and to increase computer memory.

In the medical field, nanoparticle technology is used in a variety of ways to help in preventing illness and diagnosing and treating diseases. Because of their small size, nanoparticles can be placed more easily in targeted areas and help improve scans taken through magnetic resonance imaging (MRI), a process that uses magnetic waves to view inside the body. Particles known as perfluorocarbon emulsion nanoparticles have a high acoustic reflectivity and can be used to enhance ultrasound images in fields from obstetrics to oncology.

Since gold is one of the few metals not toxic to the human body, gold nanoparticles are often used in the treatment of cancer. Cancer is the abnormal and uncontrolled growth of cells in the body. The difficulty in treating cancer is that methods used to kill cancerous cells also harm healthy ones. Using particles such as gold nanoparticles or carbon nanotubes, doctors can target specific cancer cells and coordinate treatment with the nanoparticles. The particles coating the cells can be irradiated with short pulses from a laser to create small shockwaves that destroy the cancerous cells, while leaving healthy cells intact. A similar method involves using nanoparticles with magnetic properties to target cancer with magnetic waves. The waves heat the nanoparticles and cause enough damage to kill the cells. The particles can also be used to determine if a potential cancer treatment is working, allowing doctors the ability to continue or alter treatment more quickly.

One of the reasons nanoparticles are so effective is that despite their smaller size, they actually have more surface area than larger objects with the same mass. This can be illustrated with a six-sided cube with each side measured at one centimeter. The total surface area of the cube would be six square centimeters. If that same cube were to be turned into one thousand millimeter-sized cubes, each of those smaller cubes would have a surface area of six square millimeters or sixty square centimeters. While the total mass of the cube did not change, it would now have a greater total surface area. A larger surface area means that more of the particles can come into contact with the surrounding material, increasing their reactive ability.

—Richard Sheposh

BIBLIOGRAPHY

Binns, Chris. *Introduction to Nanoscience and Nanotechnology.* John Wiley & Sons, 2010.

D'Almeida, Carolyn M., and Bradley J. Roth. "Medical Applications of Nanoparticles." *University of Michigan–Flint,* www.umflint.edu/sites/default/files/groups/Research_and_Sponsored_Programs/

MOM/c.dalmeida_b.roth_.pdf. Accessed 2 Dec. 2016.

Heiligtag, Florian J., and Markus Niederberger. "The Fascinating World of Nanoparticle Research." *Materials Today*, vol. 16, no. 7–8, July–Aug. 2013, pp. 262–71, www.sciencedirect.com/science/article/pii/S1369702113002253. Accessed 2 Dec. 2016.

Horikoshi, Satoshi, and Nick Serpone. "Introduction to Nanoparticles." *Microwaves in Nanoparticle Synthesis: Fundamentals and Applications.* Wiley-VCH, 2013, pp. 1–24.

Kulkarni, Ashish. "New Nanoparticle Reveals Cancer Treatment Effectiveness in Real Time." *Brigham and Women's Hospital*, 28 Mar. 2016, www.sciencedaily.com/releases/2016/03/160328191842.htm. Accessed 2 Dec. 2016.

"Nanotechnology Timeline." *National Nanotechnology Initiative*, www.nano.gov/timeline. Accessed 2 Dec. 2016.

"What Is a Nanoparticle?" *Horiba Scientific*, www.horiba.com/scientific/products/particle-characterization/applications/what-is-a-nanoparticle/. Accessed 2 Dec. 2016.

"What's So Special about the Nanoscale?" *National Nanotechnology Initiative*, www.nano.gov/nanotech-101/special. Accessed 2 Dec. 2016.

Nanotechnology

SUMMARY

Nanotechnology is dedicated to the study and manipulation of structures at the extremely small nano level. The technology focuses on how particles of a substance at a nanoscale behave differently than particles at a larger scale. Nanotechnology explores how those differences can benefit applications in a variety of fields. In medicine, nanomaterials can be used to deliver drugs to targeted areas of the body needing treatment. Environmental scientists can use nanoparticles to target and eliminate pollutants in the water and air. Microprocessors and consumer products also benefit from increased use of nanotechnology, as components and associated products become exponentially smaller.

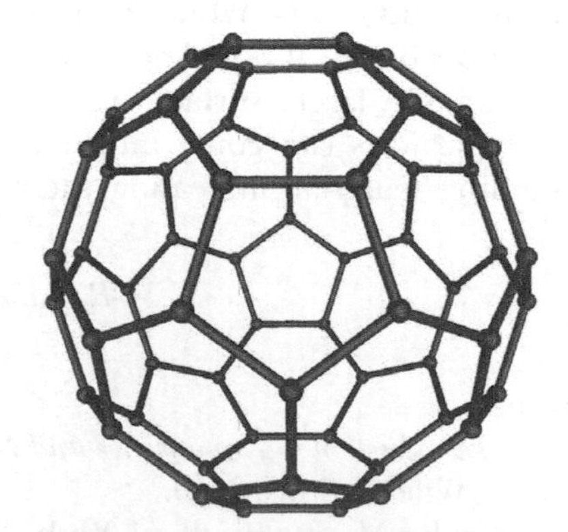

Buckminsterfullerene C60, also known as the buckyball, is a representative member of the carbon structures known as fullerenes. Members of the fullerene family are a major subject of research falling under the nanotechnology umbrella. By Mstroeck at en.wikipedia Later versions were uploaded by Bryn C at en.wikipedia.

DEFINITION AND BASIC PRINCIPLES

Nanotechnology is the science that deals with the study and manipulation of structures at the nano level. At the nano level, things are measured in nanometers (nm), or one billionth of a meter (10^{-9}). Nanoparticles can be produced using various techniques known as top-down nanofabrication, which starts with a larger quantity of material and removes portions to create the nanoscale material. Another method is bottom-up nanofabrication, in which individual atoms or molecules are assembled to create nanoparticles. One area of research involves developing bottom-up self-assembly techniques that would allow nanoparticles to create themselves when the necessary materials are placed in contact with one another.

Nanotechnology is based on the discovery that materials behave differently at the nanoscale, less than 100 nm in size, than they do at slightly larger scales. For instance, gold is classified as an inert material because it neither corrodes nor tarnishes; however, at the nano level, gold will oxidize in carbon monoxide. It will also appear as colors other than the yellow for which it is known.

Nanotechnology is not simply about working with materials such as gold at the nanoscale. It also involves taking advantage of the differences at this scale to create markers and other new structures that are of use in a wide variety of medical and other applications.

BACKGROUND AND HISTORY

In 1931, German scientists Ernst Ruska and Max Knoll built the first transmission electron microscope (TEM). Capable of magnifying objects by a factor of up to one million, the TEM made it possible to see things at the molecular level. The TEM was used to study the proteins that make up the human body. It was also used to study metals. The TEM made it possible to view particles smaller than 200 nm by focusing a beam of electrons to pass through an object, rather than focusing light on an object, as is the case with traditional microscopes.

In 1959 the noted American theoretical physicist Richard Feynman brought nanoscale possibilities to the forefront with his talk "There's Plenty of Room at the Bottom," presented at the California Institute of Technology in 1959. In this talk, he asked the audience to consider what would happen if they could arrange individual atoms, and he included a discussion of the scaling issues that would arise. It is generally agreed that Feynman's reputation and influence brought increased attention to the possible uses of structures at the atomic level.

In the 1970s scientists worked with nanoscale materials to create technology for space colonies. In 1974 Tokyo Science University professor Norio Taniguchi coined the term "nano-technology." As he defined it, nanotechnology would be a manufacturing process for materials built by atoms or molecules.

In the 1980s the invention of the scanning tunneling microscope (STM) led to the discovery of fullerenes, or hollow carbon molecules, in 1986. The carbon nanotube was discovered a few years later. In 1986, K. Eric Drexler's seminal work on nanotechnology, *Engines of Creation*, was published. In this work, Drexler used the term "nanotechnology" to describe a process that is now understood to be molecular nanotechnology. Drexler's book explores the positive and negative consequences of being able to manipulate the structure of matter. Included in his book are ruminations on a time when all the works in the Library of Congress would fit on a sugar cube and when nanoscale robots and scrubbers could clear capillaries or whisk pollutants from the air. Debate continues as to whether Drexler's vision of a world with such nanotechnology is even attainable.

In 2000 the U.S. National Nanotechnology Initiative was founded. Its mandate is to coordinate federal nanotechnology research and development. Great growth in the creation of improved products using nanoparticles has taken place since that time. The creation of smaller and smaller components—which reduces all aspects of manufacture, from the amount of materials needed to the cost of shipping the finished product—is driving the use of nanoscale materials in the manufacturing sector. Furthermore, the ability to target delivery of treatments to areas of the body needing those treatments is spurring research in the medical field.

The true promise of nanotechnology is not yet known, but this multidisciplinary science is widely viewed as one that will alter the landscape of fields from manufacturing to medicine.

HOW IT WORKS

Basic Tools. Nanoscale materials can be created for specific purposes, but there exists also natural nanoscale material, like smoke from fire. To create nanoscale material and to be able to work with it requires specialized tools and technology. One essential piece of equipment is an electron microscope. Electron microscopy makes use of electrons, rather than light, to view objects. Because these microscopes have to get the electrons moving, and because they need several thousand volts of electricity, they are often quite large.

One type of electron microscope, the scanning electron microscope (SEM), requires a metallic sample. If the sample is not metallic, it is coated with gold. The SEM can give an accurate image with good resolution at sizes as small as a few nanometers.

For smaller objects or closer viewing, a TEM is more appropriate. With a TEM, the electrons pass through the object. To accomplish this, the sample has to be very thin, and preparing the sample is time consuming. The TEM also has greater power needs than the SEM, so SEM is used in most cases, and the TEM is reserved for times when a resolution of a few tenths of a nanometer is absolutely necessary.

The atomic force microscope (AFM) is a third type of electron microscope. Designed to give a clear image of the surface of a sample, this microscope uses a laser to scan across the surface. The result is an image that shows the surface of the object, making visible the object's "peaks and valleys."

Moving the actual atoms around is an important part of creating nanoscale materials for specific purposes. Another type of electron microscope, the scanning tunneling microscope (STM), images the surface of a material in the same way as the AFM. The tip of the probe, which is typically made up of a single atom, can also be used to pass an electrical current to the sample, which lessens the space between the probe and the sample. As the probe moves across the sample, the atoms nearest the charged atom move with it. In this way, individual atoms can be moved to a desired location in a process known as quantum mechanical tunneling.

Molecular assemblers and nanorobots are two other potential tools. The assemblers would use specialized tips to form bonds with materials that would make specific types of materials easier to move. Nanorobots might someday move through a person's blood stream or through the atmosphere, equipped with nanoscale processors and other materials that enable them to perform specific functions.

Bottom-Up Nanofabrication. Bottom-up nanofabrication is one approach to nanomanufacturing. This process builds a specific nanostructure or material by combining components of atomic and molecular scale. Creating a structure this way is time consuming, so scientists are working to create nanoscale materials that will spontaneously join to assemble a desired structure without physical manipulation.

Top-Down Nanofabrication. Top-down nanofabrication is a process in which a larger amount of material is used at the start. The desired nanomaterial is created by removing, or carving away, the material that is not needed. This is less time consuming than bottom-up nanofabrication, but it produces considerable waste.

Specialized Processes. To facilitate the manufacture of nanoscale materials, a number of specialized processes are used. These include nanoimprint lithography, in which nanoscale features are stamped or printed onto a surface; atomic layer epitaxy, in which a layer that is only one atom thick is deposited on a surface; and dip-pen lithography, in which the tip of an atomic force microscope writes on a surface after being dipped into a chemical.

APPLICATIONS AND PRODUCTS

Smart Materials. Smart materials are materials that react in ways appropriate to the stimulus or situation they encounter. Combining smart materials with nanoscale materials would, for example, enable scientists to create drugs that would respond when encountering specific viruses or diseases. They could also be used to signal problems with other systems, such as nuclear power generators or pollution levels.

Sensors. The difference between a smart material and a sensor is that the smart material will generate a response to the situation encountered, while the sensor will generate an alarm or signal that there is something that requires attention. The capacity to incorporate sensors at a nanoscale greatly enhances the ability of engineers and manufacturers to create structures and products with a feedback loop that is not cumbersome. Nanoscale materials can easily be incorporated into the product.

Medical Uses. The potential uses of nanoscale materials in the field of medicine are of particular interest to researchers. Theoretically, nanorobots could be programmed to perform functions that would eliminate the possibility of infection at a wound site. They could also speed healing. Smart materials could be designed to dispense medication in appropriate doses when a virus or bacteria is encountered. Sensors could be used to alert physicians to the first stages of malignancy. There is great potential for nanomaterials to meet the needs of aging populations without intrusive surgeries requiring lengthy recovery and rehabilitation.

Energy. Nanomaterials also hold promise for energy applications. With nanostructures, components of heating and cooling systems could be tailored to control temperatures with greater efficiency. This could be accomplished by engineering the materials so that some types of atoms, such as oxygen, can pass through, while others, such as mold or moisture, cannot. With this level of control, living conditions could be designed to meet the specific needs of different categories of residents.

Extending the life of batteries and prolonging their charge has been the subject of decades of research. With nanoparticles, researchers at Rutgers University and Bell Labs have been able to better separate the chemical components of batteries, resulting in longer battery life. With further nanoscale research, it may be possible to alter the internal composition of batteries to achieve even greater performance.

Light-emitting diode (LED) technology uses 90 percent less energy than conventional, non-LED lighting. It also generates less heat than traditional metal-filament light bulbs. Nanomanufacture would make it possible to create a new generation of efficient LED lighting products.

Electronics. Moore's law states that transistor density on integrated circuits doubles about every two years. With the advent of nanotechnology, the rate of miniaturization has the potential to double at a much greater rate. This miniaturization will profoundly affect the computer industry. Computers will become lighter and smaller as nanoparticles are used to increase everything from screen resolution to battery life while reducing the size of essential internal components, such as capacitors.

SOCIAL CONTEXT AND FUTURE PROSPECTS

Whether nanotechnology will ultimately be good or bad for the human race remains to be seen, as it continues to be incorporated into more and more products and processes, both common and highly specialized. There is tremendous potential associated with the ability to manipulate individual atoms and molecules, to deliver medications to a disease site, and to build products such as cars that are lighter yet stronger than ever. Much research is devoted to using nanotechnology to improve fields such as pollution mitigation, energy efficiency, and cell and tissue engineering. However, there also exists the persistent worry that humans will lose control of this technology and face what Drexler called a "gray goo" scenario, in which self-replicating nanorobots run out of control and ultimately destroy the world.

Despite fears linked to cutting-edge technology, many experts, including nanotechnology pioneers, consider such doomsday scenarios involving robots to be highly unlikely or even impossible outside of science fiction. More worrisome, many argue, is the potential for nanotechnology to have other unintended negative consequences, including health impacts and ethical challenges. Some studies have shown that the extremely small nature of nanoparticles makes them susceptible to being breathed in or ingested by humans and other animals, potentially causing significant damage. Structures including carbon nanotubes of graphene have been linked to cancer. Furthermore, the range of possible applications for nanotechnology raises various ethical questions about how, when, and by whom such technology can and should be used, including issues of economic inequality and notions of "playing God." These risks, and the potential for other unknown negative impacts, have led to calls for careful regulation and oversight of nanotechnology, as there has been with nuclear technology, genetic engineering, and other powerful technologies.

—Gina Hagler, MBA

Bibliography

Berlatsky, Noah. *Nanotechnology.* Greenhaven, 2014.

Binns, Chris. *Introduction to Nanoscience and Nanotechnology.* Hoboken: Wiley, 2010. Print.

Biswas, Abhijit, et al. "Advances in Top-Down and Bottom-Up Surface Nanofabrication: Techniques, Applications & Future Prospects." *Advances in Colloid and Interface Science* 170.1–2 (2012): 2–27. Print.

Demetzos, Costas. *Pharmaceutical Nanotechnology: Fundamentals and Practical Applications.* Adis, 2016.

Drexler, K. Eric. *Engines of Creation: The Coming Era of Nanotechnology.* New York: Anchor, 1986. Print.

Drexler, K. Eric. *Radical Abundance: How a Revolution in Nanotechnology Will Change Civilization.* New York: PublicAffairs, 2013. Print.

Khudyakov, Yury E., and Paul Pumpens. *Viral Nanotechnology.* CRC Press, 2016.

Ramsden, Jeremy. *Nanotechnology.* Elsevier, 2016.

Ratner, Daniel, and Mark A. Ratner. *Nanotechnology and Homeland Security: New Weapons for New Wars.* Upper Saddle River: Prentice, 2004. Print.

Ratner, Mark A., and Daniel Ratner. *Nanotechnology: A Gentle Introduction to the Next Big Idea.* Upper Saddle River: Prentice, 2003. Print.

Rogers, Ben, Jesse Adams, and Sumita Pennathur. *Nanotechnology: The Whole Story.* Boca Raton: CRC, 2013. Print.

Rogers, Ben, Sumita Pennathur, and Jesse Adams. *Nanotechnology: Understanding Small Systems.* 2nd ed. Boca Raton: CRC, 2011. Print.

Stine, Keith J. *Carbohydrate Nanotechnology.* Wiley, 2016.

NETWORK INTERFACE CONTROLLER (NIC)

SUMMARY

A *network interface controller (NIC)* is a component installed in a computer that allows it to connect to a network of computers. A network interface can be a card, an adapter, or another type of device. Most interface controllers used in the twenty-first century allow computers to connect to a network wirelessly, but there are controllers that require a wired connection. Originally, most computers used a separate interface controller installed in one of the computer's ports; however, it has become more common for an interface device to be installed directly on the computer's motherboard.

BACKGROUND

Soon after personal computers began to grow in availability and popularity in the 1970s, it became necessary to find a way for them to communicate with one another. The first NICs were added to computers after they were manufactured. These were called expansion cards, and they were installed by connecting them to the computer's motherboard through a special connection port. A motherboard is an electronic circuit board built into the computer that controls its mechanical functions, such as how the different parts of the computer communicate with one another and how power is used.

The longer computers were available, the more important computer networks became; this was especially the case after the Internet became widely available in the 1980s. Even computers that were not connected to the Internet often needed to be connected to one another, such as computers in an office that needed to be integrated in a local access network (LAN) for the company's employees. As it became increasingly common for computers to be part of a network, manufacturers began incorporating network interface controllers directly into the motherboard.

When wireless Internet technology gained popularity and availability in the twenty-first century, computer manufacturers responded by making the network interfaces wireless. These allow the computer to connect to a network without being wired in through Ethernet or other cabled connections, providing flexibility in placing a desktop computer and mobility for a laptop computer. These wireless network interface controllers, often referred to as WNICs, use antennas to connect to the wireless signals. Older computers that are not equipped with a WNIC can often be adapted with an NIC dongle, or a device that plugs into one of the computer's universal serial bus (USB) ports. In some cases, a computer might have two network interface controllers.

TYPES OF INTERFACE CONTROLLERS

There are different types of network interface controllers. They are designed to fit in certain styles of computers and to function with different operating systems. In effect, the

An ATM network interface. By Barcex (Own work)

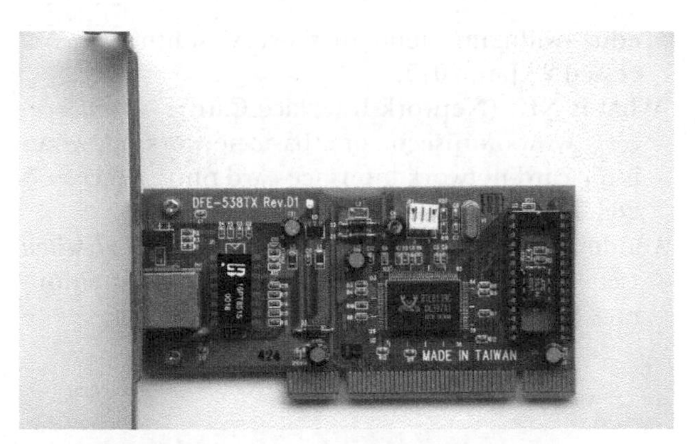

An ethernet card, a low-price network interface card. By futase_tdkr (Self-photographed)

network interface transfers and translates information passed between the computer and the larger network.

The network interface controller makes it possible for the single bits of information sent through the network to be converted to the eight-bit format that computers receive. *Bit* stands for "binary digit" and represents the smallest unit of information encoded on a computer. The network interface controller both converts the information to the correct format, depending on whether the computer is currently receiving or sending information, and arranges it in a way that the computer can understand. This is known as formatting.

For a network interface to work, the network must be able to identify it. Each device includes a media access control (MAC) address permanently installed on the interface controller's ROM card. A read-only memory (ROM) card is a memory storage device. They are common in computers as well as digital cameras, video games, and other electronic devices. The MAC address serves as a sort of key to the computer; if the computer recognizes the MAC address as valid and thinks it matches the information for the computer, it will let the packet through. MAC addresses can also be used by network administrators to limit access to a network; the network can be programmed to recognize only certain MAC addresses and only allow those addresses to access the network, or to access only certain parts of the network. For example, in a school setting, there may be one level of access for student-assigned computers, another level for computers assigned to staff, and a third level for computers that are not known to the network, or guests.

The MAC address is a 48-bit code that is unique to each specific NIC. Information is passed from computer to computer in packets; a packet is a portion of data enclosed in additional information known as wrappers. They are also sometimes called buffers. A wrapper, which can be a header or trailer attached to the data packet to identify it, serves much like the wrapper on a takeout burger or sandwich to identify what is inside. The wrapper also helps to direct the packet to its intended destination.

When information needs to go from one computer in a network to another computer, it goes first from wherever it is on the computer to a protocol stack. Protocols are the rules for how data is handled; the stack refers to the software that implements the rules. In the protocol stack, the information is turned into a packet in a wrapper that identifies it and labels it with an intended destination. A single file may require multiple packets. In that case, the NIC labels them so that the information can be read properly at its destination.

The information packet then goes through the Ethernet hardware, which adds additional identifying information to a wrapper before sending the packet on its way. Once it arrives, the network interface controller will remove the information added by the Ethernet hardware at the sending computer. The packet is then sent to the protocol stacks, which further "unwrap" the packet before sending it where it belongs in the destination computer.

Network interface controllers provide an important way for data to flow from one computer to another. They allow for security both within the computer and within the network. Without them, computers would not be able to communicate with one another.

—*Janine Ungvarsky*

BIBLIOGRAPHY

"Ethernet Protocols and Packets." *Savvius*, www.wildpackets.com/resources/compendium/ethernet/ethernet_packets. Accessed 28 Jan. 2017.

"The Network Interface Card." *Knowledge Systems Institute Graduate School*, pluto.ksi.edu/~cyh/cis370/ebook/ch02c.htm. Accessed 28 Jan. 2017.

"Network Interface Card (NIC)." *Scottish Qualifications Authority*, 21 Mar. 2010, www.sqa.org.uk/e-learning/HardOSEss04CD/page_43.htm. Accessed 28 Jan. 2017.

"Network Interface Cards (NIC)." *Teach-ICT*, www.teach-ict.com/as_a2_ict_new/ocr/A2_G063/333_networks_coms/network_components/miniweb/pg7.htm. Accessed 28 Jan. 2017.

Spanbauer, Scott. "Wired or Wireless? Choose Your Network." *PC World*, 30 Sept. 2003, www.pcworld.com/article/112560/article.html. Accessed 28 Jan. 2017.

"A 10 Gigabit Programmable Network Interface Card – An Overview." *Rice University*, www.ece.rice.edu/~willmann/teng_nics_overview.html. Accessed 28 Jan. 2017.

"What Is NIC (Network Interface Card)?" *OmniSecu.com*, www.omnisecu.com/basic-networking/what-is-nic-card-network-interface-card.php. Accessed 28 Jan. 2017.

Wonder, Dan. "The Function of the NIC." *Houston Chronicle*, smallbusiness.chron.com/function-nic-68223.html. Accessed 28 Jan. 2017.

Network Topology

SUMMARY

Network topology refers to the way computers are connected. A topology is the configuration in which the network is laid out, specifically how the computers are connected to one another and to a central point. There are four main network topology configurations and a number of variations of each. Larger networks may include several different topologies; these are often called hybrids. Network topologies make it possible for computers to communicate with one another, whether that communication takes place across a room or across the world.

BACKGROUND

Electronic computers were first available in the 1940s. Computer languages and computer chips followed in the 1950s, and the first personal computers became available in the 1970s. In the earliest days, the only way to transfer information from one computer to another was to put it on a storage device such as a tape drive or floppy disk, which was then physically carried and inserted into another computer. A Xerox employee named Robert Taylor is considered the earliest pioneer of computer networking. Taylor is also credited with inventing Ethernet, or a form of local area network (LAN). Computer networks provide a way to share information between computers.

Computer networks include computers, cabling, a network hub, network switch, or network router that can serve as a central control point for the information in the system, and peripherals such as printers and scanners. These components are laid out according to one of four computer topologies, or a variation of one of these. The word *topology* comes from the Greek *topos,* which means "place," and *logos,* which means "study."

Network Topologies. By NetworkTopologies.png: Maksim derivative work: Malyszkz (NetworkTopologies.png)

CATEGORIES OF COMPUTER TOPOLOGIES

Computer topologies generally fit into one of four major categories: bus, star, ring, or mesh. The characteristics of each determine whether it will be appropriate for use in a particular setting. When a particular application requires two or more topologies, a hybrid of multiple topologies might be used.

The bus topology is the most commonly used, in part because it is the type used for Ethernet. Ethernet is used to set up the architecture of a majority of LAN systems; therefore, the bus topology has become very common. In a bus system, sometimes called a linear bus topology, all of the computers and peripherals

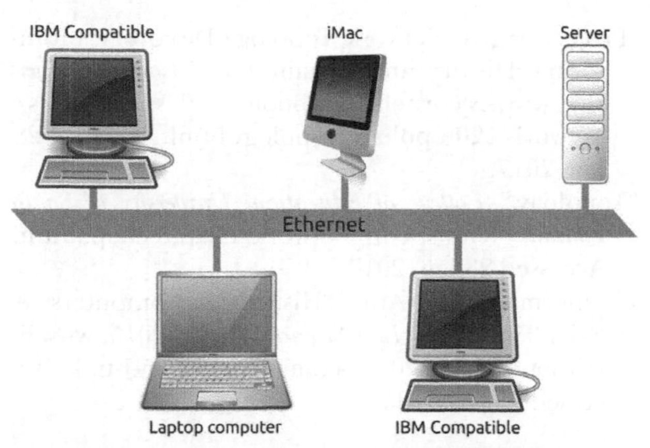

A conceptual diagram of a local area network. By T.seppelt, derivate work from File:Ethernet.png, including content of the Open Clip Art Library, by © 2007 Nuno Pinheiro & David Vignoni & David Miller & Johann Ollivier Lapeyre & Kenneth Wimer & Riccardo Iaconelli / KDE / LGPL 3, User:George Shuklin a

are connected to a single cable, which serves as a sort of spinal cord for the system. A typical system will include a file server to back up the systems, some form of printer or other peripheral device, and a number of individual computers. Each end of the cable is finished with a device called a terminator, which prevents signals from bouncing back to the connected computers.

Besides the fact that it is part of Ethernet systems, there are several advantages to bus topologies. Connecting each computer directly to the main line means that these networks use less cable than some other topologies, making them less expensive to set up. It is also relatively easy to add a computer to the network.

The same single cable that makes the bus system easier and less expensive to set up is also one of its weaknesses. If the cable fails, all computers on the system will become inoperable. The single cable also makes it difficult to identify exactly where a problem is if the system does fail. The two terminators the bus system requires also add to the cost of setting it up. Finally, the linear configuration means that this system is not a good solution for a large building; it can only work effectively with a few dozen computers.

In contrast to the linear bus system with each component connected to one line, a star topology features an individual cable that goes from each computer or peripheral directly to a central component. This central point, which is called a hub node, controls the flow of data through the system and can be a network hub, a switch, or a router; the way each computer radiates out from this central point is what gives the system the name "star." Information shared between computers goes through this hub node, whether it is traveling to another computer within the network or off to another computer through the Internet.

This central hub makes it easy to control security on the data on a star topology system. The layout makes it easy to add components up to the limits of the central hub node. While the failure of the hub will result in an outage of all the computers on the system, the failure of a single computer or other component will not affect the rest of the system and is easy to detect and repair. However, the additional cable used to connect each system and the cost of the hub adds to the cost of installation for the system.

The star topology is also limited in size depending on how many components can be added to the central hub. However, it is possible to get around this by using one of the input areas on the hub to add a line to another hub that can be the central point for additional computers. This configuration is called an expanded star or a tree topology.

In a ring topology, each computer on the system is paired with two others. All information shared from one of these computers travels around the ring—this can be clockwise or counter clockwise—to the other computers. A special device known as a token ring is often used to connect the components in this topology, which is commonly used in offices and schools. However, failure of one computer or the cable connecting them will cause all of the computers in the ring to fail as well.

Some networks are built with routes, or multiple paths that data can take in traveling from computer to computer. Systems like this, where each computer is connected directly or indirectly to all of the others, are called mesh topologies. The Internet is one such system.

—Janine Ungvarsky

BIBLIOGRAPHY

"Evolution of Computer Networks." *University of Notre Dame,* www3.nd.edu/~dwang5/courses/fall16/pdf/evolution.pdf. Accessed 28 Jan. 2017.

"Internet Hall of Fame Honors UT Austin Professor Bob Metcalfe and Alum Robert Taylor." *University of Texas at Austin,* 3 Aug. 2013, www.engr.utexas.

edu/news/7652-internet-hall-of-fame. Accessed 28 Jan. 2017.

McMillan, Robert. "Xerox: Uh, We Didn't Invent the Internet." *Wired,* 23 July 2012, www.wired.com/2012/07/xerox-internet/. Accessed 28 Jan. 2017.

"Network Topology Definition." *Linux Information Project,* 2 Nov. 2005, www.linfo.org/network_topology.html. Accessed 28 Jan. 2017.

"Network Topologies." *Pace University,* webpage.pace.edu/ms16182p/networking/topologies.html. Accessed 28 Jan. 2017.

Norman, Jeremy M., editor. *From Gutenberg to the Internet: A Sourcebook of the History of Information Technology.* HistoryofScience.com, 2005.

"Project Argus: Network Topology Discovery, Monitoring, History, and Visualization." *Cornell University,* www.cs.cornell.edu/boom/1999sp/projects/network%20topology/topology.html. Accessed 28 Jan. 2017.

"Topology." *College of Education, University of South Florida,* fcit.usf.edu/network/chap5/chap5.htm. Accessed 28 Jan. 2017.

Zimmerman, Kim Ann. "History of Computers: A Brief Timeline." *Live Science,* 8 Sept. 2015, www.livescience.com/20718-computer-history.html. Accessed 28 Jan. 2017.

Neural Engineering

SUMMARY

Neural engineering is an emerging discipline that translates research discoveries into neurotechnologies. These technologies provide new tools for neuroscience research, while leading to enhanced care for patients with nervous-system disorders. Neural engineers aim to understand, represent, repair, replace, and augment nervous-system function. They accomplish this by incorporating principles and solutions derived from neuroscience, computer science, electrochemistry, materials science, robotics, and other fields. Much of the work focuses on the delicate interface between living neural tissue and nonliving constructs. Efforts focus on elucidating the coding and processing of information in the sensory and motor systems, understanding disease states, and manipulating neural function through interactions with artificial devices such as brain-computer interfaces and neuroprosthetics.

DEFINITION AND BASIC PRINCIPLES

Neural engineering (or neuroengineering, NE) is an emerging interdisciplinary research area within biomedical engineering that employs neuroscientific and engineering methods to elucidate neuronal function and design solutions for neurological dysfunction. Restoring sensory, motor, and cognitive function in the nervous system is a priority. The strong emphasis on engineering and quantitative methods separates NE from the "traditional" fields of neuroscience and neurophysiology. The strong neuroscientific approach distinguishes NE from other engineering disciplines such as artificial neural networks. Despite being a distinct discipline, NE draws heavily from basic neuroscience and neurology and brings together engineers, physicians, biologists, psychologists, physicists, and mathematicians.

At present, neural engineering can be viewed as the driving technology behind several overlapping fields: functional electrical stimulation, stereotactic and functional neurosurgery, neuroprosthetics and neuromodulation. The broad scope of NE also encompasses neurodiagnostics, neuroimaging, neural tissue regeneration, and computational approaches. By using mathematical models of neural function (computational neuroscience), researchers can perform robust testing of therapeutic strategies before they are used on patients.

The human brain, arguably the most complex system known to humankind, contains about 10^{11} neurons and several times more glial cells. Understanding the functional neuroanatomy of this exquisite device is a sine qua non for anyone aiming to manipulate and repair it. The "neuron doctrine," pioneered by Spanish neuroscientist Santiago Ramón y Cajal, considers the neuron to be a distinct anatomical and functional unit. The extension introduced by American neuroscientist Warren S. McCullogh and American logician Walter Pitts asserts that the neuron is the basic information-processing unit of the brain. For neuroengineers, this means that a particular goal can be reached just by manipulating a cell or group of cells. One argument in favor of this view is that stimulating groups of neurons produces a regular effect. Motor activity, for example, can be induced by stimulating the motor cortex with electrodes. In addition, lesions to specific brain areas due to neurodegenerative disorders or stroke lead to more or less predictable clinical manifestation patterns.

BACKGROUND AND HISTORY

Electricity (in the form of electric fish) was used by ancient Egyptians and Romans for therapeutic purposes. In the eighteenth century, the work of Swiss anatomist Albrecht von Haller, Italian physician Luigi Galvani, and Benjamin Franklin set the stage for the use of electrical stimulation to restore movement to paralyzed limbs. The basis of modern NE is early neuroscience research demonstrating that neural function can be recorded, manipulated, and mathematically modeled. In the mid-twentieth century, electrical recordings became popular as a window into neuronal function. Metal wire electrodes recorded extracellularly, while glass pipettes probed individual cells. Functional electrical stimulation (FES) emerged with a distinct engineering orientation and the aim to use controlled electrical stimulation to restore function. Modern neuromodulation has developed since the 1970's, driven mainly by clinical professionals. The first peripheral nerve, then spinal cord and deep brain stimulators were introduced in the 1960's. In 1997, the Food and Drug Administration (FDA) approved deep brain stimulation (DBS) for the treatment of Parkinson's disease. An FES-based device that restored grasp was approved the same year.

In the 1970's, researchers developed primitive systems controlled by electrical activity recorded from the head. The U.S. Pentagon's Advanced Research Projects Agency (ARPA) supported research aimed at developing bionic systems for soldiers. Scientists demonstrated that recorded brain signals can communicate a user's intent in a reliable manner and found cells in the motor cortex the firing rates of which correlate with hand movements in two-dimensional space.

Since the 1960's, engineers, neuroscientists, and physicists have constructed mathematical models of the retina that describe various aspects of its function, including light-stimulus processing and transduction. In addition, scientists have made attempts to treat blindness using engineering solutions, such as nonbiological "visual prostheses." In 1975, the first multichannel cochlear implant (CI) was developed and implanted two years later.

HOW IT WORKS

Neuromodulation and Neuroaugmentation. Neural engineering applications have two broad (and sometimes overlapping) goals: neuromodulation and neuroaugmentation. Neuromodulation (altering nervous system function) employs stimulators and infusion devices, among other techniques. It can be applied at multiple levels: cortical, subcortical, spinal, or peripheral. Neural augmentation aims to amplify neural function and uses sensory (auditory, visual) and motor prostheses.

Neuromuscular Stimulation. Based on a method that has remained unchanged for decades, electrodes are placed within the excitable tissue that provide current to activate certain pathways. This supplements or replaces lost motor or autonomic functions in patients with paralysis. An example is application of electrical pulses to peripheral motor nerves in patients with spinal cord injuries. These pulses lead to action potentials that propagate across neuromuscular junctions and lead to muscle contraction. Coordinating the elicited muscle contractions ultimately reconstitutes function.

Neural Prosthetics. Neural prostheses (NP) aim to restore sensory or motor function—lost because of disease or trauma—by linking machines to the nervous system. By artificially manipulating the biological system using external electrical currents, neuroengineers try to mimic normal sensorimotor function.

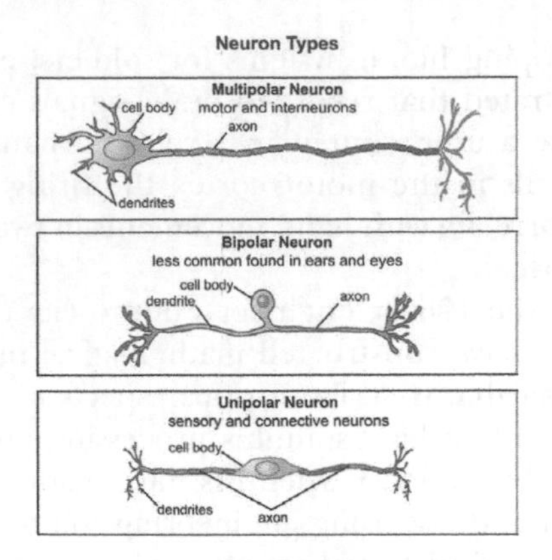

Electrodes act as transducers that excite neurons through electrical stimulation, or record (read) neural signals. In the first approach, stimulation is used for its therapeutic efficacy, for example, to alleviate the symptoms of Parkinson's disease, or to provide input to the nervous system, such as converting sound to neural input with a cochlear implant. The second paradigm uses recordings of neural activity to detect motor intention and provide input signal to an external device. This forms the basis of a subset of neural prosthetics called brain-controlled interfaces (BCI).

Microsystems. Miniaturization is a crucial part of designing instruments that interface efficiently with neural tissue and provide adequate resolution with minimal invasiveness. Microsystems technology integrates devices and systems at the microscopic and submicroscopic levels. It is derived from microelectronic batch-processing fabrication techniques. A "neural microsystem" is a hybrid system consisting of a microsystem and its interfacing neurons (be they cultured, part of brain slices, or in the intact nervous system). Technologies such as microelectrodes, microdialysis probes, fiber optic, and advanced magnetic materials are used. The properties of these systems render them suitable for simultaneous measurements of neuronal signals in different locations (to analyze neural network properties) as well as for implantation within the body.

APPLICATIONS AND PRODUCTS

Some of the most common applications of NE methods are described below.

Cochlear Implants. Cochlear implants (CI), by far the most successful sensory neural prostheses to date, have penetrated the mainstream therapeutic arsenal. Their popularity is rivaled only by the cardiac pacemakers and deep brain stimulation (DBS) systems. Implanted in patients with sensorineural deafness, these devices process sounds electronically and transmit stimuli to the cochlea. A CI includes several components: a microphone, a small speech processor that transforms sounds into a signal suitable for auditory neurons, a transmitter to relay the signal to the cochlea, a receiver that picks up the transmitted signal, and an electrode array implanted in the cochlea. Individual results vary, but achieving a high degree of accuracy in speech perception is possible, as is the development of language skills.

Retinal Bioengineering. Retinal photoreceptor cells contain visual pigment, which absorbs light and initiates the process of transducing it into electrical signals. They synapse onto other types of cells, which in turn carry the signals forward, eventually through the optic nerve and into the brain, where they are interpreted. Every neuron in the visual system has a "receptive field," a particular portion of the visual space within which light will influence that neuron's behavior. This is directly related to (and represented by) a specific region of the retina. Inherited retinal degenerations such as retinitis pigmentosa (RP) or age-related macular degeneration (AMD) are responsible for the compromised or nonexistent vision of millions of people. In these disorders, the retinal photoreceptor cells lose function and die, but the secondary neurons are spared.

Using an electronic prosthetic device, a signal is sent to these secondary neurons that ultimately causes an external visual image. A miniature video camera is mounted on the patient's eyeglasses that captures images and feeds them to a microprocessor, which converts them to an electronic signal. Then the signal is sent to an array of electrodes located on the retina's surface. The electrodes transmit the signal to the viable secondary neurons. The neurons process the signal and pass it down the optic nerve to the brain to establish the visual image.

Several different versions of this device exist and are implanted either into the retina or brain. Cortical visual prostheses could entirely bypass the retina, especially when this structure is damaged from diseases such as diabetes or glaucoma. Retinal prostheses, or artificial retinas (AR), could take advantage of any remaining functional cells and would target photoreceptor disorders such as RP. Two distinct retinal placements are used for AR. The first type slides under the retina (subretinal implant) and consists of small silicon-based disks bearing microphotodiodes. The second type would be an epiretinal system, which involves placing the camera or sensor outside the eye, sending signals to an intraocular receiver. In addition to challenges related to miniaturization and power supply, developing these systems faces obstacles pertaining to biocompatibility, such as retinal health and implant damage, and vascularization.

Functional Electrical Stimulation (FES). Some FES devices are commercialized, and others belong to clinical research settings. A typical unit includes an electronic stimulator, a feedback or control unit, leads, and electrodes. Electrical stimulators bear one or multiple channels (outputs) that are activated simultaneously or in sequence to produce the desired movement. Applications of FES include standing, ambulation, cycling, grasping, bowel and bladder control, male sexual assistance, and respiratory control. Although not curative, the method has numerous benefits, such as improved cardiovascular health, muscle-mass retention, and enhanced psychological well-being through increased functionality and independence.

Brain-Controlled Interfaces. A two-electrode device was implanted into a 1998 stroke victim who could communicate only by blinking his eyes. The device read from only a few neurons and allowed him to select letters and icons with his brain. A team of researchers helped a young patient with a spinal cord injury by implanting electrodes into his motor cortex that were connected to an interface. The patient was able to use the system to control a computer cursor and move objects using a robotic arm.

Brain-controlled interfaces (BCIs), a subset of NP, represent a new method of communication based on brain-generated neural activity. Still in an experimental phase, they offer hope to patients with severe motor dysfunction. These interfaces capture neural activity mediating a subject's intention to act and translate it into command signals transmitted to a computer (brain-computer interface) or robotic limb. Independent of peripheral nerves and muscles, BCI have the ability to restore communication and movement. This exciting technological advance is not only poised to help patients, but it also provides insight into the way neurons interact.

Every BCI has four main components: recording of electrical activity, extraction of the planned action from this activity, execution of the desired action using the prosthetic effector (actuator), and delivery of feedback (via sensation or prosthetic device).

Brain-controlled interfaces rely on four main recording modalities: electroencephalography, electrocorticography, local field potentials, and singe-neuron action potentials. The methods are noninvasive, semi-invasive, or invasive, depending on where the transducer is placed: scalp, brain surface, or cortical tissue.

The field is still in its infancy; however, several basic principles have emerged from these and other early experiments. A crucial requirement in BCI function, for example, is for the reading device to obtain sufficient information for a particular task. Another observation refers to the "transparency of action" in brain-machine interface (BMI) systems: Upon reaching proficiency, the action follows the thought, with no awareness of intermediate neural events.

Deep Brain Stimulation (DBS) and Other Modulation Methods. Deep brain stimulation of thalamic nuclei decreases tremors in patients with Parkinson's disease. It may alleviate depression, epilepsy, and other brain disorders. One or more thin electrodes, about 1 millimeter in diameter, are placed in the brain. An external signal generator with a power supply is also implanted somewhere in the body, typically in the chest cavity. An external remote control sends signals to the generator, varying the parameters of the stimulation, including the amount and frequency of the current and the duration and frequency of the pulses. The exact mechanism by which this method works is still unclear. It appears to exert its effect on axons and act in an inhibitory manner, by inducing an effect akin to ablation of target area, much like early Parkinson's treatment.

One major advantage of DBS over other previously employed methods is its reversibility and absence of structural damage. Another valuable neuromodulatory approach, the electrical stimulation of the vagus nerve, can reduce seizure frequency in patients with epilepsy and alleviate treatment-resistant depression. Transcutaneous electrical nerve stimulation (TENS) represents the most common form of electrotherapy and is still in use for pain relief. Cranial electrotherapy stimulation involves passing small currents across the skull. The approach shows good results in depression, anxiety, and sleep disorders.

Transcranial magnetic stimulation uses the magnetic field produced by a current passing through a coil and can be applied for diagnostic (multiple sclerosis, stroke), therapeutic (depression), or research purposes.

SOCIAL CONTEXT AND FUTURE PROSPECTS

Bioelectrodes for neural recording and neurostimulation are an essential part of neuroprosthetic devices. Designing an optimal, stable electrode that records long-term and interacts adequately with neural tissue remains a priority for neural engineers. The implementation of microsystem technology opens new perspectives in the field.

About 466 million people around the world suffer from hearing loss, according to the World Health Organization in 2018, mainly because sensory hair cells in the cochlea have degenerated. The only efficient therapy for patients with profound hearing loss is the CI. Improvements in CI performance have increased the average sentence recognition with multichannel devices. An exciting new development, auditory brainstem implants, show improved performance in patients with impaired cochlear nerves.

Millions of Americans have vision loss. The need for a reliable prosthetic retina is significant, and rivals the one for CI. Technological progress makes it quite likely that a functioning implant with a more sophisticated design and higher number of electrodes will be on the market soon. The epiretinal approach is promising, but providing interpretable visual information to the brain represents a challenge. In addition, even if they prove to be successful, retinal prostheses under development address only a limited number of visual disorders. Much is left to be discovered and tested in this field.

The coming years will also see rapid gains in the area of BCI. Whether they achieve widespread use will depend on several factors, including performance, safety, cost, and improved quality of life.

The advent of gene therapy, stem cell therapy, and other regenerative approaches offers new hope for patients and may complement prosthetic devices. However, many ethical and scientific issues still have to be solved.

Implanted devices are changing the way neurological disorders are treated. An unprecedented transition of NE discoveries from the research to the commercial realm is taking place. At the same time, new discoveries constantly challenge the basic tenets of neuroscience and may alter the face of NE in the coming decades. People's understanding of the nervous system, especially of the brain, changes, and so do the strategies designed to enhance and restore its function.

—Mihaela Avramut, MD, PhD

Bibliography

Blume, Stuart. *The Artificial Ear: Cochlear Implants and the Culture of Deafness.* New Brunswick, N.J.: Rutgers University Press, 2010.

DiLorenzo, Daniel J., and Joseph D. Bronzino, eds. *Neuroengineering.* Boca Raton, Fla.: CRC Press, 2008.

Durand, Dominique M. "What Is Neural Engineering?" *Journal of Neural Engineering* 4, no. 4 (September, 2006).

He, Bin, ed. *Neural Engineering.* New York: Kluwer Academic/Plenum Publishers, 2005.

Hu, Xiaoling. "Advances in Neural Engineering for Rehabilitation." *Behavioural Neurology,* vol. 2017, 2017, doi: 10.1155/2017/9240921. Accessed 28 Feb. 2018.

Katz, Bruce F. *Neuroengineering the Future: Virtual Minds and the Creation of Immortality.* Hingham, Mass.: Infinity Science Press, 2008.

Montaigne, Fen. *Medicine By Design: The Practice and Promise of Biomedical Engineering.* Baltimore: The Johns Hopkins University Press, 2006.

Numerical Analysis

SUMMARY

Numerical analysis is the study of how to design, implement, and optimize algorithms that provide approximate values to variables in mathematical expressions. Numerical analysis has two broad subareas: first, finding roots of equations, solving systems of linear equations, and finding eigenvalues; and second, finding solutions to ordinary and partial differential equations (PDEs). Much of this field involves using numerical methods (such as finite differences) to solve sets of differential equations. Examples are Brownian motion of polymers in solution, the kinetics of phase transition, the prediction of material microstructures, and the development of novel methods for simulating earthquake mechanics.

DEFINITION AND BASIC PRINCIPLES

Most of the phenomena of science have discrete or continuous models that use a set of mathematical equations to represent the phenomena. Some of the equations have exact solutions as a number or set of numbers, but many do not. Numerical analysis provides algorithms that, when run a finite number of times, produce a number or set of numbers that approximate the actual solution of the equation or set of equations. For example, since ϖ is transcendental, it has no finite decimal representation. Using English mathematician Brook Taylor's series for the arctangent, however, one can easily find an approximation of ϖ to any number of digits. One can also do an error analysis of this approximation by looking at the tail of the series to see how closely the approximation came to the exact solution.

Babylonian clay tablet YBC 7289 (c. 1800–1600 BCE) with annotations. The approximation of the square root of 2 is four sexagesimal figures, which is about six decimal figures. $1 + 24/60 + 51/60^2 + 10/60^3 = 1.41421296....$

Finding roots of polynomial equations of a single variable is an important part of numerical analysis, as is solving systems of linear equations using Gaussian elimination (named for German mathematician Carl Friedrich Gauss) and finding eigenvalues of matrices using triangulation techniques. Numeric solution of ordinary differential equations (using simple finite difference methods such as Swiss mathematician Leonhard Euler's formula, or more complex methods such as German mathematicians C. Runge and J. W. Kutta's Runge-Kutta algorithm) and partial differential equations (using finite element or grid methods) are the most active areas in numerical analysis.

BACKGROUND AND HISTORY

Numerical analysis existed as a discipline long before the development of computers. By 1800, Lagrange polynomials, named for Italian-born mathematician Joseph-Louis Lagrange, were being used for general approximation, and by 1900 the Gaussian technique for solving systems of equations was in common use. Ordinary differential equations with boundary conditions were being solved using Gauss's method in 1810, using English mathematician John Couch Adams's difference methods in 1890, and using the Runge-Kutta algorithm in 1900. Analytic solutions of partial differential equations (PDEs) were being developed by 1850, finite difference solutions by 1930, and finite element solutions by 1956.

The classic numerical analysis textbook *Introduction to Numerical Analysis* (1956), written by American mathematician Francis Begnaud Hildebrand, had substantial sections on numeric linear algebra and ordinary differential equations, but the algorithms were computed with desktop calculators. In these early days, much time was spent finding multiple representations of a problem in order to get a representation that worked best with desktop calculators. For example, a great deal of effort was spent on deriving Lagrange polynomials to be used for approximating curves. The early computers, such as the Electronic Numerical Integrator And Computer (ENIAC), built by John Mauchly and John Presper Eckert in 1946, were immediately applied to the existing numerical analysis algorithms and stimulated the development of many new algorithms as well.

Modern computer-based numerical analysis really got started with John von Neumann and Herman Goldstine's 1947 paper "Numerical Inverting of Matrices of High Order," which appeared in the *Bulletin of the American Mathematical Society*. Following that, many new and improved techniques were developed for numerical analysis, including cubic-spline approximation, sparse-matrix packages, and the finite element method for elliptic PDEs with boundary condition.

HOW IT WORKS

Numerical analysis has many fundamental techniques. Below are some of the best known and most useful.

Approximation and Error. It is believed that the earliest examples of numerical analysis were developed by the Greeks and others as methods of finding numerical quantities that were approximations of values of variables in simple equations. For example, the length of the hypotenuse of a 45-degree right triangle with side of length 1 is $\sqrt{2}$, which has no exact decimal representation but has rational approximations to any degree of accuracy. The first three elements of the standard sequence of approximations of $\sqrt{2}$ are 1.4, 1.414, and 1.4142. Another fundamental idea is error (a bound of the absolute difference between a value and its approximation); for example, the error in the approximation of $\sqrt{2}$ by 1.4242, to ten digits, is 0.000013563.

There are many examples of approximation in numerical analysis. The Newton-Raphson method, named for English physicist Sir Isaac Newton and English mathematician Joseph Raphson, is an iterative method used to find a real root of a differentiable function and is included in most desktop math packages. A function can also be approximated by a combination of simpler functions, such as representing a periodic function as a Fourier series (named for French mathematician Jean-Baptiste Joseph Fourier) of trigonometric functions, representing a piecewise smooth function as a sum of cubic splines (a special polynomial of degree 3), and representing any function as a Laguerre series (a sum of special polynomials of various degrees, named for French mathematician Edmond Laguerre).

Solution of Systems of Linear Equations. Systems of linear equations were studied shortly after the introduction of variables; examples existed in early Babylonia. Solving a system of linear equations involves determining whether a solution exists and then using either a direct or an iterative method to find the solution. The earliest iterative solutions of linear system were developed by Gauss, and newer iterative algorithms are still published. Many of the problems of science are expressed as systems of linear equations, such as balancing chemical equations, and are solved when the linear system is solved.

Finite Differences. Many of the equations used in numerical analysis contain ordinary or partial derivatives. One of the most important techniques used in numerical analysis is to replace the derivatives in an equation, or system of equations, with equivalent finite differences of the same order, and then develop an iterative formula from the equation. For example, in the case of a first-order differential equation with an initial value condition, such as $f'(x) = F[x, f(x)]$, $y_0 = f(x_0)$, one can replace the derivative by using equivalent differences, $f'(x_1) = [f'(x_1) - f(x_0)]/(y_1 - x_0)$ and solve the resulting equation for y_1. After getting y_1, one can use the same technique to find approximations for yi given xi for i greater than 2. There are many examples of using finite differences in solving differential equations. Some use forward differences, such as the example above; others use backward differences, much like antiderivatives; and still others use higher-order differences.

Grid Methods. Grid methods provide a popular technique for solving a partial differential equation with n independent variables that satisfies a set of boundary conditions. A grid vector is a vector of the independent variables of a partial differential equation, often formed by adding increments to a boundary point. For example, from (t, x) one could create a grid by adding t to t and x to x systematically.

The basic assumption of the grid method is that the partial differential equation is solved when a solution has been found at one or more of the grid points. While grid methods can be very complex, most of them follow a fairly simple pattern. First, one or more initial vectors are generated using the boundary information. For example, if the boundary is a rectangle, one might choose a corner point of the rectangle as an initial value; otherwise one might have to interpolate from the boundary information to select an initial point. Once the initial vectors are selected, one can develop recursive formulas (using techniques like Taylor expansions or finite differences) and from these generate recursive equations over the grid vectors. Adding in the information for the initial values yields sufficient information to solve the recursive equations and thus yields a solution to the partial differential equation. For example, given a partial differential equation and a rectangle the lower-left corner point of which satisfies the boundary condition, one often generates a rectangular grid and set of recursive equations that, when solved, yield a solution of the partial differential equation. Many grid methods support existence and uniqueness proofs for the PDE, as well as error analysis at each grid vector.

APPLICATIONS AND PRODUCTS

The applications of numerical analysis, including the development of new algorithms and packages within the field itself, are numerous. Some broad categories are listed below.

PACKAGES TO SUPPORT NUMERICAL ANALYSIS. One of the main applications of numerical analysis is developing computer software packages implementing sets of algorithms. The best known and most widely distributed package is LINPACK, a software library for performing numeric linear algebra developed at Stanford University by students of American mathematician George Forsythe. Originally developed in FORTRAN for supercomputers, it is now available in many languages and can be run on large and small computers alike, although it has largely been superseded by LAPACK (Linear Algebra Package). Another software success story is the development of microcomputer numerical analysis packages. In 1970, a few numerical analysis algorithms existed for muMath (the first math package for microcomputers); by the twenty-first century, almost all widely used math packages incorporated many numerical analysis algorithms.

ASTRONOMY, BIOLOGY, CHEMISTRY, GEOLOGY, AND PHYSICS. Those in the natural sciences often express phenomena as variables in systems of equations, whether differential equations or PDEs. Sometimes a symbolic solution is all that is necessary, but numeric answers are also sought. Astronomers use numeric integration to estimate the volume of Saturn a million years ago; entomologists can use numeric integration to predict the size of the fire ant population in Texas twenty years from the present. In physics, the solutions of differential equations associated with dynamic light scattering can be used to determine the size of polymers in a solution. In geology, some of the Earth's characteristics, such as fault lines, can be used as variables in models of the Earth, and scientists have predicted when earthquakes will occur by solving these differential equations.

MEDICINE. Many of the phenomena of medicine are represented by ordinary or partial differential equations. Some typical applications of numerical analysis to medicine include estimating blood flow for stents of different sizes (using fluid flow equations), doing a study across many physicians of diaphragmatic hernias (using statistical packages), and determining the optimal artificial limb for a patient (solving some differential equations of dynamics).

ENGINEERING. Engineers apply the natural sciences to real-world problems and often make heavy use of numerical analysis. For example, civil and mechanical engineering have structural finite element simulations and are among the biggest users of computer time at universities. In industry, numerical analysis is used to design aerodynamic car doors and high-efficiency air-conditioner compressors. Electrical engineers, at universities and in industry, always build a computer model of their circuits and run simulations before they build the real circuits, and these simulations use numeric linear algebra, numeric solution of ordinary differential equations, and numeric solution of PDEs.

FINITE ELEMENT PACKAGES. For some problems in numerical analysis, a useful technique for solving the problem's differential equation is first to convert it to a new problem, the solution of which agrees with that of the original equation. The most popular use of the finite element, to date, has been to solve elliptic partial PDEs. In most versions of finite element packages, the original equations are replaced by a new system of equations that agree at the boundary points. The new system of differential equations is easier to solve "inside" the boundary than is the original system and is proved to be close to the solution of the original system. Much care is taken in selecting the original finite element grid and approximating functions. Examples of finite element abound, including modeling car body parts, calculating the stiffness of a beam that needs to hold a number of different weights and simulating the icing of an airplane wing.

SOCIAL CONTEXT AND FUTURE PROSPECTS

Advances in numerical algorithms have made a number of advances in science possible, such as improved weather forecasting, and have improved life for everyone. If scientific theories are to live up to their full potential in the future, ways of finding approximations to values of the variables used in these theories are needed. The increased complexity of these scientific models is forcing programmers to design, implement, and test new and more sophisticated numerical analysis algorithms.

—George M. Whitson III, BS, MS, PhD

Bibliography

Burden, Richard L., and J. Douglas Faires. *Numerical Analysis.* 9th ed. Boston: Brooks, 2011. Print

Hildebrand, F. B. *Introduction to Numerical Analysis.* 2nd ed. New York: McGraw, 1974. Print.

Iserles, Arieh. *A First Course in the Numerical Analysis of Differential Equations.* 2nd ed. New York: Cambridge UP, 2009. Print.

Moler, Cleve B. *Numerical Computing with MATLAB.* Philadelphia: Soc. for Industrial and Applied Mathematics, 2008. Print.

Overton, Michael L. *Numerical Computing with IEEE Floating Point Arithmetic.* Philadelphia: Soc. for Industrial and Applied Mathematics, 2001. Print.

Ralston, Anthony, and Philip Rabinowitz. *A First Course in Numerical Analysis.* 2nd ed. Mineola: Dover, 2001. Print.

Strauss, Walter A. *Partial Differential Equations: An Introduction.* 2nd ed. Hoboken: Wiley, 2008. Print.

O

Objectivity (Science)

SUMMARY

Scientific objectivity is the basis on which the scientific community strives to conduct its research. Objectivity is the concept that scientists observe and document scientific truths, theories, and facts through study and experimentation while remaining neutral on the subject themselves. The scientific method, the process by which modern scientific research is performed, seeks to prevent personal and professional biases from corrupting the scientific process. This method typically employs the following steps: (1) conducting research and making an observation, (2) forming a hypothesis, (3) performing an experiment, (4) recording and analyzing the data, and (5) drawing an evidence-based conclusion. In addition, a neutral party must be able to repeat the same process using the same methods and duplicate the original results. This method helps ensure that scientists remain objective in their research and do not consciously or unconsciously distort or force inaccurate scientific conclusions on the results of their experiments.

BACKGROUND

This concept of scientific objectivity became prominent in the early nineteenth century as the scientific method was honed and research technology and methods were advanced. Before this time, scientists and naturalists sought to make genuine discoveries through studying the natural world. To capture the stages of their scientific research and observations, they created artistic representations; however, their methods tended to be based on the perspective of the researcher and did not always authentically depict the observations being made.

The development of photographic camera technology in the nineteenth century helped usher in a new era of scientific objectivity, as visual observations could be accurately recorded instead of interpreted through works of art such as drawings. In the twentieth century, the objective scientific method was further complemented by the standardized use of technological data and images that could be interpreted for many purposes in a variety of professions.

STRATEGIES TO MAINTAIN OBJECTIVITY

Scientists employ several different strategies to keep their work accurate and free of conscious or unconscious personal influence. These strategies typically include crafting an unambiguous hypothesis, ensuring that both the experimenter and the participant are unaware of who belongs to the control group and who belongs to the test group, testing a large sample, and having qualified third-party professionals analyze the results.

An additional way scientists endeavor to remain objective in their work is by using standard units and tools for measurements of quantities such as length, weight, distance, and time. Standardized tools may include rulers, thermometers, calipers, scales, and electrometers. Using standard equipment accepted by one's discipline or field of study helps eliminate the possible discrepancies that could arise from multiple scientists performing the same or similar research with differently calibrated tools. Measurements are expressed using a standard system, typically the International System of Units (SI), allowing scientists across multiple nations and disciplines to comprehend the results and repeat the scientific process, hopefully achieving the same results. This practice is employed in the science classroom for students at all levels as well. An instructor will give students a set of standard materials in a laboratory environment and ask the students to perform a certain set of steps to conduct an experiment and reach a specific result, cultivating a dedication to unbiased procedures.

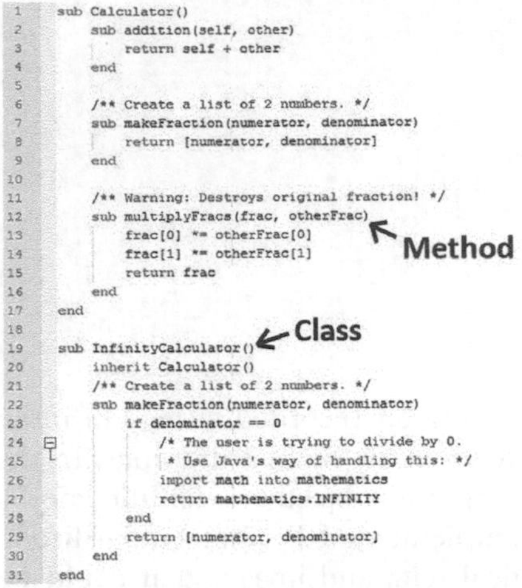

```
1  sub Calculator()
2      sub addition(self, other)
3          return self + other
4      end
5
6      /** Create a list of 2 numbers. */
7      sub makeFraction(numerator, denominator)
8          return [numerator, denominator]
9      end
10
11     /** Warning: Destroys original fraction! */
12     sub multiplyFracs(frac, otherFrac)   <- Method
13         frac[0] *= otherFrac[0]
14         frac[1] *= otherFrac[1]
15         return frac
16     end
17 end
18
19 sub InfinityCalculator()   <- Class
20     inherit Calculator()
21     /** Create a list of 2 numbers. */
22     sub makeFraction(numerator, denominator)
23         if denominator == 0
24             /* The user is trying to divide by 0.
25              * Use Java's way of handling this: */
26             import math into mathematics
27             return mathematics.INFINITY
28         end
29         return [numerator, denominator]
30     end
31 end
```

A source code example that shows classes, methods, and inheritance. By Carrot Lord (Taking a screenshot, then editing using Paint.NET)

The verification of scientific research is often carried out in the form of peer reviews, where members of the same or similar disciplines or fields of study analyze and comment on one another's findings and conclusions. These reviews are conducted in a variety of ways, most often as part of the submission process for academic journals, in which several other scientists read and evaluate the scientific process and methods used before the journal accepts and circulates the conclusions. This requirement can also be met through presentations at various scientific symposia and forums, where members of similar disciplines can directly question the researchers about their process and results. The scientific community publishes thousands of peer-reviewed academic journals and regularly holds dedicated conferences where this information can be presented and reviewed by fellow experts in the appropriate field, with each specific discipline having a variety of dedicated journals in which scientists can share their findings. Academic journals are not the same as scientific magazines, which also publish articles on scientific results; the intent of an academic journal is to provide a forum for peer-reviewed materials, while the scientific magazine provides an independent overview of ongoing scientific news and research.

The scientist's ability to remove himself or herself from an experiment or research, and the very possibility of objectivity in general, has been widely debated over the years, but those involved in the scientific community believe that objectivity should be the foundation of the research they conduct. Regardless of the specific discipline, scientific objectivity demands that researchers employ a repeatable experimentation process that can be used to verify the authenticity and value of the scientific research and studies being performed.

—L. L. Lundin, MA

BIBLIOGRAPHY

Ackermann, Robert John. *Data, Instruments, and Theory: A Dialectical Approach to Understanding Science.* Princeton: Princeton UP, 1985. Print.

Agazzi, Evandro. *Scientific Objectivity and Its Contexts.* Cham: Springer, 2014. Print.

Couvalis, George. *The Philosophy of Science: Science and Objectivity.* London: Sage, 1997. Print.

Daston, Lorraine, and Peter Galison. *Objectivity.* 2007. New York: Zone, 2010. Print.

Gauch, Hugh G., Jr. *Scientific Method in Brief.* New York: Cambridge UP, 2012. Print.

Hofmann, Angelika H. *Scientific Writing and Communication: Papers, Proposals, and Presentations.* 2nd ed. New York: Oxford UP, 2014. Print.

Roessler, Johannes, Hemdat Lerman, and Naomi Eilan, eds. *Perception, Causation, and Objectivity.* New York: Oxford UP, 2011. Print.

Vaughan, Simon. *Scientific Inference: Learning from Data.* Cambridge: Cambridge UP, 2013. Print.

OBJECT-ORIENTED PROGRAMMING (OOP)

SUMMARY

Object-oriented programming (OOP) is a type of computer programming that is based on the idea of objects, which are bundles of code. Some famous programming languages (e.g., Java, C++, and Python) are used for OOP. The bundles of code in OOP are useful because programmers can easily insert or remove them

An icon showing "OOP" (Object Oriented Programming) in web 2.0 style. By Wi-destination (Own work)

from programs and reuse them in other programs. OOP is useful to programmers because the objects can make programming less time-consuming and make coding errors less common. Although OOP has a number of benefits, it is not as flexible as some types of procedural programming.

MAJOR PRINCIPLES OF OOP

OOP is based on the idea of objects. In OOP, *objects* are bundles of code that have specific functions. All OOP objects have states and behaviors. This is similar to objects in the real world. For example, a cell phone—which is an object in real life—also has states and behaviors. The phone can be turned on or off, which are two of its states. The phone can also place calls, display the time, and more; these are all behaviors of the phone. The state of an object in OOP is stored in fields in the coding. Objects in OOP are used for many functions, and they can make programming simpler.

In OOP, objects are separated into different *classes.* Just as real-life objects can be classified by groups (e.g., tools, clothes, furniture), OOP objects can belong to different classes. The classes in OOP are blueprints that explain how objects in a particular class should be made.

Objects in real life have classes (e.g., houses), but they also have subclasses (e.g., ranches, cottages, mobile homes). Objects can also belong to *subclasses* in OOP. The subclasses are important because each member of the subclass can inherit particular traits from the main class. For instance, if the class of objects were houses and the subclass were mobile homes, the subclass would inherit certain traits, such as having rooms and a door, that are common to the main class. In the same way, a coding object that belongs to a subclass can inherit traits from its main class. This idea is called *inheritance,* and it is a fundamental concept in OOP.

Another fundamental idea people working with OOP have to understand is that idea of an *interface.* Objects in OOP have interfaces. Objects in everyday life have interfaces too. For example, a television has buttons a user can press to interact with it. A person can push the power button to turn the TV on and off. The TV's power has only two functions, so a person cannot make the TV do something it was not designed to do using the interface. A person could not change the colors on the TV screen using the power button. The TV's interface determines how people interact with the TV. In the same way, an OOP object's interface will allow the object to have specific, set functions. People can use an interface to make an object perform one of those functions. But a person cannot use the interface to make the object perform a function for which it was not designed.

A package is another concept that is important to OOP. *Packages* are spaces that organize sets of related classes and interfaces. Packages are similar to folders in a filing system. Each folder has a name and holds similar types of information. Programs can include hundreds and even thousands of classes, so it is important to use packages to organize this information.

Encapsulation, data abstraction, and polymorphism are other important concepts in OOP. *Encapsulation* is the process of putting many pieces of data into a single unit. OOP allows programmers to put these pieces together, which hides unnecessary data from users. This provides users with a clear and simple interface. *Data abstraction* is another important concept. This aspect of OOP allows users to reuse objects and even small programs over and over again without fully understanding how they work. Just as a person can use a cell phone without understanding the elements that make it function, a programmer can use objects and subprograms without understanding the code that makes them work. A final important concept in OOP is *polymorphism,* which is related to abstraction. This concept allows users to give general commands to all members of a class, even though each member of the class may execute the command in a different way. A real-life example of this is a teacher asking two students to add together

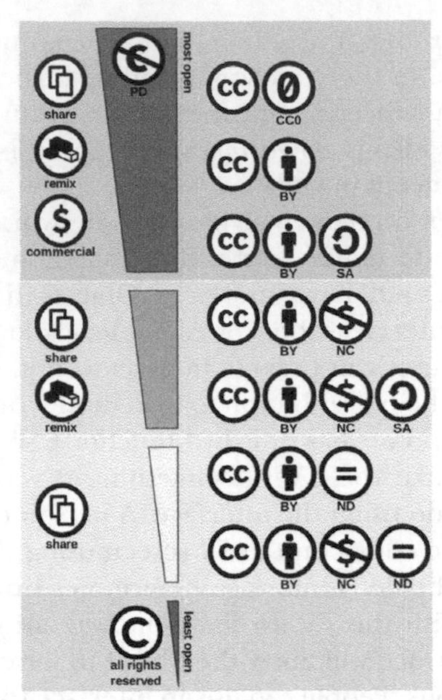

Creative commons license spectrum between public domain (top) and all rights reserved (bottom). By Creative commons (the original CC license symbols), the combined work by Shaddim and is hereby cc-by-4.0 licensed.

two numbers. One student may add the numbers together on paper. The other student may add the numbers together on a calculator. Even though the two students used different methods, both executed the action the teacher asked them to perform.

BENEFITS OF OOP

OOP is popular with many programmers because it has a number of benefits. Objects that are used in OOP can be designed and written by one programmer and adopted by another. A programmer can use objects that another programmer has already written and tested for accuracy. Once an object is written and tested, it can be easily inserted into another program using OOP. Another benefit is that objects in OOP can be removed. If programmers find that a particular object is causing program errors, the problematic object can be removed without changing or deleting the entire program.

Although OOP is useful because objects can be easily added or removed, this type of programming does not allow programmers to simply string objects together to create a useable program. Objects are tools that programmers can use and reuse in a number of ways, but they are only building blocks that can be added to sound programming structures.

—Elizabeth Mohn

BIBLIOGRAPHY

Hock-Chuan, Chua. "Java Programming Tutorial: Object-Oriented Programming (OOP) Basics." Nanyang Technological University. Nanyang Technological University. Apr. 2013. Web. 11 Aug. 2015. http://www.ntu.edu.sg/home/ehchua/programming/java/J3a_OOPBasics.html

"Lesson: Object-Oriented Programming Concepts." *Oracle.* Oracle. Web. 11 Aug. 2015. https://docs.oracle.com/javase/tutorial/java/concepts/index.html

Nakov, Svetlin. "Chapter 20. Object-Oriented Programming Principles (OOP)." *Introduction to Programming with C#/Java Books.* Svetlin Nakov and Team. http://www.introprogramming.info/english-intro-csharp-book/read-online/chapter-20-object-oriented-programming-principles/#_Toc362296569

"Object-Oriented Programming (C# and Visual Basic)." *Microsoft Developer Network.* Microsoft. Web. 11 Aug. 2015. https://msdn.microsoft.com/en-us/library/Dd460654.aspx

"Object-Oriented Programming (OOP) Definition." *TechTarget.* TechTarget. Web. 11 Aug. 2015. http://searchsoa.techtarget.com/definition/object-oriented-programming

OPEN ACCESS (OA)

SUMMARY

Open access (OA) is a term used in publishing. It refers to data, scientific findings, technical journal entries, research articles, and other primarily nonfiction literature that are available for free, with little to no restrictions regarding licenses and copyright. The vast majority of open-access materials are published

Open Access logo. By art designer at PLoS, modified by Wikipedia users Nina, Beao, and JakobVoss (http://www.plos.org/)

online since that allows for a much wider distribution at a much lower cost than physical media.

Traditionally, scholarly articles are published in journals that need subscriptions to read. If other scientists want to use findings for their own research, use or licensing fees are often required. This model persists and has expanded to include online publications. While the open-access model continues to grow in popularity, many scholars defend the traditional model. Some authors rely on the royalties from copyright and licensing fees to fund their research and simply earn living wages.

Open-access materials can gain a larger readership more quickly than articles published in traditional journals. This allows scientists to share data and discoveries efficiently, leading to more progress. Proponents of the open-access model argue that knowledge and information should be readily available to the public, and that society should be as informed as possible.

BRIEF HISTORY

By the second half of the twentieth century, some scholarly journals, believing that scientific knowledge should be shared on principle, began setting the condition that any content they published would be set in the public domain. However, the move was not financially sound for the majority of journals.

In 1989, the European Organization for Nuclear Research—CERN—oversaw the invention of the World Wide Web. While the Internet had been around for decades, the Web was a critical development in making it much more accessible to a general audience. With the growth of the Internet, several scholarly journals began publishing online. Without the costs of printing and shipping physical copies of journals, it became much more feasible to distribute publications.

The Human Genome Project set an early example for how science could benefit from open access. The project was conceived in the mid-1980s as an effort to map the sequences of genetic code that make up deoxyribonucleic acid (DNA). It officially began in 1990 and involved collaboration on a global scale. Research was carried out across twenty different laboratories, and communication between them was essential.

The project set a very important precedent in 1996. Following conflict over the patent rights of a particular finding, project leaders met to discuss strategy. They found that a lot of delays were stemming from researchers' attempts to protect their rights to their individual findings—actions that they were entitled to take, but were hindering the project. They established an entirely new set of rules for this project, agreeing to share all findings and data with all other laboratories within twenty-four hours. The decision to prioritize the advancement of the project as a whole led to accelerated progress, and it was considered completed in 2003.

The success of the Human Genome Project prompted the scientific community and publishers alike to evaluate the benefits of sharing materials more readily. In 2001, the Center for the Public Domain helped establish Creative Commons, an organization dedicated to licensing works so that they are readily accessible, while protecting the creators' interests.

TOPIC TODAY

The start of the twenty-first century was critical for the development of open access. The Internet and other communications technology saw rapid advances that helped disseminate information quickly and easily. For instance, BioMed Central emerged shortly after the turn of the century, and it became the model for open-access databases. It hosted articles from numerous journals, as well as self-published pieces, providing a single destination for those seeking the relevant information. Other online databases soon followed.

Open access saw significant opposition, however, from journal publishers and authors alike. Several of the traditional scholarly journals had developed reputations as reliable, prestigious publications. This was important to many scholars because many of them published work associated with small projects that would require substantially more funding and interest to continue researching. Many also relied on their published work to expand their networks and build their careers; therefore, if their work was in a respectable journal, it gave the authors greater visibility and respect. From their point of view, open access was an extremely successful model for the Human Genome Project, but that was a global study backed by substantial resources. A paper on a small project from a single laboratory may never stand out if it was simply shared online.

Tens of thousands of scholars rallied to help raise the profile of the open-access model. They signed what was titled "An Open Letter to Scientific Publishers" in 2001, in which they agreed to only submit and peer review materials published in open-access publications, hoping to pressure publishers to convert to this model. While few traditional publications changed their structures, the letter did help raise awareness of open-access publishing globally and created the Public Library of Science, one of the most prestigious open-access publishers.

Public institutions were major supporters during the rapid growth of open access. Libraries and universities need to pay for subscriptions to stock scholarly journals, and most welcome an alternative method of providing patrons with comparable materials without the cost. Since universities produce research as well as provide journals, their relationships with publishers are more complicated. One trend that has helped encourage more universities to adopt the open-access model is increasing collaboration. Since the late twentieth century, universities have formed more research partnerships than in the past, both with other universities and with industries.

The 2010s saw more progress for open access. Jay Bradner, working in a Harvard University laboratory, discovered a molecule in mice that appeared to have cancer-fighting potential in 2010. He published the results via open access, and after follow-up research and tests, the molecule—JQ1—proved valuable in treatments of several serious diseases. In 2013, CERN began an initiative known as SCOAP^3, the Sponsoring Consortium for Open Access Publishing in Particle Physics. This global effort helped physicists share data on a massive scale.

Discussion of open access spread outside of the scholarly community, becoming a subject in highly influential places. In the 2010s, the U.S. Congress discussed multiple cases involving it. In 2016, the Bill & Melinda Gates Foundation made a policy that all research funded by its grants must be open access, with no waiting period. The argument that open access was not only convenient but also provided a right to information became more common, particularly in the medical field.

—Reed Kalso

Bibliography

Barbour, Virginia. "How the Insights of the Large Hadron Collider Are Being Made Open to Everyone." *Phys Org,* 13 Jan. 2017, phys.org/news/2017-01-insights-large-hadron-collider.html. Accessed 14 Jan. 2017.

"Benefits of Open Access Journals." *PLOS,* www.plos.org/open-access. Accessed 14 Jan. 2017.

"Global Open Access Portal." *United Nations Educational, Scientific and Cultural Organization,* www.unesco.org/new/en/communication-and-information/portals-and-platforms/goap/. Accessed 14 Jan. 2017.

Harmon, Elliot. "Open Access Rewards Passionate Curiosity: 2016 in Review." *Electronic Frontier Foundation,* 24 Dec. 2016, www.eff.org/deeplinks/2016/12/open-access-rewards-passionate-curiosity-2016-review. Accessed 14 Jan. 2017.

______. "What If Elsevier and Researchers Quit Playing Hide-and-Seek?" *Electronic Frontier Foundation,* 16 Dec. 2015, www.eff.org/deeplinks/2015/12/what-if-elsevier-and-researchers-quit-playing-hide-and-seek. Accessed 14 Jan. 2017.

"How Did the Human Genome Project Make Science More Accessible?" *YourGenome.org,* 13 June 2016, www.yourgenome.org/stories/how-did-the-human-genome-project-make-science-more-accessible. Accessed 14 Jan. 2017.

"Open Access Week." *Open Access Week,* www.openaccessweek.org/. Accessed 14 Jan. 2017.

Suber, Peter. "Open Access Overview." *Earlham College,* 5 Dec. 2015, legacy.earlham.edu/~peters/fos/overview.htm. Accessed 14 Jan. 2017.

"Who Are We & Why This Website?" *EnablingOpenScholarship,* www.openscholarship.org/jcms/c_5012/en/home. Accessed 14 Jan. 2017.

Wilder, Richard, and Melissa Levine. "Let's Speed Up Science by Embracing Open Access Publishing." *STAT,* 19 Dec. 2016, www.statnews.com/2016/12/19/open-access-publishing/. Accessed 14 Jan. 2017.

Optical Storage

SUMMARY

Optical storage refers to a variety of technologies that are used to read and write data. It employs special materials that are selected for the way they interact with light (an optical, or visible, medium). As of 2011, most optical storage devices being manufactured are digital; however, some are analog. Both the computer and entertainment industries offer numerous Practical Applications of optical storage devices. Common optical storage applications are compact discs (CDs), digital versatile discs (DVDs), and Blu-ray discs (BDs). A variety of data can be stored optically, including audio, video, text, and computer programs. Data are stored in binary form.

DEFINITION AND BASIC PRINCIPLES

Optical storage differs from other data storage technologies such as magnetic tape, which stores data as an electrical charge. Optical disks are flat and circular and contain binary data in the form of microscopic pits, which are non-reflective and have a binary value of 0. Smooth areas are reflective and have a binary value of 1. Optical disks are both created and read with a laser beam. The disks are encoded in a continuous spiral running from the center of the disk to the perimeter. Some disks are dual layer: With these disks, after reaching the perimeter, a second spiral track is etched back to the center. The amount of data storage is dependent upon the wavelength of the laser beam. The shorter the wavelength, the greater the storage capacity (shorter-wavelength lasers can read a smaller pit on the disk surface). For example, the high-capacity Blu-ray Disc uses short-wavelength blue light. Lasers can be used to create a master disk from which duplicates can be made by a stamping process.

Optical media is more durable than electromagnetic tape and is less vulnerable to environmental conditions. With the exception of Blu-ray discs, the speed of data retrieval is considerably slower than that of a computer hard drive. The storage capacity of optical disks is significantly less than that of hard drives. Another, less common form of optical storage is optical tape, which consists of a long, narrow strip of plastic upon which patterns can be written and from which the patterns can be read back.

BACKGROUND AND HISTORY

Optical storage originated in the nineteenth century. In 1839, English inventor John Benjamin Dancer produced microphotographs with a reduction ratio of 160:1. Microphotography progressed slowly for almost a century until microfilm began to be used commercially in the 1920's. Between 1927 and 1935, more than three million pages of books and manuscripts in the British Library were microfilmed by the Library of Congress. Newspaper preservation on film had its onset in 1935 when Kodak's Recordak

This image depicts a flat view of a CD-R with interference colours. It was scanned with an HP ScanJet 4400c, and ran through ACDSee's auto-level filter. One fallback of the image is the dust fibres. It was initially saved as JPG with IrfanView at 90% qu. By User Black and White on en.wikipedia

division filmed and published the *New York Times* on thirty-five-millimeter (mm) microfilm reels.

Analog optical disks were developed in 1958 by American inventor David Paul Gregg, who patented the videodisc in 1961. In 1969, physicists at the Netherlands-based Royal Philips Electronics began experimenting with optical disks. Subsequently, Philips and the Music Corporation of America (MCA) joined forces to create the laser disc, which was first introduced in the United States in 1978. Although the laser disc achieved greater popularity in Asia and Europe than it did in the United States, it never successfully competed with VHS tape. In 1980, Philips partnered with Sony to develop the compact disc (CD) for the storage of music. A few years later, the CD had evolved into a compact disc read-only memory (CD-ROM) format, which in addition to audio, could store computer programs, text, and video. In November 1966, the digital versatile disc (DVD) format was first introduced by Toshiba in Japan; it first appeared in the United States in March, 1997; in Europe in October, 1998; and in Australia in February, 1999. A format war between two higher-capacity data storage technologies emerged in 2006 when Sony's Blu-ray and Toshiba's HD DVD players became commercially available for the recording and playback of high-definition video. Two years later, Toshiba conceded to Sony; Blu-ray was based on newer technology and had a greater storage capacity.

HOW IT WORKS

CDs, DVDs, and BDs are produced with a diameter of 120 mm. The storage capacity is dependent upon the wavelength of the laser: the shorter the wavelength, the greater the storage capacity. CDs have a wavelength of 780 nanometers (nm), DVDs have a wavelength of 650 nm, and BDs have a wavelength of 405 nm. Some disk drives can read data from a disk while others can both read and write data.

Optical System. In a disk reader, the optical system consists of pickup head (which houses a laser), a lens for guiding the laser beam, and photodiodes that detect the light reflection from disk's surface. The photodiodes convert the light into an electrical signal. An optical disk drive contains two main servomechanisms: One maintains the correct distance between the lens and the disk and also ensures that the laser beam is focused on a small area of the disk; the other servomechanism moves the head along the disk's radius, keeping the beam on a continuous spiral data path. The same servomechanism can be used to position the head for both reading and writing. A disk writer employs a laser with a significantly higher power output. It burns data onto the disk by heating an organic dye layer, which changes the dye's reflectivity. Higher writing speeds require a more powerful laser because of the decreased time the laser is focused on a specific point. Some disks are rewritable—they contain a crystalline metal alloy in their recording layer. Depending on the amount of power applied, the substance may be melted into a crystalline form or left in an amorphous form. This enables the creation of marks of varying reflectivity. The number of times the recording layer of a disk can be reliably switched between its crystalline and amorphous states is limited. Estimates range from 1,000 to 100,000 times, depending on the type of media. Some formats may employ defect-management schemes to verify data as it is written and skip over or relocate problems to a spare area of the disk. A third laser function is available on Hewlett-Packard's LightScribe disks. The laser can burn a label onto the side opposite the recording surface on specially coated disks.

Double-Layer Media. Double-layer (DL) media has up to twice the storage capacity of single-layer media. DL media have a polycarbonate first layer with a shallow groove; a first data layer, a semi-reflective layer; a second polycarbonate spacer layer with a deep groove; and a second data layer. The first groove spiral begins on the inner diameter and extends outward; the second groove starts on the outer diameter and extends inward. If data exists in the transition zone, a momentary hiccup of sound and/or video will occur as the playback head changes direction.

Disk Replication. Most commercial optical drives are copy protected; in some cases, a limited number of copies can be made. Disks produced on home or business computers can be readily copied with inexpensive (or included) software. If two optical drives are available, a disk-to-disk copy can be made. If only one drive is available, the data is first stored on the computer's hard drive and then transferred to a blank disk placed in the same read-write drive.

For copying larger numbers of disks, dedicated disk-duplication devices are available. The more expensive ones incorporate a robotic disk-handling system, which automates the process. Some products incorporate a label printer. Industrial processes are used for mass replication of more than 1,000 disks, such as DVDs, CDs, or computer programs. These disks are manufactured from a mold and are created via a series of industrial processes including pre-mastering, mastering, electroplating, injection molding, metallization, bonding, spin coating, printing, and advanced quality control.

APPLICATIONS AND PRODUCTS

Numerous applications and products are focused on optical storage. Most applications are geared toward the computer and entertainment industries.

Optical Disks. Most optical disks are read-only; however, some are rewritable. They are used for the storage of data, computer programs, music, graphic images, and video games. Since the first CD was introduced in 1982, this technology has evolved markedly. Optical data storage has in large part supplanted storage on magnetic tape. Although optical storage media can degrade over time from environmental factors, they are much more durable than magnetic tape, which loses its magnetic charge over time. Magnetic tape is also subject to wear as it passes through the rollers and recording head. This is not the case for optical media, in which the only contact with the recording surface is the laser beam. CDs are commonly used for the storage of music: A CD can hold an entire recorded album and has replaced the vinyl record, which was subject to wear and degradation. A limitation of the CD is its storage capacity: 700 megabytes of data (eighty minutes of music). The DVD, which appeared in 1996, rapidly gained popularity and soon outpaced VHS tape for the storage of feature-length movies. The DVD can store 4.7 gigabytes of data in single-layer format and 8.5 gigabytes in dual-layer format. The development of high-definition television fueled the development of higher-capacity storage media. The Blu-ray disc (BD) can store about six times the amount of data as a standard DVD: 25 gigabytes of data in single-layer format and 50 gigabytes of data in dual-layer format. The most recent evolution of the BD disc is the 3-D format; the increased storage capacity of this medium allows for the playback of video in three dimensions.

Computer Applications. Almost all computers, including basic laptops, contain one or more optical drives, most of which house a read-write drive. The drive is used to load computer programs onto a hard drive, data storage and retrieval, and entertainment. The inclusion of Blu-ray readers as well as writers (burners) is increasing; this is a result of significant drops in price of these devices as well as blank media since their introduction to the marketplace. Most internal drives for computers are designed to fit in a 5.25-inch drive bay and connect to their host via an interface. External drives can be added to a computer; they connect via a universal serial bus (USB) or FireWire interface.

Entertainment Applications. Optical disk players are a common form of home entertainment. Most play DVDs as well as CDs. As with computer applications, the presence of Blu-ray devices is rapidly increasing. Some load a single optical disk at a time; others load a magazine, which holds five or six optical disks; others have a capacity for several hundred disks. The higher-capacity players often have a TV interface in which the user can make a selection from a list of the disks contained within. Often, the list contains a thumbnail image of the CDs or DVDs. These devices contain audio and video outputs to interface with home entertainment systems. If attached to an audio-video receiver (AVR), surround-sound audio playback can be enjoyed. Some DVD players have an Ethernet interface for connection to the Internet. This allows for streaming video from companies such as Netflix and Web surfing. Many automobiles contain a CD or DVD drive. DVD video is displayed to backseat passengers. Vehicle navigation systems often contain an optical drive (CD or DVD) containing route information. Portable players are also available. These range from small devices that can be strapped on an arm to small battery-operated players similar in size to a laptop computer. Many laptop owners use their devices for viewing DVD or BD movies.

Games. Although a wealth of games can be played on a computer, a number of devices in the marketplace are designed strictly for game playing. Most of these devices attach to a television set and are interactive;

thus, the player can immerse oneself in the action. Some accommodate more than one player. Although earlier devices had proprietary cartridges for data storage, the trend is toward DVDs and BDs. Three-dimension consoles, which have added realism, are also available.

SOCIAL CONTEXT AND FUTURE PROSPECTS

Merely three decades after optical-storage devices appeared on the market, laser applications have become innumerable and ubiquitous. Most households contain one or more computers with optical drives. Many have a DVD burner. DVD and CD players are common household items. Optical disks with games that can be played on a computer or specialized gaming consoles are popular not only with children but also with adults. Criticism has been directed at gaming because some are fearful that children and even adults will devote excessive amounts of time at the expense of other, more productive activities. Many computer games are violent. The goal of these games is to kill game inhabitants; often a "kill" is graphically displayed. Critics complain that this could lead to antisocial behavior, particularly in teens.

The trend for optical storage has progressed from CD to DVD to BD. High-capacity media are being used not only for entertainment, such as high-definition movies, but also for data storage. Many computer programs are marketed in DVD format. This is cost-effective because the production cost of three or more CDs is much higher than that of a single DVD, which can hold all of the data. Another trend is toward video streaming and hard-disk storage. Music can be downloaded from a Web site onto a hard drive or solid-state memory. Movies are being streamed directly to a computer, DVD player, or gaming console from companies such as Netflix. CD and DVD "jukeboxes," which can house up to several hundred disks are giving way to hard-drive devices (either computer or stand-alone). Some automobiles can be equipped with a hard-drive player of audio and sometimes video. These devices can be programmed with playlists to suit particular musical tastes. In contrast to a jukebox, the transition between selections is swift without any annoying "clanking." These devices are commonly loaded from CDs or DVDs, which can be stored for future use in the event of a hard-drive failure.

—*Robin L. Wulffson, MD, F.A.C.O.G.*

BIBLIOGRAPHY

Bunzel, Tom. *Easy Creating CDs and DVDs.* 2d ed. Indianapolis: Que, 2005.

McDonald, Paul. *Video and DVD Industries.* London: British Film Institute, 2008.

Taylor, Jim, Mark R. Johnson, and Charles G. Crawford. *DVD Demystified.* 3d ed. New York: McGraw-Hill, 2006.

Taylor, Jim, et al. *Blu-ray Disc Demystified.* New York: McGraw-Hill, 2009.

P

Parallel Computing

SUMMARY

Parallel computing involves the execution of two or more instruction streams at the same time. Parallel computing takes place at the processor level when multiple threads are used in a multi-core processor or when a pipelined processor computes a stream of numbers, at the node level when multiple processors are used in a single node, and at the computer level when multiple nodes are used in a single computer. There are many memory models used in parallel computing, including shared cache memory for threads, shared main memory for symmetric multiprocessing (SMP) and distributed memory for grid computing. Parallel computing is often done on supercomputers that are capable of achieving processing speeds of more than 10^6 Gflop/s.

DEFINITION AND BASIC PRINCIPLES

Several types of parallel computing architectures have been used over the years, including pipelined processors, specialized SIMD machines and general MIMD computers. Most parallel computing is done on hybrid supercomputers that combine features from several basic architectures. The best way to define parallel processing in detail is to explain how a program executes on a typical hybrid parallel computer.

To perform parallel computing one has to write a program that executes in parallel; it executes more than one instruction simultaneously. At the highest level this is accomplished by decomposing the program into parts that execute on separate computers and exchanging data over a network. For grid computing, the control program distributes separate programs to individual computers, each using its own data, and collects the result of the computation. For in-house computing, the control program distributes separate programs to each node (the node is actually a full computer) of the same parallel computer and, again, collects the results. Some of the control programs manage parallelization themselves, while others let the operating system automatically manage parallelization.

The program on each computer of a grid, or node of a parallel computer, is itself a parallel program, using the processors of the grid computer, or node, for parallel processing over a high-speed network or bus. As with the main control program, parallelization at the node can be under programmer or operating system control. The main difference between control program parallelization at the top level and the node level is that data can be moved more quickly around one node than between nodes.

The finest level of parallelization in parallel programming comes at the processor level (for both grid and in-house parallel programming). If the processor uses pipelining, then streams of numbers are processed in parallel, using components of the arithmetic units, while if the processor is multi-core, the processor program decomposes into pieces that run on the different threads.

BACKGROUND AND HISTORY

In 1958, computer scientists John Cocke and Daniel Slotnick of IBM described one of the first uses of parallel computing in a memo about numerical analysis. A number of early computer systems supported parallel computing, including the IBM MVS series (1964–1995), which used threadlike tasks; the GE Multics system (1969), a symmetric multiprocessor; the ILLIAC IV (1964–1985), the most famous array processor; and the Control Data Corporation (CDC) 7600 (1971–1983), a supercomputer that used several types of parallelism.

IBM's Blue Gene/P massively parallel supercomputer. By Argonne National Laboratory's Flickr page

American computer engineer Seymour Cray left CDC and founded Cray Research in 1972. The first Cray 1 was delivered in 1977 and marked the beginning of Cray's dominance of the supercomputer industry in the 1980s. Cray used pipelined processing as a way to increase the flops of a single processor, a technique also used in many of the Reduced Instruction Set Computing (RISC), such as the i860 used in the Intel Hypercube. Cray developed several MIMD computers, connected by a high-speed bus, but these were not as successful as the MIMD Intel Hypercube series (1984–2005) and the Thinking Machines Corporation SIMD Connection Machine (1986–1994). In 2004, multi-core processors were introduced as the latest way to do parallel computing, running a different thread on each core, and by 2011, many supercomputers were based on multi-core processors.

Many companies and parallel processing architectures have come and gone since 1958, but the most popular parallel computer in the twenty-first century consists of multiple nodes connected by a high-speed bus or network, where each node contains many processors connected by shared memory or a high-speed bus or network, and each processor is either pipelined or multi-core.

HOW IT WORKS

The history of parallel processing shows that during its short lifetime (1958 through the present), this technology has taken some very sharp turns resulting in several distinct technologies, including early supercomputers that built super central processing units (CPUs), SIMD supercomputers that used many processors, and modern hybrid parallel computers that distribute processing at several levels. What makes this more interesting is that all of these technologies remain active, so a full explanation of parallel computing must include several rather different technologies.

Pipelined and Super Processors. The early computers had a single CPU, so it was natural to improve the CPU to provide an increase in speed. The earliest computers had CPUs that provided control and did integer arithmetic. One of the first improvements to the CPU was to add floating point arithmetic to the CPU (or an attached coprocessor). In the 1970s, several people, most notably Seymour Cray, developed pipelined processors that could process arrays of floating point numbers in the CPU by having CPU processor components, such as part of the floating point multiplier, operate on numbers in parallel with the other CPU components. The Cray 1, X-MP and Y-MP were the leaders in supercomputers for the next few years. Cray, and others, considered using gallium arsenide rather than silicon for the next speed improvement for the CPU, but this technology never worked. A number of companies attempted to increase the clock speed (the number of instructions per second executed by the CPU) using other techniques such as pipelining, and this worked reasonably well until the 2000's.

A number of companies were able to build a CPU chip that supported pipelining, with Intel's i860 being one of the first. While chip density increased as predicted by Moore's law (density doubles about every two years), signal speed on the chip limited the size of pipelined CPU that could be put on a chip. In 2005, Intel introduced its first multi-core processor that had multiple CPUs on a chip. Applications software was then developed that decomposed a program into components that could execute a different thread on each processor, thus achieving a new type of single CPU parallelism.

Another idea has developed for doing parallel processing at the processor level. Some supercomputers are being built with multiple graphics processing units (GPUs) on a circuit board, and some have been built with a mix of CPUs and GPUs on a board. A variety of techniques is being used to combine these processors into nodes and computers, but they are similar to the existing hybrid parallel computers.

Data Transfer. Increasing processor speed is an important part of supercomputing, but increasing the speed of data transfers between the various components of a supercomputer is just as important. A processor is housed on a board that also contains local memory and a network or bus connection module. In some cases, a board houses several processors that communicate via a bus, shared memory, or a board-level network. Multiple boards are combined to create a node. In the node, processors exchange data via a (backplane) bus or network. Nodes generally exchange data using a network, and if multiple computers are involved in a parallel computing program, the computers exchange data via a transmission-control protocol/Internet protocol (TCP/IP) network.

Flynn's Taxonomy. In 1972, American computer scientist Michael J. Flynn described a classification of parallel computers that has proved useful. He divided the instruction types as single instruction (SI) and multiple instruction (MI); and divided the data types as single data (SD) and multiple data (MD). This led to four computer types: SISD, SIMD, MISD and MIMD. SISD is an ordinary computer, and MISD can be viewed as a pipelined processor. The other classifications described architectures that were extremely popular in the 1980's and 1990's and are still in use. SIMD computers are generally applied to special problems, such as numeric solution of partial differential equations, and are capable of very high performance on these problems. Examples of SIMD computers include the ILLIAC IV of the University of Illinois (1974), the Connection Machine (1980s), and several supercomputers from China. MIMD computers are the most general type of supercomputer, and the new hybrid parallel processors can be seen as a generalization of these computers. While there have been many successful MIMD computers, the Intel Hypercube (1985) popularized the architecture, and some are still in use.

Software. Most supercomputers use some form of Linux as their operating system, and the most popular languages for developing applications are FORTRAN and C++. Support for parallel processing at the operating-system level is provided by operating-system directives, which are special commands to the operating system to do something, such as use the maximum number of threads within a code unit. At the program level, one can use blocking/unblocking message passing, access a thread library, or use a section of shared memory with the OpenMP API (application program interface).

APPLICATIONS AND PRODUCTS

Weather Prediction and Climate Change. One of the first successful uses of parallel computing was in predicting weather. Information, like temperature, humidity, and rainfall, has been collected and used to predict the weather for more than 500 years. Many early computers were used to process weather data, and as the first supercomputers were deployed in the 1970s, some of them were used to provide faster and more accurate weather forecasts. In 1904, Norwegian physicist and meteorologist Vilhelm Bjerknes proposed a differential-equation model for weather forecasting that included seven variables, including temperature, rainfall, and humidity. Many have added to this initial model since its introduction, producing complex weather models with many variables that are ideal for supercomputers, and this has led to many government agencies involved in weather prediction using supercomputers. The European Centre for Medium-Range Weather Forecasts (ECMWF) was using a CDC 6600 by 1976; the National Center for Atmospheric Research (NCAR) used an early model by Cray; and in 2009, the National Oceanic and Atmospheric Administration (NOAA) announced the purchase of two IBM 575s to run complex weather models, which improved forecasting of hurricanes and tornadoes. Supercomputer modeling of the weather has also been used to determine previous events in weather, such as the worldwide temperature of the Earth during the age of the dinosaurs, as well as to predict future phenomena such as global warming.

EFFICIENT WIND TURBINES. Mathematical models describing the characteristics of airflow over a surface, such as an airplane wing, using partial differential equations (Claude-Louis Navier, George Gabriel Stokes, Leonhard Euler, and others) have existed for more than a hundred years. The solution of these equations for a variable, such as the lift applied to an airplane wing for a given airflow, has used computers since their invention, and as soon as supercomputers appeared in the 1970s, they were used to solve these types of problems. There also has been great interest in using wind turbines to generate electricity. Researchers are interested in designing the best blade to generate thrust without any loss due to vortices, rather than developing a wing with the maximum lift. The EOLOS Wind Energy Research Consortium at the University of Minnesota used the Minnesota Supercomputer Institute's Itasca supercomputer to perform simulations of airflow over a turbine (three blades and their containing device) and as a result of these simulations was able to develop more efficient turbines.

MULTISPECTRAL IMAGE ANALYSIS. Satellite images consist of large amounts of data. For example, Landsat 7 images consists of seven tables of data, where each entry in a table represents a different magnetic wavelength (blue, green, red, or thermal-infrared) for a 30-meter-square pixel of the Earth's surface. A popular use of Landsat data is to determine what a particular set of pixels represents, such as a submerged submarine. One approach used to classify Landsat pixels is to build a backpropagation neural network and train it to recognize pixels (determining the difference between water over the submarine and water that is not). A number of neural networks has been implemented on supercomputers over the years to identify multispectral images.

BIOLOGICAL MODELING. Many applications of supercomputers involve the modeling of biological processes on a supercomputer. Both continuous modeling, involving the solution of differential equations, and discrete modeling, finding selected values of a large set of values, has made use of supercomputers., Dr. Dan Siegal-Gaskins of the Ohio State University built a continuous model of cress cell growth, consisting of seven differential equations and twelve unknown factors, to study why some cress cells divide into trichomes, a cell that assists in growth, as opposed to an ordinary cress cell. The model was run on the Ohio Supercomputer Center's IBM 1350 Cluster, which has 9,500 core CPUs and a peak computational capability of 75 teraflops. After running a large number of models, and comparing the results of the models to the literature, Siegal-Gaskins decided that three proteins were actively involved in determining whether a cell divided into a trichome or ordinary cells. Many examples of discrete modeling in biology using supercomputers are also available. For example, many DNA and protein-recognition programs can only be run on supercomputers because of their computational requirements.

ASTRONOMY. There are many applications of supercomputers to astronomy, including using a supercomputer to simulate events in the future, or past, to test astronomical theories. The University of Minnesota Supercomputer Center simulated what a supernova explosion originating on the edge of a giant interstellar molecular gas cloud would look like 650 years after the explosion.

SOCIAL CONTEXT AND FUTURE PROSPECTS

Parallel processing can be used to solve some of societies' most difficult problems, such as determining when, and where, a hurricane is going to hit, thus improving people's standard of living. An interesting phenomenon is the rapid development of supercomputers for parallel computing in Europe and Asia. While this will result in more competition for the United States' supercomputer companies, it should also result in a wider use in the world of supercomputers, and improve the worldwide standard of living.

Supercomputers have always provided technology for tomorrow's computers, and technology developed for today's supercomputers will provide faster processors, system buses, and memory for future home and office computers. Looking at the growth of the supercomputer industry in the beginning of the twenty-first century and the state of computer technology, there is good reason to believe that the supercomputer industry will experience just as much growth in the 2010s and beyond. If quantum computers become a reality in the future, that will only increase the power and variety of future supercomputers.

—George M. Whitson III, BS, MS, PhD

Bibliography

Culler, David, Jaswinder Pal Singh, and Anoot Gupta. *Parallel Computer Architecture: A Hardware/Software Approach.* San Francisco: Kaufmann, 1999. Print.

Hwang, Kai, and Doug Degroot, eds. *Parallel Processing for Supercomputers and Artificial Intelligence.* New York: McGraw, 1989. Print.

Kirk, David, and Wen-mei W. Hwu. *Programming Massively Parallel Processors: A Hands-On Approach.* Burlington: Kaufmann, 2010. Print.

Rauber, Thomas, and Gudula Rünger. *Parallel Programming: For Multicore and Cluster Systems.* New York: Springer, 2010. Print.

Varoglu, Sevin, and Stephen Jenks. "Architectural support for thread communications in multi-core processors." *Parallel Computing* 37.1 (2011): 26–41. Print.

Pattern Recognition

SUMMARY

Pattern recognition is a branch of science concerned with identifying patterns within any type of data, from mathematical models to visual and auditory information. Applied pattern recognition aims to create machines capable of independently identifying patterns and using patterns to perform tasks or make decisions. The field involves cooperation between statistical analysis, mechanical and electrical engineering, and applied mathematics. Pattern recognition technology emerged in the 1970's from work in advanced theoretical mathematics. The development of the first pattern recognition systems for computers occurred in the 1980's and early 1990's. Pattern recognition technology is found both in complex analyses of systems such as economic markets and physiology and in everyday electronics such as personal computers.

KEY TERMS AND CONCEPTS

- **Algorithm:** Rule or set of rules that describe the steps taken to solve a certain problem.
- **Artificial Intelligence:** Field of computer science and engineering seeking to design computers that have the capability to use creative problem-solving strategies.
- **Character Recognition:** Programs and devices used to allow machines and computers to recognize and interpret data symbolized with numbers, letters, and other symbols.
- **Computer-Aided Diagnosis (CAD):** Field of research concerned with using computer analysis to more accurately diagnose diseases.
- **Machine Learning:** Field of mathematics and computer science concerned with building machines and computers capable of learning and developing systems of computing and analysis.
- **Machine Vision:** Field concerned with developing systems that allow computers and other machines to recognize and interpret visual cues.
- **Neural Network:** Patterns of neural connections and signals occurring within a brain or neurological system.
- **Pattern:** Series of distinguishable parts that has a recognizable association, which in turn forms a relationship among the parts.

DEFINITION AND BASIC PRINCIPLES

Pattern recognition is a field of science concerned with creating machines and programs used to categorize objects or bits of data into various classes. A pattern is broadly defined as a set of objects or parts that have a relationship. Recognizing patterns therefore requires the ability to distinguish individual parts, identify the relationship between the parts, and remember the pattern for future applications. Pattern recognition is closely linked to machine learning and artificial intelligence, which are fields concerned with creating machines capable of learning new information and using it to make decisions. Pattern recognition is one of the tools used by learning machines to solve problems.

One example of a pattern recognition machine is a computer capable of facial recognition. The computer first evaluates the face using a visual sensor and then divides the face into parts, including the eyes, lips, and nose. Next, the system assesses the relationship between individual parts, such as the distance between the eyes, and notes features such as the length of the lips. Once the machine has evaluated an individual's face, it can store the data in memory and later compare the information against other facial scans. Facial recognition machines are most often used to confirm identity in security applications.

BACKGROUND AND HISTORY

The earliest work on pattern recognition came from theoretical statistics. Early research concentrated on creating algorithms that would later be used to control pattern recognition machines. Pattern recognition became a distinct branch of mathematics and statistics in the early 1970's. Engineer Jean-Claude Simon, one of the pioneers of the field, began publishing papers on optical pattern recognition in the early 1970's. His work was followed by a number of researchers, both in the United States and abroad. The first international conference on pattern recognition was held in 1974, followed by the creation of the International Association for Pattern Recognition in 1978.

In 1985, the United States spent $80 million on the development of visual and speech recognition systems. During the 1990's, pattern recognition systems became common in household electronics and were also used for industrial, military, and economic applications. By the twenty-first century, pattern recognition had become a robust field with applications ranging from consumer electronics to neuroscience. The field continues to evolve along with developments in artificial intelligence and computer engineering.

HOW IT WORKS

Pattern recognition systems are based on algorithms, or sets of equations that govern the way a machine performs certain tasks. There are two branches of pattern recognition research: developing algorithms for pattern recognition programs and designing machines that use pattern recognition to perform a function.

Obtaining Data. The first step in pattern recognition is to obtain data from the environment. Data can be provided by an operator or obtained through a variety of sensory systems. Some machines use optical sensors to evaluate visual data, and others use chemical receptors to detect molecules or auditory sensors to evaluate sound waves.

Most pattern recognition computers focus on evaluating data according to a few simple rules. For example, a visual computer may be programmed to recognize only red objects and to ignore objects of any other color, or it may be programmed to look only at objects larger than a target length.

Translating Data. Once a machine has intercepted data, the information must be translated into a digital format so that it can be manipulated by the computer's processor. Any type of data can be encoded as digital information, from spatial relationships and geometric patterns to musical notes.

A character recognition program is programmed to recognize symbols according to their spatial geometry. In other words, such a program can distinguish the letter *A* from the letter *F* based on the unique organization of lines and spaces. As the machine identifies characters, these characters are encoded as digital signals. A user can then manipulate the digital signals to create new patterns, using a computer interface such as a keyboard.

A character recognition system that is familiar to many computer users is the spelling assistant found on most word-processing programs. The spelling assistant recognizes patterns of letters as words and can therefore compare each word against a preprogrammed list of words. If a word is not recognized, the program looks for a word that is similar to the one typed by the user and suggests a replacement.

Memory and Repeated Patterns. In addition to recognizing patterns, machines must also be able to use patterns in problem solving. For example, the spelling assistant program allows the computer to compare patterns programmed into its memory against input given by a user. Word processors can also learn to identify new words, which become part of the machine's permanent memory.

Advanced pattern recognition machines must be able to learn without direct input from a user or engineer. Certain learning robots, for instance, are programmed with the capability to change their own programming according to experience. With repeated exposure to similar patterns, the machines can become faster and more accurate.

APPLICATIONS AND PRODUCTS

Computer-Aided Diagnosis. Pattern recognition technology is used by hospitals around the world in the development of computer-aided diagnosis (CAD) systems. CAD is a field of research concerned with using computer analysis to more accurately diagnose disease. CAD research is usually conducted by radiologists and also involves participation from computer and electrical engineers. CAD systems can be used to evaluate the results taken from a variety of imaging techniques, including radiography, computed tomography (CT), magnetic resonance imaging (MRI), and ultrasound. Radiologists can use CAD systems to evaluate disorders affecting any body system, including the pulmonary, cardiac, neurologic, and gastrointestinal systems.

At the University of Chicago Medical Center, the radiology department has obtained more than seventy patents for CAD systems and related technology. Among other projects, specialists have been working on systems to use CAD to evaluate potential tumors.

Speech Recognition Technology. Some pattern recognition systems allow machines to recognize and respond to speech patterns. Speech recognition computers function by recording and analyzing sound waves; individual speech patterns are unique, and the computer can use a recorded speech pattern for comparison. Speech recognition technology can be used to create security systems in which an individual's speech pattern is used as a passkey to gain access to private information. Speech recognition programs are also used to create dictation machines that translate speech into written documents.

Neural Networks. Many animals, including humans, use pattern recognition to navigate their environments. By examining the way that the brain behaves when confronted with pattern recognition problems, engineers are attempting to design artificial neural networks, which are machines that emulate the behavior of the brain.

Biological neural networks have a complex, nonlinear structure, which makes them unpredictable and adaptable to new problems. Artificial neural networks are designed to mimic this nonlinear function by using sets of algorithms organized into artificial neurons that imitate biological neurons. The artificial network is designed to be adaptive, so that repeated exposure to similar problems creates strong connections among the artificial neurons—similar to memory in the human brain.

Although artificial neural networks have just begun to emerge, they could potentially be used for any pattern recognition application from economic analysis to fingerprint identification. Many engineers have come to believe that artificial neural networks are the future of pattern recognition technology and will eventually replace the linear algorithms that have been used most frequently.

Military Applications. Pattern recognition is one of the most powerful tools in the development of military technology, including both surveillance and offensive equipment. For example, the Tomahawk cruise missile, sometimes called a smart bomb, is an application of pattern recognition used for offensive military applications. The missile uses a digital scene area matching correlation (DSAMC) system to guide the missile toward a specific target identified by the pilot. The missile is equipped with sensors, an onboard computer, and flight fins that can be used to adjust its trajectory. After the missile is fired, the DSAMC adjusts the missile's flight pattern by matching images from its visual sensors with the target image.

IMPACT ON INDUSTRY

Academic projects in pattern recognition receive funding from a variety of public and private granting agencies. In the United States, grants for pattern recognition technology have come from the National Institutes of Health, the National Science Foundation, and a variety of other sources. The CEDAR research group at the University of Buffalo has received funding from the U.S. Postal Service to develop technology using pattern recognition for document analysis. In October, 2009, Con Edison received $136 million from the U.S. Department of Energy for research into the development of an automated pattern recognition program that would be used to prevent blackouts and power failures.

In addition, numerous private and public companies have made investments in pattern recognition technology because of its numerous marketable applications. As the use of pattern recognition systems expands globally, researchers and engineers will have increasing opportunities to find funding from foreign investors.

SOCIAL CONTEXT AND FUTURE PROSPECTS

Pattern recognition technology has become familiar to many consumers. Voice-activated telephones, fingerprint security for personal computers, and character analysis in word-processing software are just a few of the many applications that affect daily life. Advances in medicine, military technology, and economic analysis are further examples of how pattern recognition has come to shape the development of society.

Projects that may represent the future of pattern recognition technology include the development of autonomous robots, space exploration, and evaluating complex dynamics. In the field of robotics, pattern recognition is being used in an attempt to create robots with the ability to locate objects in their environment. The National Aeronautics and Space Administration has begun research to create probes capable of using pattern recognition to find objects or sites of interest on alien landscapes. Combined with research on artificial neural networks, automated systems may soon be capable of making complex decisions based on the recognition of patterns.

—Micah L. Issitt, BS

Bibliography

Brighton, Henry, and Howard Selina. *Introducing Artificial Intelligence.* Edited by Richard Appignanesi. 2003. Reprint. Cambridge, England: Totem Books, 2007.

Frenay, Robert. *Pulse: The Coming Age of Systems and Machines Inspired by Living Things.* 2006. Reprint. Lincoln: University of Nebraska Press, 2008.

Marques de Sá, J. P. *Pattern Recognition: Concepts, Methods and Applications.* New York: Springer Books, 2001.

McCorduck, Pamela. *Machines Who Think: A Personal Inquiry into the History and Prospects of Artificial Intelligence.* 2d ed. Natick, Mass.: A. K. Peters, 2004.

Singer, Peter Warren. *Wired for War: The Robotics Revolution and Conflict in the Twenty-first Century.* New York: Penguin Press, 2009.

Theodoridis, Sergio, and Konstantinos Koutroumbas. *Pattern Recognition.* 4th ed. Boston: Academic Press, 2009.

PHOTOGRAMMETRY

SUMMARY

Photogrammetry is a technical field that combines elements of art and science. Its practitioners blend photographic images with data recorded and measured from electromagnetic energy and other sources. The results offer information about objects and environmental features that could not be obtained from a single source.

Photogrammetry often involves aerial or satellite photographs and provides detailed images used for surveying, mapping, or creating electronic models. Photogrammetry is especially useful for creating images of large areas and for analyzing areas that would be dangerous for direct study by human surveyors.

BACKGROUND

Photographic technology did not exist until the early part of the nineteenth century, but the concepts involved in photogrammetry go back centuries. The ancient Greeks and Egyptians created art and building plans that made use of the ideas of *perspective*, or a viewpoint that includes accurate representations of height, width, depth, and relative closeness of objects, as well as *projective geometry,* or the study of the properties of objects as they appear.

Projective geometry differs from basic geometry. For example, basic geometry views a square as a flat object with four equal sides and corners formed by ninety-degree angles. Projective geometry examines the properties of the square from the perspective of a

Low altitude aerial photograph for use in Photogrammetry - Location Three Arch Bay, Laguna Beach CA. By WPPilot (Own work)

particular point, such as one corner. From this point of view, the square appears distorted, with some sides appearing longer than others and the corners not appearing as right angles.

The great artist and inventor Leonardo da Vinci (1452–1519) described the concept of photogrammetry in 1480 without naming it as such. A number of da Vinci's contemporaries in both the artistic and scientific worlds continued to research the visual and mathematical components of what would later become known as photogrammetry. By 1759, Swiss mathematician Johan Lambert (1728–1777) had determined a way to calculate the specific point from which an image was created by working backward from the image's perspective.

The invention of photography in the early 1800s provided new ways to apply the idea of object perspective. In 1840, Dominique Francois Jean Arago (1786-1853), a French *geodesist* (a scientist who measures and monitors the planet to determine the exact location of specific points) first practiced photogrammetry using early photographs known as *daguerreotypes.* In 1883, Guido Hauck (1845-1905) and his associates established the relationship between photogrammetry and projective geometry. Prussian architect Albrecht Meydenbauer (1834-1921) coined the term "photogrammetry" in 1867.

RESULTS AND USES

The photogrammetry process produces three results: photographic products, computational results, and maps. A photographic product is a photograph that has been created by enhancing and creating a composite of one or more original photographs. The original image is printed and made into a *diapositive,* or image reproduced on glass or a clear film, such as a slide. The photo may also be *rectified,* or converted to a horizontal image, which makes an aerial view look as if it were taken from ground level. When the rectified image is further enhanced to remove distortion resulting from camera orientation and adjusted to accurately reflect the distances between objects within it, it is called an *orthoimage.* Orthoimages can be further manipulated to allow viewing from various angles. Because they are digital, they can easily be shared and used in different software applications.

Photogrammetry can also be used to calculate, plot, and compare two or more points, such as areas of terrain covered by a highway construction project. The computational results gained through photogrammetry allow planners to determine elevations and distances. For example, planners can project the amount of dirt and rock to remove, the ending length of a bridge or road, and other details needed to accurately estimate projections.

The most common use of photogrammetry is the creation of maps. By combining and manipulating photos, the photogrammeter can create a topographical map that shows geographical features from different angles, digital terrain maps that depict elevations and other details, or special-use maps that focus on one specific aspect of an area, such as a waterway or roadway.

TOOLS AND PROCESS

Early photogrammeters often relied on simple cameras, hot air balloons, and manual calculation methods to generate their results. By the time the U.S. Geological Survey began using photogrammetry for topographical mapping in 1904, special survey cameras were available to capture panoramic views.

The development of air flight improved the ability to take aerial photos. Additionally, photogrammeters developed special tools for their trade, such as the *stereocomparator* devised by Carl Pulfrich (1858-1929) to compare two points. Frederick Vivian Thompson (1880-1917) created a *stereoplotter* that could plot contours from a photo taken on land.

Initially, photogrammeters followed a simple process. For example, they might take aerial photos from a hot air balloon and then perform calculations manually. By the 1960s, however, photogrammeters began applying more advanced technologies and analytical techniques to their study.

Contemporary photogrammeters have access to photographs from planes and satellites, and the advantage of digital photography, which improves resolution and allows for easier image manipulation and sharing. Specialists have access to computer assisted drafting (CAD) technology and digital terrain modeling (DTM) to create three-dimensional images. They can use software to digitally rectify images and produce orthoimages from multiple photos. They can use light detecting and ranging (LIDAR) equipment to scan the ground to detect and produce features not easily visible by the eye, such as depressions in large grassy areas.

Photogrammeters can employ remote sensing to capture information based on the color bands generated by satellite imagery to help detect and define details of a terrain by color, brightness, or other characteristics. They can create realistic three-dimensional representations of objects or terrain even when it is difficult or dangerous for human observers. Photogrammetry can be used to study or to help plan expeditions to far of parts of Earth, such as Antarctica, and to other celestial bodies in the solar system.

—*Janine Ungvarsky*

Bibliography

Duraiswami, Ramani. "Projective Geometry." University of Maryland Institute of Advanced Computer Studies, University of Maryland. Web. 3 Mar 2016. http://www.umiacs.umd.edu/~ramani/cmsc828d/ProjectiveGeometry.pdf

"History of Photogrammetry." The Center for Photogrammetric Training. Department of Spatial Sciences, Curtain University. Web. 3 Mar 2016. http://spatial.curtin.edu.au/local/docs/HistoryOfPhotogrammetry.pdf

"Orthophoto Imagery." Pima County, Arizona. Web. 3 Mar 2016. http://webcms.pima.gov/cms/one.aspx?pageId=23722

"Photogrammetry Surveys." CalTrans Survey Manual 2006. California Department of Transportation. Web. 3 Mar 2016. http://www.dot.ca.gov/hq/row/landsurveys/SurveysManual/13_Surveys.pdf

Pillai, Anil Narendran. "A Brief Introduction to Photogrammetry and Remote Sensing." GIS Lounge. 12 July 2015. Web. 3 Mar 2016. https://www.gislounge.com/a-brief-introduction-to-photogrammetry-and-remote-sensing/

Pneumatics

SUMMARY

Pneumatics technology generates and uses compressed gas, normally air, to provide energy to mechanical equipment performing work. Pneumatic technology can be used in large-scale industrial applications, such as a drill for blasting rock, or on a much smaller scale, such as to move a prosthetic device such as an artificial leg. Pneumatics is based on the fact that all gases can be compressed, although the compression affects the volume and temperature. Air, the primary gaseous medium, is in endless supply, inexpensive, and environmentally safe.

DEFINITION AND BASIC PRINCIPLES

Pneumatics is a technology resulting from the molecular and chemical composition of a gas, a medium that can move machinery. The machinery can be very large, such as a marine hovercraft, or very small, such as a portable ventilator for a human being. Pneumatics is half of a technology labeled as fluid power, the other half being hydraulics. Hydraulics uses many liquids because liquids are also dependent on pressure and temperature for their volume, although not as much as gases. In scientific terms, the word "fluid" applies to both liquids and gases. Different laws govern the behavior of each medium.

Women workers in ordnance shops, Midvale Steel and Ordnance Co., Nicetown, Pa. Hand chipping with pneumatic hammers. 1918. National Archives and Records Administration

Gases are molecular compounds. Molecules in a fluid are in constant motion. Matter in the gaseous state occupies a volume about one thousand times larger than the same amount of matter in the liquid state. The molecules in a gas are also much farther apart. Because they have no inherent shape, they can be easily confined.

Air is the most common gas used in pneumatics. It is composed of nitrogen (78.08 percent), oxygen (20.95 percent), argon (0.93 percent), and several other elements in trace amounts on the order of parts per million. Carbon dioxide and water vapor are also present in varying and unpredictable amounts. Because water can be damaging to a system, most pneumatic systems have a component for drying the air. An important value is the weight of a one-inch-square column of air from sea level to the beginning of outer space (thermosphere). Expressed in a variety of international measuring systems, the unit one atmosphere (atm) weighs is 101.3 kilopascals (kPa, named for French scientist Blaise Pascal), 14.7 pounds per square inch (psi), 760 torr (named for Italian physicist Evangelista Torricelli), or 29.9 inches of mercury (mmHg), so called because 1 atm can support a column of mercury that high.

BACKGROUND AND HISTORY

Pneumatics is derived from Latin words meaning "pertaining to air." The blacksmith's bellows pumping a fire is a primitive example. Windmills in twelfth-century central Europe exhibited the technology's first large-scale possibility. Development was dependent on discoveries made in the seventeenth and eighteenth centuries about gas properties, especially how the volume of a gas is affected by pressure and temperature. The Italian physicist Evangelista Torricelli invented the mercury barometer in 1643 and then measured atmospheric pressure. He also formed a vacuum by inverting a mercury-filled tube. The first vacuum pump was invented by German scientist Otto von Guericke.

Boyle's law, developed by Irish chemist Robert Boyle, states that the volume of a gaseous substance at constant temperature is inversely proportional to the pressure it is under. French physicist Jacques Charles contributed the insight that the volume of a given mass of gaseous substance increases or decreases 1/273 for each degree Celsius of temperature change; this is known as Charles's law. French chemist Joseph Louis Gay-Lussac discovered the relationship between volume and temperature. He devised a hydrometer and calculated the volume of a gaseous compound is equal to or less than the sum of the volumes of the combined gases. Amedeo Avogadro, an Italian physicist, determined that at the same temperature and pressure, equal volumes of different gases contain equal numbers of molecules.

Once the gaseous state was understood, gases could be manipulated to do work, and the field of pneumatics was possible.

HOW IT WORKS

Pneumatics works by compressing a quantity of air (or some other gas), filtering it to make sure it is clean, drying it, and sending it through a system with one or more valves. The compressed air is delivered to a mechanical apparatus that can then move a force through a distance. The air is an energy-transfer medium.

Pneumatic System Components. The major components of a pneumatic system include a compressor, an air storage tank (receiver), an air dryer, and a main line filter, all connected by a series of hoses. These sections are followed by an air-conditioning section composed of a filter, regulator, and lubricator; then a controlling section made up of a directional control valve and a speed controller; and, finally, the actuator or operating system that the pneumatic system was designed to power.

This basic system is used in large factories where truck snow-plowing blades are spray painted. Such a system would be automated for mass production and include a timing apparatus, usually using some sort of digital logic, to control when and how much paint was sprayed. The paint is atomized and mixed with the pressurized air as it is released.

On a smaller scale, everyday handheld tools, such as a drill hammer for blasting slabs of concrete or an air hammer for pounding nails, use the same principle. Such tools have less need, and thus less capacity, for cleaning and removing moisture from the air sucked into the tool. The compressed air pressure supplied for pneumatic tools is normally about 90 pounds per square inch gauge (psig).

Valves. Valves control the way the gas is used, stopped, or directed. They must function under a range of temperatures. Control or proportional valves in a process system are power-operated mechanisms able to alter fluid flow. A pneumatic valve actuator adjusts valve position by making the air pressure either linear or rotary motion. Ball valves provide the shutoff capability. Gas valves are specialized to control the flow of another medium, such as natural gas. A pressure-relief valve is a self-actuated safety valve that relieves pressure. A butterfly valve controls the flow of air or a gas through a circular disk or vane by turning the valve's pivot axis at right angles to the directing flow in the pipe.

Gaseous Medium. The gas or energy medium traveling through the system is unique in its behavior. Charles's law states that a volume of a gas is directly proportional to the temperature. Boyle's law states the volume of a gas is inversely proportional to the pressure applied to it. If the amount of gas is calculated in moles (expressed as n), which is an amount of a substance that contains exactly $6.02214129 \times 10^{23}$ atoms or molecules of that substance; the temperature is in degrees Celsius (expressed as T); the volume of the gas is in liters (expressed as V); and the pressure is in atmospheres (expressed as p), then calculations can be made as to what happens to that gas changing one or more of those variables using the formula $pV = nRT$. R is known as the molar gas constant and has the same value for all gases. When changes are made to the conditions of a gas, the calculation can be made to find out what will happen to any one or two values of the gas in question. This is expressed $pV/T = pV/T$. This is one of the most important formulas in pneumatics.

Pneumatic System Categories. One way of categorizing pneumatic systems is whether the actuator that the air drives is rotor or reciprocating. Compressed air enters a rotating compressor and pushes on the vanes, rotating a central shaft or spindle. Tools such as drills or grinding wheels use this system. In a reciprocating piston compressor, the air enters a cylinder, expands, and forces the piston to move. In a reciprocating tool, the piston also has a return stroke, which is actuated by compressed air on the other end of the piston or a spring action. A riveting hammer is a tool using this technique.

Another way to categorize pneumatic devices is as either portable or rock drills. Portable devices include buffers, drills, screwdrivers, wrenches, and paint mixers. This category is powered by a rotary-vane type of air motor. Rock drills or percussion hammers are composed of high-carbon steels. In this type, the compressed air drives a piston down onto a loosely held drill inside a cylinder.

APPLICATIONS AND PRODUCTS

Pneumatic systems have positive and negative features. Potentially, pneumatics can be used in any situation where mechanical work needs to be done. Its overarching desirable quality is a boundless supply of free air. The desirable features include stability under temperature changes, cleanliness (leaks do not cause contamination), and work at high speeds measured in feet per second (fps); the tools stall when they are overloaded with pressure and are therefore inherently safe. Since pneumatics is so versatile, it can be used across a broad spectrum of industries. The undesirable features are: a constant actuator velocity is not possible; under normal conditions, usually 100 psi, the force is limited to between 4,500 and 6,700 pounds; and the gas must be prepared by removing dirt and humidity as these degrade a system.

Heavy Loads. This is the traditional area. Factories often deal in products weighing tons and under extreme temperatures. Pneumatic systems can open heavy or hot doors. They can unload hoppers (containers usually with sloping floors used to carry bulk

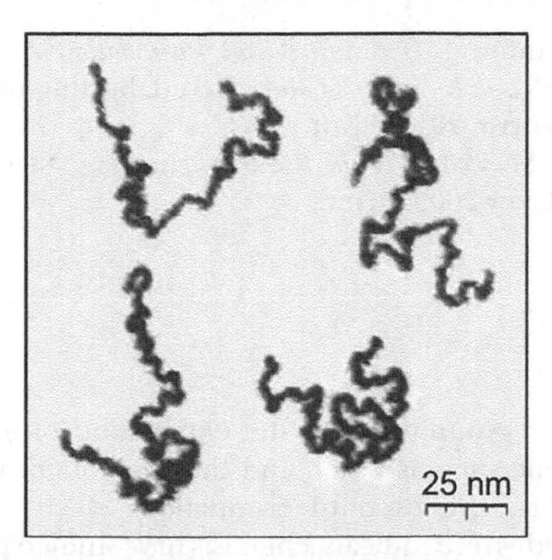

Appearance of real linear polymer chains as recorded using an atomic force microscope on a surface, under liquid medium. Chain contour length for this polymer is ~204 nm; thickness is ~0.4 nm. By Yurko (Own work)

goods) in industries such as construction, steelmaking, and chemical. Slab molding machines are usually lifted and moved with pneumatic systems. Road or large flat-surface construction is usually rammed or tamped with them. The capability of moving heavy loads proves the efficiency of pneumatic systems.

Spraying. Pneumatic valves are capable of directing air and then atomizing liquids into fine droplets for this application. Large- and small-scale surfaces needing coverage include agriculture, where crops are sprayed with chemicals or water, and painting, as in automobile, motorcycle, or bicycle manufacturing.

Repetition. Repetitive motion has been used since the first mass production and has grown with production of ever-larger quantities in ever-shorter periods of time. Pneumatics is advantageous because of its low cost of operation and maintenance. Both large-scale and small-scale tools are also examples. Both handheld and robotic riveters can secure bolts and screws at very fast speeds. Pneumatic systems are ideal for holding maneuvers where pieces are clamped while other work is accomplished as in: woodworking, furniture making, gluing, heat sealing, or welding plastics. Guillotine blades for slicing precise quantities of a material by the millions are a classic example of this capability. Bottling and filling machines and household cleaners also use the systems.

Breaking and Smashing. The jackhammer, used for road repairs, is a pneumatic tool.

Medical Devices. Pneumatics is being used more in medical devices. When the dentist, drill or polisher in hand, steps on a small plate, a pneumatic system is activated that drives the instrument in a circular motion at a very high speed. Pneumatic systems are excellent for applications in patient ventilators. Ventilators can either take over breathing for an extremely sick patient or assist a patient with some breathing ability.

Robotics. Robotics for work as well as entertainment is another application of pneumatics. Most movable monsters or lifelike creatures in theme parks or museums are operated pneumatically.

SOCIAL CONTEXT AND FUTURE PROSPECTS

Energy awareness, conservation, and environmental protection are forces that will allow pneumatics to remain a promising field into the future. Despite the commercial size of the field of pneumatics, few people know about it even though they use its technology daily.

From its inception, the industry has not changed much because the basic system remains the same. What does change is that component providers continue to improve on the efficiency and materials, as well as the advances made by computerization and miniaturization of parts. Hospital ventilators at one time took up as much room as a hospital bed. Now, they can be placed in a belt pack and the patient sent home.

The field of nanotechnology works with unusual matter located between molecules in the size range of 0.000000001 meter. It allows pneumatic component parts such as pistons and seals to be manufactured with ultraprecision.

—Judith L. Steininger, BA, MA

Bibliography

Cundiff, John S. *Fluid Power Circuits and Controls: Fundamentals and Applications.* Boca Raton: CRC, 2001. Print.

Daines, James R. *Fluid Power: Hydraulics and Pneumatics.* 2nd ed. Tinley Park: Goodheart, 2013. Print.

Fleischer, Henry. *Manual of Pneumatic Systems Optimization.* New York: McGraw, 1995. Print.

Johnson, Olaf A. *Fluid Power: Pneumatics.* Chicago: Amer. Technical Soc., 1975. Print.

Parr, Andrew. *Hydraulics and Pneumatics: A Technician's and Engineer's Guide.* 3rd ed. Burlington: Butterworth, 2011. Print.

Rabie, M. Galal. *Fluid Power Engineering.* New York: McGraw 2009. Print.

POLYMER SCIENCE

SUMMARY

Polymer science is a specialized field concerned with the structures, reactions, and applications of polymers. Polymer scientists generate basic knowledge that often leads to various industrial products such as plastics, synthetic fibers, elastomers, stabilizers, colorants, resins, adhesives, coatings, and many others. A mastery of this field is also essential for understanding the structures and functions of polymers found in living things, such as proteins and deoxyribonucleic acid (DNA).

DEFINITION AND BASIC PRINCIPLES

Polymers are very large and often complex molecules constructed, either by nature or by humans, through the repetitive yoking of much smaller and simpler units. This results in linear chains in some cases and in branched or interconnected chains in others. The polymer can be built up of repetitions of a single monomer (homopolymer) or of different monomers (heteropolymer). The degree of polymerization is determined by the number of repeat units in a chain. No sharp boundary line exists between large molecules and the macromolecules characterized as polymers. Industrial polymers generally have molecular weights between ten thousand and one million, but some biopolymers extend into the billions.

Chemists usually synthesize polymers by condensation (or step-reaction polymerization) or addition (also known as chain-reaction polymerization). A good example of chain polymerization is the free-radical mechanism in which free radicals are created (initiation), facilitating the addition of monomers (propagation), and ending when two free radicals react with each other (termination). A general example of step-reaction polymerization is the reaction of two or more polyfunctional molecules to produce a larger grouping, with the elimination of a small molecule such as water, and the consequent repetition of the process until termination.

Besides free radicals, chemists have studied polymerizations utilizing charged atoms or groups of atoms (anions and cations). Physicists have been concerned with the thermal, electrical, and optical properties of polymers. Industrial scientists and engineers have devoted their efforts to creating such new polymers as plastics, elastomers, and synthetic fibers. These traditional applications have been expanded to include such advanced technologies as biotechnology, photonics, polymeric drugs, and dental plastics. Other scientists have found uses for polymers in such new fields as photochemistry and paleogenetics.

BACKGROUND AND HISTORY

Nature created the first polymers and, through chemical evolution, such complex and important macromolecules as proteins, DNA, and polysaccharides. These were pivotal in the development of increasingly multifaceted life-forms, including Homo sapiens, who, as this species evolved, made better and better use of such polymeric materials as pitch, woolen and linen fabrics, and leather. Pre-Columbian Native Americans used natural rubber, or *cachucha*, to waterproof fabrics, as did Scottish chemist Charles Macintosh in nineteenth century Britain.

The Swedish chemist Jöns Jakob Berzelius coined the term "polymer" in 1833, though his meaning was far from a modern chemist's understanding. Some scholars argue that the French natural historian Henri Braconnot was the first polymer scientist since, in investigating resins and other plant products, he created polymeric derivatives not found in nature. In 1836, the Swiss chemist Christian Friedrich Schönbein reacted natural cellulose with nitric and sulfuric acid to generate semisynthetic polymers.

In 1843 in the United States, hardware merchant Charles Goodyear accidentally discovered "vulcanization" by heating natural rubber and sulfur, forming a new product that retained its beneficial properties in cold and hot weather. Vulcanized rubber won prizes at the London and Paris Expositions in 1850, helping to launch the first commercially successful product of polymer scientific research.

In the early twentieth century, the Belgian-American chemist Leo Baekeland made the first totally synthetic polymer when he reacted phenol and formaldehyde to create a plastic that was marketed under the name Bakelite. The nature of this and other synthetic polymers was not understood until the 1920's and 1930's, when the German chemist Hermann Staudinger proved that these plastics (and other polymeric materials) were extremely long molecules built up from a sequential catenation of basic units, later called monomers. This enhanced understanding led the American chemist Wallace Hume Carothers to develop a synthetic rubber, neoprene, that had numerous applications, and nylon, a synthetic substitute for silk. Synthetic polymers found wide use in World War II, and in the postwar period the Austrian-American chemist Herman Francis Mark founded the Polymer Research Institute at Brooklyn Polytechnic, the first such facility in the United States. It helped foster an explosive growth in polymer science and a flourishing commercial polymer industry in the second half of the twentieth century.

HOW IT WORKS

After more than a century of development, scientists and engineers have discovered numerous techniques for making polymers, including a way to make them using ultrasound. Sometimes these techniques depend on whether the polymer to be synthesized is inorganic or organic, fibrous or solid, plastic or elastomeric, crystalline or amorphous. How various polymers function depends on a variety of properties, such as melting point, electrical conductivity, solubility, and interaction with light. Some polymers are fabricated to serve as coatings, adhesives, fibers, or thermoplastics. Scientists have also created specialized polymers to function as ion-exchange resins, piezoelectrical devices, and anaerobic adhesives. Certain new fields have required the creation of specialized polymers like heat-resistant plastics for the aerospace industry.

Condensation Polymerization. Linking monomers into polymers requires the basic molecular building blocks to have reaction sites. Carothers recognized that most polymerizations fall into two broad categories, condensation and addition. In condensation, which many scientists prefer to call step, step-growth, or stepwise polymerization, the polymeric chain grows from monomers with two or more reactive groups that interact (or condense) intermolecularly, accompanied by the elimination of small molecules, often water. For example, the formation of a polyester begins with a bifunctional monomer, containing a hydroxyl group (OH, oxygen bonded to hydrogen) and a carboxylic acid group (COOH, carbon bonded to an oxygen and an OH group). When a pair of such monomers reacts, water is eliminated and a dimer formed. This dimer can now react with another monomer to form a trimer, and so on. The chain length increases steadily during the polymerization, necessitating long reaction times to get "high polymers" (those with large molecular weights).

Addition Polymerization. Many chemists prefer to call Carothers's addition polymerization chain, chain-growth, or chain-wise polymerization. In this process, the polymer is formed without the loss of molecules, and the chain grows by adding monomers repeatedly, one at a time. This means that monomer concentrations decline steadily throughout the polymerization, and high polymers appear quickly. Addition polymers are often derived from unsaturated monomers (those with a double bond), and in the polymerization process the monomer's double bond is rearranged in forming single bonds with other molecules. Many of these polymerizations also require the use of catalysts and solvents, both of which have to be carefully chosen to maximize yields. Important examples of polymers produced by this mechanism are polyurethane and polyethylene.

APPLICATIONS AND PRODUCTS

Since the start of the twentieth century, the discoveries of polymer scientists have led to the formation of hundreds of thousands of companies worldwide that manufacture thousands of products. In the United States, about 10,000 companies are manufacturing plastic and rubber products. These and other products exhibit phenomenal variety, from acrylics to zeolites. Chemists in academia, industry, and governmental agencies have discovered many applications for traditional and new polymers, particularly in such modern fields as aerospace, biomedicine, and computer science.

ELASTOMERS AND PLASTICS. From its simple beginnings manufacturing Bakelite and neoprene, the plastic and elastomeric industries have grown rapidly in the quantity and variety of the polymers their scientists and engineers synthesize and market. Some scholars believe that the modern elastomeric industry began with the commercial production of vulcanized rubber by Goodyear in the nineteenth century. Such synthetic rubber polymers as styrene-butadiene, neoprene, polystyrene, polybutadiene, and butyl rubber (a copolymer of butylene and isoprene) began to be made in the first half of the twentieth century, and they found extensive applications in the automotive and other industries in the second half.

Although an early synthetic plastic derived from cellulose was introduced in Europe in the nineteenth century, it was not until the twentieth century that the modern plastics industry was born, with the introduction of Bakelite, which found applications in the manufacture of telephones, phonograph records, and a variety of varnishes and enamels. Thermoplastics, such as polyethylene, polystyrene, and polyester, can be heated and molded, and billions of pounds of them are produced in the United States annually. Polyethylene, a low-weight, flexible material, has many applications, including packaging, electrical insulation, housewares, and toys. Polystyrene has found uses as an electrical insulator and, because of its clarity, in plastic optical components. Polyethylene terephthalate (PET) is an important polyester, with applications in fibers and plastic bottles. Polyvinyl chloride (PVC) is one of the most massively manufactured synthetic polymers. Its early applications were for raincoats, umbrellas, and shower curtains, but it later found uses in pipe fittings, automotive parts, and shoe soles.

Carothers synthesized a fiber that was stronger than silk, and it became known as nylon and led to a proliferation of other artificial textiles. Polyester fibers, such as PET, have become the world's principal man-made materials for fabrics. Polyesters and nylons have many applications in the garment industry because they exceed natural fibers, including cotton and wool, in such qualities as strength and wrinkle resistance. Less in demand are acrylic fibers, but, because they are stronger than cotton, they have had numerous applications by manufacturers of clothing, blankets, and carpets.

OPTOELECTRONIC, AEROSPACE, BIOMEDICAL, AND COMPUTER APPLICATIONS. As modern science and technology have expanded and diversified, so, too, have the applications of polymer science. For example, as researchers explored the electrical conductivity of various materials, they discovered polymers that have exhibited commercial potential as components in environmentally friendly battery systems. Transparent polymers have become essential to the fiber optics industry. Other polymers have had an important part in the improvement of solar-energy devices through products such as flexible polymeric film reflectors and photovoltaic encapsulants. Newly developed polymers have properties that make them suitable for optical information storage. The need for heat-resistant polymers led the U.S. Air Force to fund the research and development of several such plastics, and one of them, polybenzimidazole, has achieved commercial success not only in aerospace but also in other industries as well.

Following the discovery of the double-helix structure of DNA in 1953, a multiple of applications followed, starting in biology and expanding into medicine and even to such fields as criminology. Nondegradable synthetic polymers have had multifarious medical applications as heart valves, catheters, prostheses, and contact lenses. Other polymeric materials show promise as blood-compatible linings for cardiovascular prostheses. Biodegradable synthetic polymers have found wide use in capsules that release drugs in carefully controlled ways. Dentists regularly take advantage of polymers for artificial teeth, composite restoratives, and various adhesives. Polymer scientists have also contributed to the acceleration of computer technology since the 1980's by developing electrically conductive polymers, and, in turn, computer science and technology have enabled polymer scientists to optimize and control various polymerization reactions.

SOCIAL CONTEXT AND FUTURE PROSPECTS

Barring a total global economic collapse or a cataclysmic environmental or nuclear-war disaster, the trend of expansion in polymer science and engineering, well-established in the twentieth century, should continue throughout the twenty-first. As polymer scientists created new materials that contributed to twentieth-century advances in such areas as transportation, communications, clothing, and health, so they are well-positioned to meet the challenges that will dominate the twenty-first century in such areas as energy, communications, and the health of humans and the environment. Many observers have noted the increasing use of plastics in automobiles, and polymer scientists will most likely help to create lightweight-plastic vehicles of the future. The role of polymer science in biotechnology will probably exceed its present influence, with synthesized polymers to monitor and induce gene expression or as components of nanobots to monitor and even improve the health of vital organs in the human body. Environmental scientists have made the makers of plastics aware that many of their products end up as persistent polluters of the land and water, thus fostering a search that will likely continue throughout the twenty-first century for biodegradable polymers that will serve both the needs of advanced industrialized societies and the desire for a sustainable world.

—Robert J. Paradowski, MS, PhD

BIBLIOGRAPHY

Carraher, Charles E., Jr. *Giant Molecules: Essential Materials for Everyday Living and Problem Solving.* 2d ed. Hoboken, N.J.: John Wiley & Sons, 2003.

_______. *Introduction to Polymer Chemistry.* 2d ed. Boca Raton, Fla.: CRC Press, 2010.

Ebewele, Robert O. *Polymer Science and Technology.* Boca Raton, Fla.: CRC Press, 2000.

Morawetz, Herbert. *Polymers: The Origins and Growth of a Science.* Mineola, N.Y.: Dover, 2002.

Painter, Paul C., and Michael M. Coleman. *Essentials of Polymer Science and Engineering.* Lancaster, Pa.: DEStech Publications, 2009.

Scott, Gerald. *Polymers and the Environment.* Cambridge, England: Royal Society of Chemistry, 1999.

Seymour, Raymond B., ed. *Pioneers in Polymer Science: Chemists and Chemistry.* Boston: Kluwer Academic Publishers, 1989.

PROBABILITY AND STATISTICS

SUMMARY

Probability and statistics are two related fields covering the science of collecting, measuring, and analyzing information in the form of numbers. Both probability and statistics are branches of applied mathematics. Probability focuses on using numeric data to predict a future outcome. Statistics incorporates theory into the gathering of numerical data and the drawing of accurate conclusions. Because nearly all fields in applied science rely on the analysis of numbers in some way, probability and statistics are one of the most diverse areas in terms of subjects and career paths. Statisticians also practice in areas of the academic world outside of science and throughout industry.

DEFINITION AND BASIC PRINCIPLES

Probability and statistics are two interconnected fields within applied mathematics. In both fields, principles of scientific theory are applied to the analysis of groups of data in the form of numbers. The main objective of probability and statistics is to ask and answer questions about data with as much accuracy as possible.

Defining "probability" can be a challenge, as multiple schools of thought exist. In one view, held by a group of scholars known as frequentists, probability is defined as the likelihood that a statement about a set of data will be true in the long run. Frequentists focus on the big picture, specifically at the collective outcome of multiple experiments conducted over time, rather than on specific data items or outcomes.

A pair of dice, symbolic of probability. By Gaz at en.wikipedia

In contrast, scholars known as Bayesians prefer to start with a probability-based assumption about a set of data, then test to see how close the actual data come to the initial assumption. On both sides of the debate, probabilists are seeking to understand patterns in data to predict how a population might behave in the future.

Statistics is a field with a broader scope than probability, but in some ways, it is easier to define. The academic discipline of statistics is based on the study of groups of numbers in three stages: collection, measurement, and analysis. At the collection stage, statistics involves issues such as the design of experiments and surveys. Statisticians must answer questions such as whether to examine an entire population or to work from a sample. Once the data are collected, statisticians must determine the level of measurement to be used and the types of questions that can be answered with validity based on the numbers.

No matter how rigorous an individual study might be, statistical findings are often met with doubt by scholars and the general public. A quote repeated often (and mistakenly attributed to former British prime minister Benjamin Disraeli) is "There are three kinds of lies: lies, damned lies, and statistics."

BACKGROUND AND HISTORY

Probability has been a subject of interest since dice and card games were first played for money. Gambling inspired the first scholarly discussions of probability in the sixteenth and early seventeenth centuries. The Italian mathematician Gerolamo Cardano wrote *Libellus de ratiociniis in ludo aleae* (*The Value of All Chances in Games of Fortune,* 1714) in about 1565, although the work was not published until 1663. In the mid-1600's, French mathematicians Blaise Pascal and Pierre de Fermat discussed principles of probability in a series of letters about the gambling habits of a mutual friend.

The earliest history of a statistical study is less clear, but it is generally thought to involve demographics. British scholar John Graunt studied causes of mortality among residents of London and published his findings in 1662. Graunt found that statistical data could be biased by social factors, such as relatives' reluctance to report deaths due to syphilis. In 1710, John Arbuthnot analyzed the male-female ratio of babies born in Britain since 1629. His findings—that there were more males than females—were used to support his argument in favor of the existence of a divine being.

A third branch of statistics, the design of experiments and the problem of observational error, has its roots in the eighteenth century work of German astronomer Tobias Mayer.

However, a paper by British theologian Thomas Bayes published in 1764 after his death is considered a turning point in the history of probability and statistics. Bayes dealt with the question of how much confidence could be placed in the predictions of a mathematical model based on probability. The convergence between probability and statistics has increased over time. The development of modern computers has led to major advances in both fields.

HOW IT WORKS

In terms of scope, probability and statistics are some of the widest, most diverse fields in applied mathematics. If a research project involves items that must be counted or measured in some way, statistics will be part of the analysis. It is common to associate statistics with research in the sciences, but an art historian tracking changes in the use of color in painting from one century to another is just as likely to use a form of statistical analysis as a biologist working in a laboratory. Similarly, probability is used by anyone relying on numbers to make an educated guess about events in the future.

The word "statistic" can refer to almost any number connected to data. When statistics are discussed as a discipline, though, there is a multistep process that most projects follow: definition and design, collection, description and measurement, and analysis and interpretation.

Definition and Design. Much of the scholarship in the field of statistics focuses on data definition and the design of surveys and experiments. Statistical projects must begin with a question, such as "Does grade-point average in high school have an effect on income level after graduation?" In the data definition phase, the statistician chooses the items to be studied and measured, such as grades and annual earnings. The next step is to define other factors, such as the number of people to be studied, the areas in which they live, the years in which they graduated from high school, and the number of years in which income will be tracked. Good experimental design ensures that the rest of the project will gather enough data, and the right data, to answer the question.

Collection. Once the data factors have been defined, statisticians must collect them from the population being studied. Experimental design also plays a role in this step. Statistical data collection must be thorough and must follow the rules of the study. For example, if a survey is mailed to one thousand high school graduates and only three respond, more data must be collected before the survey's findings can be considered valid. Statisticians also must ensure that collected data are accurate by finding a way to check the reliability of answers.

Description and Measurement. Collected data must be stored, arranged, and presented in a way that can be used by statisticians to form statistics and draw conclusions. Grade-point averages, for example, might be compared easily if all the survey participants attended the same school. If different schools or grading systems were used, the statistician must develop rules about how to convert the averages into a form that would allow them to be compared. Once these conversions are made, the statistician would decide whether to present the data in a table, chart, or other form.

Analysis and Interpretation. In terms of statistical theory, the most complex step is the analysis and interpretation of data. When data have been collected, described, and measured, conclusions can be drawn in a number of ways—none of which is right in every case. It is this step in which statisticians must ask themselves a few questions: What is the relationship between the variables? Does a change in one automatically lead to a change in the other? Is there a third variable, or a lurking variable, not covered in the study that makes both data points change at the same time? Is further research needed?

One of the most common methods used for statistical analysis is known as modeling. A model allows the statistician to build a mathematical form, such as a formula, based on ideas. The data collected by the study can then be compared with the model. The results of the comparison tell a story that supports the study's conclusions. Some models have been found to be so innovative that they have earned their creators awards such as the Nobel Prize. However, even the best models can have flaws or can fail to explain actual data. Statistician George Box once said, "All models are wrong, but some models are useful."

Prediction. Probability deals with the application of statistical methods in a way that predicts future behavior. The goal of many statistical studies is to establish rules that can be used to make decisions. In the example, the study might find that students who achieve a grade-point average of 3 or higher earn twice as much, as a group, as their fellow students whose averages were 2.9 or lower. As an academic discipline, probability offers several tools, based on theory, that allow a statistician to ask questions such as: How likely is a student with a grade-point average of 3.5 to earn more than $40,000 per year?

For both statistics and probability, one of the primary objectives is to ask mathematical questions about a smaller population to find answers that apply to a larger population.

APPLICATIONS AND PRODUCTS

It would be nearly impossible to find a product or service that did not rely on probability or statistics in some way. In the case of a cup of coffee, for example, agricultural statistics guided where the coffee beans were grown and when they were harvested. Industrial statistics controlled the process by which the beans were roasted, packaged, and shipped. Statistics even influenced the strength to which the coffee was brewed—whether in a restaurant, coffeehouse, or a home kitchen. Probability played a role in each step as well. Forecasts in weather and crop yields, pricing on coffee bean futures contracts, and anticipated caffeine levels each had an effect on a single brewed cup.

One way to understand the applications and products of probability and statistics is to look at some general categories by function. These categories cover professional fields that draw some of the highest concentrations of professionals with a statistical background.

Process Automation. One of the broadest and most common ways in which statistical methods are applied is in process automation. Quality control is a leading example. When statistical methods are applied to quality control, measures of quality such as how closely products match manufacturing specifications can be translated into numbers and be tracked. This approach allows manufacturers to ensure that their products meet quality standards such as durability and reliability. It also verifies that products meet physical standards, including size and weight. Quality control can be used to evaluate service providers as well as manufacturers. If measures of quality such as customer satisfaction can be converted into numerical data, statistical methods can be applied to measure and increase it. One well-known quality control application, Six Sigma, was developed by the Motorola Corporation in the early 1980's to reduce product defects. Six Sigma relies on both probability and statistical processes to meet specific quality targets.

Another field in which process automation is supported by statistical analysis is transport and logistics. Transport makes up a significant amount of the cost of manufacturing a product. It also plays a major role in reliability and customer satisfaction. A manufacturer must ensure that its products will make the journey from one place to another in a timely way without being lost or damaged. To keep costs down, the method of transport must be as inexpensive as possible. Statistical methods allow manufacturers to calculate and choose the best transportation options (such as truck versus rail) and packaging options (such as cardboard versus plastic packaging) for their products. When fuel costs rise, the optimization of logistics becomes especially important for manufacturers. Probability gives manufacturers tools such as futures and options on fuel costs.

Biostatistics. Statistics are used in the biological sciences in a variety of ways. Epidemiology, or the study of disease within a population, uses statistical techniques to measure public health problems. Statistics allow epidemiologists to measure and document the presence of a specific illness or condition in a population and to see how its concentration changes over time. With this information, epidemiologists can use probability to predict the future behavior of the health problem and recommend possible solutions.

Other fields in the biological sciences that rely on statistics include genetics and pharmaceutical research. Statistical analysis has played a key role in the Human Genome Project. The amount of data generated by the effort of mapping human genes could be analyzed only with complex statistical processes, some of which are still being fine-tuned to meet the project's unique needs. Probability analyses allow geneticists to predict the influence of a gene on a trait in a living organism.

Pharmaceutical researchers use statistics to build clinical trials of new drugs and to analyze their effects. The use of statistical processes has become so widespread in pharmaceutical research that it is extremely difficult to obtain approvals from the U.S. Food and Drug Administration (FDA) without it. The FDA publishes extensive documentation guiding researchers through the process of complying with the agency's statistical standards for clinical trials. These standards set restrictions on trial factors ranging from the definition of a control group (the group against which the drug's effects are to be measured) to whether the drug effectively treats the targeted condition.

Spatial Statistics. Understanding areas of space requires the analysis of large amounts of data. Spatial statistics are used in fields such as climatology, agricultural science, and geology. Statistical methods provide climatologists with specialized tools to model the effects of factors such as changes in air temperature or atmospheric pollution. Meteorology also depends on statistical analysis because its data are time based and often involve documenting repeated events over time, such as daily precipitation over the course of several years. One of the best-known applications of probability to a field of science is weather forecasting.

In agricultural science, researchers use statistics and probability to evaluate crop yields and to predict success rates in future seasons. Statistics are also used to measure environmental impact, such as the depletion of nutrients from the soil after a certain kind of crop is grown. These findings guide recommendations, based on probabilistic techniques, about what kinds of crops to grow in seasons to come. Animal science relies on statistical analysis at every stage, from genetic decisions about the breeding of livestock to the environmental and health impacts of raising animals under certain kinds of farming conditions.

Geologists use statistics and probability in a wide range of ways. One way that draws a significant amount of interest and funding from industry is the discovery of natural resources within the Earth. These efforts include the mining of metals and the extraction and refining of products such as oil and gasoline. Mining and petroleum operations are leading employers of statisticians and probabilists. Statistical processes are critical in finding new geologic sites and in measuring the amount and quality of materials to be extracted. To be done profitably, mining and petroleum extraction are functions that must be carried out on a large scale, assisted by sizable amounts of capital and specialized equipment. These functions would not be possible without sophisticated statistical analysis. Statistics and probability are also used to measure environmental impact and to craft effective responses to disasters such as oil spills.

Risk Assessment. As a science, statistics and probability have their roots in risk assessment—specifically, the risk of losing money while gambling. Risk assessment remains one of the areas in which statistical analysis plays a chief role. A field in which statistical analysis and risk assessment are combined at an advanced level is security. Strategists are beginning to apply the tools of statistics and game theory to understanding problems such as terrorism. Although terrorism was once regarded as an area for study in the social sciences, people are developing ways to control and respond to terrorist events based on statistics. Probability helps strategists take lessons from terrorism in one political context or world region and apply them to another situation that, on the surface, might look very different.

Actuarial science is one of the largest and most thoroughly developed fields within risk assessment. In actuarial science, statistics and probability are used to help insurance and financial companies answer questions such as how to price an automobile insurance policy. Actuaries look at data such as birth rates, mortality, marriage, employment, income, and loss rates. They use this data to guide insurance companies and other providers of financial products in setting product prices and capital reserves.

Quantitative finance uses statistical models to predict the financial returns, or gains, of certain types of securities such as stocks and bonds. These models can help build more complex types of securities known as derivatives. Although derivatives are often not well understood outside of the finance industry, they have a powerful effect on the economy and can influence everyday situations such as a bank's ability to loan money to a person buying a new home.

Survey Design and Execution. Surveys use information gathered from populations to make statements about the population as a whole. Statistical methods ensure that the surveys gather enough high-quality data to support accurate conclusions. A survey of an entire population is known as a census. One prominent example is the United States Census, conducted every ten years to count the country's population and gather basic information about each person. Other surveys use data from selected individuals through a process known as sampling. In nearly all surveys, participants receive questions and provide information in the form of answers, which are turned into mathematical data and analyzed statistically.

Aside from government censuses, some of the most common applications of survey design are customer relationship management (CRM) and consumer product development. Through the gathering of survey data from customers, companies can use customer relationship management to increase the effectiveness of their services and identify frequent problems to be fixed. In creating and introducing new products, most companies rely on data gathered from consumers through Internet and telephone surveys and on the results of focus groups.

SOCIAL CONTEXT AND FUTURE PROSPECTS

There is an increasing need for professionals with a knowledge of probability and statistics. The growth of the Internet has led to a rapid rise in the amount of information available, both in professional and in private contexts. This growth has created new opportunities for the sharing of knowledge. However, much of the information being shared has not been filtered or evaluated. Statistics, in particular, can be fabricated easily and often are disseminated without a full understanding of the context in which they were generated.

Statisticians are needed to design experiments and studies, to collect information on the basis of sound research principles, and to responsibly analyze and interpret the results of their work. Aside from a strong knowledge of this process, statisticians must be effective communicators. They must consider the ways in which their findings might be used and shared, especially by people without a mathematical Background. Their results must be presented in a way that will ensure clear understanding of purpose and scope.

Although overall job growth for statisticians is predicted to remain steady, fields likely to see a higher rate of growth are those involving statistical modeling. An increase in computer software and other tools to support modeling has fueled higher demand for professionals with a familiarity in this area. Statistical modeling is useful in many contexts related to probability. These contexts range from the analysis of clinical trial data for pharmaceutical products to the forecasting of monetary gains and losses in a portfolio of complex financial instruments. Career growth prospects, particularly at the entry level, are most attractive for candidates with a Background both in statistics and in their applied fields.

The number of jobs for actuaries is expected to grow more rapidly than for statisticians in general. However, the popularity of actuarial science as an area of study for college undergraduates may produce a surplus of qualified applicants. Insurers are expected to be the primary employers of actuaries for the foreseeable future, but demand for actuaries is growing among consulting firms. As the tools for modeling the impact of large-scale disasters become more sophisticated, companies in many industries will seek expertise in protecting themselves against risk—a trend that may heighten the need for actuarial expertise.

—Julia A. Rosenthal, MS

Bibliography

Black, Ken. *Business Statistics: For Contemporary Decisionmaking.* 8th ed. Malden: Wiley, 2014. Print.

Boslaugh, Sarah. *Statistics in a Nutshell.* 2nd ed. Sebastopol: O'Reilly, 2013. Print.

Enders, Walter, and Todd Sandler. *The Political Economy of Terrorism.* New York: Cambridge UP, 2005. Print.

Fung, Kaiser. *Numbers Rule Your World: The Hidden Influence of Probabilities and Statistics on Everything You Do.* New York: McGraw, 2010. Print.

Mlodinow, Leonard. *The Drunkard's Walk: How Randomness Rules Our Lives.* New York: Pantheon, 2008. Print.

Nisbet, Robert, John Elder, and Gary Miner. *Handbook of Statistical Analysis and Data Mining Applications.* Burlington: Elsevier, 2009. Print.

Ott, R. Lyman, and Michael T. Longnecker. *An Introduction to Statistical Methods and Data Analysis.* Belmont: Brooks, 2010. Print.

Sharpe, Norean R., Richard De Veaux, and Paul F. Velleman. *Business Statistics: A First Course.* 2nd ed. New York: Pearson, 2013. Print.

Takahashi, Shin. *The Manga Guide to Statistics.* San Francisco: No Starch, 2009. Print.

Programming Languages for Artificial Intelligence

SUMMARY

Artificial intelligence (AI) is a branch of computer science that creates computers and software with the thinking capacity of humans. The key to the capabilities of these computers is the programming language used in the process. There are two major programming languages for AI, *LISP* and *PROLOG.* LISP is more widely used than PROLOG, but both are respected and recognized as invaluable parts of this scientific discipline.

HISTORY AND FUNCTIONALITY OF LISP

Computers are programmed by humans to "think." The ability of a computer to perform its function is only as strong as the language used to instruct it. Two different programming languages have evolved for use in artificial intelligence, and computer scientists have noticed promise in both.

A language called LISP is the older of the two. LISP—which stands for List Processor—was developed in 1958 by an American computer scientist named John McCarthy, who also coined the term *artificial intelligence.* The foundation for LISP was an older programming language known as Information Processing Language (IPL), developed by Allen Newell, Herbert Simon, and Cliff Shaw in 1956. This language was useful in the early days of computing, but it was highly complicated. A simpler language was needed to enable programmers to understand it better. McCarthy altered IPL by adding basic principles from what is known as *lambda calculus,* a branch of mathematical logic that focuses on variables and substitution. An advantage of LISP is that it enables a data management mechanism known as *garbage collection* in the AI "thinking" process. Items considered "garbage" in the world of AI are units of memory created during the running of an AI operation that serve as temporary holders of information. While they are useful for a time, they eventually outlive that usefulness. LISP is able to identify this used memory and discard it. Unlike IPL, LISP is able to assign storage space for information effectively and conservatively.

LISP is the most commonly used programming language for AI in the United States. The most crucial part of LISP's functionality is an element called the *interpreter.* The interpreter takes statements presented to it by users and decides the value of the expression through a process known as *evaluation.* The evaluation process depends on three factors: the identity of the symbols used in a statement, or what the words or figures presented mean at their most basic level; the associations (e.g. other terms, or lists of terms) with each symbol in a statement; and the list of possible functions that might be associated with a statement. Lists are central to LISP, and yet they are not necessarily the same items we would typically call lists; they are more collections of possible functions or equations centering on a term or symbol. It is significant that the programmer in LISP gives the computer commands by constantly defining new functions and new equations. This process could potentially result in the growth of ability in the machine—as it might with a human, engaged in the act of intellectual exploration.

HISTORY AND FUNCTIONALITY OF PROLOG

The other major programming language for artificial intelligence is called PROLOG. The name comes from the phrase "programming in logic." PROLOG was first developed in 1973 by French computer scientist Alain Colmerauer at the University of Aix-Marseilles. Robert Kowalski, an American computer scientist at the University of Edinburgh in Scotland, developed it further. The chief strength of PROLOG is its ability to implement logical processes. If two statements are entered, the program can determine if there is a logical connection between them. One reason PROLOG is able to do this is that programmers give the computer a set of facts to work with before beginning to enter statements. These facts might be descriptive statements about people or objects, such as "the car is green," or about relationships between people or objects, "Sam is the father of Carla." The facts can also be rules concerning the people or objects, such as "Sam is the grandfather of Emma is true if Sam is the father of Carla is true and Carla is the mother of Emma is true." This gives the program a basic framework to answer questions. People who use PROLOG do not necessarily have to know its code. Embedded in the language is a program called *ProtoThinker* (PT-Thinker), which translates normal statements into language a computer driven by PROLOG understands.

Another characteristic that distinguishes PROLOG from other programming languages is that it is declarative. To use this language, the programmer must first decide what the ultimate goal is they want to achieve. The goal is then expressed in PROLOG, and the software determines how that goal can best be met. This allows a considerable programming burden to be taken off the back of the programmer. Rather than telling the computer what to do, the PROLOG user lets the computer determine what to do. The computer does this by sorting through the information it has been given previously to find a match for the request made of it by the user.

The structures native to PROLOG have aspects in common with many searchable databases. PROLOG recognizes words in lowercase and recognizes basic numbers. Among other uses, PROLOG is used in the programming of more intelligent machines and certain trainable robots.

—*Max Winter*

BIBLIOGRAPHY

Copeland, Jack. "What Is Artificial Intelligence?" *AlanTuring*. AlanTuring, 2000. Web. 5 Oct. 2015. http://www.alanturing.net/turing_archive/pages/reference%20articles/what_is_AI/What%20is%20AI05.html.

Philips, Winfred. "Introduction to Logic for Proto-Thinker." *Mind*. Consortium on Cognitive Science Instruction, 2006. Web. 6 Oct. 2015. http://www.mind.ilstu.edu/curriculum/protothinker/logic_intro.php.

"Introduction to PROLOG." *Mind*. Consortium on Cognitive Science Instruction, 2006. Web. 5 Oct. 2015. http://www.mind.ilstu.edu/curriculum/protothinker/prolog_intro.php.

Wilson, Bill. "Introduction to PROLOG Programming." *CSE*. Bill Wilson/U of New South Wales, 26 Feb. 2012. Web 5 Oct. 2015. http://www.cse.unsw.edu.au/~billw/cs9414/notes/prolog/intro.html.

PROPORTIONALITY

SUMMARY

A proportion is a statement that two ratios are equal. For example, 2/7 = 6/21 is a proportion. Proportions also show a relationship between two ratios.

Proportions are very important in everyday thinking, for example, realizing that it has taken you 15 minutes to drive 10 miles can allow you to predict that it will take you 1 hour to drive 40 miles at that same rate. Proportional thinking is also used to determine the better value when buying food items, as when comparing a 20-ounce box of cereal that costs $2.00 with a 28-ounce box that costs $3.25. In this example, the larger box is more expensive by 50%, yielding 30 ounces for $3.00. The relative cost, $2.00 for 20 ounces, is in proportion to $3.00 for 30 ounces. By comparing this new ratio ($3.00 for 30 ounces) with the 28-ounce box, it becomes apparent that the $2.00 box is the better deal (more cereal for less money). A more traditional way of solving this problem is to find a unit rate for each box of cereal. In this case we get:

$2.00/20 ounces = 10 cents/ounce

$3.25/28 ounces = 11.6 cents/ounce

It is important to note that in the above situation the two boxes ware not in proportion regarding the ratio of cost to amount. If they were, then neither of the boxes would have been a better deal. However, because the ratio of the 20-ounce box was less than that of the 32-ounce box, the ratios were not in proportion, and one was a better deal.

RATIOS

A ratio is a relationship between two quantities. Ratios convey the relative sizes of the two quantities in a particular situation. For example, a particular room could contain a ratio of 5 men to 7 women. We usually write this ratio as 5:7, or 5/7. A rate is a ratio that can be extended to a broad range of situations, such as 70 miles per hour. There are several ways of articulating a ratio with words; the ratio a/b could be said as "*a* to *b*," "*a* per *b*," "*a* for *b*," "*a* for each *b*," "for every *b* there are *a*," "the ratio of *a* to *b*," and "*a* is to *b*."

CONSTANT OF PROPORTIONALITY

A key aspect of proportional situations (when the two ratios are equal) is identifying what is known as the constant of proportionality. This is a ratio that describes the way in which two quantities that are in proportion are related. Hourly wages provide most workers with experience of a constant of proportionality. For example, a rate of $10.00 per hour, 10 is the constant of proportionality that relates the two quantities of hours worked and money earned. Table 1

Hours Worked	Money Earned
1.5	$10.00
2.5	$20.00
3.5	$35.00
4.5	$40.00

Table 1. Income for hours worked at a rate of $10.00 per hour. © EBSCO

shows various amounts of money made by working various amounts of hours, and it is clear that the two columns are related by a factor of 10. This shows how the two quantities are in proportion.

Another way of seeing this is to take any related pair of numbers in the table (e.g., $40 and 4 hours) and find their ratio—the result will always be the same, in this case 10, the constant of proportionality.

Many situations allow the use of proportional thinking. These are situations where the ratios between two quantities change in a constant manner (in other words, where a constant of proportionality exists). For example, a 16-ounce bottle that is 30% alcohol in water has a ratio of alcohol to water that is 3/7 (30% alcohol, 70% water). Removing a spoonful would leave the ratio as 3/7, as this mixture is assumed to be uniformly mixed. Note that 3/7 is the constant of proportionality in this situation. If an office has a total computer supply of 9 Macs and 15 PCs, the ratio of Macs to PCs is 3/5. If 8 computers were removed from one of the cubicles, it would not be possible to tell the ratio of Macs to PCs for this selection without knowing if the computers in the cubicle are in the same proportion as those in the office as a whole. Therefore, while proportional thinking can be used in many daily situations, there are some situations for which proportional thinking may lead to false conclusions.

QUALITATIVE PROPORTIONAL THINKING

Proportional thinking can also be used in situations that do have specific quantities. These are often called qualitative proportional situations, as the comparisons are made without the use of numbers. For example, suppose Bill and George have two pieces of wood, and that Bill's piece is longer. They each need to hammer nails into the board in a straight line so that the nails are equally spaced apart. If Bill has fewer nails to hammer than George, then qualitative proportional thinking can be used to tell which board will have the nails farther apart. Specifically, because Bill has fewer nails and a longer piece of wood, he would need to space them out more than those on George's board. This idea of more for less is a common way of thinking about these kinds of situations, which are not proportional, but use proportional thinking.

—*David Slavit*

BIBLIOGRAPHY

Brousseau, Guy, Virginia Warfield. *Teaching Fractions Through Situations: A Fundamental Experiment.* New York: Springer, 2013.

Ercole, Leslie K., Manny Frantz, and George Ashline. "Multiple Ways to Solve Proportions." *Mathematics Teaching in the Middle School* 16.8 (2011): 482-490.

Lobato, Joanne, and Amy B. Ellis. *Developing Essential Understanding of Ratios, Proportions, and Proportional Reasoning for Teaching Mathematics: Grades 6–8.* Reston, VA: National Council of Teachers of Mathematics, 2010.

PUBLIC-KEY CRYPTOGRAPHY

SUMMARY

Public-key cryptography is a form of data encryption technology used to protect confidential information. *Data encryption* refers to using a code as a key to protect important information and limit access to a certain user or group of users. Public-key cryptography, also called public-key encryption, uses an *asymmetric algorithm,* or mathematical formula, that applies one key to encrypt the information and another one to decode it. Data is more secure when the encryption key does not have to be shared.

BACKGROUND

Ciphers and codes are two ways of protecting data so that it can remain secret. A *code* is a list of prearranged substitutes—one thing replaces another. One simple form involves simply substituting a number for a letter; for instance, if A=1, B=2, etc., *cat* becomes "3-1-20." Yo decode a message, one uses the key to replace the substitutes with the original letters, words, or phrases. Every party who will be sharing the message needs the key to code or decode messages.

Ciphers are more complicated and use a mathematical algorithm to both substitute for the original information and multiply that result by another number that is the key to the algorithm. Using the example above, for instance, and encrypting it with a cipher with a key of 3, means multiplying "3-1-20" by 3 to get "9-3-60." To decipher the message, the recipient needs the specific key, or algorithm, that works for the cipher. The recipient applies the key in reverse to decode the message. Real ciphers use complicated mathematical keys.

The earliest known cipher is the Caesar cipher, which originated in the sixth century BCE. It is a symmetric cipher, meaning both parties use the same code to send and receive messages that either can read. For many centuries, this was the only type of cipher that existed. As technology increased to allow information to be shared faster and more easily between places, it became necessary to change codes more frequently. A code that fell into the wrong hands could be used to decode secret messages and also allowed an enemy to encrypt and send fake messages. To keep information safe, users were forced to develop more complicated codes and to change them often. This created administrative problems to manage and track multiple keys.

NEW METHOD

Cryptologists—those who work with codes and ciphers—realized that the way to solve the problem was to have a separate way to encrypt and decrypt the messages. This asymmetrical method means that data coded with the public key can only be decrypted with the private key, and vice versa. The public key could even be widely available and still protect the information, because it can only be accessed by someone with the private key. Meanwhile, those with the private key do not have to worry about replacing the key to protect sensitive information. Public-key cryptography protects that data and can be used to determine if the information is coming from an authentic source, since only a holder of the private key can send a message. This is the basis of a digital signature in electronic data transmissions.

This asymmetric encryption form was first explored secretly by British intelligence services in the early 1970s. Around the same time, Stanford mathematicians Whitfield Diffie (1944—) and Martin Hellman (1945—) proved it was theoretically possible to develop an asymmetric ciphering method that uses two keys. In 1977, MIT mathematicians Ronald L. Rivest (1947—), Adi Shamir (1952—), and Leonard M. Adleman (1945—) devised an algorithm based on factoring that allows for the first public-key encryption coding. Their method, known as the Rivest-Shamir-Adleman encryption, or RSA, created a special algorithm that can be used with multiple keys to encrypt information.

HOW IT WORKS

Factoring is the secret behind public-key cryptography. Factoring a number means identifying the two prime numbers that can be multiplied together to produce the original number. The process of factoring very large numbers is difficult and slow, even when computers are used. This makes the encryption secure. The person, company, or other entity that wants to encrypt a message chooses—or has a computer choose—two prime numbers that are each more than 100 digits long and multiplies them together. This becomes the key and can be made public without endangering the code because it is so difficult to factor it back to the two prime numbers with which the person or entity started. It can be and often is published freely—the key is made public—and can be used by parties to send coded messages that only the originator of the code can read.

For example, Company A wants to be able to receive coded messages from its customers. Company A uses a computer to choose two large prime numbers—x and y—and multiplies them to get z. Company A then shares z on its website. Customer B can now take his message and the key z and enter them into the RSA algorithm. The algorithm will encode the message, which can now be sent to Company A. The key to unlocking it is x and y, which only Company A knows, so the company can decipher the message. The message remains secure because factoring z to x and y is difficult and time consuming.

This process is used in countless computer transactions every day, often without the person encrypting the message even realizing it is taking place. The method was so transformative for computer cryptography that the Association for Computing Machinery awarded the ACM Turing Award and $1 million in prize money to Diffie and Hellman in March 2016.

—*Janine Ungvarsky*

Bibliography

Bright, Peter. "Locking the Bad Guys Out with Asymmetric Encryption." *ARSTechnica.* Condé Nast. 12 Feb. 2013. Web. 4 March 2016. http://arstechnica.com/security/2013/02/lock-robster-keeping-the-bad-guys-out-with-asymmetric-encryption/

"Cryptography Defined/Brief History." *The College of Liberal Arts.* University of Texas at Austin. Web. 4 March 2016. http://www.laits.utexas.edu/~anorman/BU.S.FOR/course.mat/SSim/history.html

Mann, Charles C. "A Primer in Public-Key Encryption." *The Atlantic.* The Atlantic Monthly Group. Sept. 2002. Web. 4 March 2016. http://www.theatlantic.com/magazine/archive/2002/09/a-primer-on-public-key-encryption/302574/

Niccolai, James. "As Encryption Debate Rages, Inventors of Public Key Encryption Win Prestigious Turing Award." *PCWorld.* IDG Consumer & SMB. 2 March 2016. Web. 4 March 2016. http://www.pcworld.com/article/3039911/encryption/as-encryption-debate-rages-inventors-of-public-key-encryption-win-prestigious-turing-award.html

"Public Key Cryptography." *PC Magazine Encyclopedia.* Ziff Davis, LLC. Web. 4 March 2016. http://www.pcmag.com/encyclopedia/term/40522/cryptography

"Understanding Public Key Cryptography." *Microsoft Tech Net.* Microsoft. Web. 4 March 2016. https://technet.microsoft.com/en-us/library/aa998077%28v=exchg.65%29.aspx

Python

SUMMARY

Python is a popular general-purpose dynamic programming language that allows multiple paradigms. Both structured programming and object-oriented programming are supported in full, while functional programming, aspect-oriented programming, metaprogramming, logic programming, and other paradigms are supported either in specific features or by extensions. Introduced in 1991, Python has a passionate user community. Its key strengths are its readability, flexibility, and large standard library.

HISTORY AND BACKGROUND

Python was developed by Dutch programmer Guido van Rossum as a hobby programming project, and named for the British comedy troupe Monty Python. In the Python community, van Rossum is designated Benevolent Dictator for Life, the traditional title given to creators or other development leaders of open-source projects. In an unusual arrangement, van Rossum was allowed to spend half of his work hours developing Python while employed at Google from 2005 to 2012. By that point, Python had become one of the most popular programming languages, consistently among the most frequently mentioned in job listings, and van Rossum's guidance was something that benefited the community at large.

The goals of Python's design were to be easy to learn, which meant code that was both easily readable and intuitive so that it was easy to remember; open source (publicly accessible and modifiable), in order to encourage a community of developers; and optimized for short development times. Some of the syntax in Python reflects an influence from the C programming language, because of its widespread familiarity and influence in the programming world, while many of the functional programming features resemble those of the Lisp and Haskell languages.

Further Python development occurred principally via Python Enhancement Proposals (PEP). The PEP process collects proposals for new features and other design decisions, as well as recording comments on those proposals made by members of the Python community. Each PEP is numbered sequentially. The programming philosophy of Python was developed by 1999, when PEP 20 was released, called the Zen of Python, consisting of twenty design aphorisms, including "Beautiful is better than ugly," "Explicit is better than implicit," "Special cases aren't special enough to break the rules," and "In the face of ambiguity, refuse the temptation to guess." Most of the aphorisms were written by Python community member Tim Peters, who also developed Timsort, a Python sorting algorithm.

A key element of Python's design is its extensibility. The core of the language is small enough to be embedded in applications to create a programmable interface. The language's flexibility emerges from the combination of its lightweight design and its large standard library, which provides users with many possibilities in extensions. The strong preferences for a certain programming aesthetic in the Python community has led to the complimentary adjective *pythonic*, usually meaning both minimalist and readable.

Python is the main programming language for the Raspberry Pi project, a single board computer the size of a credit card, priced between $5 and $35 depending on the model. The Raspberry Pi was developed to teach computer science in developing countries, but is also a famously flexible device for computer hobbyists, thanks in part to the possibilities of Python.

—Bill Kte'pi, MA

Bibliography

Goriunova, Olga. *Fun and Software: Exploring Pleasure, Paradox, and Pain in Computing.* New York: Bloomsbury, 2016. Print.

Grus, Joel. *Data Science from Scratch: First Principles with Python.* Sebastopol: O'Reilly, 2015. Print.

Hoare, C. A. R. *Communicating Sequential Processes.* Upper Saddle River: Prentice, 1985. Print.

Lubanovic, Bill. *Introducing Python: Modern Computing in Simple Packages.* Sebastopol: O'Reilly, 2014. Print.

Lutz, Mark. *Learning Python.* Sebastopol: O'Reilly, 2015. Print.

Ramalho, Luciano. *Fluent Python.* Sebastopol: O'Reilly, 2015. Print.

Roscoe, A.W. *Theory and Practice of Concurrency.* Upper Saddle River: Prentice, 1997. Print.

Weikum, Gerhard, and Gottfried Vossen. *Transactional Information Systems: Theory, Algorithms, and the Practice of Concurrency Control and Recovery.* Burlington: Morgan, 2001. Print.

Q

Quantum Computing

SUMMARY

Quantum computing is a type of computing that uses ideas gained from quantum physics to make a computer run faster than traditional computers. Quantum computing is a relatively new field, though the idea has been around for several decades. Today, researchers, physicists, and others are working to create viable quantum computing technology that can be used by large corporations and governments. However, quantum technology is still in its infancy, and quantum computers do not work well enough to be used in everyday settings. Current quantum computers require very specific environments. Scientists are working to make quantum computers more reliable as they strive to understand quantum physics and quantum computing components.

QUANTUM PHYSICS AND QUANTUM COMPUTING

Quantum computing is based on the field of quantum physics, and the field of quantum physics must advance for quantum computing to progress. At the quantum level, physics has different rules than it does in the regular world. For example, quantum physicists have found a strange phenomenon called superposition, in which particles can exist in multiple states at once. Because of superposition, quantum particles can play different roles at one time.

Traditional computers use bits, which are represented by zeros and ones. A string of ones and zeros holds information that makes up numbers, letters, and so on in traditional computing. Quantum computing works differently. This type of computing uses qubits rather than bits. A bit can exist as only a one or a zero. A qubit can exist as a one, a zero, or both at the same time. This is possible because of superposition. This phenomenon allows qubits to be more versatile than bits, and it allows quantum computers to have exponentially more computing power than regular computers.

CURRENT STATE OF QUANTUM COMPUTING

In the 1980s and 1990s, people first start considering the possibility of quantum computing. Experts pointed out that some of the computing predications made in the early days of computing (e.g., Alan Turing's concepts about the future of computing in the 1950s) were based on classical ideas about physics. These same experts noted that advances in quantum physics could change the future of computing. For many years, quantum computing was only theoretical. After some important theories were developed, scientists and researchers had the groundwork they needed to start to develop quantum computers. Even today the quantum computers that exist are experimental and do not fully function.

Quantum computing is very much an evolving field with many questions that still need to be answered. Some technology professionals question the effect that quantum computing will have on technology in general. Some experts believe that the progress in traditional computing is slowing down. Some of these same people believe that quantum computing is the force that will help propel scientific advances in the future, as humans reach the limits of what traditional technology can do. Most experts agree it will take scientists many years to develop an operational quantum computer.

One problem that is holding back the advancement of quantum computing is that quantum computers require very specific environments to work. Quantum computers must be kept at temperatures approaching absolute zero, and any changes to the environment can affect the computer. Scientists also encase the computers in extremely thick shielding to ensure no changes to the environment occur, as these changes would interrupt the quantum processing. One of the most challenging problems in the field of quantum computing is controlling the quantum system and finding practical ways to keep the systems protected.

Another problem with developing quantum computing is the very same thing that makes it work: quantum physics. If researchers try to investigate how the qubits work, the qubits will stop acting as both ones and zeros at the same time. Scientists want to better understand qubits and the reason why they work the way they do, but current methods and technology are not advanced enough to give them the answer.

POSSIBLE FUTURE OF QUANTUM COMPUTING

Quantum computing could potentially change the future of technology. Encryption methods that are all but impossible to break with current technology could be cracked easily in the age of quantum computing. In addition, quantum computers could create scientific models that are more advanced than anything that has come before. Such models could help scientists answer important questions, including questions about the beginning of the universe. Although quantum computing could change people's lives, even experts find it difficult to give exact examples of ways quantum computing could be utilized. While experts speculate that these major advances in technology would change life, they are unsure of what effect these changes could have on people.

Because quantum computing is seen as a potential game-changing technology, researchers from both the public and private sectors are investigating the field. Several universities have groups that are researching the technology, and research labs around the world have been trying to tackle the problem of quantum computing. Venture capitalists and governments are also putting resources into quantum computing projects. Major technology companies such as Microsoft and Google have even invested in quantum computing research.

—*Elizabeth Mohn*

Bibliography

DeAngelis, Stephen F. "Closing In on Quantum Computing." *Wired.* Conde Nast. Web. 6 Aug. 2015. http://www.wired.com/insights/2014/10/quantum-computing-close/

Dickerson, Kelly. "Here's Why We Should Be Really Excited about Quantum Computers." *Business Insider.* Business Insider, Inc. 17 Apr. 2015. Web. 6 Aug. 2015. http://www.businessinsider.com/why-be-excited-about-quantum-computers-2015-4

Galchen, Rivka. "Dream Machine." *New Yorker.* Conde Nast. 2 May 2011. Web. 6 Aug. 2015. http://www.newyorker.com/magazine/2011/05/02/dream-machine

"A Little Bit, Better." *The Economist.* The Economist Newspaper. 20 June 2015. Web. 6 Aug. 2015. http://www.economist.com/news/science-and-technology/21654566-after-decades-languishing-laboratory-quantum-computers-are-attracting

"Scientists Achieve Critical Steps to Building First Practical Quantum Computer." *Phys.org.* Phys.org. 29 Apr. 2015. Web. 6 Aug. 2015. http://phys.org/news/2015-04-scientists-critical-quantum.html

Wadhwa, Vivek. "Quantum Computing Is About to Overturn Cybersecurity's Balance of Power." *The Washington Post.* The Washington Post. 11 May 2015. Web. 6 Aug. 2015. http://www.washingtonpost.com/news/innovations/wp/2015/05/11/quantum-computing-is-about-to-overturn-cybersecuritys-balance-of-power/

R

R

SUMMARY

In the field of statistics, R is a free software package used in the analysis of data. Other software packages exist (such as SAS or SPSS), but they are company based. R is managed by a core team of expert individuals with whom its scope is extremely broad. On the downside, however, learning how to work with R presents a steep learning curve. When it comes to looking for help, there is a forum, but its managers require that one's question has been thoroughly researched without success before being presented; otherwise, the query is dismissed.

HISTORY AND BACKGROUND

In the early 1980s the statistical programming language S was created. A commercial product based on S, called S-Plus, was released and enjoyed widespread success. Then Ross Ikaha and Robert Gentleman of the University of Auckland, New Zealand, wrote a simplified version of S to use as a teaching tool and named it R.

With R, it is always possible to do more on the results of a statistical procedure. Where R is based on a formal computer language, it is very dynamic in its uses. Analysis of even moderately complicated data is best approached via ad-hoc statistical model building, and here R comes into its own. In addition, R does not require everything to be written in its own language and so it allows for a mix of languages to be used in a single program, depending on what language is best for the desired functionality.

In 1995 statistician Martin Mächler urged Ikaha and Gentleman to release the source code for R under the General Public Licence, which allows for its free use and distribution. At the same time the Linux operating system promulgated an upsurge in open source software. R was a good fit for those who wished to use Linux for statistical programming. At this time a forum was set up to discuss and explore bugs in the system and in the development of R.

By 1997 an international core team was created and had expanded. Its members collaborate via the Internet to control the development of R. As of 2016 there are two presidents of the board, two members at large, two auditors, and thirty-two ordinary members, elected by a majority vote of the general assembly. New members are evaluated based on nonmonetary contributions, such as new code and other efforts, to the R project.

An educational version of R was released on May 3, 2016, version 3.3.0, while version 3.2.5 of the language proper was released on April 14, 2016.

R is viewed by many as a statistics system, though it is better understood as providing an environment within which statistical procedures can be implemented. R's uses can easily be extended by the use of packages, or sets of code that allow for new sets of tasks with set parameters. There are eight packages provided from the R distribution, while many more can be found through the Comprehensive R Archive Network (CRAN) family of Internet sites.

—Martin P. Holt, MSc

Bibliography

Bloomfield, Victor. *Using R for Numerical Analysis in Science and Engineering.* Boca Raton: CRC, 2014. Print.

Crawley, Michael, J. *Statistics: An Introduction using R.* Medford: Wiley, 2014. Print.

Dayal, Vikram. *An Introduction to R for Quantitative Economics: Graphing, Simulating and Computing.* New York: Springer, 2015. Print.

Hothorn, Torsten, and B. S. Everitt. *A Handbook of Statistical Analyses Using R.* 3rd ed. Boca Raton: CRC, 2014. Print.

Machlis, Sharon. "Beginner's Guide to R." *Computerworld.* Computerworld, 6 June 2013. Web. 22 Aug. 2016.

Nash, John C. *Nonlinear Parameter Optimization Using R Tools.* Medford: Wiley, 2014. Print.

Stowell, Sarah. *Using R for Statistics.* New York: Apress, 2014. Print.

Sun, Changyou. *Empirical Research in Economics: Growing up with R.* Starkville: Pine Square, 2015. Print.

RATE

SUMMARY

Rate as a concept is a summary measure that conveys the idea of risk over time, where risk is the probability of occurrence of the event of interest. (Risk is calculated as the number of events divided by the number of people at risk.) The idea of a summary measure is that of a statistic which collapses the information available from several data values into a single value. Another summary measure underlying rate is relative risk: the ratio of the risk (as defined above) of a given event in one group of subjects exposed to a particular disease of interest compared to that of another group not exposed. This leads to relative risk being used as a synonym for risk ratio, or more generally, for odds ratios and rate ratios.

Rate is a measure of the frequency of occurrence of an item of interest. It consists of the number of events occurring in a specific period, per unit of time. Time is therefore an essential element of this concept. As this frequency can change over time, rate is time specific. Typically, the essential components of a rate are 1) a numerator, 2) a denominator, 3) a duration, and 4) a multiplier.

PRACTICAL EXAMPLES OF RATES

Many practical examples of rates exist, prominently statistics relating to life and death of the population at large: birth rates, fertility, rates and death rates. Among the latter are the crude mortality rate, age-specific mortality rates, and age-standardized mortality rates. The infant mortality rate is the number of deaths of children under one year of age divided by the number of live births. The perinatal mortality rate is the number of stillbirths and deaths in the first week of life divided by the total births. Almost all child deaths are preventable, so child mortality is of great concern in some populations (but not in others, where attention shifts to adult mortality). Adult mortality is concerned with deaths between the ages of 15 and 59 years, ignoring geriatric mortality. Adult mortality is of great interest as ages from 15 years old to 59 years of age are expected to be the most healthy period of life. Death is easy to identify in nearly all cases, and records of the date of death are generally available. Therefore, mortality statistics are considered to be reliable and are used across the world.

Fundamentally, there is the crude death rate. This is the number of deaths in 1 year divided by the mid-year population, all multiplied up by 1,000. The main aim in calculating a rate is to allow comparisons of groups, times, and so on. A rate, however, is valid as a tool only for stable conditions over the period for which it is calculated. Otherwise, the mid-year population can be fallacious.

—Martin P. Holt, MSc

BIBLIOGRAPHY

Bureau of Labor Statistics. Office of Survey Methods Research. "Wage Estimates by Job Characteristic: NCS and OES Program Data." By Michael Lettau and Dee Zamora. N.p.: n.p., 2013.

Forbes, Catherine, Merran Evans, Nicholas Hastings, and Brian Peacock. *Statistical Distributions.* Hoboken, NJ: Wiley, 2011.

Freedman, David, Robert Pisani, and Roger Purves. *Statistics.* 4th ed. London: Norton, 2011.

Glantz, Stanton A. *Primer of Biostatistics.* 7th ed. New York,: McGraw, 2011.

Replication

SUMMARY

When used in computing, the term *replication* refers to the duplication of databases and database elements. Database replication is a complex process that involves copying data found on one server and distributing the copies to additional servers over a local area network (LAN) or a wide area network (WAN). Replication allows multiple people to have access to the same set of data and prevents data loss in the event of software or hardware failure. Replication systems may be configured into fan-in systems, fan-out systems, or multimaster systems.

The process of database replication involves several parts. Any piece of data stored in a database is called a *database object.* A database object simultaneously stored in multiple databases or locations is called a *replication object.* Replication objects related to each other are then sorted into *replication groups.* Any place a replication group exists is referred to as a *replication site.* A *master site* is any replication site that stores a full copy of the database and can modify the contained data. Sites that are not able to modify the data they receive are called *snapshot sites.*

DATABASE REPLICATION SYSTEMS

Database replication is the act of copying data stored on one server and sending it to another location. Replication systems are set up for many purposes. Most commonly, database replication allows people at several locations to read identical sets of data. In some cases, database replication allows several users to modify a single data set. This dataset then updates information held at many other shared locations. Finally, database replication may be used to back up data in the case of a hardware or software failure.

One of the most basic replication scenarios involves a single computer sharing information with another computer. For example, Computer A contains several replication groups. Someone on Computer B needs information from one of Computer A's replication groups. Once the two computers have been connected through LAN or WAN, specialized software sends a copy of the data from Computer A to Computer B. Whenever someone changes the data on Computer A, the software will send an updated copy to Computer B. Because Computer A stores the entire database and can make modifications to it, Computer A is considered a master site. Because Computer B cannot modify the data it receives, Computer B is a snapshot site.

Database replication becomes more complicated as more sites are added to the replication network. Two of the most commonly utilized configurations of multiple replication sites are fan-out systems and fan-in systems. In a *fan-out system,* one master site sends data to several snapshot sites. This configuration is used whenever up-to-date information is required at multiple locations, such as franchise stores that require frequently updated price catalogues. In a *fan-in system,* multiple computers send data to a single master site, which compiles all the data into a locally stored database.

Fan-in systems are used to collect information and to store it in a single, easily accessed location. Additionally, such systems are used to back up important information. In this circumstance, many computers periodically send data to a secure server. The server then creates an organized database of all that data. If the contributing computers ever fail, the data they stored can be retrieved from the server.

Many replication networks contain several master sites. In these *multimaster systems,* all the master sites are constantly communicating. Whenever someone changes a replication object at one master site, the computer sends an updated version of the object to every other master site in the system. The other master systems receive the updated object and use it to replace the old object. By constantly maintaining this process, all the master sites in a multimaster system should have exactly the same files at any given time. Each of these master sites can still distribute files to several snapshot sites.

Multimaster systems are used in any situation where databases absolutely need to be available at any time. Even if one master site is rendered nonoperational, the other master sites in the system can be configured to automatically begin distributing updated replication groups to the nonoperational master group's snapshot sites. If the master sites in a multimaster system are stored in separate locations, it is extremely unlikely that the database will become completely inaccessible. Many multimaster systems are comprised of several smaller fan-in and fan-out systems.

Finally, data replication can take place synchronously or asynchronously. When a replication system is configured to replicate *synchronously*, the system is set to update any modified replication objects on other systems as soon as they are changed. Synchronous replication is useful because it eliminates any chance of data being lost before being shared with other systems. However, synchronous replication requires a powerful connection between computers and is very demanding on networks. The farther apart two computers are, the more difficult synchronous replication is to maintain. In contrast, when a replication system is set to update *asynchronously*, it does not update in real time. Instead, it keeps a log of any changes made at a master site and updates other sites with the changes at predetermined intervals. While asynchronous systems are far easier to maintain, they come with the possibility that data could be lost before being transferred to other replication sites.

—Tyler Biscontini

Bibliography

Bradbury, Danny. "Remote Replication: Comparing Data Replication Methods." *ComputerWeekly.com*. TechTarget. Web. 29 Dec. 2014. http://www.computerweekly.com/feature/Remote-replication-Comparing-data-replication-methods

"Database Replication." *UMBC Computer Science and Engineering*. University of Maryland, Baltimore County. Web. 29 Dec. 2014. http://www.csee.umbc.edu/portal/help/oracle8/server.815/a67781/c31repli.htm#12888

"Data Replication." *Documentation.commvault.com*. CommVault Systems Inc. Web. 29 Dec. 2014. http://documentation.commvault.com/hds/release_8_0_0/books_online_1/english_us/features/data_replication/data_replication.htm#Fan-In

"Replication." *Webopedia.com*. Quinstreet Enterprise. Web. 29 Dec. 2014. http://www.webopedia.com/TERM/R/replication.html

Robotics

SUMMARY

Robotics is an interdisciplinary scientific field concerned with the design, development, operation, and assessment of electromechanical devices used to perform tasks that would otherwise require human action. Robotics applications can be found in almost every arena of modern life. Robots, for example, are widely used in industrial assembly lines to perform repetitive tasks. They have also been developed to help physicians perform difficult surgeries and are essential to the operation of many advanced military vehicles. Among the most promising robot technologies are those that draw on biological models to solve problems, such as robots whose limbs and joints are designed to mimic those of insects and other animals.

DEFINITION AND BASIC PRINCIPLES

Robotics is the science of robots—machines that can be programmed to carry out a variety of tasks independently, without direct human intervention. Although robots in science fiction tend to be androids or humanoids (robots with recognizable human forms), most real-life robots, especially those designed for industrial use, do not resemble humans physically. Robots typically consist of at least three parts: a mechanical structure (most commonly a robotic arm) that enables the robot to physically affect either itself or its task environment; sensors that gather information about physical properties such as sound, temperature, motion, and pressure; and some kind of processing system that transforms data from the robot's sensors into instructions about what actions to perform. Some devices, such as the

Two robot snakes. The snake on the left has 64 motors (with 2 degrees of freedom per segment), the snake on the right has 10.

search-engine bots that mine the Internet daily for data about links and online content, lack mechanical components. However, they are nevertheless often considered robotic because they can perform repeated tasks without supervision.

Many robotics applications also involve the use of artificial intelligence. This is a complex concept with a shifting definition, but in its most basic sense, a robot with artificial intelligence possesses features or capabilities that mimic human thought or behavior. For example, one aspect of artificial intelligence involves creating parallels to the human senses of vision, hearing, or touch. The friendly voices at the other ends of customer-service lines, for example, are increasingly likely to be robotic speech-recognition devices capable not merely of hearing callers' words but also of interpreting their meanings and directing the customers' calls intelligently.

More advanced artificial intelligence applications give robots the ability to assess their environmental conditions, make decisions, and independently develop efficient plans of action for their situations—and then modify these plans as circumstances change. Chess-playing robots do this each time they assess the state of the chessboard and make a new move. The ultimate goal of artificial intelligence research is to create machines whose responses to questions or problems are so humanlike as to be indistinguishable from those of human operators. This standard is the so-called Turing test, named after the British mathematician and computing pioneer Alan Turing.

BACKGROUND AND HISTORY

The word "robot" comes from a Czech word for "forced labor" that the Czech writer Karel apek used in his 1921 play *R.U.R.* about a man who invents a humanlike automatic machine to do his work. During the 1940s, as computing power began to grow, the influential science-fiction writer Isaac Asimov began applying "robotics" to the technology behind robots. The 1950s saw the development of the first machines that could properly be called robots. These prototypes took advantage of such new technologies as transistors (compact, solid-state devices that control electrical flow in electronic equipment) and integrated circuits (complex systems of electronic connections stamped onto single chips) to enable more complicated mechanical actions. In 1959, an industrial robot was designed that could churn out ashtrays automatically. Over the ensuing decades, public fascination with robots expanded far beyond their actual capabilities. It was becoming clear that creating robots that could accomplish seemingly simple tasks—such as avoiding obstacles while walking—was a surprisingly complex problem.

During the late twentieth century, advances in computing, electronics, and mechanical engineering led to rapid progress in the science of robotics. These included the invention of microprocessors, single integrated circuits that perform all the functions of computers' central processing units; production of better sensors and actuators; and developments in artificial intelligence and machine learning, such as a more widespread use of neural networks. (Machine learning is the study of computer programs that improves their performance through experience.)

Cutting-edge robotics applications are being developed by an interdisciplinary research cohort of computer scientists, electrical engineers, neuroscientists, psychologists, and others, and combine a greater mechanical complexity with more subtle information processing systems than were once possible. Homes may not be populated with humanoid robots with whom one can hold conversations, but mechanical robots have become ubiquitous in industry. Also, unmanned robotic vehicles and planes are essential in warfare, search-engine robots crawl the World Wide Web every day collecting and analyzing data about Internet links and content, and robotic surgical tools are indispensable in health care. All this is evidence of the extraordinarily broad range of problems robotics addresses.

HOW IT WORKS

Sensing. To move within and react to the conditions of task environments, robots must gather as much information as possible about the physical features of their environments. They do so through a large array of sensors designed to monitor different physical properties. Simple touch sensors consist of electric circuits that are completed when levers receive enough pressure to press down on switches. Robotic dogs designed as toys, for example, may have touch sensors in their backs or heads to detect when they are being petted and signal them to respond accordingly. More complex tactile sensors can detect properties such as torque (rotation) or texture. Such sensors may be used, for example, to help an assembly-line robot's end effector control the grip and force it uses to turn an object it is screwing into place.

Light sensors consist of one or more photocells that react to visible light with decreases in electrical resistance. They may serve as primitive eyes, allowing unmanned robotic vehicles, for example, to detect the bright white lines that demarcate parking spaces and maneuver between them. Reflectance sensors emit beams of infrared light, measuring the amounts of that light that reflect back from nearby surfaces. They can detect the presence of objects in front of robots and calculate the distances between the robots and the objects—allowing the robots either to follow or to avoid the objects. Temperature sensors rely on internal thermistors (resistors that react to high temperatures with decreases in electrical resistance). Robots used to rescue human beings trapped in fires may use temperature sensors to navigate away from areas of extreme heat. Similarly, altimeter sensors can detect changes in elevation, allowing robots to determine whether they are moving up or down slopes.

Other sensor types include magnetic sensors, sound sensors, accelerometers, and proprioceptive sensors that monitor the robots' internal systems and tell them where their own parts are located in space. After robots have collected information through their sensors, algorithms (mathematical processes based on predefined sets of rules) help them process that information intelligently and act on it. For example, a robot may use algorithms to help it determine its location, map its surroundings, and plan its next movements.

Motion and Manipulation. Robots can be made to move around spaces and manipulate objects in many different ways. At the most basic level, a moving robot needs to have one or more mechanisms consisting of connected moving parts, known as links. Links can be connected by prismatic or sliding joints, in which one part slides along the other, or by rotary or articulated joints, in which both parts rotate around the same fixed axis. Combinations of prismatic and rotary joints enable robotic manipulators to perform a host of complex actions, including lifting, turning, sliding, squeezing, pushing, and grasping. Actuators are required to move jointed segments or robot wheels. Actuators may be electric or electromagnetic motors, hydraulic gears or pumps (powered by compressed liquid), or pneumatic gears or pumps powered by pressurized gas. To coordinate the robots' movements, the actuators are controlled by electric circuits.

Motion-description languages are a type of computer programming language designed to formalize robot motions. They consist of sets of symbols that can be combined and manipulated in different ways to identify whole series of predefined motions in which robots of specified types can engage. Motion-description languages were developed to simplify the process of manipulating robot movements by allowing different engineers to reuse common sets of symbols to describe actions or groups of actions, rather than having to formulate new algorithms to describe every individual task they want robots to perform.

Control and Operation. A continuum of robotic control systems ranges from fully manual operation to fully autonomous operation. On the one hand, a human operator may be required to direct every movement a robot makes. For example, some bomb disposal robots are controlled by human operators working only a few feet away, using levers and buttons to guide the robots as they pick up and remove the bombs. On the other side of the spectrum are robots that operate with no human intervention at all, such as the KANTARO—a fully autonomous robot that slinks through sewer pipes inspecting them for damage and obstructions. Many robots have control mechanisms lying somewhere between these two extremes.

Robots can also be controlled from a distance. Teleoperated systems can be controlled by human operators situated either a few centimeters away, as in robotic surgeries, or millions of miles away, as in outer space applications. "Supervisory control" is a term given to teleoperation in which the robots themselves are capable of performing the vast majority of their tasks independently; human operators are present merely to monitor the robots' behavior and occasionally offer high-level instructions.

ARTIFICIAL INTELLIGENCE. Three commonly accepted paradigms, or patterns, are used in artificial intelligence robotics: hierarchical, reactive, and hybrid. The hierarchical paradigm, also known as a top-down approach, organizes robotic tasks in sequence. For example, a robot takes stock of its task environment, creates a detailed model of the world, uses that model to plan a list of tasks it must carry out to achieve a goal, and proceeds to act on each task in turn. The performance of hierarchical robots tends to be slow and disjointed since every time a change occurs in the environment, the robot pauses to reformulate its plan. For example, if such a robot is moving forward to reach a destination and an obstacle is placed in its way, it must pause, rebuild its model of the world, and begin lurching around the object.

In the reactive (or behavioral) paradigm, also known as a bottom-up approach, no planning occurs. Instead, robotic tasks are carried out spontaneously in reaction to a changing environment. If an obstacle is placed in front of such a robot, sensors can quickly incorporate information about the obstacle into the robot's actions and alter its path, causing it to swerve momentarily.

The hybrid paradigm is the one most commonly used in artificial intelligence applications being developed during the twenty-first century. It combines elements of both the reactive and the hierarchical models.

APPLICATIONS AND PRODUCTS

INDUSTRIAL ROBOTS. In the twenty-first century, almost no factory operates without at least one robot—more likely several—playing some part in its manufacturing processes. Welding robots, for example, consist of mechanical arms with several degrees of movement and end effectors in the shape of welding guns or grippers. They are used to join metal surfaces together by heating and then hammering them, and produce faster, more reliable, and more uniform results than human welders. They are also less vulnerable to injury than human workers. Another common industrial application of robotics is silicon-wafer manufacturing, which must be performed within meticulously clean rooms so as not to contaminate the semiconductors with dirt or oil. Humans are far more prone than robots to carry contaminants on them.

Six major types of industrial robots are defined by their different mechanical designs. Articulated robots are those whose manipulators (arms) have at least three rotary joints. They are often used for vehicle assembly, die casting (pouring molten metal into molds), welding, and spray painting. Cartesian robots, also known as gantry robots, have manipulators with three prismatic joints. They are often used for picking objects up and placing them in different locations, or for manipulating machine tools. Cylindrical robots have manipulators that rotate in a cylindrical shape around a central vertical axis. Parallel robots have both prismatic and rotary joints on their manipulators. Spherical robots have manipulators that can move in three-dimensional spaces shaped like spheres. SCARA (Selective Compliant Assembly Robot Arm) robots have two arms connected to vertical axes with rotary joints. One of their arms has another joint that serves as a wrist. SCARA robots are frequently used for palletizing (stacking goods on platforms for transportation or loading).

SERVICE ROBOTS. Unlike industrial robots, service robots are designed to cater to the needs of individual people. Robopets, such as animatronic dogs, provide companionship and entertainment for their human owners. The Sony Corporation's AIBO (Artificial Intelligence roBOt) robopets use complex systems of sensors to detect human touch on their heads, backs, chins, and paws, and can recognize the faces and voices of their owners. They can also maintain their balance while walking and running in response to human commands. AIBOs also function as home-security devices, as they can be set to sound alarms when their motion or sound detectors are triggered. Consumer appliances, such as iRobot Corporation's robotic vacuum cleaner, the Roomba, and the robotic lawn mover, the RoboMower, developed by Friendly Robotics, use artificial intelligence approaches to safely and effectively maneuver around their task environments while performing repetitive tasks to save their human users time.

Even appliances that do not much resemble public notions of what robots should look like often contain robotic components. For example, digital video recorders (DVRs) such as TiVos, contain sensors, microprocessors, and a basic form of artificial intelligence that enable them to seek out and record programs that conform to their owners' personal tastes. Some cars can assist their owners with driving tasks such as parallel parking. An example is the Toyota Prius, which offers an option known as Intelligent Parking Assist. Other cars have robotic seats that can lift elderly or disabled passengers inside. In the 2010s, many companies actively explored technologies for autonomous cars (self-driving vehicles), with the goal of making mass-produced models available to the public that would function flawlessly on existing roads. Several prototypes were introduced, although much controversy surrounded the potential hazards and liabilities of such technology.

Many companies or organizations rely on humanoid robots to provide services to the public. The Smithsonian National Museum of American History, for example, has used an interactive robot named Minerva to guide visitors around the museum's exhibits, answering questions and providing information about individual exhibits. Other professional roles filled by robots include those of receptionists, floor cleaners, librarians, bartenders, and secretaries. At least one primary school in Japan even experimented with a robotic teacher developed by a scientist at the Tokyo University of Science. However, an important pitfall of humanoid robots is their susceptibility to the uncanny valley phenomenon. This is the theory that as a robot's appearance and behavior becomes more humanlike, people will respond to it more positively—but only up to a point. On a line graph plotting positive response against degree of human likeness, the response dips (the "uncanny valley") as the likeness approaches total realism but does not perfectly mimic it. In other words, while people will prefer a somewhat anthropomorphic robot to an industrial-looking one, a highly human-like robot that is still identifiably a machine will cause people to feel revulsion and fear rather than empathy.

MEDICAL USES. Robotic surgery has become an increasingly important area of medical technology. In most robotic surgeries, a system known as a master-slave manipulator is used to control robot movements. Surgeons look down into electronic displays showing their patients' bodies and the robots' tool tips. The surgeons use controls attached to consoles to precisely guide the robots' manipulators within the patients' bodies. A major benefit of robotic surgeries is that they are less invasive—smaller incisions need to be made because robotic manipulators can be extremely narrow. These surgeries are also safer because robotic end effectors can compensate for small tremors or shakes in the surgeons' movements that could seriously damage their patients' tissues if the surgeons were making the incisions themselves. Teleoperated surgical robots can even allow surgeons to perform operations remotely, without the need to transport patients over long distances. Surgical robots such as the da Vinci system are used to conduct operations such as prostatectomy, cardiac surgery, bariatric surgery, and various forms of neurosurgery.

Humanoid robots are also widely used as artificial patients to help train medical students in diagnosis and procedures. These robots have changing vital signs such as heart rates, blood pressure, and pupil dilation. Many are designed to breathe realistically, express pain, urinate, and even speak about their conditions. With their help, physicians-in-training can practice drawing blood, performing cardiopulmonary resuscitation (CPR), and delivering babies without the risk of harm to real patients.

Other medical robots include robotic nurses that can monitor patients' vital signs and alert physicians to crises and smart wheelchairs that can automatically maneuver around obstacles. Scientists are also working on developing nanorobots the size of bacteria that can be swallowed and sent to perform various tasks within human bodies, such as removing plaque from the insides of clogged arteries.

Robot Exploration and Rescue. One of the most intuitive applications of robotic technology is the concept of sending robots to places too remote or too dangerous for human beings to work in—such as outer space, great ocean depths, and disaster zones. The six successful manned Moon landings of the Apollo program carried out during the late 1960s and early 1970s are dwarfed in number by the unmanned robot missions that have set foot not only on the Moon but also on other celestial bodies, such as planets in the solar system. The wheeled robots Spirit and Opportunity, for example, began analyzing material samples on Mars and sending photographs back to Earth in 2004. Roboticists have also designed biomimetic robots inspired by frogs that take advantage of lower gravitational fields, such as those found on smaller planets, to hop nimbly over rocks and other obstacles.

Robots are also used to explore the ocean floor. The Benthic Rover, for example, drags itself along the seabed at depths up to 2.5 miles below the surface. It measures oxygen and food levels, takes soil samples, and sends live streaming video up to the scientists above. The rover is operated by supervisory control and requires very little intervention on the part of its human operators.

Rescue robots seek out, pick up, and safely carry injured humans trapped in fires, under rubble, or in dangerous battle zones. For example, the U.S. Army's Bear (Battlefield Extraction-Assist Robot) is a bipedal robot that can climb stairs, wedge itself through narrow spaces, and clamber over bumpy terrain while carrying weights of up to three hundred pounds.

Military Robots. Militaries have often been among the leading organizations pioneering robotics, as robots have the potential to complete many military tasks that might otherwise prove dangerous to humans. While many projects, such as bomb-removal robots, have proven highly useful and have been widely accepted, other military applications have proven more controversial. For example, many observers and activists have expressed concern over the proliferation of drones, particularly unmanned aerial vehicles (UAVs) capable of enacting military strikes such as rocket or missile launches while going virtually undetected by radar. The U.S. government has used such technology (including the Predator drone) to succesfully destroy terrorist positions, including in remote territory that would otherwise be difficult and dangerous to access, but there have also been notable examples of misidentified targets and civilian collateral damage caused by drone strikes. Opponents of drones argue that the potential for mistakes or abuse of their capabilities is dangerous.

SOCIAL CONTEXT AND FUTURE PROSPECTS

In the twenty-first century, the presence of robots in factories all over the world is taken for granted. Meanwhile, robots are also increasingly entering daily life in the form of automated self-service kiosks at supermarkets, electronic lifeguards that detect when swimmers in pools are struggling, and cars whose robotic speech-recognition software enables them to respond to verbal commands. A science-fiction future in which ubiquitous robotic assistants perform domestic tasks such as cooking and cleaning may not be far away, but many technological limitations must be overcome for that to become a reality. For example, it can be difficult for robots to process multipart spoken commands that have not been preprogrammed—a problem for researchers in the artificial intelligence field of natural language processing. However, the voice-activated personal assistant software in smartphone and computer operating systems, such as Apple's Siri, shows how such technology is rapidly evolving.

Robots that provide nursing care or companionship to the infirm are not merely becoming important parts of the health care industry but may also provide a solution to the problem increasingly faced by countries in the developed world—a growing aging population who need more caretakers than can be found among younger adults. Another area where robots can be particularly useful is in performing dangerous tasks that would otherwise put a human's life at risk. The use of robotics in the military remains a growing field—not only in the controversial use of combat drones, but also for tasks such as minesweeping. Robots are also more and more heavily used in space exploration. In the future, robots may be used in other dangerous fields as well; during the 2014 ebola outbreak, roboticists discussed the possibility of using robots in future medical emergencies of the type for purposes such as disinfecting contaminated areas.

There are also concerns about the growing use of robots to perform tasks previously performed by humans, however. As robotics technology improves and becomes less expensive, companies may well turn to cheap, efficient robots to do jobs that are typically performed by immigrant human labor, particularly in such areas as agriculture and manufacturing. Meanwhile, some observers are concerned that the rise of industrial and professional service robots is already eliminating too many jobs held by American workers. Many of the jobs lost in the 2008–2009 recession within the struggling automotive industry, for example, are gone for good, because costly human workers were replaced by cheaper robotic arms. However, the issue is more complicated than that. In certain situations, the addition of robots to a factory's workforce can actually create more jobs for humans. Some companies, for example, have been able to increase production and hire additional workers with the help of robot palletizers that make stacking and loading their products much faster.

Safety concerns can sometimes hinder the acceptance of new robotic technologies, even when they have proven to be less likely than humans to make dangerous mistakes. Robotic sheep shearers in Australia, for example, have met with great resistance from farmers because of the small risk that the machines may nick a major artery as they work, causing the accidental death of a sheep. And while fully autonomous vehicles, including self-driving cars, are seen by many automotive and technology companies alike as a major area of innovation, crashes and other accidents by several prototypes in the 2010s drew significant concerns from regulators and the general public. It is critical in the field of robotics to not only develop the technology necessary to make a design a reality, but to understand the cultural and economic landscape that will determine whether a robot is a success or failure.

—*M. Lee, BA, MA*

Bibliography

Dinwiddie, Keith. *Basic Robotics.* Cengage Learning, 2016.

Faust, Russell A., ed. *Robotics in Surgery: History, Current and Future Applications.* New York: Nova Science, 2007. Print.

Floreano, Dario, and Claudio Mattuissi. *Bio-Inspired Artificial Intelligence: Theories, Methods, and Technologies.* Cambridge: MIT P, 2008. Print.

Gutkind, Lee. *Almost Human: Making Robots Think.* 2006. Reprint. New York: Norton, 2009. Print.

Jones, Joseph L. *Robot Programming: A Practical Guide to Behavior-Based Robotics.* New York: McGraw-Hill, 2004. Print.

"Our Friends Electric." *Economist.* Economist Newspaper, 7 Sept. 2013. Web. 14 Nov. 2014.

Popovic, Marko B. *Biomechanics and Robotics.* Boca Raton: CRC, 2013. Print.

Samani, Hooman. *Cognitive Robotics.* CRC Press, 2016.

Siciliano, Bruno, et al. *Robotics: Modeling, Planning, and Control.* London: Springer, 2009. Print.

Siciliano, Bruno, and Oussama Khatib, eds. *The Springer Handbook of Robotics.* Springer, 2016.

Springer, Paul J. *Military Robots and Drones: A Reference Handbook.* Santa Barbara: ABC-CLIO, 2013. Print.

Ruby

SUMMARY

Ruby is a general-purpose computer programming language that, while originating as an object-oriented scripting language, supports multiple programming paradigms, or styles. An object-oriented programming language consists of small pieces of code that are treated as a single unit, or "object," each of which has specified states and behaviors, belongs to a specified class of objects, and is able to be added, deleted, or copied with ease. Ruby is also open source, meaning that the code can be freely altered, copied, and distributed.

Ruby has developed a reputation as one of the easiest general-purpose languages to learn, and because of its close relationship to prominent languages like Perl and Lisp, it is often recommended to beginning programmers as their first language. As a

Yukihiro Matsumoto, creator of the Ruby programming language. By Cep21

general-purpose language, it is suitable for most programming applications. However, in practice, professional programmers tend to use it in two main areas: system administration tools and web programming. In the latter area, Ruby is the language in which the web application framework Ruby on Rails is written.

HISTORY AND BACKGROUND

Ruby was developed in 1993 by Japanese programmer Yukihiro Matsumoto, who was attracted to the idea of an object-oriented scripting language but was dissatisfied with the extant version of Perl and did not consider Python a true scripting language. The first public version of Ruby was released in 1995, and a stable Ruby 1.0 launched in December 1996.

The first several releases were popular mainly in Japan. Interest in Ruby spread to the English-speaking world by 1998 and grew significantly following the publication of the first English-language Ruby programming manual in 2000.

The development of Ruby on Rails, or "Rails" for short, is often credited with the enduring popularity of Ruby, as well as its widespread association with web programming. Ruby on Rails 1.0 was released in 2005 and shipped with Mac OS X in 2007. Rails is one of the most popular web frameworks and its popularity drove awareness of Ruby in the mid-2000s. However, it also overshadowed the many other uses of Ruby, such as writing command-line tools and narrow-focused but mathematically-rich programs like those used for music composition.

RubyGems.org hosts libraries of Ruby programs (known as "gems") for the RubyGems package manager, which helps developers to download, install, and maintain pieces of Ruby code that others have created. The first RubyGems project was begun by Ryan Leavengood in 2001 but lacked momentum. A new RubyGems project was begun in 2003 by Rich Kilmer, Chad Fowler, David Black, Paul Brannan, and Jim Weirch, who sought Leavengood's permission to use the RubyGems name. By that time, the English-language Ruby community had overtaken the original Japanese-language Ruby community.

There are several alternate implementations of Ruby, including JRuby and Rubinius, which run on virtual machines created using the Java and C++ programming languages, respectively. The Ruby Game Scripting System allows RPG Maker software users to design their own role-playing games.

—Bill Kte'pi, MA

Bibliography

Black, David A. *The Well-Grounded Rubyist.* Shelter Island: Manning, 2014. Print.

Carlson, Lucas, and Leonard Richardson. *Ruby Cookbook.* Sebastopol: O'Reilly Media, 2015. Print.

Flanagan, David, and Yukihiro Matsumoto. *The Ruby Programming Language.* Sebastopol: O'Reilly Media, 2008. Print.

Fulton, Hal, and André Arko. *The Ruby Way.* 3rd ed. Boston: Pearson Education, 2015. Print.

Goriunova, Olga, ed. *Fun and Software: Exploring Pleasure, Paradox, and Pain in Computing.* New York: Bloomsbury, 2016. Print.

Olsen, Russ. *Eloquent Ruby.* Boston: Pearson Education, 2011. Print.

Shaw, Zed A. *Learn Ruby the Hard Way.* 3rd ed. Upper Saddle River: Pearson Education, 2015. Print.

Yukihiro Matsumoto. *Ruby in a Nutshell.* Trans. David L. Reynolds Jr. Sebastopol: O'Reilly Media, 2002. Print.

S

Scale Model

SUMMARY

A scale model is a representation of a figure, structure or other phenomenon that maintains the overall shape and appearance of the original but alters (usually reducing) its dimensions in terms of length, width, height and so on. A common example of a scale model is seen in the architectural models used to represent buildings that are planned for construction. A scale model is usually intended to represent a real world object in miniature, in order to study the object in ways that would be either impossible or extremely difficult or inconvenient if the original object were used. For example, one might create a scale model of a skyscraper in a computerized, virtual reality environment in order to test how the building stands up to extreme conditions such as earthquakes and hurricanes. This would allow one to learn about how the building could best be designed without having to first construct the building (at great expense), test its behavior under destructive conditions, and then have to rebuild it in order to incorporate the improvements that had been discovered during testing. Architects use scale models for similar purposes, because the scale model allows them to see what a building will look like without having to go to the expense and effort of building it in real life.

WORKING WITH SCALE MODELS

One of the most important features that must be kept in mind when one is constructing a scale model is to keep measurements consistently in proportion to one another. The word "scale" in the phrase scale model refers to this idea, which assigns a particular ratio to all measurements the model seeks to reproduce. For example, if a scale model of an amusement park is constructed to a scale of 1 inch to 100 feet, that means that each linear inch of distance on or within the scale model is meant to represent a distance of one hundred feet in the real world. In this way, an amusement park that actually takes up thousands of square feet can be modeled in miniature on a tabletop.

If the scale model is to provide useful information, however, all of its parts must be constructed according to the same scale. In the example of the scale model of the amusement park, the model would look very strange if distances were given at a scale of 1 inch to every 100 feet but buildings were constructed to a scale of 1 inch to every ten feet—the scale model amusement park would be crowded with enormously oversized buildings, making it difficult if not impossible to extract any useful information from the inspection of the model. Some types of scale models have challenges beyond those presented by the need to maintain linear proportionality. For example, scale models of automobile prototypes are often used to test how aerodynamic a particular design for a car is, so in addition to the concern for the size of the model to be accurately proportional, designers must also ensure that its material properties behave in proportion to their real world counterparts.

—Scott Zimmer, MLS, MS

Bibliography

Adam, John A. *Mathematics in Nature: Modeling Patterns in the Natural World.* Princeton, NJ: Princeton UP, 2003.

Bay-Williams, Jennifer M, and William R. Speer. *Professional Collaborations in Mathematics Teaching and Learning: Seeking Success for All.* Reston, VA: National Council of Teachers of Mathematics, 2012.

Ceccato, Cristiano, Lars Hesselgren, Mark Pauly, Helmut Pottmann, and Johannes Wallner. *Advances in Architectural Geometry.* Wien: Springer, 2010.

Cook, William. *In Pursuit of the Traveling Salesman: Mathematics at the Limits of Computation.* Princeton, NJ: Princeton UP, 2012.

Dunn, Fletcher, and Ian Parberry. *3D Math Primer for Graphics and Game Development.* Boca Raton, FL: CRC, 2011.

Kutz, Jose N. *Data-Driven Modeling & Scientific Computation: Methods for Complex Systems & Big Data.* Oxford: Oxford UP, 2013.

Roth, Richard, and Stephen Pentak. *Design Basics.* Boston: Cengage, 2013.

Serway, Raymond A., John W. Jewett, and Vahé Peroomian. *Physics for Scientists and Engineers with Modern Physics.* Boston: Cengage, 2014.

SCIENTIFIC CONTROL

SUMMARY

A *scientific control* is the component of a scientific experiment that is not altered so that the test subjects that are changed may later be compared to it. Scientists may employ as many controls as necessary in their experiments. Controls are vital to the success of experiments because they act as standards that indicate whether the results of the experiment are positive or negative.

In an experiment testing the potency of a certain drug, for instance, the people who receive the drug are called the experimental group, while the control, or control group, is the set of people who are not given the drug. After the experiment, the effects of the drug on the experimental group are compared to the condition of the control group to determine what exactly the drug did. Controls interact in different ways with the other components of an experiment—the independent variables, dependent variables, and constants—to produce results that answer the questions scientists proposed before executing the experiment.

BACKGROUND

The scientific control is only one of the vital parts of an experiment. Controls, independent variables, dependent variables, and constants must all be present for an experiment to be considered scientifically valid. Each component plays a unique role and is important to the success of the experiment in its own way.

Independent variables are the parts of an experiment that are not changed by the other parts. They alter other parts of an experiment by being applied to them. Independent variables can be certain amounts of substances given to or applied to test subjects, unchanging forces such as speed or pressure, or the age of an individual being tested. These substances, forces, and ages are considered independent variables because they stay the same throughout the experiment.

Dependent variables, conversely, are the components of an experiment that are influenced by other factors. Dependent variables are the changes that scientists measure in the experiment. While an unchanging force may be the independent variable, the different objects upon which the force is acting are the dependent variables, since each object will react differently to the force.

Controls are the parts of an experiment that act as the basis of comparison to other experimental subjects. The experimental groups are altered due to the application of a force or other change, but controls are not altered in the same way. Scientists observe

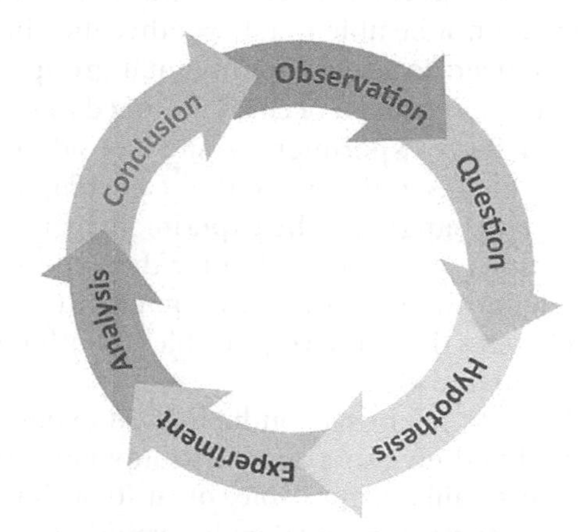

Scientific controls are part of the scientific method. By Thebiologyprimer (Own work)

If fertilizer is given to half of identical growing plants and there are differences between the fertilized group and the unfertilized "control" group, these differences may be due to the fertilizer. Forest & Kim Starr

what becomes of all their test subjects to determine how the experimental groups have been affected by the changes they introduced.

Finally, constants are the factors that never change throughout an experiment. It is important that they remain the same because they could interfere with the results of the experiment if they were to change. For instance, a test subject's gender, height, and weight may all be constants in an experiment. If any of these changed during a scientific test, the dependent variable could change with them, thus making the conclusions of the experiment invalid.

IMPACT

Controls are necessary to producing accurate results in a scientific experiment because, without them, scientists would not be able to tell whether the changes they introduced to their experimental groups actually caused the results to occur. They need to include controls in their experiments—and not administer the same changes to them—so they can compare the effects of the changes on the experimental groups to the conditions of the controls after the experiment. This will determine whether it was in fact the administered changes that were responsible for influencing the experimental groups.

It is best for scientists conducting an experiment to ensure that their experimental groups and control groups are as alike as possible, or at least that they share some character traits or life experiences such as being the same age or holding the same job. These commonalities will prove useful because the personal differences that led to the varied results between the experimental group and the control group will be highlighted.

For instance, scientists who want to test whether a weight loss medication works will first assemble two similar groups of people, an experimental group and a control group. The scientists will then administer the same dosage of the medication to each person in the experimental group, where the uniform dosage is the independent variable. The control group will not receive the medication at all. The scientists will then determine how much weight, if any, each person in the experimental group lost. The different amounts are the dependent variables. The constants in the experiment are the individuals' ages, heights, and weights. These factors must remain the same throughout the test so the scientists can determine how the medication interacts with people who possess certain characteristics.

The importance of the control is apparent after the experiment has been completed. If the experimental group lost weight but the control group did not, the scientists can conclude that the weight loss medication may be effective. This cannot be declared as a definite fact yet because other factors may also have influenced the individuals' weight loss. The scientists would then need to perform more experiments with the medication to prove whether it truly was responsible for the subjects' weight loss.

If both the experimental group and the control group lost weight, the scientists would need to determine what actions each group took to do so. If both groups ate healthier diets and exercised and lost similar amounts of weight, for example, the scientists might conclude that it was these actions, and not the drug, that caused the weight loss.

—Michael Ruth

Bibliography

Bradford, Alina. "Empirical Evidence: A Definition." *Live Science*, 24 Mar. 2015, www.livescience.com/21456-empirical-evidence-a-definition.html. Accessed 19 Jan. 2017.

______. "Science & the Scientific Method: A Definition." *Live Science*, 30 Mar. 2015, www.livescience.com/20896-science-scientific-method.html. Accessed 19 Jan. 2017.

Cothron, Julia H., et al. *Science Experiments by the Hundreds.* Kendall/Hunt Publishing Company, 1996, 12–13.

Crowder, C.D. "Explaining the Control in a Science Experiment." *Bright Hub Education,* 1 Jan. 2013, www.brighthubeducation.com/science-fair-projects/107152-what-is-the-control-in-a-science-experiment/. Accessed 19 Jan. 2017.

Fitz-Gibbon, Carol T., and Lynn Lyons Morris. *How to Design a Program Evaluation.* Sage Publications, 1987, 25–27.

Harris, William. "How the Scientific Method Works." *HowStuffWorks.com,* science.howstuffworks.com/innovation/scientific-experiments/scientific-method6.htm. Accessed 19 Jan. 2017.

Privitera, Gregory J. *Research Methods for the Behavioral Sciences.* 2nd ed., Sage Publications, 2016, 264–265.

Santos, Kay. "What Components Are Necessary for an Experiment to Be Valid?" *Seattle Post-Intelligencer,* education.seattlepi.com/components-necessary-experiment-valid-3630.html. Accessed 19 Jan. 2017.

SCRATCH

SUMMARY

Scratch is a visual, event-driven programming language developed at the Massachusetts Institute of Technology (MIT) Media Lab and released to the public for free with few restrictions. Scratch uses visual objects called sprites, either created in Scratch's included editor or imported from other files or a webcam, in order to create animation, games, puzzles, visualizations, interactive presentations, and other visually driven multimedia programs. Scratch is primarily intended as a teaching tool, and is equally useful as a beginner's programming language for students or as a simple programming language for teachers to use to create educational tools, programs, and presentations for use in their classes.

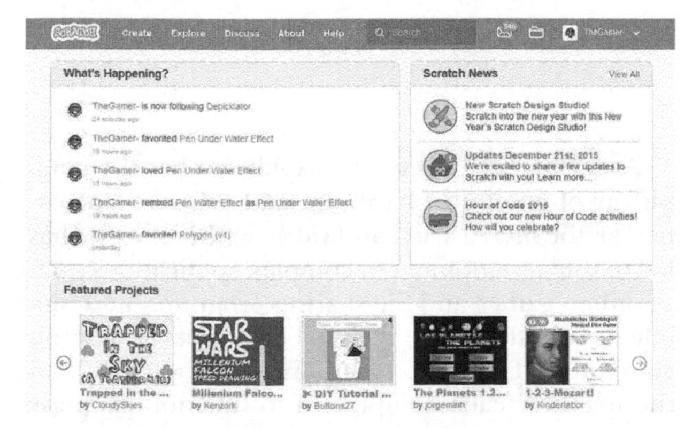

The homepage of Scratch.mit.edu. By LLK group / Scratch Team (https://scratch.mit.edu)

HISTORY AND BACKGROUND

The MIT Media Lab is a research laboratory at the Massachusetts Institute of Technology (MIT), founded in 1985 and known for its interdisciplinary work and technological social engagement efforts (the One Laptop per Child nonprofit grew out of a Media Lab project, for example). The Media Lab is divided into several working groups. Scratch was designed by Mitchel Resnick, director of the Media Lab's Lifelong Kindergarten group, which focuses on projects inspired by kindergarten-style play such as wooden blocks and finger painting. Kindergarten-style learning is based on play and exploration, driven not by reading and lectures but by designing, creation, expression, and experimentation. Scratch was designed as a programming language meant to both teach programming through this style of learning and facilitate the creation of other learning tools in this style.

The underlying educational theory behind Scratch's design is constructionism, which says that learning involves students assembling mental models as a strategy for understanding the world—itself an extension of developmental psychologist Jean Piaget's constructivist theory, describing the way children learn through experience. Scratch was developed for young people, especially students aged eight to sixteen, and after several years of development by the Media Lab and the consulting company Playful Invention Company, it was launched publically in 2007.

Script for Hello, World! in Scratch. Mberry at English Wikipedia

"Scratching" is a programming term meaning to reuse existing code for a new purpose, and is derived from the practice of record-scratching in turntablism. The Scratch community encourages users to "remix" programs by downloading projects from public spaces and combining them with others, or using parts of them in different combinations and contexts. The language was designed with the online community in mind, and the slogan "Imagine, Program, Share," emphasizes both the play philosophy and the remixing and communal-collaborative nature of Scratch. The official Scratch website hosted by the Media Lab puts up frequent design studio challenges to stimulate creativity and communication in the community.

By default, Scratch is organized with a stage area on the left hand side of the screen that shows the results of the program in process. In addition to the sprite editor built into the program, Scratch comes with a library of sprites that also stores previously created sprites. New sprites can be imported from files, clip art, or photos taken with cameras. The conventions of the language are designed to ease the transition from Scratch programming to programming in languages such as Java, Python, or Ruby.

Resnick also collaborated on ScratchJr, released in 2014. Inspired by Scratch, it is a simpler programming language intended for younger children and released as a free app for iOS in 2014, Android in 2015, and Chromebooks in 2016.

—Bill Kte'pi, MA

BIBLIOGRAPHY

"About Scratch." *Scratch.* MIT Media Lab, 2016. Web. 23 Aug. 2016.

Benson, Pete. *Scratch.* Ann Arbor: Cherry Lake, 2015. Print.

Kafai, Yasmin B., and Quinn Burke. *Connected Code: Why Children Need to Learn Programming.* Cambridge: MIT P, 2014. Print.

Marji, Majed. *Learn to Program with Scratch: A Visual Introduction to Programming with Games, Art, Science, and Math.* San Francisco: No Starch, 2014. Print.

Sweigart, Al. *Scratch Programming Playground: Learn to Program by Making Cool Games.* San Francisco: No Starch, 2016. Print.

Van Pul, Sergio, and Jessica Chiang. *Scratch 2.0 Game Development.* Birmingham: Packt, 2014. Print.

Woodcock, Jon. *Coding Games in Scratch.* New York: DK, 2016. Print.

SELF-MANAGEMENT

SUMMARY

In a self-managing computer system all of the operations are run and coordinated autonomously without intervention in the process from humans. Such systems are also known as "autonomic systems," a phrase adapted from the medical term "autonomic nervous system," the area of the human body that automatically controls the respiratory system and the body's heart rate. In theory, autonomic computer systems should function similarly.

As technology has grown rapidly, so has the complexity of computer hardware and software. In particular, the increase in bandwidth on the Internet has led to the development of complex applications such as online games and file-sharing programs, and mobile devices that serve the same functions as desktop computers. Different computing technologies and the need for each component to function properly as a whole system has created problems that experts hope will be solved by the emergence of self-managing computer systems.

BACKGROUND

Computer-industry experts expect the market for tablet computers alone to grow by 38 percent each year for the foreseeable future. The number of smartphones also continues to grow at a rapid pace, while other computing devices such as laptop and desktop computers, though not exhibiting the same type of growth, continue to be important consumer devices. Therefore, there is an increasing need, and shortfall, for computer technicians to manage these devices. However, autonomic computer systems could be a solution to bridge the gap between the growing number of devices and those able to manage them.

The biggest advocate and financial backer of autonomic computing has been IBM, which spearheaded the autonomic computing initiative (ACI) in 2001. IBM derived four goals that developers of autonomic computers should have: self-configuration, or the ability to optimize the various aspects of a system to help them function harmoniously; self-healing, or the ability to diagnose and solve problems automatically when errors occur in the system; self-optimization, or the ability to distribute resources to each component of a system so that each receives at least a minimum amount of resources; and self-protection, or the ability to ensure that outside effects such as viruses do not harm the system.

A typical example of the development of self-management in computers is peer-to-peer, or file-sharing, networks. Typically, applications that facilitate file sharing require a great deal of bandwidth to exchange information between one connection and another. Thus, there exists the possibility of bandwidth overload and subsequent failure to exchange the files. This failure was a common occurrence in early peer-to-peer applications, since they used what is called "random neighbor communication." By contrast, updated versions of these programs use what is called "structured overlay networks," which have the ability to use self-correcting measures if a failure is imminent.

EXAMPLES OF AUTONOMIC COMPUTING

The most fundamental idea related to autonomic computing is that machines can "learn," meaning they can gain knowledge and skills outside their programming and do not require a human to inject new skills into them. Since the end of the twentieth century, IBM has been the foremost technology company in its efforts to develop machine learning and associated technology. In the 1990s, IBM introduced Deep Blue, which was touted as a chess-playing machine that could defeat world champions. The company claimed that the machine could analyze over 200 million positions per second. In 1997, Deep Blue defeated Garry Kasparov, world chess champion, in a six-game contest. Considered by many to be the greatest chess player of all time, Kasparov was embarrassed by his loss to Deep Blue to such a degree that he accused IBM of intervening during critical moments of the matches, implying the company was cheating.

In 2011, IBM introduced "Watson," a computer system that used machine learning to defeat human contestants on the popular television quiz show *Jeopardy!*. Watson had the processing power of 500 gigabytes, meaning it could read and digest a million books in a matter of seconds. To enable this processing power, IBM engineers outfitted Watson with $3 million worth of hardware and uploaded nearly the entire contents of Wikipedia onto its servers. Watson was then matched with two of the most successful contestants ever on *Jeopardy!*. While Watson was able to defeat its human adversaries handily, its inability to interpret some questions as a human would was exposed. For example, when prompted with a question in the category of "U.S. cities," Watson inexplicably answered "Toronto."

The most ambitious modern technological effort at autonomic computing is the long-held dream of a self-driving car. While autonomous cars were conceived as early as the 1920s, not until the 2010s did they become a practical reality. As of late 2016, five U.S. states—California, Florida, Michigan, Nevada, and Virginia—as well as Washington, DC, have allowed self-driving cars to be tested on their roads. However, a June 2016 incident severely hampered the public perception of self-driving cars as a safe means of transportation. While driving on a highway in Florida, Joshua Brown, a business owner, was killed when his Tesla automobile, outfitted with the Tesla Autopilot system, failed to apply its brakes when a semitrailer truck merged in front of him from the right lane. Still, research into self-driving technology continued, with several companies expressing plans to introduce autonomous vehicles.

Though most experts believe that self-management is the inevitable next wave in computing, for some, the technology is fraught with ethical issues. Some fear that artificial intelligence will replace humans in the workforce or will even exert influence or control over people. Though the latter may be somewhat far-fetched, these ethical dilemmas are part of the conversation related to self-managing computer systems.

—Kevin Golson

Bibliography

"An Architectural Blueprint for Autonomic Computing." 3rd ed. IBM, June 2005. PDF file.

Cong-Vinh, Phan. *Formal and Practical Aspects of Autonomic Computing and Networking: Specification, Development, and Verification.* Hershey: Information Science Reference, 2012. Print.

Dargie, Waltenegu. *Context-Aware Computing and Self-Managing Systems.* Boca Raton: CRC, 2009. Print.

Kephart, Jeffrey O., and David M. Chess. "The Vision of Autonomic Computing." IEEE Computer Society, Jan. 2003. PDF file.

Lalanda, Philippe, Julie McCann, and Ada Diaconescu. *Autonomic Computing: Principles, Design and Implementation.* London: Springer, 2014. Print.

Michalski, Ryszard S., Jaime G. Carbonell, and Tom M. Mitchell. *Machine Learning: An Artificial Intelligence Approach.* New York: Springer, 2013. Print.

Nilsson, Nils J. *Principles of Artificial Intelligence.* San Francisco: Morgan Kaufmann, 2014. Print.

Zimmer, Ben. "Is It Time to Welcome Our New Computer Overlords?" *Atlantic.* Atlantic Monthly Group, 17 Feb. 2011. Web. 23 Aug. 2016.

Semantic Web

SUMMARY

The *Semantic Web*, which is sometimes called Web 3.0 or the Web of data, is a framework that allows different applications and programs to share data such as dates, numbers, formulas, and more. The Semantic Web is an extension of the World Wide Web (WWW), but it does not serve the same purpose as the WWW. While both the Semantic Web and the WWW aim to make knowledge and information available to people, the WWW achieves this goal by presenting users with countless documents that they can read and learn from, and the Semantic Web classifies individual pieces of data in a manner that allows computers and people to use the data in a variety of ways.

Much of the data people use each day is not actually part of the WWW. The goal of the Semantic Web is to make as much data as possible accessible using architecture that is similar to the WWW architecture (e.g., URIs and URLs). Most of the data on the WWW is there for humans to read, but the data on the Semantic Web is there for computers and humans to link to, use, and categorize. For example, one day people could use the Semantic Web to find a doctor appointment (data provided by the doctor's office) and their own schedule (data provided by the calendars on their computers or smart phones). The Semantic Web is often compared to a Web of data, with data being pulled from many different places.

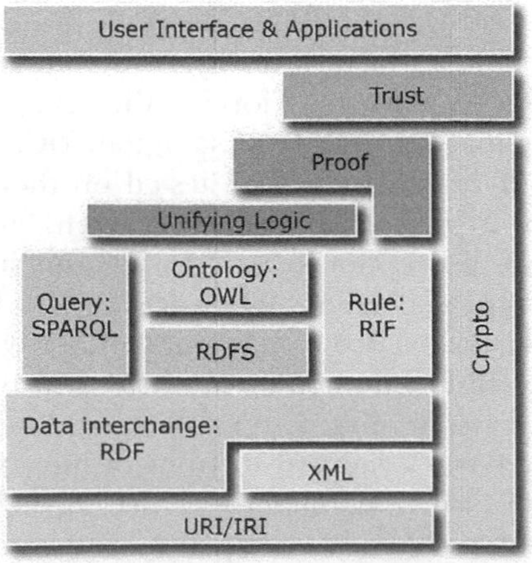

Semantic Web Stack. By W3C

HISTORY OF THE SEMANTIC WEB

The Semantic Web is different from the conventional Web, but it is not totally separate from it. The Semantic Web is an extension of the conventional Web and

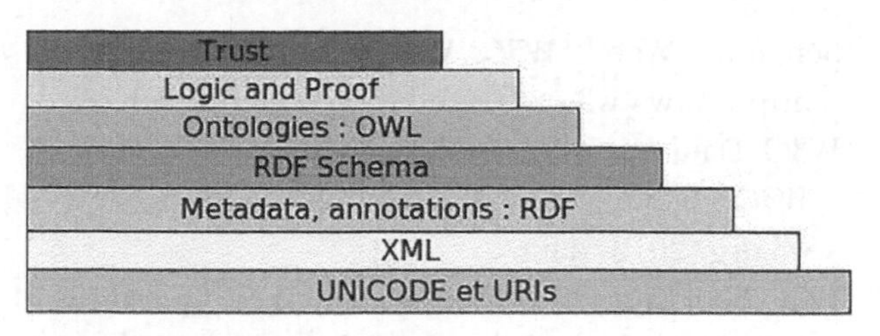

Architecture of semantic web.

is often used in conjunction with it. The history of the Semantic Web also is closely tied to the WWW. The founder of the WWW, Tim Berners-Lee, was also influential in the development of the Semantic Web. In 2001, Lee and others published an article in Scientific American that discussed the Semantic Web and its possibilities. That same year, the World Wide Web Consortium (W3C; a group with volunteers and paid workers who develop Web standards) launched the Semantic Web Activity, which helped create the groundwork for the Semantic Web. The group saw the Semantic Web as being similar to the WWW in that people would publish pages of data in the same way they published websites. However, the Semantic Web became more important in the nackground of the Web, and few people publish large stores of data.

GOALS OF THE SEMATIC WEB

The Sematic Web has a number of goals. The most important goal is to make data easy to access. Another goal of the Sematic Web is to increase data sharing to make it possible for people and technology to find relationships among data. In addition, the Semantic Web tries to make data available so it can be used to create models of problems in everyday life and try to help solve them. Another way the data might be used is to build machines that are able to "think." The Semantic Web has the potential to change the way people use the Internet. However, the current Semantic Web runs mostly in the background of the WWW and has not changed the way everyday Internet users experience the Web.

BUILDING BLOCKS OF THE SEMANTIC WEB

The building blocks that make up the WWW and the Semantic Web are program languages that help users to write software. Two of the most important building blocks for the Semantic Web are eXtensible Markup Language (XML) and the Resource Description Framework (RDF). Data on the Semantic Web often are categorized and organized so that humans and devices can use them more efficiently. XML and RDF are two important ways people organize data on the Semantic Web.

The WC3 recommends XML as one possible language to use on the conventional Web and the Semantic Web. XML is a markup language, like the popular HTML language. Unlike HTML, XML allows people to develop their own tags to categorize data. For example, a person using HTML has to use official HTML tags, but a person using XML can devise any tags (e.g., <firstname>, <zipcode>, <birthdate>, <age>, etc.). XML is an important building block of the Semantic Web because it allows users to code data in more useful ways than some other languages.

Although XML helps people structure data, it does not give the data meaning. Another important building block, RDF, helps give the data on the Semantic Web meaning. RDF encodes pieces of data with information that is similar to the subject, the verb, and the object of a sentence. This three-part structure can be used to describe much of the data computers hold and process. For example, RDF could be used to indicate that a name on a page is the author of a text. The meaning given to the data by the RDF coding could help people find this information on the Semantic Web. People can choose to use either XML or RDF depending on the goals they want to achieve.

Another important language in the development of the Semantic Web is structured query language (SQL). SQL is a type of query language that allows people to manage and gather information from databases. The Semantic Web, in one way, can be seen as a huge database of information. Therefore, people can use SQL to access, organize, and manage the data on the Semantic Web.

—Elizabeth Mohn

BIBLIOGRAPHY

Berners-Lee, Tim, James Hendler, and Ora Lassila. "The Semantic Web." *Scientific American*. Scientific American, a Division of Nature America, Inc. May 2001. Web. 4 Aug. 2015. www-sop.inria.fr/acacia/cours/essi2006/Scientific%20American_%20

Feature%20Article_%20The%20Semantic%20Web_%20May%202001.pdf

Hitzler, Pascal, Markus Krotzsch, and Sebastian Rudolph. *Foundations of Semantic Web Technologies.* Boca Raton, FL: Taylor & Francis Group, 2010. Print.

"Introduction to the Semantic Web." *Cambridge Semantics.* Cambridge Semantics. Web. 4 Aug. 2015. http://www.cambridgesemantics.com/semantic-university/introduction-semantic-web

"Semantic Web." *W3C.* W3C. Web. 6 Aug. 2015. http://www.w3.org/standards/semanticweb/

"W3C Data Activity: Building the Web of Data." *W3C.* W3C. Web. 6 Aug. 2015. http://www.w3.org/2013/data/

"W3C Semantic Web Frequently Asked Questions." *W3C.* W3C. Web. 6 Aug 2015. http://www.w3.org/RDF/FAQ

SEQUENCE

SUMMARY

A sequence is an ordered list of mathematical objects, such as numbers, that may be infinite or finite and allows for repetition. A an important basic concept in mathematics, a sequence can take many forms and have various characteristics. They form the basis for series.

Finite sequences, or strings, have a definite number of terms with them, such as (1, –2, 7, π) and (0, $\sqrt{2}$, 0). Infinite sequences, or streams, contain an infinite number of terms, such as the sequence of natural numbers: (1, 2, 3, 4, 5...). Although both sequences and sets are collections of elements, they differ in two main ways. First, the order in which terms appear is significant in a sequence but not in a set. For example, while (1, 2) and (2, 1) represent different sequences, {1, 2} and {2, 1} are simply two ways of describing the set containing 1 and 2. Secondly, terms may be counted multiple times within a sequence, but only once within a set. The sequences (0, 0, 1, 2) and (0, 1, 2) are distinct, but the sets {0, 0, 1, 2} and {0, 1, 2} are equivalent.

NAMED SEQUENCES

Sequences are frequently indexed by listing each term in the form of a_n, where the subscript n is the position of the term within the sequence. In other words, a sequence is a function that maps n to a_n. As a result, the first term can be represented by a_1 (or a_0) the second by a_2 (or a_1), and so on. To avoid having to list out every term, a_n can be defined in terms of an explicit formula. For example, the sequence of natural numbers can be represented by $a_n = n$.

Certain sequences show up so often in pure or applied mathematics that they are given names. For instance, arithmetic sequences are sequences of the form $a_n = cn + b$, where c and b are constants. Geometric sequences have the form $a_n = d \bullet g^n$, where d and g are constants. Yet another type of sequence is a recursive sequence, in which terms are defined based on previous terms of the sequence. A notable historic example of a recursive sequence is the Fibonacci sequence, named after Leonardo Fibonacci of Pisa. In this sequence, each term is defined as the sum of the two previous terms, which can be expressed as follows:

$$a_x = a_{n-1} + a_{n-2}, \text{ where } a_1 = 1 \text{ and } a_2 = 1$$

The Fibonacci sequence is one example of an infinite sequencc. When working with infinite sequences, it is often useful to find the limit of the sequence, the value a_n approaches as n becomes arbitrarily large. If a_n gets increasingly close or equal to a particular value k as n becomes larger, the sequence is said to "converge to k." If no such limit exists, the sequence is deemed divergent. One example of a convergent sequence is defined by $a_n = 1/n$, because as n approaches infinity, a_n approaches 0. Conversely, the Fibonacci Sequence and the sequence defined by $a_n = n$ are divergent, since a_n approaches infinity as n does.

APPLICATIONS OF SEQUENCES

Sequences can be applied to situations both in everyday life and in more advanced mathematical topics. Because any list of ordered numbers is a sequence, everything from prime numbers to population growth can be represented as a sequence. There is even a growing field called sequence analysis that uses sequences to model historical events.

One of the most important applications of sequences in higher mathematics are the summations of their terms, which are called series. Like sequences, series can be either finite or infinite, and infinite series can be either divergent or convergent. Convergent series are extremely useful in a variety of areas, but most importantly they can be used to calculate integrals that are impossible to find using other methods.

—Matthew LeBar, Serena Zhu, and Daniel Showalter, PhD

BIBLIOGRAPHY

Blanchard, Philippe, Felix Bühlmann, and Jacques-Antoine Gauthier, eds. *Advances in Sequence Analysis: Theory, Method, Applications.* New York: Springer, 2014. Print.

Boyer, Carl B. "The Fibonacci Sequence." *A History of Mathematics.* 2nd ed. New York: Wiley, 1991. 255–56. Print.

Epp, Susana S. *Discrete Mathematics and Applications.* Boston: Cengage, 2011. Print.

Grabiner, Judith V. *The Origins of Cauchy's Rigorous Calculus.* Mineola: Dover, 2011. Print.

SERIES

SUMMARY

A series is the sum of all the mathematical objects, or terms, in a sequence, which may be numbers or functions. Series may be finite or infinite. A key aspect of mathematics, they are also important to related fields such as computer science and physics.

Mathematically, computers can really only add, multiply, and compare two numbers. Thus, a natural question is how can a computer evaluate a function such as sin x which cannot be expressed as a finite number of additions and multiplications. Likewise, how can a computer deal with numbers that have a non-terminating decimal expansion?

The answer to such questions lies in infinite series. A series is an infinite sum of the numbers a_x, where k ranges from one to infinity. This sum is represented by $\sum_{k=1}^{\infty} a_k$.

If the numbers a_k get small enough, then successive terms contribute less to the sum. Hence, there is the possibility that the series will converge to a finite number. In this case, we can approximate the value of the sum by looking at the partial sum, $\sum_{k=1}^{x} a_k$ for sufficiently large values of n. Partial sums are an important tool in determining the convergence of series.

GEOMETRIC AND P-SERIES

Two of the more important series are the geometric series and the p-series. The geometric series is of the form $\sum_{k=0}^{\infty} x^k$. This converges to $\frac{1}{1-x}$ for $|x| < 1$ and diverges otherwise. The p-series is of the form $\sum_{k=1}^{\infty} \frac{1}{k^p}$. This converges for $p > 1$ and diverges otherwise. In particular, if $p = 1$, then this is called the harmonic series.

There are a number of tests that can determine the convergence of series that are neither a geometric series nor a p-series. The test for divergence states that if $\lim_{k\to\infty} a_k \neq 0$, then the series $\sum_{k=1}^{\infty} a_k$ diverges. The integral test compares the series against the corresponding improper integral. The direct comparison test states that:

If $\sum_{k=1}^{\infty} c_k$ is a convergent series and $0 < a_k < c_k$, then $\sum_{k=1}^{\infty} a_k$ is likewise convergent.

If $\sum_{k=1}^{\infty} d_k$ is a divergent series and $0 < d_k < a_k$, then $\sum_{k=1}^{\infty} a_k$ is likewise divergent.

The limit comparison test states that if $\lim_{k\to\infty} {a_k}/{b_k} = L$, where $0 < L < \infty$, then $\sum_{k=1}^{\infty} a_k$ and $\sum_{k=1}^{\infty} b_k$ both converge or both diverge. The ratio test computes $\lim_{k\to\infty} \left| {a_{k+1}}/{a_k} \right| = L$ then uses this number to determine convergence in the following way:

If $L < 1$, then $\sum_{k=1}^{\infty} a_k$ converges.
If $L > 1$, then $\sum_{k=1}^{\infty} a_k$ diverges.
If $L = 1$, then the test is inconclusive.
-+The root test instead computes

$$\lim_{k\to\infty} \sqrt[k]{|a_k|} = L,$$

but has the same conclusions as the ratio test. Finally, the alternating series test states that if $a_k > a_{k+1}$, $a_k > 1$, and $\lim_{k\to\infty} a_k =$, then the series $\sum_{k=1}^{\infty}(-1)^{k_d}$ converges.

If $\sum_{k=1}^{\infty}|a_k|$ converges, then $\sum_{k=1}^{\infty} a_k$ is said to be absolutely convergent. With the exception of the Alternating Series Test, all of the above tests look for absolute convergence. If $\sum_{k=1}^{\infty}|a_k|$ diverges but $\sum_{k=1}^{\infty} a_k$ converges, then the series is said to be conditionally convergent. Since absolute convergence is preferable to conditional convergence, the Alternating Series Test is usually the last one implemented.

POWER SERIES

Series can be used to approximate functions such as sine, we can employ series. A power series for a function $f(x)$ centered at c is the function

$$f(x)=\sum_{k=0}^{\infty} f_k (x-c)^k$$

A power series will converge for all x in an interval I (which may consist of a single point) and diverge for all x outside of I. This interval is called the interval of convergence. The radius of convergence is half of the width of this interval. The radius of convergence can usually be computed using the ratio test or the root test. However, neither test will give information about convergence at the endpoints of the interval. Hence, another test must be used for the endpoints.

The coefficients for the above power series can be computed by

$$\frac{f^{(k)}(c)}{k!},$$

where $f^k(c)$ is the kth derivative of $f(x)$ evaluated at c. This is called the Taylor series for $f(x)$.

If $c = 0$, then this is called the Maclaurin series. When this method is used on the function $\frac{1}{1-x}$, this gives the geometric series $\sum_{k=1}^{\infty} a_k$. Other important power series (and their respective intervals of convergence) include:

Table 1. Various power series. © EBSCO

$e^x=\sum_{k=0}^{\infty}\frac{x^k}{k!}$	$-\infty < x < \infty$	$\sin x=\sum_{k=0}^{\infty}\frac{(-1)^k x^{2k+1}}{(2k+1)!}$	$-\infty < x < \infty$
$\cos x=\sum_{k=0}^{\infty}\frac{(-1)^k x^{2k}}{(2k)!}$	$-\infty < x < \infty$	$\tan^{-1} x=\sum_{k=0}^{\infty}\frac{(-1)^k x^{2k+1}}{(2k+1)}$	$\lvert x\rvert \le 1$
$In(x+1)=\sum_{k=1}^{\infty}\frac{(-1)^{k-1} x^k}{k}$	$-1 < x \le 1$	$(1+x)^m=\sum_{k=0}^{\infty}\binom{m}{k} x^k$	$\lvert x\rvert<1$\

—*Robert A. Beeler, PhD*

Bibliography

Hass, Joel, Maurice Weir, and George B. Thomas. *Thomas' Calculus.* 13th ed. Boston: Pearson, 2014. Print.

Hirschman, Isadore Isaac. *Infinite Series.* Belmont: Cengage, 2014. Print.

Knopp, Konrad. *Theory and Application of Infinite Series.* Mineola: Dover, 1990. Print.

Wade, William R. *An Introduction to Analysis.* 4th ed. Boston: Pearson, 2009. Print.

Set Notation

SUMMARY

In broadest terms, a set is a collection of things, with the individual things within the set called either elements or members of the set. Mathematicians usually speak of sets as a collection of numbers, but a set may be defined with ordinary objects. Sets are usually referenced by an italicized capital letter, with the object in the set listed between a pair of curly brackets. A set of objects in a living room, *L*, might be defined as L = {television, sofa, coffee table, bookcase, lamp}. In this case, the set L is a finite set that contains five elements. This set notation is used throughout mathematics to describe the elements of sets and how sets relate to each other.

Rather than listing out individual elements within a set, some sets contain all elements that fit a given rule or criterion. An equation with a solution of $y > 4$ can be written using set notation as $\{y|y > 4\}$, which is read as "the set of all y, such that y is greater than 4." This is an infinite set, because it contains an infinite number of elements.

THE USE OF SET NOTATION

Set notation is ubiquitous in mathematics, because it provides mathematicians with a way to discuss relationships among groups of numbers. Set notation can be used in algebra and calculus to describe sets of numbers. Functions are defined in terms of a mapping from one set of inputs (the domain) to a set of outputs (the range). In statistics and probability, set notation is often used to designate the sample space of interest.

Having defined a pair of sets A and B, the mathematician is then able to use set notation to designate different ways of combining and relating sets to each other. For example, A could be a subset of B, which indicates that every element of A is also an element of B. Set notation for indicating that A is a subset of B is $A \subset B$.

The use of sets developed into the field of set theory under Georg Cantor (1845–1918). By defining sets that have common features, it is possible for mathematicians to deeply explore fundamental relationships within mathematics. The importance of this approach to understanding deep truths about mathematics becomes more obvious when you consider some relevant sets of numbers.

$\mathbb{N}$: the set of all counting numbers
$\mathbb{Z}$: the set of all integers
$\mathbb{Q}$: the set of all rational numbers
$\mathbb{R}$: the set of all real numbers

These sets all contain an infinite number of elements, but the question that Cantor posed was built around the idea of whether all infinite sets contained the same number of infinities. For example, there are an infinite number of integers, but between any two integers are also an infinite number of rational numbers. Cantor wanted to know if this meant that the collection of rational numbers was a larger infinity than the collection of integers, and he applied set theory to this problem. Ultimately, his exploration of these sets led to a revolution in how mathematicians viewed the concept of infinity.

—*Andrew Zimmerman Jones, MS*

Bibliography

Bagaria, Joan. "Set Theory." Palo Alto, CA: Stanford University, 8 Oct. 2014. Web. 3 Jan. 2015. http://plato.stanford.edu/entires/set-theory/

Berlinski, David. "Sets." *Infinite Ascent: A Short History of Mathematics.* New York, NY: Modern Library, 2005. Print.

Devlin, Keith. J. *The Joy of Sets: Fundamentals of Contemporary Set Theory.* 2nd ed. New York: Springer, 1993. Print.

Elwes, Richard. "Cardinal Numbers." *Math in 100 Key Breakthroughs.* New York: Quercus, 2013. Print.

Strogatz, Steven. "The Hilbert Hotel." *The Joy of X.* Boston: Houghton, 2012. Print.

Wallace, David Foster. *Everything and More: A Compact History of Infinity.* New York: Norton, 2010.

SIRI

SUMMARY

Siri is an artificial intelligence program released by Apple on its iPhone 4S in 2011. The program has been updated and included on each subsequent model of iPhone as well as on several other Apple products since that time. Siri is a virtual assistant that uses voice commands to perform actions such as reading and replying to text messages, creating events and reminders in the calendar, opening programs, suggesting restaurants, and searching the Internet while interacting with the user. First given a voice by American artist Susan Bennett, who has since risen to fame on social media, Siri gained a male voice option in 2013.

HISTORY AND BACKGROUND

Siri started as a standalone application (app) added to the Apple App Store in early 2010 by Siri Inc. Siri Inc.'s cofounders—Dag Kittlaus, Adam Cheyer, and Tom Gruber—developed the technology after having worked for the Defense Department on the Cognitive Assistant That Learns and Organizes (CALO) project, which was part of the Personal Assistant That Learns (PAL) program. They used the basics of what they put together working on CALO to create Siri, which they called more of a "do engine" than a search engine. When they released Siri on Apple's App Store, they had plans to also release the app on Android; they even started a deal with Verizon

Siri Remote and old Apple Remote. By Dandumont (Own work)

Screenshot of the Siri interface in macOS Sierra. By Apple Inc. (Upload: WitherOrNot

to make it a default app on Android phones. Three weeks after Siri was released on the App Store, however, Steve Jobs, chairman and chief executive officer of Apple, met with Kittlaus and made a deal to acquire Siri Inc. for a reported $150 to $250 million.

Siri has gone through many updates and changes since it was acquired by Apple. Many of the original features, such as being able to buy plane tickets and movie tickets as well as to make restaurant reservations just with the app, were discarded. However, Siri has learned how to understand and communicate in the language of every country that the iPhone is released in and retained some of the app's original snappy remarks.

When Siri was first released as a default on all iPhones, Apple had somewhat of a monopoly on the market. Soon, others such as Google, Samsung, and some smaller app developers released their own versions of virtual assistants. Since early 2016, several other virtual assistants have dominated the market. Amazon created Alexa, Windows released Cortana, Google developed Google Now, and Samsung worked to improve S-Voice. Siri's popularity grew and became available on other Apple products, including the iPad Pro, the iPad Air, the iPad (3rd generation) or later, the iPad mini, and the iPod touch (5th generation) or later. It could also be found on the Apple Watch and Apple TV.

Artificial intelligence is a constantly growing and changing technology with virtual assistants such as Siri taking the lead on development. The complex algorithms used to distinguish human language and interpret meaning are getting better and better, with the end goal being that the virtual assistant may soon be able to know what a user wants before the user even knows.

—Rebecca Orner

Bibliography

Bosker, Bianca. "Siri Rising: The Inside Story of Siri's Origins—And Why She Could Overshadow the iPhone." *Huffington Post*, 22 Jan. 2013, www.huffingtonpost.com/2013/01/22/siri-do-engine-apple-iphone_n_2499165.html. Accessed 9 Dec. 2016.

Dwoskin, Elizabeth. "Siri's Creators Say They've Made Something Better That Will Take Care of Everything for You." *Washington Post*, 4 May 2016, www.washingtonpost.com/news/the-switch/wp/2016/05/04/siris-creators-say-theyve-made-something-better-that-will-take-care-of-everything-for-you/?utm_term=.f669c8c9f507. Accessed 9 Dec. 2016.

Lafrance, Adrienne. "Why Do So Many Digital Assistants Have Feminine Names?" *Atlantic*, 30 Mar. 2016, www.theatlantic.com/technology/archive/2016/03/why-do-so-many-digital-assistants-have-feminine-names/475884/. Accessed 9 Dec. 2016.

"How Apple's Siri Got Her Name." *The Week*, 29 Mar. 2012, theweek.com/articles/476851/how-apples-siri-got-name. Accessed 9 Dec. 2016.

Martin, James A. "Virtual Assistant Faceoff: Alexa, Cortana, Google Assistant and Siri." *CIO*, 5 Oct. 2016, www.cio.com/article/3127736/software/virtual-assistant-faceoff-alexa-cortana-google-assistant-and-siri.html. Accessed 9 Dec. 2016.

Reuters. "Apple Enhances Siri but Still Trails in Artificial Intelligence Race." *NBC News*, 14 June 2016, www.nbcnews.com/tech/tech-news/apple-enhances-siri-still-trails-artificial-intelligence-race-n591786. Accessed 9 Dec. 2016.

"Siri." *Apple*, www.apple.com/ios/siri/. Accessed 9 Dec. 2016.

Stern, Joanna. "Apple's Siri: A Lot Smarter, but Still Kind of Dumb." *Wall Street Journal*, 20 Sept. 2016, www.wsj.com/articles/apples-siri-a-lot-smarter-but-still-kind-of-dumb-1474390910. Accessed 9 Dec. 2016.

Smart City

SUMMARY

Recent trends in urban planning and development have contributed to the rise of the concept of the *smart city*. The term is used in various ways and does not have a universally accepted definition, but one widely cited conceptualization was developed by European Smart Cities, an initiative of the Vienna University of Technology in Austria. According to the European Smart Cities model, a smart city has six key "smart" features—smart economy, smart mobility, smart environment, smart people, smart living, and smart governance. In this context, *smart* refers to progressive, inclusive, sustainable, and forward-thinking policies that leverage technology, human capital, and social responsibility.

Connectivity technologies are also poised to play a major role in the ongoing move toward building smart cities. With the Internet of Things (IoT) continuing to develop at a rapid pace, technology is playing an increasingly prominent and essential role in the development of smart urban spaces.

BACKGROUND

In 2012, the United Nations (UN) reported that more than half of Earth's population was living in cities. In a related trend, European Smart Cities notes that globalization is driving economic changes with widespread impacts. Among other effects, globalization and increasing levels of urbanization are forcing cities in both developed and developing countries to explore new ways to remain competitive. Meanwhile, as technology continues to develop at an unprecedented rate, new applications that could positively affect urban life are emerging. Thus, while the "smart city" concept has been defined in many ways, the vast majority of theoretical models circle back to these two core principles: competitiveness and the transformative power of technology.

Cities that remain competitive tend to have stronger economies, lower unemployment rates, and offer a better overall quality of life than cities that lag behind. The ability to innovate also gives cities a competitive edge, which is why human and social capital

Smart City Roadmap. By Cybersecurity101 (Own work)

investments are so important to the development of smart cities. Most smart city conceptualizations also focus on sustainability, with technology viewed as a core means of achieving this goal. Finally, some models emphasize the importance of civic participation in governance and the need for city residents and businesses to make equitable use of available resources.

Sustainability has become a particularly pertinent issue. Governments around the world have become increasingly focused on minimizing the environmental toll of human activity, especially in urban areas. According to statistics presented at the 2013 International Forum on Knowledge Asset Dynamics (IFKAD), cities consume as much as 80 percent of the world's energy, with higher levels of energy consumption and transportation-related pollution being associated with lower-density urban areas. Thus, increasing urban population density and decreasing reliance on less efficient forms of transportation are salient topics in the smart city sphere. To boost urban population density, cities are looking to build "up" rather than "out" by building more mid-rise and high-rise residential buildings rather than continually expanding the reaches of suburbs and city limits. On the transportation front, smart cities look to decrease the use of single-occupancy vehicles by encouraging residents and commuters to explore the growing number of available alternatives.

TOPIC TODAY

It is possible to explore current and developing trends in modern urban planning using the six key "smart" features cited in the European Smart Cities model. From an economic standpoint, one of the features that differentiates smart cities is the implementation of mixed-use land development policies. Mixed-use land development designates specific urban spaces for various uses, including a blend of commercial, cultural, industrial, institutional, residential, and retail applications. One of the key advantages of mixed-use development is that it makes cities and neighborhoods far more walkable, which helps fight traffic congestion, improves community safety, and reduces pollution. Researchers and experts have also noted that information and communication technologies (ICTs) are influencing practically every aspect of contemporary urban economies, including changes in land use trends.

However, the mobility aspect of smart cities is the area in which ICTs are widely considered to have the greatest potential for positive change. Examples of smart mobility technologies include traffic light and traffic flow sensors that ease gridlock and congestion, especially during peak periods, and public transportation and parking apps that help users plan journeys using mass transit and spend less time searching for parking spots. These applications, along with the rising trends toward car-sharing and ride-sharing programs, are expected to continue to improve and advance as IoT technologies proliferate in urban areas.

From an environmental standpoint, energy efficiency is one of the smart city's primary goals. IoT technologies are poised to bring about major shifts in urban power generation and distribution strategies. According to one proposal, public and industrial buildings would supplement their use of grid-supplied electricity by generating their own using wind and solar power technologies to capture energy and IoT applications to store excess electricity for later use. Internet-connected sensor networks would monitor individual and municipal power generation and distribution systems in real time, with some projections indicating that such initiatives could cut energy costs by 25 percent or more.

Beyond driving innovation, the inhabitants of a smart city have a major role to play in the ongoing effort to improve the quality of life for urban residents. The European Smart Cities model cites four key characteristics of smart city residents: educated, engaged in lifelong learning, multicultural, and

Amsterdam is hailed for using smart city technologies, such as upgrading street lamps to dim based on pedestrian usage. By Massimo Catarinella (Own work)

open-minded. While these notions have drawn some criticism for being too abstract, they nevertheless reveal a possible path forward for cultivating a more civically engaged population base.

Living in a smart city has implications that extend beyond using ICTs to increase efficiency and accessibility. In designating cities as "smart," European Smart Cities evaluates criteria including the presence of cultural facilities, leisure facilities, quality housing, and educational infrastructure. High levels of safety, security, and social cohesion are also considered favorable, and cities that make concerted efforts to improve these characteristics generate higher scores due to their strong associations with improved quality of life.

Finally, on the topic of governance, smart cities display high levels of cooperation between citizens and municipal officials and between residents and the local business community. Smart cities aim to share key resources to the greatest possible degree, and the European Smart Cities model cites social service availability, government transparency, and high levels of political awareness and public engagement as desirable elements.

—Jim Greene, MFA

BIBLIOGRAPHY

Albino, Vito, et. al. "Smart Cities: Definitions, Dimensions, and Performance." *International Forum on Knowledge Asset Dynamics,* 2013. www.ifkad.org/Proceedings/2013/papers/session7/103.pdf. Accessed 23 Dec. 2016.

Batty, M., et. al. "Smart Cities of the Future." *The European Physical Journal: Special Topics,* vol. 214, no. 1, Nov. 2012, pp. 481–518.

"European Smart Cities." *Vienna University of Technology,* 2015, www.smart-cities.eu/?cid=-1&ver=4. Accessed 23 Dec. 2016.

"Five ICT Essentials for Smart Cities." *Escher Group,* www.eschergroup.com/files/8914/4491/8222/Smart_City_Planning.pdf. Accessed 23 Dec. 2016.

Pennington, James. "4 Ways Smart Cities Will Make Our Lives Better." *World Economic Forum,* 10 Feb. 2016, www.weforum.org/agenda/2016/02/4-ways-smart-cities-will-make-our-lives-better/. Accessed 23 Dec. 2016.

"Readings on Smart Cities." *IEEE Smart Cities,* smartcities.ieee.org/articles-publications/ieee-xplore-readings-on-smart-cities.html. Accessed 23 Dec. 2016.

Vinod Kumar, T.M., editor. *Smart Economy in Smart Cities.* Springer Singapore, 2017.

Zanella, A., et. al. "Internet of Things for Smart Cities." *IEEE Internet of Things Journal,* vol. 1, no. 1, 14 Feb. 2014, pp. 22–32.

SMART HOMES

SUMMARY

Smart-home technology encompasses a wide range of everyday household devices that can connect to one another and to the Web. This connectivity allows owners to program simple daily tasks and, in some cases, to control device operation from a distance. Designed for convenience, smart homes also hold the promise of improved independent living for elderly people and those with disabilities.

THE EMERGING FIELD OF HOME AUTOMATION

Smart homes, or automated homes, are houses in which household electronics, environmental controls, and other appliances are connected in a network. Home automation is a growing trend in the 2010s that allows owners to monitor their homes from afar, automate basic home functions, and save money in the long run on utility payments and other costs. Creating a smart home generally involves buying a hub and then various smart devices, appliances, plugs, and sensors that can be linked through that hub to the home network. Setting up a smart home is costly and requires strong network infrastructure to work well.

CREATING THE HOME NETWORK

Not all smart devices are designed to connect in the same way. Wi-Fi enabled devices connect to Wi-Fi networks, which use super-high frequency (SHF) radio signals to link devices together and to the Internet via routers, modems, and range extenders. Other smart devices have Bluetooth connectivity. Bluetooth also uses SHF radio waves but operates over shorter ranges. Assorted Bluetooth devices may be connected to a smart-home network for remote in-home control but cannot be operated outside the home. Essentially, a Wi-Fi network is necessary for Internet-based control.

More advanced options for smart-home networking involve mesh networks. In these networks, each node can receive, repeat, and transmit signals to all others in the network. Mesh networks may be wired, with cables connecting the nodes. More often they are wireless, using Bluetooth or Wi-Fi to transmit radio signals between nodes. Smart-home network company Insteon has created products linked into dual-mesh networks. Such networks use a dual-band connection, with both wired and wireless connections between the nodes. Dual-mesh networks are designed to avoid electronic interference from outside signals coming from sources such as microwave ovens and televisions. In a dual-mesh network, interruption to the power lines entering the home or to the home's wireless network will not prevent signal from reaching a networked device. Mesh networks also work with devices from different manufacturers, while other types of networks may not be compatible with all of an owner's devices.

CONNECTING SMART DEVICES AND UTILITIES

Different types of devices can be connected to smart-home networks. Smart light switches and dimmers can be used to control existing light fixtures, turning lights on or off or connecting them to a scheduler for timed activation. A smart-home owner might program their lights to coincide with the sunrise/sunset cycle, for example. The use of smart plugs also allows for the control of existing, ordinary devices. Smart kitchen appliances and entertainment units are becoming available as well.

Other devices can be used to monitor home utilities. For instance, smart water sensors can detect moisture or leaks. Smart locking devices placed on doors and windows can alert the owner when a door or window is unsecured and allow them to lock and unlock doors and windows remotely. Sensors and controls can measure heat within each room and adjust thermostat settings. Smart-home owners can use programmable thermostats to lower temperatures during sleep periods and to raise heat levels before they wake. While many thermostats have timers, automated systems also allow owners to adjust heat and energy consumption from their smartphone or computer, even when away from the home. Some even feature machine learning, in which the device adjusts its behavior over time in response to repeated usage patterns. By shutting off air conditioning, heat, or electricity when not needed, automated monitoring and control can save owners large amounts of money in utility costs each year.

GOALS AND CHALLENGES

Smart-home technology is part of a technological movement called the Internet of things (IoT). IoT is based on the idea that devices, vehicles, and buildings will all someday be linked into a wireless network and collect and communicate data about the environment. This concept relies on ubiquitous computing, the practice of adding microprocessors and Internet connectivity to basic devices and appliances. Ubiquitous computing aims to make data processing a constant background activity embedded in daily life.

While still in its infancy, smart-home technology is intended to enhance security and convenience for homeowners. A home could be configured, for instance, so that motion sensors detecting activity in the morning open window shades, turn on a coffee maker, and raise the heat. Although still largely a hobbyist industry, home automation is also beginning to help elderly people and those with disabilities. Features such as voice activation, mobile apps, and touch screens make it easier for those owners to control their environment and live independently.

Consumers have been slow to adopt smart-home technology for several reasons. One is fear that manufacturers will use such devices to collect and sell data about their personal habits. Another is fear that the security of such systems could be hacked and lead to safety issues such as burglary. The multiple users' preferences, power dynamics, and rapid shifts in schedule that present in family life pose further challenges for smart-home systems. With the debut of wearable digital devices like smart watches, some companies are experimenting with linking wearable and smart-home devices. Wearable devices that can record biometric data could be used to verify a homeowner's identity for increased security and personalization, for example.

—*Micah L. Issitt*

Bibliography

Clark, Don. "Smart-Home Gadgets Still a Hard Sell." *Wall Street Journal.* Dow Jones, 5 Jan. 2016. Web. 12 Mar. 2016.

Glink, Ilyce. "10 Smart Home Features Buyers Actually Want." *CBS News.* CBS Interactive, 11 Apr. 2015. Web. 12 Mar. 2016.

Gupta, Shalene. "For the Disabled, Smart Homes Are Home Sweet Home." *Fortune.* Fortune, 1 Feb. 2015. Web. 15 Mar. 2016.

Higginbotham, Stacey. "5 Reasons Why the 'Smart Home' Is Still Stupid." *Fortune.* Fortune, 19 Aug. 2015. Web. 12 Mar. 2016.

Taylor, Harriet. "How Your Home Will Know What You Need Before You Do." *CNBC.* CNBC, Jan 6 2016. Web. 11 Mar. 2016.

Vella, Matt. "Nest CEO Tony Fadell on the Future of the Smart Home." *Time.* Time, 26 June 2014. Web. 12 Mar. 2016.

Smart Label

SUMMARY

A *smart label* is any form of product packaging or identification that uses technology to provide more information than could be listed on the product's paper or plastic packaging. Smart labels interact with other technology to provide information to consumers, help track and manage inventory, or provide antitheft protection. Quick response (QR) codes, surveillance tags, and radio frequency identification (RFID) tags are all examples of smart labels. The technology is used by manufacturers, retailers, shipping companies, hospitals, libraries, and others to gather and share information, but it is not without controversy.

BACKGROUND

Prior to the mid-1970s, retailers and manufacturers who wanted to track their products relied on manual systems. Change began to occur in June 1974, when a ten-pack of Wrigley's Gum purchased in Ohio became the first product to be scanned with a bar code scanner

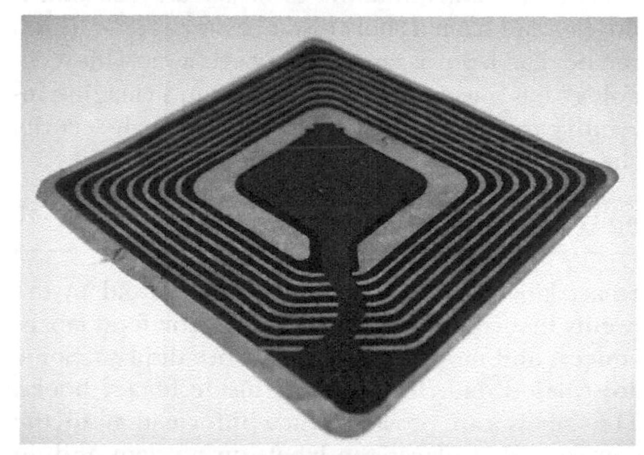

A generic RFID chip. Maschinenjunge assumed (based on copyright claims).

in the first system that would go on to widespread use. Efforts had been underway for several decades to find a way to create a uniform method to streamline the process of choosing and/or paying for merchandise, but it was not until the 1970s that the right format and the technology to implement it were put together to create the first Universal Product Code (UPC).

UPC codes use a system of thick and thin lines that can be customized for each product. Once scanned, the scanning system interprets those lines and returns the price to the register while also updating the inventory for the product. This was very helpful to manufacturers and stores, but it did not provide any useful information to consumers.

The next innovation, the 2D bar code, was developed to hold more information. These codes, sometimes known as quick response, or QR, codes, capture information in two dimensions—both horizontally, like traditional bar codes, and vertically. As a result, they are able to hold exponentially more information. UPC codes capture twenty characters of data. QR codes can hold more than seven thousand characters. When used in conjunction with a smartphone, a QR code can direct a consumer to a web page or display additional data right on the phone.

Other forms of smart label technology include electronic article surveillance (EAS) codes, such as those used in libraries and stores to code merchandise. These codes can be deactivated at the checkout counter by special equipment. They can sound an alarm if items are removed from the premises without being deactivated. A radio frequency ID tag, or RFID, uses a microchip to code information about items and a small antenna to broadcast it so that it can be read from a short distance away. Some RFIDs can be read from as far as thirty yards away. This technology has a wide range of uses, from managing inventory to allowing people to pass through security checkpoints.

PRACTICAL APPLICATIONS

Smart labels are nearly impossible to avoid in the twenty-first century. They are present on food labels, concert and movie posters, and store displays. Some are carried daily in wallets or inside library books. The labels can provide much information to the company that places the labels on an item and, in many cases, to the consumer as well.

A matrix barcode.

QR codes are the most visible of the smart technologies. The small squares of coding appear on a wide range of products and sometimes are even on store displays, on posters, or in newspapers and magazines. Some can be read by simply using the smartphone's camera; others require an app, which often can be downloaded for free. QR codes can direct the person "reading" them to a website with additional information. A food manufacturer may include additional nutrition information and information about the source of the food product. A concert or movie poster might include a bit of video, such as an excerpt of a song or a movie trailer. Codes on store displays or in magazines can play commercials right on smartphones, while codes in newspapers may connect the reader to a website with updated information on the story being read. All of these help companies share more information with consumers.

Other smart labels can be used to help track merchandise and packages as they move from one site to another. The code is scanned as the product or package reaches each stage of its journey, allowing the producer and, in cases of shipped packages, the customer to know where the package is as it travels. This technology is widely used by postal services and delivery companies such as UPS and FedEx to avoid loss and provide accountability during the shipping process.

New technology for these processes is being developed all the time. For instance, companies are working with new thin plastic film that can encode smart technology so that more types of items can be labeled. Technology is also available that can detect temperatures where the product is being transported, allowing manufacturers of perishable products such as milk and yogurt to ensure their products remain safe as they travel.

Additional forms, such as RFID, allow items "tagged" with the technology to be read from a distance. This technology allows swipe reading of credit and debit cards. U.S. Homeland Security uses RFID technology at border crossings to "read" identity documents to help speed passage across borders. Retail stores and many other businesses use the technology to keep track of people and objects, and new uses are found regularly.

While smart labels provide a wealth of information for both companies and consumers, concerns about their usage do exist. Some worry that the ease with which labels such as RFID tags can be read makes it easy for thieves to steal personal information. Organizations concerned about the health and safety of food raise concerns that food companies will use the codes to get around posting information about their products that can be perceived as negative. For example, groups concerned about foods that contain genetically modified organisms (GMOs) worry that companies will get around requirements to label these products by putting the information in smart labels.

A 2015 study by the Mellman Group indicated that 88 percent of respondents preferred key information, such as a product's GMO status, to be listed on product packaging. The same study found that only 17 percent of respondents had ever "read" a code on a package. Nevertheless, the Grocery Manufacturer's Association predicts that by 2020, about 80 percent of all food products will carry a QR code.

—Janine Ungvarsky

Bibliography

"10 Reasons Why GMO Smart Label Isn't 'Smart' at All." *EcoWatch,* www.ecowatch.com/10-reasons-why-gmo-smart-label-isnt-smart-at-all-1882130207.html. Accessed 21 Dec. 2016.

"About Us: SmartLabel™ Working for You." *Smartlabel.org,* www.smartlabel.org/about/about-us. Accessed 21 Dec. 2016.

Fish, Eric. "Smart Labels—The Next Big Thing in Packaging?" *Flexible Packaging Magazine,* 11 Jan. 2016, www.flexpackmag.com/articles/87805-smart-labels-the-next-big-thing-in-packaging. Accessed 21 Dec. 2016.

Kavis, Mike. "The Smart Labels That Will Power the Internet of Things." *Forbes,* 17 Feb. 2015, www.forbes.com/sites/mikekavis/2015/02/17/the-smart-labels-that-will-power-the-internet-of-things/#1a9c18112690. Accessed 21 Dec. 2016.

"New SmartLabel™ Initiative Gives Consumers Easy Access to Detailed Product Ingredient Information." *Grocery Manufacturers Association,* 2 Dec. 2015, www.gmaonline.org/news-events/newsroom/new-smartlabel-initiative-gives-consumers-easy-access-to-detailed-ingredien/. Accessed 21 Dec. 2016.

"Radio Frequency Identification (RFID): What Is It?" *U.S. Department of Homeland Security,* 20 Aug. 2015, www.dhs.gov/radio-frequency-identification-rfid-what-it. Accessed 21 Dec. 2016.

Rouse, Margaret. "Smart Label." *IoT Agenda,* internetofthingsagenda.techtarget.com/definition/smart-label. Accessed 21 Dec. 2016

Watson, Elaine. "80% of Packaged Groceries Will Feature SmartLabel within Five Years, Predicts GMA." *Food Navigator U.S.A,* 2 Dec. 2015, www.foodnavigator-usa.com/Manufacturers/80-of-foods-will-feature-SmartLabel-in-five-years-predicts-GMA. Accessed 21 Dec. 2016.

Weightman, Gavin. "The History of the Bar Code." *Smithsonian.com,* 23 Sept. 2015, www.smithsonianmag.com/innovation/history-bar-code-180956704/. Accessed 21 Dec. 2016.

Smartphones, Tablets, and Handheld Devices

SUMMARY

Handheld devices are very small computers. They are very common today and easily fit in one or both hands. Many handheld devices have touch screens. Devices with touch screens do not need keyboards to enter information. Some handheld devices make phone calls or send and receive messages. Some search the Internet and receive e-mails. Some play music. Others allow people to buy and read electronic books, or e-books. Advanced handheld devices do all of these tasks and more.

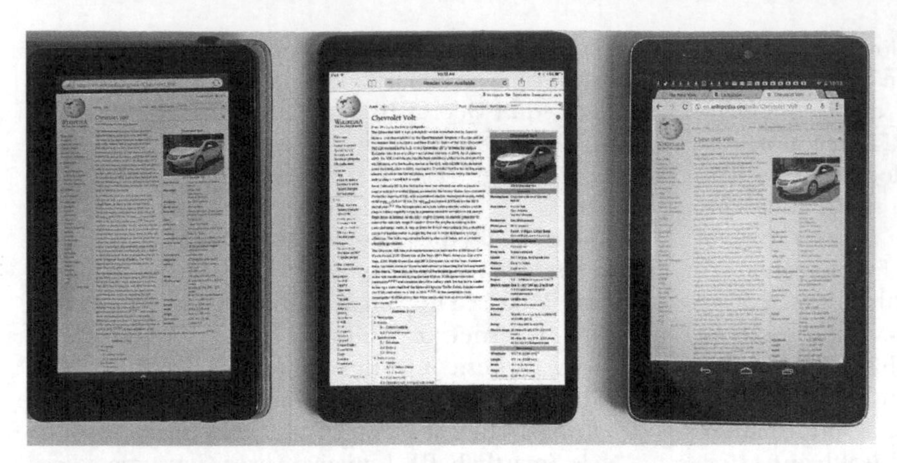

Comparison of several mini tablet computers: Amazon Kindle Fire (left), iPad Mini (center), and Google Nexus 7 (right). By Mariordo (Mario Roberto Durán Ortiz).

CELL PHONES AND LANDLINES

Cellular phones, or cell phones, are among the most well-known handheld devices. Cell phones are phones that can make phone calls from almost anywhere. In the past, people could make phone calls only from landlines. Landlines are phones inside a person's home. Landlines connect to wires that stretch from a person's house to the road. These wires connect to phone cables. The cables carry the sound of people's voices across great distances.

Cell phones do not need wires to make calls. Cell phone companies build towers in many places. Each tower serves a cell, or small area. The towers send and receive signals to and from nearby cell phones. People often travel from one cell to another while talking on their cell phones. Cell phones simply switch to the closest tower so people can keep talking as they move.

Cell phones may seem like modern inventions. However, the first cell phone call was made in 1973. Cell phones became more popular in the 1980s. Early cell phones were large and weighed more than two pounds. Smaller cell phones arrived in the 1990s. By 2010 many people used cell phones more than they used landlines.

OTHER HANDHELD DEVICES

Cell phones are just one of many handheld devices. Personal digital assistants (PDAs) are handheld organizers. PDAs first appeared in the early 1990s. They are like electronic day planners and address books. People use them to schedule appointments and to save contact information. People connect PDAs to computers to move information back and forth.

Portable media players are devices that store and play sound, pictures, and video files. They are small enough to carry easily.

E-book readers or e-readers, are devices that let people save and read books whenever they want. The first e-book readers arrived in the late 1990s. However, people had to connect them to a computer to save books on them. They were large and could save only a few books. E-readers have improved since then. Today they are thin and light. They can save hundreds of books.

SMARTPHONES AND TABLETS

Smartphones and tablets are two more popular handheld devices. Smartphones are a type of cell phone. People can use smartphones to make phone calls. They can use them to take pictures, play games, and listen to music. Smartphones can also connect to the Internet. Most smartphones have touch screens. Smartphones perform the tasks of many handheld devices at once.

Tablets are thin, handheld computers. Tablets are larger than smartphones but smaller than laptop computers. They perform the same tasks as computers and smartphones. However, most tablets cannot make phone calls. Tablets usually have touch screens rather than keyboards.

TRENDS IN HANDHELD DEVICES

A trend is something that is popular for a short time. Trends in handheld devices change quickly. New devices keep being made. Some devices are special for being the first or the most popular of their kind, and other companies often make similar devices to compete with them.

The IBM Simon was the first smartphone. It was created in 1993 before the word "smartphone" existed. It was a PDA and a cell phone. It even had a touch screen. However, it was large and expensive. It never became popular.

The PalmPilot was one of the most popular PDAs. It first appeared in the mid-1990s. Later PDAs used the PalmPilot as a model.

The BlackBerry was first sold in the early 2000s. It was another early smartphone. The BlackBerry was a PDA and a cell phone, and it could send and receive e-mails. It became very popular among businesspeople.

The iPod was introduced in 2001. It quickly became one of the most successful portable media players of the twenty-first century.

The Kindle is an e-reader. The Kindle began to be sold in 2007. It was not the first e-reader, but it became popular very quickly. The Kindle is small and light. It can save hundreds of books. It produces a very sharp screen image. Reading text on the Kindle is like reading text on paper. The Kindle is also wireless, which means it can connect to the Internet without wires.

The iPhone is a smartphone. It was first sold in 2007. The iPhone was considered the best smartphone when it was first sold. The iPhone has been improved several times since then. It remains one of the most popular smartphones.

The iPad was first sold in 2010. The iPad is a tablet. It quickly became one of the most popular tablets.

—*Lindsay Rock Rohland*

BIBLIOGRAPHY

Lebert, Marie. "eBooks: 1998 – The First eBook Readers." Project Gutenberg News. Project Gutenberg News, 16 July 2011. Web. 3 June 2014. http://www.gutenbergnews.org.

Wright, Rob. "The Evolution of the Smartphone in 7 Releases." CRN. Channel Company, 3 Apr. 2013. Web. 3 June 2014. http://www.crn.com/slide-shows/mobility.

SOFT ROBOTICS

SUMMARY

Soft robotics is the development of robots composed of soft components. In nature, soft body parts are adaptable; the same reasoning applies to robotics. Whereas traditional robotic components are rigid links in a mechanical chain and move in mathematically predictable ways, soft robotics may be soft materials that contain some rigid components, or may even be made entirely of soft components. They can conform themselves to their environment and adapt to different situations. Soft robotics also describes the variable resistance of the control systems of the robots.

Development of soft robotics began around 2006. Applications potentially include the biomedical field, the service industry, manufacturing of delicate materials, and surgical procedures.

BACKGROUND

Although clockwork automatons have been around since at least the Middle Ages, robotics developed primarily to serve industry. George Devol designed the first programmable robot in 1954. He and Joseph Engelberger produced the prototype, the Unimate #001, in 1959. They focused on creating a robotic arm that would perform tasks that put human workers at risk. The prototype was installed in a General Motors

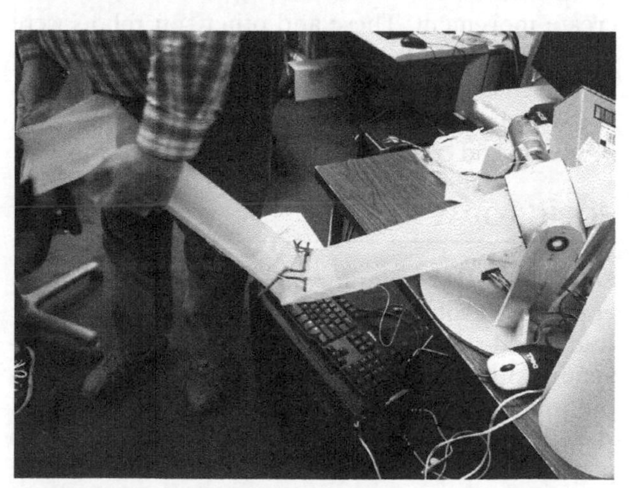

An inflatable robotic arm. By Z22 (Own work)

die-casting plant in Trenton, New Jersey. Unimate took heated die-castings from machines and performed welds. Industrial robots became common in the auto industry and other manufacturing industries within a short time, and grew increasingly complex with the advancement of computer chips. The U.S. space program developed robotic rovers, robotic arms, and experimental robotic astronauts. Other applications include search-and-rescue robots and consumer products such as vacuum cleaners, personal assistant devices, and toys such as robotic animals and building kits.

Traditional robots, with hard components, are not suitable for all applications, however. While industrial robots can pick up hard objects, they are less capable of handling delicate items or items that are not uniform shapes and sizes. They must have multiple controllable joints and force sensing components to handle such items. Robots meant to interact with humans and other living things may have difficulty with some tasks. Their ability to adapt is limited by their rigid components.

Twenty-first-century soft robotics developers responded with a variety of approaches. Researchers at Harvard University developed a series of soft robots using pneumatic actuation, meaning they use pistons inside of hollow cylinders. Pressure moves the piston along its axis, creating linear force, and a spring-back force on the opposite side returns the piston to its original position. Harvard's silicone-based soft robots use a pneumatic network, which inflates and deflates various parts of the soft body to create movement. These and other soft robots were often inspired by living creatures, including caterpillars, starfish, and octopuses. An octobot developed at Harvard University has no batteries, circuit boards, or other rigid components at all—it is entirely soft.

TOPIC TODAY

Soft robotics is developing in two primary areas. One approach involves robots built with rigid links, like traditional robots, with control systems for the interaction with the robot's environment, such as gripping abilities. In simple terms, these robots have rigid skeletal structures and soft exteriors. The other approach involves robots primarily or entirely made of soft materials, which experience vast shape changes in operation. Many of these soft-bodied robots are made of materials that can adjust to various degrees of stiffness.

A soft under-actuated-robot that mimics a tuna fish. By Pablo Valdivia y Alvarado pablov@mit.edu (Own work www.dedoux.com)

One development in gripping technology uses soft components to handle the objects, while much of the robot is composed of solid components. A soft sack made of elastic or another material, for example, can be filled with sand. Sand is granular, and when close together, the grains jam together. The sack of sand is connected to a pneumatic component, which allows a gas such as air into the sand. The robot lowers the sack onto the object to be gripped, and vacuums the gas out of the sand. The granular jamming envelops the object; to release it, the robot puts the gas back into the sack of sand, which unjams the grains, and the sack no longer conforms to the object. In addition to gripping objects of inconsistent size and shape, such granular jamming components can also pick up multiple items at once and place them, in the same configuration, in another location.

Flexible, soft components can be constructed to move in many ways. They can expand, contract, bend, and twist with adjustments to pressurized fluid or gas. For example, a tube can be constructed with a smooth side and a reinforced ribbed side. When unpressurized, the tube is straight; under pressure, the ribbed side expands, forcing the smooth side into a curve. The degree of curvature is determined by the amount of pressure.

Soft robotics development has been aided by polymer science and in particular 3-D printing. Some 3-D printers can create items out of soft, polymer-based materials. This enables roboticists to design and create any shape using soft materials, including silicone, that were previously impossible to build. The Harvard octobot is one example of this use. The team individually 3-D printed each component needed inside the soft robot body, including fuel storage and power components. Hydrogen peroxide is used as liquid fuel. A reaction using platinum inside the octobot creates gas, which inflates the eight arms. A microfluidic logic circuit controls the movement.

Soft robotics offers a wide range of possibilities in many fields. Some researchers are working on rubbery robots that can be used to perform surgery. These devices would operate more gently around soft tissues and organs, reducing risk of injuries such as punctured arteries and organs that have occurred using traditional rigid robotic systems. Researchers are also exploring the possibility of developing a device that would surround a damaged or malfunctioning heart and rhythmically contract to mechanically aid the organ in pumping blood. Other potential uses of soft robotics include robotic gloves to help patients recovering from strokes to grip objects, prosthetic devices, and soft robots that can squeeze through small, irregular gaps during rescue operations.

—Josephine Campbell

Bibliography

Brown, Eric, et al. "Universal Robotic Gripper Based on the Jamming of Granular Material." *Proceedings of the National Academy of Sciences of the United States of America,* 2 Nov. 2010, www.ncbi.nlm.nih.gov/pmc/articles/PMC2973877/. Accessed 12 Jan. 2017.

Burrows, Leah. "The First Autonomous, Entirely Soft Robot." *Harvard Gazette,* 24 Aug. 2016, news.harvard.edu/gazette/story/2016/08/the-first-autonomous-entirely-soft-robot/. Accessed 12 Jan. 2017.

Greenemeier, Larry. "Soft Touch: Squishy Robots Could Lead to Cheaper, Safer Medical Devices." *Scientific American,* 24 Sept. 2013, www.scientificamerican.com/article/soft-robotics-biomedical-surgery/. Accessed 12 Jan. 2017.

"Help NASA Create Better Vision for Robonaut." *National Aeronautics and Space Administration,* robotics.nasa.gov/news_feature.php?id=561. Accessed 12 Jan. 2017.

Laschi, Cecilia, and Matteo Cianchetti. "Soft Robotics: New Perspectives for Robot Bodyware and Control." *Frontiers in Bioengineering and Biotechnology,* 30 Jan. 2014, journal.frontiersin.org/article/10.3389/fbioe.2014.00003/full. Accessed 12 Jan. 2017.

"The Many Uses of Soft and Flexible Robots." *Association for Advancing Automation,* 31 Aug. 2016, www.a3automate.org/many-uses-soft-flexible-robots/. Accessed 12 Jan. 2017.

Scott, Clare. "TU Delft Students Develop New Technique for 3D Printing Soft Robotics." *3DPrint.com,* 9 Nov. 2016, PERLINK "https://3dprint.com/154931/tu-delft-3d-printed-soft-robotics/" 3dprint.com/154931/tu-delft-3d-printed-soft-robotics/. Accessed 12 Jan. 2017.

Shen, Helen. "Meet the Soft, Cuddly Robots of the Future." *Nature,* 3 Feb. 2016, www.nature.com/news/meet-the-soft-cuddly-robots-of-the-future-1.19285. Accessed 12 Jan. 2017.

"Soft Robotics." *Harvard Biodesign Lab,* biodesign.seas.harvard.edu/soft-robotics. Accessed 12 Jan. 2017.

"Soft Robotics." *Jaeger Lab,* jfi.uchicago.edu/~jaeger/group/Soft_Robotics/Soft_Robotics.html. Accessed 12 Jan. 2017.

"Unimate: The First Industrial Robot." *Robotic Industries Association,* www.robotics.org/joseph-engelberger/unimate.cfm. Accessed 12 Jan. 2017.

Solar Cell

SUMMARY

A *solar cell,* or photovoltaic cell, is a device capable of converting energy from the Sun into electricity. Humans have relied on the Sun's rays for light and warmth since the beginning of time, but the first solar cell capable of harnessing the Sun's energy to produce electricity was not introduced until 1954. Since then, scientists and researchers have continued to create better, more efficient solar cells to power

A conventional crystalline silicon solar cell (as of 2005). Electrical contacts made from busbars (the larger silver-colored strips) and fingers (the smaller ones) are printed on the silicon wafer. © EBSCO

everything from calculators and watches to spacecraft and power plants. In 2016, scientists developed a solar cell constructed from an inexpensive, easy-to-use material that demonstrated efficiency levels rivaling some of the best solar cells on the market.

BACKGROUND

French scientist Edmond Becquerel discovered the photovoltaic effect in 1839. During his experiments, Becquerel noticed that certain materials produced an electrical voltage when exposed to light. Other scientists, such as Willoughby Smith, Charles Fritts, and Albert Einstein, built on his ideas over time. The biggest breakthrough, however, occurred in 1954 when three scientists at Bell Labs—Daryl Chapin, Calvin Fuller, and Gerald Pearson—developed the first silicon photovoltaic cell.

Silicon, by itself, is not a great conductor of electricity. A silicon atom has fourteen electrons in three electron shells. An electron shell is the path, or orbit, that electrons follow around an atom's nucleus. In a silicon atom, the first shell, closest to the nucleus, contains two electrons and is completely full. The second shell contains eight electrons and is completely full. The third, and outermost, shell contains four electrons and is only half full. To fill its outer shell, a silicon atom will share electrons with another nearby silicon atom, thereby forming a bond. As a result, pure silicon has no free electrons to move about, making it a poor conductor of electricity. Adding energy to silicon from an outside source, such as heat, can cause a few electrons to break free and move around. Called free carriers, these electrons create an electrical current, but it is too weak to generate a significant amount of electricity.

Fuller, Pearson, and Chapin discovered that adding certain impurities—that is, atoms of other materials—to silicon can cause it to become positively charged (P-type) or negatively charged (N-type). P-type silicon is missing electrons, while N-type silicon has extra electrons. A solar cell contains both types. The point where the two types of silicon meet is called the P-N junction. Adding energy—in the form of light from the Sun—knocks loose spare electrons from the N-type silicon. These free carriers try to move to the P-type silicon to fill the gaps where electrons are missing. The P-type silicon then becomes negatively charged, and the N-type silicon becomes positively charged. As a result, an electrical field forms. This field controls the movement of electrons within the solar cell, which creates an electric current capable of powering a device.

PRACTICAL APPLICATIONS

Not long after the Bell scientists' breakthrough in 1954, photovoltaic cells began to be used to power satellites in Earth's orbit. In 1958, for example, the Vanguard I satellite was launched into space carrying a small array of solar cells to power its radios. Subsequent satellites launched that year also had systems powered by photovoltaic technology. Indeed, silicon solar cells have played a significant role in space applications since their inception; they are like little power plants, turning sunlight into electricity hundreds of miles above Earth. Even today, space agencies around the world depend on solar cells to power satellites, spacecraft, and the International Space Station (ISS).

On Earth, solar cells may be used to power everything from calculators and wristwatches to entire photovoltaic power plants. The Agua Caliente Solar Project outside Yuma, Arizona, is one of the largest photovoltaic power plants in the world. The plant covers 2,400 acres and comprises 5.2 million ground-mounted solar modules, which convert enough energy from the Sun to produce electricity for about 100,000 homes. The plant uses no water to generate electricity, makes no noise, and releases no emissions

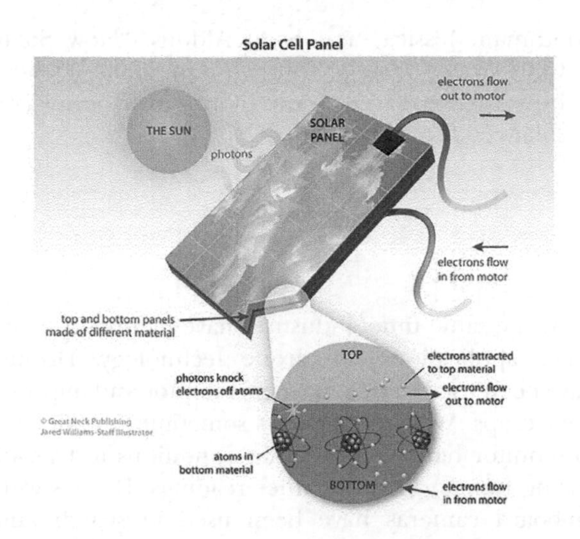

Solar Cell Panel: The solar cell panel collects energy from the sun and can provide electric energy.

into the air. Estimates suggest that the amount of carbon dioxide displaced by the Agua Caliente Solar Project is equivalent to taking 40,000 cars off the road each year.

Rising greenhouse gas emissions and dwindling fossil fuel resources have led to increased interest in more sustainable energy resources, including solar. Solar energy's benefits are numerous. Since solar energy comes from the Sun, it is available all over the world. Unlike fossil fuel supplies, the supply of solar energy is inexhaustible. While power plants that burn fossil fuels to produce electricity release greenhouse gases into the air, solar cells do not cause pollution as they harness energy from the Sun. Additionally, solar cells require very little maintenance.

Despite its many advantages, solar energy does have some drawbacks. For example, a power plant like Agua Caliente occupies a massive amount of space. The processes involved in the manufacture of solar cells contribute to pollution. Sunlight is available only at certain times of day. In addition, although technologies exist to store energy produced by solar systems for future use (such as at night, when sunlight is unavailable), these technologies can be quite expensive. Thus, using solar cells to harness the Sun's energy is more expensive than burning fossil fuels.

Nevertheless, scientists continue to research ways to make solar energy more efficient and less expensive. Manufacturing solar cells from cheaper materials is one way to accomplish this goal. Most solar cells on the market are composed of silicon, and their levels of efficiency vary considerably. Solar cells made from the purest silicon available are quite costly, and they have a maximum efficiency of about 25 percent—that means they are able to convert about 25 percent of the Sun's energy into electricity. In 2009, a team of scientists from Japan introduced a new type of solar cell made from a material called perovskite. Like silicon solar cells, perovskite solar cells are able to turn energy from sunlight into electricity. However, perovskite solar cells are both less expensive and easier to manufacture than traditional silicon solar cells. Moreover, by 2016, scientists had managed to improve the efficiency of perovskite solar cells from 4 percent to about 22 percent—nearly equal to silicon solar cells. They were hopeful that affordable, efficient perovskite solar cells would be ready for market as early as 2017.

—Lindsay Rohland

BIBLIOGRAPHY

"Agua Caliente Solar Project." *First Solar*, www.firstsolar.com/en/About-Us/Projects/Agua-Caliente-Solar-Project. Accessed 10 Nov. 2016.

"Chemists Find Key to Manufacturing More Efficient Solar Cells." *Phys.org*, 23 Sept. 2016, phys.org/news/2016-09-chemists-key-efficient-solar-cells.html. Accessed 10 Nov. 2016.

"The History of Solar." *U.S. Department of Energy*, www1.eere.energy.gov/solar/pdfs/solar_timeline.pdf. Accessed 10 Nov. 2016.

"How Do Solar Cells Work?" *Physics.org*, www.physics.org/article-questions.asp?id=51. Accessed 10 Nov. 2016.

Maehlum, Mathias Aarre. "Solar Energy Pros and Cons." *Energy Informative*, 12 May 2014, energyinformative.org/solar-energy-pros-and-cons/. Accessed 10 Nov. 2016.

"Major Advance in Solar Cells Made from Cheap, Easy-to-Use Perovskite." *Phys.org*, 8 Nov. 2016, phys.org/news/2016-11-major-advance-solar-cells-cheap.html. Accessed 10 Nov. 2016.

Perlin, John. "The Invention of the Solar Cell." *Popular Science*, 22 Apr. 2014, www.popsci.com/article/science/invention-solar-cell. Accessed 10 Nov. 2016.

"Solar Arrays." *NASA*, www.nasa.gov/mission_pages/station/structure/elements/solar_arrays.html#.WDMs1bIrKUk. Accessed 10 Nov. 2016.

Toothman, Jessica, and Scott Aldous. "How Solar Cells Work." *HowStuffWorks*, 1 Apr. 2000, science.howstuffworks.com/environmental/energy/solar-cell.htm. Accessed 10 Nov. 2016.

SPACE DRONE

SUMMARY

A *space drone* is an unmanned spacecraft. It may be piloted from a location on Earth or through global positioning satellite (GPS) technology. Drones are also known as *unmanned serial vehicles* (UAV) or *remotely piloted serial systems* (RPAS). They offer a number of advantages over manned flight, including safety and cost, and can be designed for a number of specific purposes. The U.S. Air Force has launched several flights of a space drone called the X-37B Orbital Test Vehicle and the U.S. National Aeronautics and Space Administration (NASA) conducts ongoing research into several types of drones that can be used for exploratory missions in outer space.

BACKGROUND

Experimentation with unmanned flying devices began almost as soon as manned flight was accomplished in 1903. During World War I, the U.S. Navy tried and failed to develop unmanned biplanes meant to bomb enemy territory (Sifton). When America entered World War II, the U.S. Navy used B-24 bombers rigged with explosives and remote control guidance with limited success: the planes required a pilot to get them to flying altitude and then parachute out before the plane complete its mission with remote guidance. Many crashed or exploded early, killing the pilot.

The military continued to make some use of unmanned craft for various purposes. However, it was not until computerized technology became available in the last two decades of the twentieth century that drones became reliable enough to be a major factor in military operations. Drones were used for surveillance in the Middle East and other areas of interest during the 1990s. After the September 11, 2001, terrorist attacks, armed drones were used to locate and kill suspected terrorists.

At the same time, industries have identified many other applications for drone technology. Drones have been used to help farmers monitor and improve their crops. Weather agencies sometimes use drones to monitor hazardous weather conditions instead of risking human lives to gather readings. Drones with on-board cameras have been used in search and rescue operations to fly over remote areas and to assist with three-dimensional mapping efforts. Some businesses are also considering drones for making local deliveries, and individual hobbyists enjoy using drones to capture aerial views of everything from indoor celebrations to landscapes. Drones are ideal for many of these uses because they are much lower in cost to build and operate than a manned device, and they do not put human lives at risk during operation. This makes them useful for adaptation to space missions as well.

PRACTICAL APPLICATIONS

While organizations such as NASA that conduct missions in outer space have used unmanned vehicles for decades, most were either land vehicles, such as the Mars Rover, or transport vehicles that carried

First stage of droneship Falcon 9 Flight 21 descending over the floating landing platform, January 17, 2016. By SpaceX Photos (First stage of Jason-3 rocket)

The SpaceX Dragon cargo spaceship is grappled by the International Space Station's Canadarm2. By NASA

cargo to an orbiting space station. In the twenty-first century, several agencies have worked on unmanned vehicles that more specifically fit the definition of a drone, or a vehicle that can self-navigate using GPS or be controlled by a pilot who is outside the line of sight of the craft.

In 2010, the U.S. Air Force launched the X-37B Orbital Test Vehicle. Built by the Boeing Corporation, X-37B looks like a smaller version of the space shuttle NASA used for thirty years before retiring it in 2011. Very little is known about the craft or its missions. It is a reusable craft, like the shuttle before it, and when it returned to Earth from its first eight-month mission it became the first U.S. spacecraft to land itself on a runway.

The X-37B is about twenty-seven feet (or nine meters) in length, most of which is taken up by its guidance systems and fuel supply. It is believed to have a cargo area about the size of a pickup truck bed, but very little is known about the craft or its missions. Since its first mission, the X-37B has had several additional flights, including one that was two years in length that ended in 2014. Many observers have developed theories about what the X-37B has been doing while in flight, but the Air Force and Boeing have shared few details.

NASA is also working on space drones to use in exploratory missions to Mars and other possible space destinations. One, a small space boomerang-shaped plane tentatively called Prandtl-m, weighs about two and a half pounds and could fly over the Martian surface to map far more territory than is possible with a land-based device. Researchers at NASA are also working on space drones called *extreme access flyers*, or EAFs. These devices would launch from another larger craft that would land on the planet or asteroid being explored and serve as a base for the EAFs. One prototype EAF space drone is a quadcopter. It resembles a four-legged platform or table of about five feet square (about one and a half meters) with four projecting arms, each with a small fan-like propeller at the end. These space drones would be able to travel to remote areas of space bodies that cannot be reached with land vehicles such as rovers, and would be able to recharge their batteries and other fuel sources from the base lander.

NASA scientists face some challenges getting these drones operational because they will have to be designed to fly in less gravity than Earth's and in thinner air. Despite these challenges, space drones have the potential to provide many benefits for agencies such as NASA. Space drones are much less expensive to deploy than manned craft. They can work without rest and can endure long voyages in space without the physical and mental health concerns that are factors in human explorers. Using drones also eliminates the risk of human life from space exploration while still providing quality information that can help scientists prepare for manned missions.

—*Janine Ungvarsky*

Bibliography

Atherton, Kelsey D. "NASA is Testing a Drone for Mars." *Popular Science,* 6 July 2015, www.popsci.com/. Accessed 10 Dec. 2016.

Crilly, Rob. "Top-Secret U.S. Space Drone Returns to Earth after Two-Year Orbit." *The Telegraph,* 18 Oct. 2014, "http://www.telegraph.co.uk/news/science/space/11171389/Top-secret-U.S.-space-drone-returns-to-Earth-after-two-year-orbit.html" www.telegraph.co.uk/news/science/space/11171389/Top-secret-U.S.-space-drone-returns-to-Earth-after-two-year-orbit.html. Accessed 10 Dec. 2016.

Howell, Elizabeth. "What is a Drone?" *Space.com,* 2 June 2015, www.space.com. Accessed 10 Dec. 2016.

Lewin, Sarah. "Drones in Space! NASA's Wild Idea to Explore Mars." *Space.com,* 5 Aug. 2015, .com/30155-nasa-drones-on-mars-video.html" www.space.com/30155-nasa-drones-on-mars-video.html. Accessed 10 Dec. 2016.

Murdoc, Shelby. "The Air Force's Space Drone Has Been in Orbit for Over 500 Days — And Its Mission Is Classified." *Business Insider,* 31 Oct. 2016, w.businessinsider.com/air-force-space-drone-in-orbit-500-days-2016-10" www.businessinsider.com/air-force-space-drone-in-orbit-500-days-2016-10. Accessed 10 Dec. 2016.

Murphy, Mike. "NASA Is Working on Drones That Can Fly in Space." *Quartz,* 31 July 2015. qz.com/469334/nasa-is-working-on-drones-that-can-fly-in-space/. Accessed 10 Dec. 2016.

"NASA Creating Robotic Drones for Future Space Exploration." *RT,* 5 Aug. 2015, www.rt.com/usa/311603-nasa-robotic-space-drones/. Accessed 10 Dec. 2016.

Sifton, John. "A Brief History of Drones." *The Nation,* 7 Feb. 2012, www.thenation.com/. Accessed 10 Dec. 2016.

Speech Recognition

SUMMARY

Speech recognition refers to the technology behind the ability of some computers, software programs, and electronic devices to recognize, interpret, and respond to human speech. It is also referred to as speech recognition technology (SRT), automatic speech recognition (ARS), and speech processing. The term speech recognition also refers to the branch of computer linguistics that deals with the study and development of speech recognition equipment and programs.

Speech recognition is different from voice recognition, which refers to technology that specifically determines the unique patterns of individual voices. Voice recognition is often used as a means to identify specific individuals for security purposes, but it can also be part of the way a speech recognition program "learns" to improve its responses.

There are many uses for speech recognition, including automated phone systems used by businesses, talk-to-text software that enables people to speak instructions to a computer or to dictate text for conversion to a written format, and verbal control of devices such as television remotes and handheld electronics.

HISTORY

The earliest forms of speech recognition technology could only understand numbers. Bell Laboratories created the "Audrey" system in 1952; it could identify numbers spoken by a single voice. Researchers continued to refine the technology, and by 1962 a system called "Shoebox" was developed that could understand sixteen spoken words.

These initial forays into speech recognition were built around words limited to four vowels and nine consonants, but they were enough to attract interest and funding from the U.S. Department of Defense. By 1976, researchers at Carnegie Mellon University had developed "Harpy," with the ability to understand about 1,100 words. That technological leap was made possible in part by improved search technology that enabled speech recognition programs to find the spoken words in its programming more quickly.

In the 1980s, the capabilities of speech recognition technology continued to grow because of the application of a hidden Markov model to search protocols. Technology applying this statistical method was able to determine whether certain sounds were words, rather than merely matching up sounds to known patterns. This enabled the development of some applications for businesses and dictation programs for

Electrodes used in subvocal speech recognition research. By Dominic Hart, NASA

home use that were able to work with thousands of words. Since the technology dealt with words one at a time, the programs were time-consuming to use; however, it worked well enough for businesses to begin adopting voice responsive automated answering services by the middle of the 1990s.

The ability to integrate speech recognition programs with powerful computer search engines such as Google enabled the technology to take a giant step forward in the early 2000s. Designers were able to use the search engine database, including the millions of searches done by users, instead of a predetermined database. This allowed the speech recognition programming to have access to many more possibilities for predicting and determining spoken words. It also allowed for the programming to incorporate the physical location of a speaker in the speech interpretation process based on information from searches the user had conducted. For example, the speech recognition technology may more readily recognize the name of a specific local restaurant if the user has already searched for the same place using the search engine. At the same time, the increased use of mobile devices created a market for applications and programs using speech recognition.

HOW IT WORKS

Speech recognition technology has two phases. In the first, the speech sounds are processed and turned into numeric values that represent different vocal sounds. In many cases, the technology also identifies and ignores sounds that are not determined to be part of the voice, such as Background noise. These numeric values are then compared to databases to help determine what words have been spoken. These databases include an acoustic model, a lexicon, and a language model.

The acoustic model includes all the sounds used in a language. Some applications of speech recognition include voice recognition and can be "trained" to recognize an individual's voice and the words used most frequently. This allows the technology to comprehend the unique way the person speaks, increasing accuracy of its interpretations and responses.

The lexicon is a list of words and the way they correspond to the numeric values determined in the first step of the process. The lexicon also helps the speech recognition program determine how to pronounce the words.

The language model helps the speech recognition program with correctly combining words to form grammatically correct sentences or phrases. Different forms of speech recognition technology have varying levels of language support. For instance, a phone system designed to direct a caller to one of a handful of departments in a small car dealership can have a restricted database that focuses on words callers are most likely to use, such as "service" or "sales." Limiting a system to a set linguistic content in this way can help it respond quicker and more accurately to a caller's input.

FUTURE APPLICATIONS

Both the technology behind speech recognition and the demand for it continue to grow in the early part of the twenty-first century. Experts theorize that as more people become comfortable with giving verbal directions to cell phones, tablets, and other mobile devices, they will come to expect to have this ability in other places as well. It has already become part of other small devices such as video games and television remotes, and automobile manufacturers have begun to include technology that allows drivers to verbally make phone calls and send texts, search for directions, change radio stations, and perform other simple tasks. Experts anticipate that consumers will soon be able to buy appliances that respond to voice commands, and that typing could become a thing of the past as voice-to-text speech recognition improves to the point where the speed and accuracy top what can be accomplished via a keyboard.

—Janine Ungvarsky

Bibliography

Blunsom, Phil. "Hidden Markov Models." *Utah State University Computer Science Department.* PDF. Web. 8 Mar. 2016. http://digital.cs.usu.edu.

"How Speech Recognition Works." *Microsoft.* Microsoft. Web. 8 Mar. 2016. https://msdn.microsoft.com.

Juang, B.H. and Lawrence R. Rabiner. "Automatic Speech Recognition – A Brief History of the Technology Development." *University of California, Santa Barbara.* The Regents of the University of California. PDF. Web. 8 Mar. 2016. http://www.ece.ucsb.edu.

Oremus, Will. "I Didn't Type This Article." *Slate.* The Slate Group. 23 Apr. 2014. Web. 8 Mar. 2016. http://www.slate.com.

Pinola, Melanie. "Speech Recognition through the Decades: How We Ended Up with Siri." *PCWorld.* IDG Consumer & SMB. Web. 8 Mar. 2016. http://www.pcworld.com.

"Speaker Recognition." *U.S. National Science and Technology Council Subcommittee on Biometrics and Identity Management.* PDF. Web. 8 Mar. 2016. http://www.biometrics.gov/Documents/SpeakerRec.pdf

Stem-and-Leaf Plots

SUMMARY

A stem-and-leaf plot is a diagram portraying the distribution of a set of data. It is an effective tool to visualize a set of data in exploratory data analysis.

PRACTICAL APPLICATIONS

Historically, the stem-and-leaf plot originated from the ideas of the British statistician Sir Arthur Lyon Bowley (1869–1957). It gained popularity in applications after the work of John Wilder Tukey (1915–2000) on exploratory data analysis. Today, the stem-and-leaf plot, typically introduced with histograms, pie charts, and dot diagrams, is documented in most introductory statistics textbooks.

Consider an example analyzing gasoline prices. Suppose that the prices of the nine gasoline stations in the east side of a city are {$3.81, $3.72, $3.71, $3.62, $3.59, $3.74, $3.53, $3.81, $3.63} and the prices in the north side (on the same day) are {$3.73, $3.78, $3.82, $3.54, $3.89, $3.84, $3.86}. At first glance, it seems there is not much difference between the two data sets. However, we can analyze the price distributions of the two data sets.

Notice that all the prices in the two data sets are in the three dollar range. Thus, to compare the two sets of prices is to compare the final two digits of the prices, and we move our attention to the cent values. For the east side, the data become {81, 72, 71, 62, 59, 74, 53, 81, 63} and the data for the north side become {73, 78, 82, 54, 89, 84, 86}. A stem-and-leaf plot presents this data graphically.

East Side		North Side	
Stem	Leaf	Stem	Leaf
5	3 9	5	4
6	2 3	6	
7	1 2 4	7	2 8
8	1 1	8	2 4 6 9

The two plots show that the eastside gasoline prices are almost evenly distributed, while the prices in the north side are skewed to the upper $3.70 to $3.80 range.

—*John Tuhao Chen*

Bibliography

Tukey, J. W. *Exploratory Data Analysis: Past, Present, and Future.* Technical Report No. 302 (Series 2), Department of Statistics. Princeton: Princeton UP, 1993.

Utt, Jessica M., and Robert F. Heckard. *Mind on Statistics.* Stamford, CT: Cengage, 2015.

Wall, Jennifer J., and Christine C. Benson. "So Many Graphs, So Little Time." *Mathematics Teaching in the Middle School* 15.2 (2009): 82-91.

Structured Query Language (SQL)

SUMMARY

Structured Query Language (SQL; pronounced "S-Q-L" or "sequel") is an interactive programming language that allows users to find and change information in databases. SQL is a language that manages and accesses information from databases, yet it is not comprehensive enough to create programs. First developed in the 1970s, SQL has long been the most popular query language used to access information. SQL has been enhanced over the years to keep up with changing technology. However, new technology has made other query languages more popular for some database management.

HISTORY OF SQL

Before the development of SQL, companies and organizations used database management systems (DBMS) to build and maintain databases. Because of the nature of the systems, computer programmers were generally the people who used the DBMS. If a business needed to access specific types of information from a database (e.g., monthly income figures by company sector), a programmer had to write a program to access that information. Each piece of information the company needed to access would have to be found using unique programming. This process was often time-consuming and expensive.

Because the process was inefficient, programmers developed query languages, which allowed individuals to make requests to databases. Many of these programs were developed during the 1970s. SQL was one such language and was developed in 1974 by IBM. SQL soon became one of the most popular query languages.

Even though SQL made querying databases more efficient, it did not replace the need for computer programming when locating information in databases. Programmers still develop programs to generate reports and compile important data. Often, these programs use SQL. The programmers can use SQL in the programs in one of three ways: through embedded SQL, through SQL modules, and through call-level interface (CLI). The most popular way to use SQL in programs is embedded SQL, meaning that SQL statements are incorporated into another language (e.g., C or COBOL) used by the programmer. SQL modules are made up of groups of procedures, with each procedure containing one single SQL command. CLIs are functions that send messages to the DBMS and receive information from the DBMS.

HOW SQL WORKS

A *query* is a single request made to a database. SQL helps people make these requests to the database. SQL is made up of about sixty commands, all of which

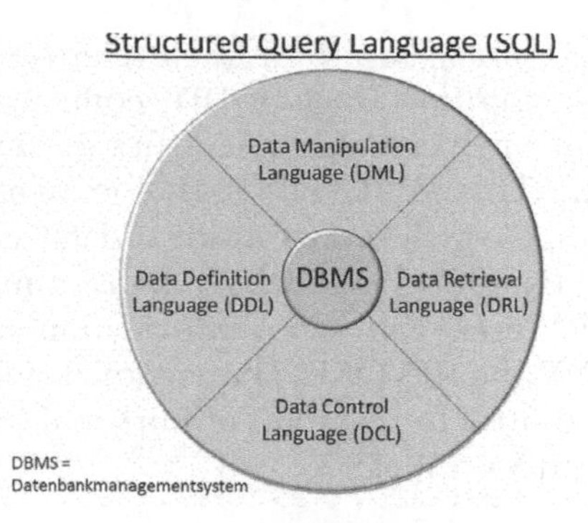

Structured Query Language. By Bagok (Own work)

make different requests to databases. Although SQL has many different commands that help it perform a number of functions, it is not a complete language. That means that it cannot create entire programs. Instead, it has specific functions that it can perform in already existing programs. That is why many SQL requests are embedded in another host language.

The DBMS is the system that actually processes a query sent by SQL. When SQL sends a query, the DBMS first parses the statement. That means that the DBMS breaks down the request into smaller parts and determines exactly what the query is asking. The DBMS usually identifies any misspellings or language errors during the parsing step. Next, the DBMS compares the information in the query with the information in the database to make sure the information being requested exists. This step is called *validation*. After validation, the DBMS creates a plan to access the information requested by the query. Then the DBMS optimizes the plan to ensure it is using the most efficient process to access the information. Although this optimization step can take some time to complete, it is very important. Some queries search for multiple pieces of data in multiple tables, and finding the most efficient way to do the search will save time in the end. Finally, the DBMS executes the plan and gathers the requested information. Although this is the general process that a DBMS uses to execute a SQL query, some systems do the steps in different orders or take different amounts of time to complete each step.

SQL commands are the part of the language that explains what a query is supposed to do or find. Four of the most basic SQL commands are DELETE (to delete existing data), SELECT (to request existing data), UPDATE (to change existing data), and INSERT (to insert new data). Four different types of SQL commands exist: *data definition language (DDL)*, *data manipulation language (DML), data control language (DCL),* and *transactional language.* DDL commands are used to create or change database structures and can include CREATE and ALTER. DML commands are used to change or select data inside a database and can include INSERT and DELETE. DCL commands control which users can access information and include GRANT and REVOKE. Transactional language commands control logical units of work and include COMMIT and SAVEPOINT.

SQL AND NOSQL DATABASES

For a long time, SQL was the standard language for building and working with databases. NoSQL databases, however, have also become popular. NoSQL software allows people to create databases with more flexibility and more options. Because of this, NoSQL is becoming a prominent force in database creation. Part of NoSQL's increase in popularity is due to the emergence of big data; as more companies are relying on big data, more of them are turning to NoSQL systems because of the systems' flexibility. Although NoSQL is becoming more popular, it can have problems working with SQL technology. As a result, some businesses still need to use SQL in their database management.

—*Elizabeth Mohn*

BIBLIOGRAPHY

"Guide to NoSQL Databases: How They Can Help Users Meet Big Data Needs." *TechTarget.* TechTarget. Web. 6 Aug. 2015. http://searchdatamanagement.techtarget.com/essentialguide/Guide-to-NoSQL-databases-How-they-can-help-users-meet-big-data-needs#guideSection1

"Introduction to Structured Query Language (SQL) – Part 1." *University of Delaware.* University of Delaware. Web. 6 Aug. 2015. http://www.udel.edu/evelyn/SQL-Class1/SQLclass1All.html

"Processing a SQL Statement." *Microsoft: Developer Network.* Microsoft. Web. 6 Aug. 2015. https://msdn.microsoft.com/en-us/library/ms713599(v=vs.85).aspx

"Structured Query Language (SQL)." *BusinessDictionary.com.* WebFinance, Inc. Web. 6 Aug. 2015. http://www.businessdictionary.com/definition/structured-query-language-SQL.html

"Structured Query Language (SQL)." *Microsoft: Developer Network.* Microsoft. Web. 6 Aug. 2015. https://msdn.microsoft.com/en-us/library/ms714670(v=VS.85).aspx

"SQL Modules." *Microsoft: Developer Network.* Microsoft. Web. 6 Aug. 2015. https://msdn.microsoft.com/en-us/library/ms709311(v=VS.85).aspx

STUXNET

SUMMARY

Stuxnet is a worm—a type of computer virus—that was first discovered in June 2010. Stuxnet quickly became one of the most publicized computer viruses in history because it was used to monitor and control computers that were involved in Iran's nuclear program. Though no one has ever claimed responsibility for creating and installing Stuxnet, many computer experts believe that a nation or multiple nations developed the program. Some analysts believe that the United States and Israel created the worm and installed it on the Iranian computers to slow the country's nuclear capability.

HOW STUXNET WORKS

Stuxnet is a worm that can monitor and control the computers on which it is installed. It generally targets industrial complexes such as plants, dams, banks, and processing centers. The program was designed specifically to attack industrial control systems built by Siemens AG, a German engineering firm. Stuxnet attacks software programs that run Supervisory Control and Data Acquisition (SCADA) systems. These systems are used in many different industrial applications, including water treatment centers and power generation facilities.

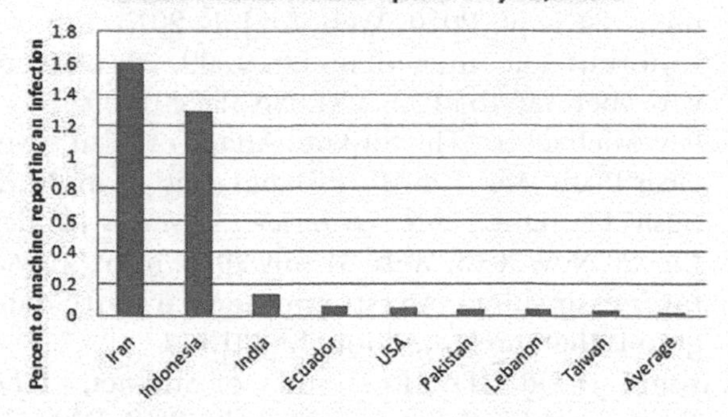

Stuxnet saturation. By Llorenzi (Own work)

The worm is very complex and uses advanced techniques to avoid being detected by malware detection software (which helps identify, isolate, and delete viruses and other malware) and by humans. The program is unique because it locates and takes advantage of four different vulnerabilities in the software it is hacking. Previously, the cybersecurity community was unaware of all four of these vulnerabilities, demonstrating that the worm is extremely advanced and required many resources to produce.

Although Stuxnet is one of the most famous computer worms in history, it has not yet been used to attack personal computers. Instead, it has been used to target larger industrial complexes. Nevertheless, computer security programs still attempt to protect personal computers against the virus.

IDENTIFICATION OF STUXNET

The Stuxnet virus is most famous for being used on computers in Iran, specifically computers that helped with uranium enrichment in that country. The Stuxnet virus was originally placed on a USB drive. Experts believe that the virus was delivered to the Iranian Natanz nuclear plant through a worker using a USB drive. The worm infected computers and machines at Natanz and then traveled to other sites. Worms can move to different computers through networks.

The Stuxnet virus eventually installed itself on computers that controlled the plant's centrifuges. It then sped up the centrifuges and made enriching uranium more difficult. The increased spinning of the centrifuges was minimal so that the workers at the plant were not alarmed by the changes. Nevertheless, the increase made the centrifuges wear down much more quickly than they would have on their own. The worm also learned as much as it could about the Natanz nuclear plant and the Iranian nuclear program in general.

In 2010, the International Atomic Energy Agency (IAEA) was touring a uranium enrichment center in Iran and was confused by the high rates of failure it noticed with the center's centrifuges. A few months later, a computer security company in Belarus located malicious files on the Iranian computers. These malicious files were part of the Stuxnet virus, and the virus had been causing the unprecedented failure rates with the centrifuges as well as other problems such as computers shutting down. The Stuxnet virus had been used as a weapon to slow the uranium enrichment process in Iran. By the time the worm had been detected on Iranian computers, Stuxnet had already caused about one-fifth of Iran's centrifuges to break down.

EFFECTS OF STUXNET

No person, group, or country admitted to creating and using the Stuxnet worm; however, many experts believe that a program as complex as Stuxnet had to be created with the help of at least one national government. Many experts also believe that the United States and Israel worked together to form and release Stuxnet in Iran. Some military experts believe that the use of Stuxnet helped change modern warfare. Stuxnet was the first computer virus used as a weapon, and many experts believe that it opened the door for cyber warfare to become a large part of international conflicts.

Although the Stuxnet worm was originally released to infect Iranian computers and slow down Iranian uranium enrichment, the worm is now a threat to all different companies, organizations, and infrastructures. Stuxnet has the potential to cripple power grids and stop water treatment centers. The Stuxnet worm or a similar program could greatly affect developed countries, including the United States. In fact, the American company Chevron admitted that its computer system was infected with the worm. Other organizations' computers may also have the worm, but it has not been detected yet. The Stuxnet worm has been located on many thousands of computers in Iran, Indonesia, Pakistan, Australia, the United Kingdom, the United States, and other

Siemens Simatic S7-300. By Ulli1105 (Own work)

countries. Although Stuxnet is potentially a powerful weapon, if an organization detects the malware, it can usually remove it and continue its operations as it normally would.

One result of the Stuxnet attack on cybersecurity in general is that it has made cybersecurity experts question the use of outside contractors. In general, large plants and organizations hire contractors to help with specific issues. Many times, however, these contractors are not cybersecurity experts. Because of their lack of security knowledge, these contractors can be targeted and used to gain access to important systems and sensitive information.

—*Elizabeth Mohn*

BIBLIOGRAPHY

"Factbox: What Is Stuxnet?" *Reuters.* Thomson Reuters. 24 Sept. 2010. Web. 31 July 2015. http://www.reuters.com/article/2010/09/24/us-security-cyber-iran-fb-idU.S.TRE68N3PT20100924

Kelley, Michael B. "The Stuxnet Attack on Iran's Nuclear Plant Was 'Far More Dangerous' than Previously Thought." *Business Insider.* Business Insider Inc. 20 Nov. 2013. Web. 31 July 2015. http://www.businessinsider.com/stuxnet-was-far-more-dangerous-than-previous-thought-2013-11

Kushner, David. "The Real Story of Stuxnet." *IEEE Spectrum.* IEEE Spectrum. 26 Feb. 2013. Web. 31 July 2015. http://spectrum.ieee.org/telecom/security/the-real-story-of-stuxnet

Schneier, Bruce. "The Story Behind the Stuxnet Virus." *Forbes.* Forbes.com LLC. 7 Oct. 2010. Web. 30 July 2015. http://www.forbes.com/2010/10/06/iran-nuclear-computer-technology-security-stuxnet-worm.html

"The Stuxnet Worm." *Norton by Symantec.* Symantec Corporation. Web. 31 July 2015. http://us.norton.com/stuxnet

Zetter, Kim. "An Unprecedented Look at Stuxnet, the World's First Digital Weapon." *Wired.* Condé Nast. 3 Nov. 2014. Web. 31 July 2015. http://www.wired.com/2014/11/countdown-to-zero-day-stuxnet/

SUPERCOMPUTER

SUMMARY

Supercomputers are large mainframe computers that are extremely fast and powerful. The high-performing machines are used primarily for scientific and engineering applications. They are used for code breaking, extremely complex mathematical calculations, weather prediction, and a variety of other purposes.

BACKGROUND

Both supercomputers and traditional computers are composed of a select number of key parts. All computers contain some form of central processing unit, or CPU. The CPU acts as the brain of the computer. It

The Cray Y 190A Supercomputer at the NASA Ames Research Center, Mountain View, California. © NASA

Cray 2 Supercomputer for the Numerical Aerodynamic Simulator at the NASA Ames Research Center, Mountain View California. © NASA

performs all of the machine's calculations. The hard disk drive, commonly called a hard drive, serves as long-term storage. Random access memory (RAM) serves as short-term storage. Other specialized parts allow the machine to display a graphical interface, or to process sound.

All of these parts are attached to a motherboard. The bus, a specific part of the motherboard made for communication, connects the RAM and drives to the motherboard, allowing them to communicate with the processor. The motherboard also allows input devices, such as keyboards, mice, and scanners, as well as output devices, such as printers, to communicate with the processor.

Most computers utilize some form of operating system. The operating system is the most important software, or programming, that runs on a computer. It interfaces directly with the computer hardware and allows the user to give the computer commands. Without an operating system, a computer user would have to know coding languages to utilize a computer. There are many different operating systems available to consumers.

EVOLUTION OF THE SUPERCOMPUTER

The first supercomputer was called Colossus. It was built by the British during World War II to function as a code breaking machine. Composed of massive numbers of vacuum tubes and digital switches, Colossus was the first electronic computer. The massive machine could read more than five thousand characters a second. This was a revolutionary amount of computing power for the time. Colossus was used to break the incredibly complex Lorenz SZ42 cipher, a secret code utilized by Adolf Hitler's commanding officers.

In the 1950s, companies began producing specialized computer models for general sale. The world realized that computers could be applicable for a wide variety of tasks, and computers quickly became smaller and cheaper. While the average consumer was still uninterested in purchasing a computer, many businesses wanted computer technology. However, the mass-produced computers built for these corporations were several orders of magnitude less powerful than the computers built for government agencies. The divide between consumer-oriented computer technologies and government-owned supercomputers began to grow. That divide only increased over the next decade.

In the early 1960s, the United States government requisitioned and funded the development of the IBM 7030 and the Rand UNIVAC LARC. These were two of the first machines intentionally built as supercomputers. They were drastically more powerful than any machine available to consumers and were intended for use in national defense calculations.

The IBM 7030 and the Rand UNIVAC LARC set a precedent in the computer development world that continued into the modern era. Governments and corporations with massive amounts of resources funded the creation of extremely powerful supercomputers that utilized new technology. As supercomputers are not mass-produced, none of this technology was directly available to consumers. However, inevitably, the parts used in creating supercomputers eventually became smaller in size and cheaper to manufacture. At this point, they were used in consumer-friendly machines, driving the consumer electronics industry forward.

In the late 1960s, the supercomputer market was primarily limited to one company, the Control Data Corporation (CDC). The U.S. government funded this company's growth, paying for the development of multiple supercomputers throughout its existence. Many of the company's engineers left to form a new company, Cray Research. Together, the two companies dominated the supercomputer market until the late 1970s. These companies spurred the development of the mass-produced central processing unit. Before this development, most computer processors were designed specifically for each computer model.

A mass-produced CPU allowed more computers to be released in shorter periods, and allowed more adaptable, varied models to be built.

In the early 1980s, the supercomputer industry saw a shift in construction. Instead of being built around a single, extremely powerful processor, several companies began to create supercomputers that contained multiple independent computers linked together by high-speed cables. These multi-CPU machines could be scaled to meet specific tasks. As processors and other computer parts became faster, supercomputers became drastically more powerful. Soon, each new supercomputer model was many times more powerful than its predecessors were. By the mid-1990s, multi-CPU supercomputers had completely erased single-CPU supercomputers. Over the next decades, supercomputer networks grew larger, eventually including dozens of powerful, high-speed servers.

While supercomputers are not common, modern supercomputers are no longer the exclusive property of governments. One of the most well-known modern supercomputers, named Watson, was built by the technology giant IBM. Named after IBM's founder—Thomas J. Watson—the supercomputer combines complex artificial intelligence–based programming with a massive and powerful network of more than 90 servers. Watson is capable of calculating more than 80 trillion operations per second, boasts more than 15 terabytes of RAM, and contains more than 2,800 processor cores. Watson is capable of a type of primitive learning. Additionally, when asked a question, Watson can search its massive databases and formulate an answer on its own.

Watson is capable of independently analyzing massive amounts of data. In 2011, Watson defeated two champion contestants on the game show *Jeopardy!* It was able to parse questions, search out relevant data, and provide correct answers faster than either human contestant. Watson-like machines have been commissioned for use by health care professionals and law firms.

—*Tyler Biscontini*

Bibliography

Anthony, Sebastian. "A History of Supercomputers." *Extreme Tech*, 10 Apr. 2012, www.extremetech.com/extreme/125271-the-history-of-supercomputers. Accessed 5 Dec. 2016.

Bell, Gordon. "Supercomputers: The Amazing Race (A History of Supercomputing, 1960–2020)." *Microsoft*, Nov. 2014, www.research.microsoft.com/en-us/um/people/gbell/MSR-TR-2015-2_Supercomputers-The_Amazing_Race_Bell.pdf. Accessed 5 Dec. 2016.

"Colussus: Birth of the Digital Computer." *Crypto Museum*, www.cryptomuseum.com/crypto/colossus/index.htm. Accessed 5 Dec. 2016.

"A Computer Called Watson." *IBM 100*, www-03.ibm.com/ibm/history/ibm100/us/en/icons/watson/. Accessed 5 Dec. 2016.

Graham, Susan L., et al., editors. *Getting Up to Speed: The Future of Supercomputing.* National Academies Press, 2005.

"How Computers Work: The CPU and Memory." *University of Rhode Island*, homepage.cs.uri.edu/faculty/wolfe/book/Readings/Reading04.htm. Accessed 5 Dec. 2016.

Kershner, Kate. "What Are Supercomputers Currently Used For?" *HowStuffWorks*, computer.howstuffworks.com/supercomputers-used-for.htm. Accessed 5 Dec. 2016.

"New Chinese Supercomputer Named World's Fastest System on Latest TOP500 List." *Top 500*, 20 June 2016, www.top500.org/news/new-chinese-supercomputer-named-worlds-fastest-system-on-latest-top500-list. Accessed 5 Dec. 2016.

T

Turing Test

SUMMARY

The *Turing test* is a test of computers and artificial intelligence (AI) that is supposed to show when computers are capable of thinking. The test gets its name from mathematician and computer pioneer Alan Turing, who suggested the basis for the test in an academic paper in the 1950s. Since the 1990s, people have used the Turing test to assess computer programs and determine how advanced AI has become. Technology has changed rapidly since Alan Turing first described his test, but a computer did not beat the Turing test until 2014. Although the Turing test is held each year, it has limitations in assessing the advancements computers have made over time.

ALAN TURING AND TURING TEST

Alan Turing first described his hypothetical test in a paper called "Computing Machinery and Intelligence," which he published in October 1950. Turning was employed at Manchester University, where he was working on a then-state-of-the-art computer. He was in charge of creating programs for the computer, but he was also very interested in ideas about what he called "intelligent machines." Turing believed that computers would change dramatically over time and that eventually humans would create machines that could think. His paper laid out some of his views about intelligent machines, and it proposed what would become the Turing test.

Turing suggested that computers could eventually be programed to have an intelligence similar to that of humans. The "imitation game," as Turing called it, would be a test in which a judge communicates with one human and one computer. The judge then has to determine which was the machine imitating a human and which was the human. Turing believed that eventually a computer would be able to fool many judges. To Turing, this test would prove the computer's ability to think in ways that are similar to the ways humans think.

HOW THE TURING TEST WORKS

What became known as the Turing test had been discussed among technology experts for decades after Turing's paper. In 1990, New York businessman Hugh Loebner started an annual competition that was set up like a Turing test to help determine the speed of AI sophistication. He offered a prize of $100,000 to any person who created a computer program that could pass the test. The Loebner Prize has been offered regularly since the 1990s, but it took more than twenty years for a computer to win. Turing himself believed that computers would be "intelligent" enough to beat the test by the year 2000; however, a computer did not earn the Loebner Prize until more than a decade later.

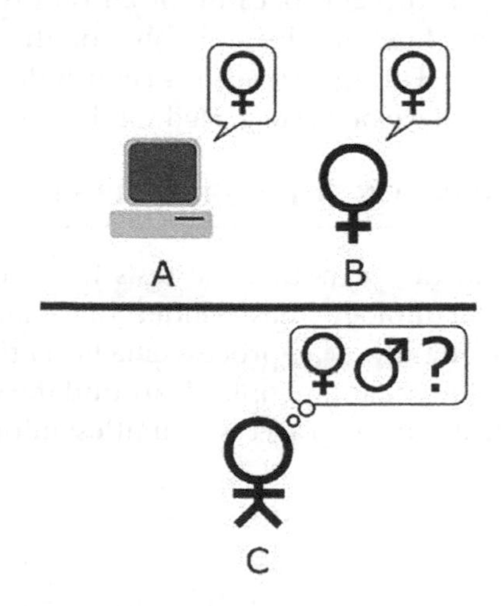

The Original Imitation Game Test, in which player A is replaced with a computer. The computer is tasked with the role of pretending to be a woman, while player B must attempt to assist the interrogator, who must determine which is male and which is female. By Bilby (Own work)

Each year for the Loebner Prize, the group organizing the test hires judges and human controls. The judges have different talents and may include, for example, language experts or psychology experts. The judges are separated from the computers and the human controls, sometimes called "confederates," throughout the test. The judges chat with different human controls and a computer for five minutes each. During that time, the human controls attempt to prove to the judges that they are actually human. At the same time, the computer programs entered into the test try to fool the judges into thinking that they are human. To win the test, the computer has to fool the judges into thinking it is human more than 30 percent of the time.

In June 2014, a computer finally won the Loebner Prize. The computer program that won was called Eugene Goostman, and it simulated a thirteen-year-old boy who was not a native English speaker. Eugene, like other computer programs entered in the contest, is a chatbot. Chatbots can carry on conversations with human respondents. Although Eugene tricked 33 percent of the judges into thinking it was actually a human, some AI experts thought that the Eugene program had used some trickery to win the test. Since this chatbot was supposed to be young and a nonnative speaker, the judges may have attributed any strange responses or odd grammar to these facts. In the past, some chatbots that failed the Turing test seemed human-like during a majority of the test, but they failed in the end because of an odd response. Though the Loebner Prize is one of the famous models of the Turing test, other—slightly different—formats also have been suggested for Turing tests.

LIMITATIONS OF THE TURING TEST

Throughout the years that the tests have been performed, technology has undergone incredible changes. Computers can process data faster than ever before, and they allow people all around the world to communicate and gain access to limitless information in a matter of seconds. Nevertheless, most modern computers would fail the Turing test because the elements of technology that have advanced the most have not been the aspects that make computers more like humans. Some AI experts worry about the value placed on the Turing test. They believe that the real advances being made in AI are being overlooked by the test. These critics argue that the test focuses on factors that might not be as important in AI as other factors. The chatbots that are entered in the annual Loebner Prize are not as useful to humans as smartphones and robots that can help assemble products. Advancements in these types of technology are, at least for now, probably more important to humans than advances in chatbots.

—Elizabeth Mohn

BIBLIOGRAPHY

"Computer AI Passes Turing Test in 'World First.'" *BBC.* BBC. 9 June 2014. Web. 30 July 2015. http://www.bbc.com/news/technology-27762088

Hern, Alex. "What Is the Turing Test? And Are We All Doomed Now?" *The Guardian.* Guardian News and Media Limited. 9 June 2014. Web. 30 July 2015. http://www.theguardian.com/technology/2014/jun/09/what-is-the-alan-turing-test

Hodges, Andrew. "Alan Turing and the Turing Test." *Alan Turing: The Enigma.* Andrew Hodges. Web. 30 July 2015. http://www.turing.org.uk/publications/testbook.html

Reingold, Eyal. "The Turing Test." *University of Toronto.* University of Toronto. Web. 30 July 2015. http://www.psych.utoronto.ca/users/reingold/courses/ai/turing.html

Sharkey, Noel. "Alan Turing: The Experiment That Shaped Artificial Intelligence." *BBC.* BBC. 21 June 2012. Web. 30 July 2015. http://www.bbc.com/news/technology-18475646

"Turing Test." *Webopedia.* QuinStreet Inc. Web. 30 July 2015. http://www.webopedia.com/TERM/T/Turing_test.html

U

UNIX

SUMMARY

UNIX is a large family of computer operating systems stemming from the original UNIX source code developed at Bell Labs in the early 1970s. After that source code was initially released for a small fee, numerous developers began creating their own versions of UNIX for both personal use and commercial sale. Soon, universities, computer companies, research institutions, and even government bodies began using UNIX operating systems and further improving the software's functionality and technological capabilities. Since that time—and with decades of continuous development—UNIX and UNIX-like operating systems have been used to run everything from mainframes and supercomputers to personal computers, tablets, and smartphones. Some of the most widely used UNIX and UNIX-like operating systems include Linux, Solaris, and Mac OS X. Today, the UNIX trademark is held by the Open Group, an industry standards consortium with more than five hundred member organizations, including Apple, Hewlett-Packard, and NASA.

THE GENESIS OF UNIX

The road to UNIX began in the late 1960s when the American Telephone & Telegraph Co. (AT&T) started a collaborative project with the Massachusetts Institute of Technology (MIT) that was aimed at creating an interactive time-sharing system called the "Multiplexed Information and Computing Service," or "Multics" for short. The proposed purpose of Multics was to allow multiple users to connect with a computer from remote terminals as a means of accessing e-mail, working with text documents, and more. AT&T funded the Multics program for five years before realizing that the project was unrealistically ambitious and withdrawing from the development effort. After AT&T's departure from the Multics program, administrators at the company's Bell Labs development facility suspended all work on operating systems. This was an especially disappointing decision for Bell Labs researchers Ken Thompson and Dennis Ritchie, who, though resigned to Multics' infeasibility, recognized the potential value of operating systems to the future of computing. Undeterred by their superiors' reluctance to explore this realm any further, Thompson, Ritchie, and other researchers continued their efforts to develop an operating system in secret.

DEVELOPMENT

After the cancellation of the Multics program, Thompson started writing a computer game called *Space Travel* in his spare time. Although he initially wrote the program for the GE-645 computer, Thompson found that it ran poorly and was too expensive for the CPU time. Eventually, he rewrote *Space Travel* to run on a PDP-7 microcomputer. This exercise subsequently led Thompson to begin creating an operating system for the PDP-7 that he largely

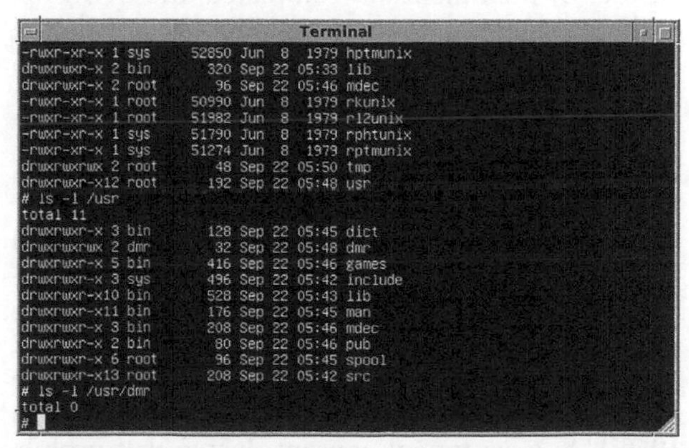

Version 7 Unix, the Research Unix, ancestor of all modern Unix systems (Wwikipedia). By Huihermit (Own work)

completed over the summer of 1969. Thompson's colleagues jokingly dubbed the operating system "Un-multiplexed Information and Computing Service," or "Unics" as a way of poking fun at the fact that it was essentially a weakened version of Multics. Sometime later, this pun was reworked into UNIX, which became the operating system's official name.

Eventually, Thompson and Ritchie realized that the already obsolete PDP-7 was no longer sufficient for the development of UNIX. Also aware that Bell Labs' management would still be reluctant to commit any funds to an operating system project, the pair told administrators that they needed a new machine to help them create text editing and formatting tools, which, by implication, would require an operating system. In the end, management acquiesced and ordered the team a PDP-11 in May 1970. When the new machine was up and running, Thompson and his colleagues developed their text formatter and used it to convince Bell Labs to supply them with an even newer and more capable PDP-11, with which they could continue to work covertly on UNIX. Although Thompson's operating system was in a constant state of evolution, the first complete version of UNIX debuted in November 1971.

While it did not include a *graphical user interface* (GUI), the inaugural version of UNIX represented a major breakthrough in computing. Among its most influential innovations was its hierarchical file system, which allowed users to place individual files in separate directories, or folders, that could then be stored within other directories. The first edition of UNIX also included the line-oriented text editor *ed*, as well as a number of basic games like chess and tic-tac-toe. By the time it was completed, the UNIX operating system was advanced enough to include the time-sharing capabilities originally intended to be a part of the aborted Multics project.

UNIX GOES PUBLIC

UNIX did not leave the confines of Bell Labs until 1974, when a paper on its design and implementation was published in *Communications of the ACM*. This paper generated a great deal of interest in UNIX and resulted in a deluge of requests for copies of the software. However, because AT&T was bound by a 1956 U.S. government consent decree that prohibited the company from selling any products outside

HP9000 workstation running HP-UX, a certified Unix operating system. By Thomas Schanz (Own work)

of telephones and telecommunications technologies, it could only sell UNIX by releasing the source code under license to interested buyers for a minimal fee. As a consequence of this unique sales model, AT&T was unable to provide any form of support for UNIX. This forced early users to work together to fix bugs, write new program code, and make other general improvements. The primary clearinghouse for this work was the Usenix user group, where UNIX users could exchange ideas, software, and fixes with one another.

Within just a few years of its public debut, UNIX was in use at universities, research facilities, computer companies, and other institutions across the country and around the world. Because users had the freedom to alter the original source code as they chose, many versions and alternate iterations of UNIX quickly appeared as well. From the late 1970s to the early 1990s, numerous vendors created and marketed their own versions of UNIX and UNIX-like operating systems, including Linux and, later, Mac OS X. In 1993, Novell, Inc. purchased the rights to UNIX from AT&T. Later, Novell resold the rights to an industry standards consortium that eventually called itself the Open Group.

—*Jack Lasky*

BIBLIOGRAPHY

"About Us." *The Open Group*. The Open Group. Web. 18 Feb. 2016. http://www.opengroup.org/aboutus

"History and Timeline." *The Open Group*. The Open Group. Web. 16 Feb. 2016. http://www.unix.org/what_is_unix/history_timeline.html

Love, Paul, Joe Merlino, Jeremy C. Reed, Craig Zimmerman, and Paul Weinstein. *Beginning Unix*. Indianapolis, IN: Wiley Publishing, Inc., 2005. Print.

"The Open Group Membership Page." *The Open Group*. The Open Group. Web. 18 Feb. 2016. http://reports.opengroup.org/membership_report_all.pdf

Toomey, Warren. "The Strange Birth and Long Life of Unix." *IEEE Spectrum*. IEEE Spectrum. 28 Nov. 2011. Web. 16 Feb. 2016. http://spectrum.ieee.org/computing/software/the-strange-birth-and-long-life-of-unix/

"What Is Unix?" *Indiana University Knowledge Base*. The Trustees of Indiana University. 14 Dec. 2015. Web. 16 Feb. 2016. https://kb.iu.edu/d/agat

V

Video Game Design and Programming

SUMMARY

Video game design and programming combines artistic and computer programming skills to create products for a variety of gaming platforms, including smartphones, personal computers (PCs), and consoles such as the Xbox and PlayStation. Whereas the products of video game design and programming were once primarily electronic amusements for children and adolescents, a number of factors have resulted in video games being used for other purposes, such as medical or military training, fitness, and physical rehabilitation. The popularity of video games has also resulted in the growth of game design and programming curricula in education.

DEFINITION AND BASIC PRINCIPLES

Video game design and programming is the field of computer science involved in the creation of video games to be played across a variety of electronic and digital platforms. This field combines artistic skills such as computer graphics, writing, and traditional drawing with computer programming skills to produce games ranging from casual games, a term denoting simple games designed for quick diversions, to elaborate games featuring immersive worlds, multiple levels, complex story lines and character development, and a fair amount of skill development for play.

Since their early history, video games have grown from being simple, entertaining diversions that could be experienced on occasion (such as arcade machines found in restaurants, bars, and arcades in the 1970s) to full-scale entertainment systems that are playable from almost any type of consumer electronics device with audio-visual capabilities. The mass adoption of video games in contemporary culture has resulted in the creation of a highly influential industry that rivals the sound-recording and motion-picture industries and may even surpass those industries in terms of profitability and user bases.

Since the mid-1990s, the widespread popularity of video games has given rise to game design and programming curricula in higher education. There are also game design programs in some high schools, but these are not as common.

BACKGROUND AND HISTORY

The first video games were created by computer scientists and programmers after World War II. However, it was not until the 1960s and 1970s that the combination of graphics and audio capabilities produced what the modern user would recognize as a standard video game. For many years, the earliest games, such as *Spacewar!*, *Pong*, and *Tank*, provided a generic model for many copycat games. However, by the mid-1980s, many genres of video games had been created, including first-person shooters, adventure games, role-playing games (RPGs), and even the earliest versions of online games, most of which were adaptations of popular board or pencil-and-paper games.

The massive growth of this industry since the 1980s is in part because of the continuous improvements made to the hardware and software with which games are created and on which they are played. Personal computers and communications devices (smartphones, for example) follow the rule of Moore's law. According to the law, which is named after Gordon E. Moore, the cofounder of Intel, the power and speed of computer chips doubles about every eighteen to twenty-four months. Thus, a computer that sold for a thousand dollars in the 1990s has far less processing power and speed than a computer selling for the same price in the 2010s. This increase in processing power allows games to be created using far more elaborate programming and design tools than were used in the past. Thus, video games have become far more realistic than before, although the cost has remained roughly the same.

HOW IT WORKS

The development of video games can be divided into several phases, from initial concept to design and programming to final product. Each of these phases has a number of processes.

CONCEPT. The initial idea for a video game is known as the concept. The design document, whether a single sheet of paper or a complicated, multipart document, will spell out all the important information that the entire design and programming team will follow for the duration of the game's development. The simplicity of the initial concept will not necessarily have any impact on how complex the final product will be, but it can guide certain decisions along the way. These include the size of the design team, the tools and resources to be used, the platform for which the game is being designed, and who the target audience will be. Early phases of game development involve decisions about the intended plot and style of the game, the language(s) of the game, the design of the game world and environments, and the characters that will populate the game world. Much of this is guided by designers and programmers as they oversee game development during each of its phases.

DESIGN AND PROGRAMMING. Once a concept has been envisioned, the next phases involve the production of the game in the desired video game format and platform, such as a video game console or PC. This involves many different steps with many levels of expertise: programming of game elements for the chosen platform, selection of the game engine or software used to program the game and the modeling and other tools to create characters and environments, and creation of the earliest models of the game so it may be tested. The roles of each development team member at this point are fairly fixed, and they work together to make sure that each element of the game matches the other elements to ensure proper game play (all experiences during a player's interaction with game systems). Major considerations during this phase include determining which game engine or software and audio and graphics tools to use. The programming languages commonly used in game design include C++, Java, and Python. Some companies make development kits available to budding game designers. At some point in this process, a version of the game is tested to ensure that software defects, also known as bugs, are eliminated and that the game works as intended.

CRUNCH TIME AND POSTPRODUCTION. Once a video game is near its final, or completion, phase, it is adjusted by designers and programmers and undergoes further testing and quality assurance. Although the game is continuously tested and adjusted by programmers and designers throughout the development phases, it is up to professional game testers to see if the game works from a user's perspective. A marketing team is usually brought in at this level to prepare for the launch and release of the game. As the video game industry is a multibillion-dollar industry, these later phases can be just as important as the early phases of design and development.

APPLICATIONS AND PRODUCTS

The games that existed before the boom in popularity of video games in the late 1970s—the era of video arcades and home systems—were relatively simple programs that had little compatibility beyond the specific gaming systems for which they were designed, such as the large, boxy coin-operated machines found at arcades. Therefore, these games were considered novelty pastimes that were enjoyed in specific places at specific times and by a fairly small percentage of the population. However, the evolution of video games, like that of motion pictures, is based on continuous adjustments to the technology used to create and consume them. The consequence of these improvements over time is a shift in the way that video games are perceived and an expansion of the market for these games to include consumers other than traditional gamers, who were typically male adolescents. Such stereotypes no longer apply, however. According to the *2016 Sales, Demographic, and Usage Data* from the Entertainment Software Association, in 2015, 65 percent of U.S. households owned a device to play video games and 48 percent of U.S. households owned a dedicated game console. Of video game consumers, 59 percent were male and 41 percent were female; 27 percent were under the age of eighteen years, 29 percent were aged eighteen to thirty-five years, and 44 percent were over the age of thirty-five. This shift in demographics has prompted a significant diversification of the video game industry in terms of platforms, content, and design.

Video Games for All Ages. The primary purpose of video game design and programming is the creation of consumer goods for entertainment. Video games, once primarily entertainment targeted at male adolescents and young men, are increasingly played by people of all ages and are being adapted to train and educate people as well as to entertain them. In the 1990s and the early twenty-first century, many new applications were created that not only elevated the gaming experience but also enabled video games to be used for purposes other than pure entertainment. Video games have become valuable tools in diverse areas such as user interfaces, storytelling, training and education, the arts, marketing, medicine, physical rehabilitation, and the military.

Beginning in the late 1980s and early 1990s, with the advent of video game production houses—and video game franchises such as *Doom, Final Fantasy,* and *Super Mario Bros.*, which generated millions of dollars—game designers and programmers have constantly pushed the boundaries of computer science and graphic arts. Innovations in video game design and programming include the use of smartphones and personal electronic devices as important new platforms for gaming and the introduction of new systems designed to appeal to nontraditional gamers, such as the Nintendo Wii. These innovations reflect the creativity of game makers in carving out new markets for themselves.

Innovative Platforms. Video game programmers and designers have increasingly developed games to run on different platforms and use novel controllers, thereby drawing in consumers who historically have not been regularly associated with video gaming. In the case of handheld devices such as the Apple iPhone, Android smartphones, and BlackBerries, consumers recognize the utility of having a device that can be used not only for telecommunications but also for entertainment. Extended battery life, ease of purchasing mobile games, and user-friendly interface designs are also factors in drawing new users.

Similarly, the simple addition of a wireless, motion-sensitive game controller called the Wii Remote to

Awesomenauts screenshot, developed by Ronimo Games and released in 2012. The game allows for split screen multiplayer gameplay.

the Nintendo Wii resulted in a new gaming environment for users and spawned a number of new game designs and concepts. Users could now control the game with physical gestures, prompting the development of games that encourage physical activity. The Wii Fit, introduced in 2008, contains a balance board that allows users to engage in yoga, aerobics, strength training, and balance games. The interactive nature of these fitness games has been noted by health and physical rehabilitation professionals, resulting in a burgeoning research field in which more and more innovative gaming technologies are being applied to serious health issues.

Games as Educational Tools. An important new field for video game designers and programmers is serious games, or those with a primary purpose other than pure entertainment. The idea of a video game environment being used to train pilots and soldiers (two of the more common occupations long associated with gaming) is not new, but the increasing realism, control, and complexity of the tools available for training make video games more and more attractive for a variety of applications. In the medical and health fields, for example, motion-sensitive controllers are useful for training physicians in surgical techniques and aiding in the rehabilitation of physical mobility in elderly patients, as well as those with limited physical movement because of congenital or age-related conditions.

Serious games have been used to teach people about such social issues as the Arab-Israeli conflict and the guerrilla war in Darfur (started in 2003). For example, in the game *Darfur Is Dying* (2006), players could take on the identities of members of a refugee family to learn more about the hazards of everyday life in that region. Other serious games deal with peak oil scenarios, climate change, and a host of other situations that could benefit from more public awareness. In particular, the relationship between game design and military recruitment highlights a possible trend in serious games and occupational fields, in that in addition to scenario simulations, games may increasingly become a part of recruitment and training.

SOCIAL CONTEXT AND FUTURE PROSPECTS

To understand why the video game industry has grown as rapidly as it has, one must look at the far-ranging applications that each new generation of game design and programming technology makes possible. Video games have a wider range of uses than ever before and additional Practical Applications are likely to be developed as gaming increasingly becomes an important part of everyday life for millions of people around the world. Future prospects include immersive environments that expand game playing into levels of interaction and sensory experience that previously could be found only in science fiction. Other possibilities include powerful, yet relatively inexpensive displays and controllers, as well as tools that allow almost anyone to create his or her own video games on a typical home computer.

Video gaming, once considered a pastime for adolescents, is becoming popular among people of all ages. According to the Entertainment Software Association, the average age for a video game player was thirty-five in 2015. In 2015, women and men were buying games with equal frequency, and 52 percent of game players asserted that computer and video games gave them the best value for their money compared to movies and music. Video games are increasingly being used for nontraditional activities, such as health and fitness regimens. These trends point to a broad change in the acceptance level of video gaming across all age groups.

—Craig Belanger, MST, BA

BIBLIOGRAPHY

Adams, Ernest. *Break into the Game Industry: How to Get a Job Making Video Games.* McGraw-Hill/Osborne, 2003.

Adams, Ernest. *Fundamentals of Game Design.* 2nd ed., New Riders, 2010. Print.

Cox, Kate. "It's Time to Start Treating Video Game Industry Like the $21 Billion Business It Is." *Consumerist,* 9 June 2014, consumerist.com. Accessed 22 Dec. 2014.

Entertainment Software Assoc. *2016 Sales, Demographic, and Usage Data: Essential Facts about the Computer and Video Game Industry.* Entertainment Software Association, 2016, essentialfacts.theesa.com/Essential-Facts-2016.pdf. Accessed 31 Oct. 2016.

Isbister, Katherine. *How Games Move Us: Emotion by Design.* MIT Press, 2016.

Kamenetz, Anya. "Why Video Games Succeed Where the Movie and Music Industries Fail." *Fast Company,* 7 Nov. 2013, www.fastcompany.com. Accessed 22 Dec. 2014.

Koster, Raph. *A Theory of Fun for Game Design.* Paraglyph, 2004.

Lecky-Thompson, Guy W. *Video Game Design Revealed.* Charles River Media, 2007.

Marcovitz, Hal. *Video Games.* Lucent, 2010.

Rogers, Scott. *Level Up!: The Guide to Great Video Game Design.* 2nd ed., Wiley, 2014.

Schell, Jesse. *The Art of Game Design: A Book of Lenses.* 2nd ed., CRC Press, 2014.

Thompson, Jim, Barnaby Berbank-Green, and Nic Cusworth. *Game Design: Principles, Practice, and Techniques.* Quarto, 2007.

VIRTUAL REALITY

SUMMARY

The applied science of virtual reality (VR) engages in the design and engineering of and research related to special immersive interactive computer systems. These virtual reality systems synthesize environments, or worlds, which are simulations of reality that are usually rendered using three-dimensional computer images, sounds, and force feedbacks. Virtual reality applications are used for pilot and astronaut training, entertainment, communication, teleoperation, manufacturing, medical and surgery training, experimental psychology, psychotherapy, education, science, architecture, and the arts. This technology, which submerges humans into altered environments and processes, intensifies experience and imagination, thereby augmenting research and education. Virtual reality training systems can simplify and improve manufacturing and maintenance, while simultaneously reducing risk exposure.

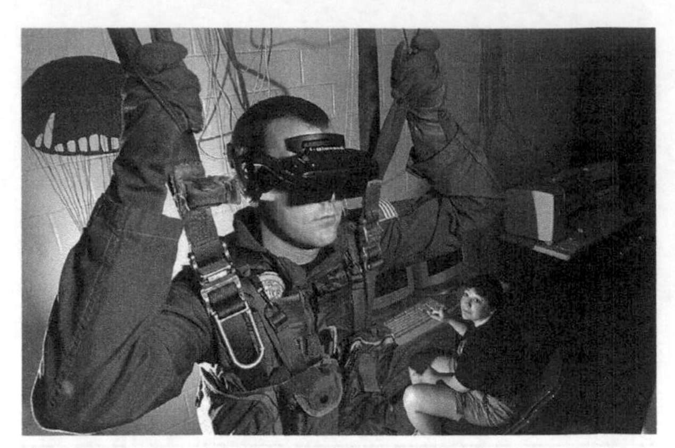

U.S. Navy personnel using a VR parachute trainer.

DEFINITION AND BASIC PRINCIPLES

A virtual reality system is an interactive technology setup (software, hardware, peripheral devices, and other items) that acts as a human-to-computer interface and immerses its user in a computer-generated three-dimensional environment. Virtual reality is the environment or world that the user experiences while using such a system. Although the term "virtual" implies that this simulated world does not actually exist, the term "reality" refers to the user's experience of the simulated environment as being real. The more that the senses are involved in a compelling fashion, the more genuine the perceived experience will be, and the more intense the imagination. Most virtual reality systems stimulate the senses of sight, hearing, touch, and other tactile-kinesthetic sense perceptions, such as equilibrioception, torque, and even temperature. Less often they include smell, and with existing technology, they exclude taste. Virtual reality must be almost indistinguishable from reality in some applications such as pilot training, but it can often differ significantly from the real world in, for example, games.

Virtual reality in a narrow sense (a computer-generated simulation that exists virtually but not materially) is not the same as augmented reality (enhanced reality) or telepresence (in the sense of teleconferencing). Augmented reality technology improves the perception of and supplements the knowledge about existing entities or processes (highlighting data of interest while abstracting the less important information). Telepresence (as teleconferencing) refers to a remote virtual re-creation of a real situation (for example, to enable audio and visual interactions between people at diverse places). However, a wider notion of virtual reality includes notions of both augmented reality and telepresence.

BACKGROUND AND HISTORY

The idea of simulated reality is often traced back to the ancient Greek philosopher Plato's allegory of the cave in *Politeia* (fourth century BCE; *Republic*, 1701). In the allegory, spectators observe images (shadows) of objects on a cave wall that they take for real objects. The term "virtual" is derived from the Latin word "virtue" (which means goodness or manliness). "Virtual" then means existing in effect or in essence but not in actuality. The notion of virtual reality can be traced back to the French theater director, actor, playwright, and illustrator Antonin Artaud who described theater as *la réalite virtuelle* in his influential *Théâtre et son double* (1938; *The Theatre and Its Double*, 1958). Computer scientist and artist Myron W. Krueger coined the technical term "artificial reality" in his book of the same title, published in 1983. Computer scientist and artist Jaron Lanier popularized the notion of virtual reality as a technical term in the 1980s. Many artists, science-fiction authors, and directors incorporated concepts of computer-generated, simulated, or augmented reality in their creations. One of the most prominent examples is the holodeck, an advanced form of virtual reality featured on the television program *Star Trek: The Next Generation* (1987–1994).

Technology. In the late 1960s, computer scientist Ivan Sutherland created the first head-mounted device, which was capable of tracking the user's viewing direction. In the 1970s, Sutherland and David Evans developed a computer graphic scene generator. In the 1970s and 1980s, force feedback was incorporated into tactile input devices such as gloves and interactive wands. Lanier and Thomas Zimmerman developed sensing gloves, which recognized finger and hand movements.

This kind of virtual reality technology was intended to improve flight simulators and applications for astronaut training. In 1981, the National Aeronautics and Space Administration (NASA) combined commercially available Sony liquid crystal display (LCD) portable television displays with special optics for a prototype stereo-vision head-mounted device called the Virtual Visual Environment Display (VIVED). NASA then created the first virtual reality system, which combined a host computer, a graphics computer, a noncontact user position-tracking system, and VIVED.

Scott Fisher and Elizabeth Wenzel developed hardware for three-dimensional virtual sound sources in 1988. Also in the late 1980s, Lanier built a virtual reality system for two simultaneous users that he named RB2 (Reality Built for Two). Fisher incorporated sound systems, head-mounted device technology, and sensor gloves into one system called the Virtual Interactive Environment Workstation (VIEW), also used by NASA. The first conference on virtual reality, "Interface for Real and Virtual Worlds," was held in Montpellier, France, in March, 1992. That same year, scientists, engineers, and medical practitioners assembled in San Diego, California, for the "Medicine Meets Virtual Reality Conference." In September, 1993, the Institute for Electrical and Electronics Engineers (IEEE) organized its first virtual reality conference in Seattle.

HOW IT WORKS

To create a realistic computer-generated world, several high-end technologies must be integrated into a single virtual reality system. This kind of high-end system is used at university, military, governmental, and private research laboratories. Adequate computing speed and power, fast image and data processors, broad bandwidth, and sophisticated software are essential. Other requirements include high-tech input-output devices or effectors (such as head-mounted devices), three-dimensional screens, surround-sound systems, and tactile devices (such as wired gloves and suits, tracking systems, and force feedback devices, including motion chairs and multidirectional treadmills).

Input. The input devices used for virtual reality systems—sensing gloves, trackballs, joysticks, wands, treadmills, motion sensors, position trackers, voice recognizers, and biosensors—are typically more complex than those used for personal computers. Biosensors recognize eye movement, muscle activity, body temperature, pulse, and blood pressure, all of which are vital to surgical applications. Position trackers and motion sensors identify and monitor the user's position and movements. The tracking systems used in virtual reality systems are mechanical, optical, ultrasonic, or magnetic devices. Steering wheels, joysticks, or wands are used for pilot training and games. Sophisticated devices for research and experiments (such as those for molecular modeling) offer six-degrees-of-freedom input. Such input devices allow the computer to adjust the virtual environment according to the data received from the user. When motion sensors and position trackers detect the user's movement, the data are processed by the computer in real time, and the display, also in real time, has to accurately render the image (such as the interior of a building). If slow data processing creates a time lag, the user may experience simulator sickness or motion sickness (nausea or dizziness), especially if the user's senses register conflicting data. In a virtual reality parachute training session, for example, equilibrioception might be in conflict with visual perception if movement is represented faster by the display than by the force feedback system.

Output. Output devices are intended to stimulate as many senses as possible for a high degree of immersion. The important visual output devices are head-mounted devices, LCDs, and projectors and screens. Developers compete to make these displays the most immersive. All displays must be able to render three-dimensional images. Other output devices are for sound and touch. Sound systems can assist in conveying the impression of three-dimensional space. Force feedback systems give the user the sense of physical resistance (essential in surgical virtual reality systems), torque, tilt, and vibration (appreciated for games and essential for pilot training).

The CAVE (Cave Automatic Virtual Environment) is a surround-sound, surround-screen, projection-based room-sized virtual reality system developed by the Electronic Visualization Laboratory at the University of Illinois at Chicago (the name is trademarked by the University of Illinois Board of Regents). Users put on lightweight stereo glasses and walk around inside the room, interacting with virtual objects. One user is an active viewer, controlling the projection, and the others are passive viewers, but all users can communicate while in the CAVE. The system was designed to help with visualization of scientific concepts.

HARDWARE AND SOFTWARE. Standard personal computers have both limited memory capacity and limited performance capability for running professional virtual reality applications. Therefore, high-end hardware and software have been developed for specific purposes such as games, research programs, and pilot, combat infantry, and medical training. Computer languages such as C++, C#, Java, and Quest3D are used for programming virtual reality software. Virtual reality computers handle tasks such as data input and output and the interaction, integration, and recomposition of all the data required in virtual environment management. Because a virtual reality system needs high computing, processing, and display speeds, the virtual reality computer architecture can at times use several computers or multiple processors.

APPLICATIONS AND PRODUCTS

AEROSPACE AND MILITARY. One of the early applications of virtual reality was in pilot training. Modern flight simulators are convincingly close to real flight experiences, although the simulation of acceleration and zero gravity are still challenges. In the military, virtual reality applications are used not only by pilots, paratroopers, and tank drivers but also by battle strategists and combat tacticians to enhance safety training and analyze battle maneuvers and positions. These military applications make targeting more precise, thereby reducing human casualties and collateral damage. Virtual reality systems are also used to evaluate new weapons systems. In space exploration, virtual reality systems help astronauts prepare for zero-gravity activities such as the repair of solar panels on the outside of a spaceship. Virtual reality training systems give the user the opportunity to review and evaluate specific sequences in a training session or the entire session.

ENTERTAINMENT AND GAMES. The military took commercial games and adapted them to create flight simulators and other professional virtual reality applications, which, in turn, were adapted to use as games. Virtual reality game applications and the virtual games industry are not common in gaming arcades, but they are making significant headway into the home and in the mobile entertainment market, with an expected consumer spending of over $5 billion in software, hardware, and accessories during 2016. Despite Sony announcing in early 2016 a new Playstation VR headset, industry analysts also noted that the gaming industry appeared less interested than the motion picture industry in developing and incorporating VR technology.

Some companies lease virtual reality game equipment (such as training applications for golf, racing, and shooting) to customers for entertainment or to corporations for internal team-bonding experiences. Some virtual reality applications enable users to journey through fantastic and futuristic worlds. The common equipment for virtual reality games—depending on the quality and level of sophistication of the games—includes head-mounted devices, several LCD screens, tracking systems, omnidirectional treadmills for walking, force feedback or rumbling seats (or platforms) for flight and race simulations, batons for tennis, and guns, wands, or sticks for shooting. The more sensory feedback provided, the more immersive the virtual reality game application.

EDUCATION. The more senses that are involved in the process of learning and training, the more a student is able to become engaged in the subject matter and the better the educational impact, especially when it comes to learning skills or practical content. The employment of a virtual reality system enables data and images to become interactive, colorful, three-dimensional, and accompanied by sound. For example, a visitor to a virtual museum is able to virtually touch the artifacts, taking them from their shelves, turning them around, and viewing them from different perspectives. In an immersive experience, a user can witness, seemingly first-hand, historical events such as famous battles as they take place. Virtual reality applications exist in almost every area of education, including sports (such as golfing), sciences (such as astronomy, physics, biology, and chemistry), the humanities (such as history), and vocational training (such as medical procedures and mechanical engineering techniques).

Art. Although classic artworks can be immersive, few of them are interactive. People are not allowed to touch exhibits in most art galleries and museums. In 1993, the Solomon R. Guggenheim Museum became the first major museum to dedicate an entire exhibition to virtual reality art. The exhibition featured virtual reality installations from Jenny Holzer, Thomas Dolby, and Maxus Systems International. Virtual reality technology enables artists to blur or combine genres (such as music, graphic arts, and video) and involves the viewer in the creation of art. Every viewer or user of the artwork perceives the work in a different way depending on his or her input into the virtual reality work of art. A virtual reality art exhibit can be programmed to produce sound or visual feedback according to, for example, a visitor's footsteps or voice input from an audience. Artworks thus become interactive, and viewers become cocreators of the art. VR technology developed in 2015 is utilized by multimedia artists and animators for content creation and making three-dimensional storyboards

Science, Engineering, and Design. Virtual reality systems can advance the creative processes involved in science, engineering, and design. Results can be tested, evaluated, shared, and discussed with other virtual reality users. Chemists at the University of North Carolina, Chapel Hill, used virtual reality systems for modeling molecules, and similar systems have been used to "observe" atoms. Buildings, automobiles, and mechanical parts can be designed on computers and evaluated using three-dimensional modeling tools and visualization techniques in a virtual reality environment. An architect can take a client on a virtual walk through a building before it is constructed, or the aerodynamics of a new automobile design can be evaluated before manufacturing a model or prototype. Virtual reality applications also can simulate crash tests. The use of virtual reality applications reduces cost, waste, and risk.

Business. Business applications include stores on the Internet featuring virtual showrooms and 360-degree, three-dimensional views of products. Another important application is teleconferencing. In contrast to the traditional conference call, virtual reality applications may permit multisensory evaluations of new products. Virtual reality systems also help network, combine, and display data from diverse sources to analyze financial markets and stock exchanges. In these situations, virtual reality applications serve as decision support systems.

Medicine, Therapy, and Rehabilitation. Experienced surgeons as well as physicians in training use virtual reality systems to practice surgery. The images for such training programs are taken from x-rays, computed tomography (CT) scans, and magnetic resonance imaging (MRI). Virtual operations can be recorded and repeated as many times as desired, in contrast to practice operations on animals or human corpses, which cannot be repeated. Force feedback gloves give practitioners the realistic feel and touch needed, for example, to determine how much force is needed for certain incisions.

Virtual reality environments are being developed for use in training patients as well as doctors. Virtual human limbs are being prototyped and studied for use in training patients to use and become comfortable with prosthetic limbs, as well as for patient rehabilitation. Some of the virtual prosthetics in development appear as avatars in a virtual reality environment and are able to accept both kinematic and neural control inputs from the patient.

In psychotherapy, virtual reality applications are often used an alternative therapy to help treat clients with disorders such as phobias, anxiety, and PTSD (post-traumatic stress disorder). Clients often undergo desensitization treatment, which is especially useful in treating phobias and PTSD. Virtual reality allows clients who fear enclosed spaces (claustrophobia), dirt and germs (mysophobia), or snakes (ophidiophobia) to gradually confront their fears without being exposed to the actual condition or object. Virtual reality applications can also help clinicians understand certain psychological problems better by enabling them to experience what the client or the patient experiences. For example, a psychiatrist, psychologist, or counselor may take a virtual bus ride in the role of a schizophrenic client, during which they experience some simulated symptoms of this disorder, such as distorted images viewed through the windows of the bus or strange voices that seem to appear from nowhere.

SOCIAL CONTEXT AND FUTURE PROSPECTS

Some experts believe that virtual reality, like the computer and the Internet, soon will become a commonplace and indispensable part of everyday life, but others do not see the potential for such a wide implementation. Most agree, however, that once computing speed and power, broad bandwidth, and peripheral systems become more affordable for average consumers, virtual reality will be more widely used. Beyond science, education, and other professional applications, the entertainment industry is thought most likely to want affordable virtual reality innovations. Critics of virtual reality applications point to personal and societal risks such as isolation, desocialization, and alienation, but advocates emphasize the technology's proven potential for augmenting people's lives. Both critics and advocates agree that experiences in virtual reality can alter people's perceptions of and responses to the real world. These changes are intentional and welcome in most cases but sometimes they take place in an unintended and potentially dangerous manner. Airplane pilots have reportedly made mistakes that could be linked to the limitations of training with a flight simulator, which, for example, is incapable of realistically simulating acceleration. However, virtual reality applications are valuable in highly technical and precise areas of medicine, industry, business, and research and are likely to gain more users, despite the potential risks and the expense.

—Roman Meinhold, MA, PhD

Bibliography

Al-Jumaily, A., and R. A. Olivares. "Bio-Driven System-Based Virtual Reality for Prosthetic and Rehabilitation Systems." *Signal, Image and Video Processing* 6.1 (2012): 71–84. *Inspec.* Web. 10 Mar. 2015.

Bailey, David Evans. "Ten Cool Applications for Virtual Reality That Aren't Just Games." *The Conversation.* The Conversation U.S., 22 Mar. 2016. Web. 24 June 2016.

Burdea, Grigore C., and Philippe Coiffet. *Virtual Reality Technology.* 2nd ed. Hoboken: Wiley-Interscience, 2003. Print.

Craig, Alan B., William R. Sherman, and Jeffrey D. Will. *Developing Virtual Reality Applications. Foundations of Effective Design.* Burlington: Morgan Kaufmann, 2009. Print.

Gaudiosi, John. "Virtual Reality Video Game Industry to Generate $5.1 Billion in 2016." *Fortune.* Time, 5 Jan. 2016. Web. 26 June 2016.

Heim, Michael. *Virtual Realism.* New York: Oxford UP, 2000. Print.

Levin, Mindy F., Emily A. Keshner, and Patrice L. (Tamar) Weiss. *Virtual Reality for Physical and Motor Rehabilitation.* New York: Springer, 2014. *eBook Collection (EBSCOhost).* Web. 10 Mar. 2015.

Putrino, David, et al. "A Training Platform for Many-Dimensional Prosthetic Devices Using a Virtual Reality Environment." *Jour. of Neuroscience Methods* (2014). *ScienceDirect.* Web. 10 Mar. 2015.

Sherman, William R., and Alan B. Craig. *Understanding Virtual Reality: Interface, Application and Design.* San Francisco: Morgan Kaufmann, 2003. Print.

Vince, John. *Introduction to Virtual Reality.* New York: Springer, 2004. Print.

"VR Futures: Where Will Virtual Reality Take You?" *EandT.* Institution of Engineering and Technology, 15 Mar. 2016. Web. 26 June 2016.

Z

Z3

SUMMARY

The *Z3* was the world's first ever automatic, program-controlled computer. Built in the early 1940s by German civil engineer Konrad Zuse, the Z3 represented a significant breakthrough in the early development of computers. While the Z3 was primarily designed to complete general mathematical calculations like other computers of its time—including those created by such computer science giants as Howard Aiken, George Stibitz, and Alan Turing—it worked differently. The Z3 was unique because it was programmable and could complete its calculations automatically. For Zuse, who completed his machine in the midst of World War II and with little support from the embattled German government, the Z3's successful debut in 1941 was the culmination of years of work in computer science and the construction of two earlier models, the Z1 and Z2. Although it was certainly primitive by contemporary standards, the Z3 set the stage for the development of modern computers and helped usher in the digital age.

BACKGROUND

As a young man, Zuse was primarily interested in art, but he also had a marked talent for engineering. Eventually, he chose to pursue a career as an engineer and enrolled at the Technical University of Berlin. After graduating in 1935, Zuse briefly worked as a structural engineer with the Henschel Aircraft Company, but soon left this job in favor of focusing on a different effort. During his time as a student and an actual engineer in the field, Zuse found the complex mathematical calculations that engineering required to be extremely slow and tedious. Determined to find a way to make completing such calculations easier, he turned his attention to computer science. Ultimately, Zuse decided to try his hand at building his own computer.

Setting up a makeshift laboratory in his parents' Berlin apartment and gathering a team of volunteer college friends to help with the effort, Zuse set to work on creating a calculating computer based on a binary system in which numeric values were represented using two different symbols (typically 1 and 0). Zuse's initial work led to the Z1, a mostly mechanical computer built with a memory unit and an arithmetic unit. Although the Z1 worked when it was completed in 1938, it did not work particularly well. Frequent problems with the gears and levers used for transmitting signals in the arithmetic unit meant that the Z1 would only work for short periods before breaking down. Upon realizing that electric wires could be used to carry signals more efficiently, Zuse quickly began making plans for a second computer.

In developing what would become the Z2, Zuse chose to include not only electrical wiring, but also telephone relays—small, electrically driven mechanical switches—in his design. These relays helped to shorten processing times and made the Z2 more efficient than its predecessor. Despite this

Replica of the Zuse Z3 in the Deutsches Museum in Munich, Germany. Venusianer at the German language Wikipedia

improvement, however, the Z2 still proved to be unstable. Regardless, Zuse, who was drafted into the German army for a time during this period, demonstrated the Z2 at the German Aeronautics Research Institute (DLV) in 1940. Impressed with his work, DLV officials offered to provide Zuse with partial funding for his next project. With this support, Zuse returned home and began work on the Z3.

EVOLUTION OF THE Z3

Once again working from his living room laboratory, Zuse built the Z3 with the help of his volunteer team. Work on the machine was completed on December 5, 1941. In terms of its design, the Z3 shared many similarities with its predecessors, but also had some key differences. Like the Z1 and Z2, the Z3 used discarded filmstrips for input instead of the punched paper tape that most other early computers used at the time. Also like Zuse's earlier models, the Z3 had a keyboard with four decimal places for entering data and an electric lamp used to display output. Unlike the Z1 and Z2, however, the Z3 was built with a much larger number of electromechanical relays—around 2,600 in total. About 1,400 of these relays were used in the memory unit alone. The Z3's arithmetic unit was composed of two different devices that separately handled different parts of numbers. Using this approach, Zuse was able to construct an arithmetic unit that could add, subtract, multiply, divide, and even find square roots. To make the Z3 as fast as possible, Zuse also programmed in a number of common multiplication problems. In part because of this, the Z3 was able to multiply two numbers in about four or five seconds and add two numbers together in anywhere from one-fourth to one-third of a second. For its time, the Z3 was a fast, cutting-edge computer.

While the Z3 had remarkable abilities and represented an important step forward in the evolution of computing, it could not otherwise be considered a success. When Zuse presented his newest machine at the DLV, officials there felt that because it had only limited memory space, the Z3 was not significantly more useful for their purposes than other existing methods of calculation. With the DLV showing no apparent interest in his work, Zuse was forced to take the Z3 back to his parents' apartment for storage. Unfortunately, the original Z3 was later destroyed during an Allied bombing raid on Berlin in 1944.

Zuse Z3 computer with Finder relays. By Dksen (Own work)

Following the DLV's rejection of the Z3, Zuse quickly began working on his next project, the Z4. Although his efforts were temporarily disrupted as Germany fell in the waning stages of World War II, Zuse eventually got the Z4 working. For a time, it was the only operational computer anywhere on the European mainland. Zuse finally found success after the war's end, establishing his own computer-manufacturing company, Zuse KG, which was absorbed into Siemens AG in 1967. In spite of his later successes, however, Zuse's crowning achievement remained the Z3. The Z3, as the first working program-controlled computer that could complete mathematic calculations, played an important role in the evolution of computer technology and contributed significantly to the development of modern computers and other similar high-tech devices. In fact, the Z3 may have been a more capable computer than Zuse himself ever realized. Two years after Zuse's death in 1995, researchers using a newly built replica of the original machine proved that the Z3 could solve any computable math problem if it was allowed enough time.

—*Jack Lasky*

BIBLIOGRAPHY

Abbany, Zulfikar. "Konrad Zuse and the Digital Revolution He Started with the Z3 Computer 75 Years Ago." *Deutsche Welle*, 5 Nov. 2016, www.dw.com/en/konrad-zuse-and-the-digital-revolution-he-started-with-the-z3-computer-75-years-ago/a-19249238. Accessed 10 Nov. 2016.

Brown, Mike. "Konrad Zuse's Z3, the World's First Programmable Computer, Was Unveiled 75 Years Ago." *Inverse*, 12 May 2016, www.inverse.

com/article/15542-konrad-zuse-s-z3-the-world-s-first-programmable-computer-was-unveiled-75-years-ago. Accessed 10 Nov. 2016.

Igarashi, Yoshihide, et al. *Computing: A Historical and Technical Perspective.* CRC Press, 2014.

Janusz, Stefan. "The Known and Unknown Pioneers of Modern Computing." *Science Node,* 20 June 2012, sciencenode.org/feature/known-and-unknown-pioneers-modern-computing.php. Accessed 10 Nov. 2016.

Lerner, K. Lee, and Brenda Wilmoth Lerner, editors. "Zuse, Konrad." *Computer Sciences.* Macmillan Reference U.S.A, 2013.

Lindsey, Victor. "Konrad Zuse." *Great Lives from History: Inventors & Inventions.* Edited by Alvin K. Benson, Salem Press, 2010.

Merrin, George. "The Computer Age." *University Observer,* 8 Nov. 2016, www.universityobserver.ie/science/the-computer-age/. Accessed 10 Nov. 2016.

Salz Trautman, Peggy. "A Computer Pioneer Rediscovered, 50 Years On." *New York Times,* 20 Apr. 1994, www.nytimes.com/1994/04/20/news/20iht-zuse.html. Accessed 10 Nov. 2016.

ZOMBIE

SUMMARY

A *zombie,* also called a bot, refers to any computer remotely under the control of another operator without the original computer user's consent. Zombie computers are often created by viruses and other malicious software called malware. The primary user does not often know that his or her computer has become a zombie. The computer still functions normally but usually works at a slower rate. Zombie computers are often used for illegal activities, such as spreading malware and hacking.

BACKGROUND

Malware is malicious programming installed without a computer user's knowledge or approval. Malware includes computer viruses, Trojan horses, spyware, adware, rootkits, and ransomware, and it is often difficult to remove. Many types of malware are used to turn an uninfected computer into a zombie. In fact, any code capable of installing additional files can be used to create a zombie computer.

While many of these malware labels are often used interchangeably, they all refer to subtly different types of code. A computer virus refers to any piece of self-replicating code that causes damage to the computer. A Trojan horse is a type of malware that is disguised as or attached to a legitimate computer program. They trick computer users into downloading the malware and then install themselves onto a computer. Spyware tracks the user's Internet history, keystrokes, or other forms of sensitive information and sends it to a third party. Adware displays large amounts of unauthorized advertisements on a computer, often with the intent of stealing credit card information. Rootkits are subtle, difficult to remove malware that allow unauthorized users access to a computer. Ransomware encrypts the contents of a computer's hard drive and only gives the user the required tools to restore the files once a large monetary payment has been made to the attacker.

Experts advise computer users to regularly run and update reputable antivirus software on their computers to detect malware. They also advise computer users to download only files from websites they trust and to never open e-mail attachments that look suspicious. If reputable antivirus software fails to remove symptoms of malware, most experts advise contacting a professional for help.

BOTNETS AND ZOMBIE ARMIES

Zombies are primarily used to create specialized networks called botnets, or zombie armies. In these circumstances, the controlling computer, often called the host computer, is not in constant contact with the zombie computers. Instead, the zombie computers are programmed to listen constantly for commands from the host computer. The host computer will

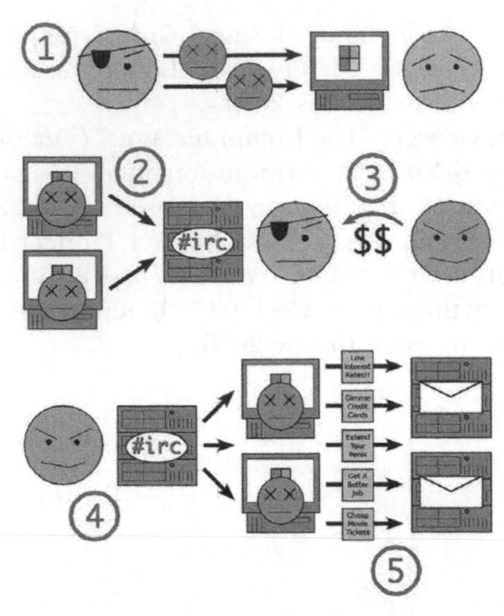

A diagram of the process by which spammers use zombie (virus-infected) computers to send spam. The original uploader was Bash at English Wikipedia.

issue commands whenever the attacker sees fit, and the zombie computers will follow those commands.

The Sub7 Trojan horse and the Pretty Park virus created the first botnets in 1999. They were the first programs to infect large numbers of computers with a remote access tool, allowing one central computer to access many other computers at once. Sub7 also offered remote keylogging tools, allowing the botnet controller to harvest massive amounts of personal information. Within a few years, various botnets had been created. Spybot, developed in 2003, was the first botnet exclusively built for datamining, the process of collecting personal information in large quantities.

Soon after the development of Spybot, criminals began to realize botnets' potential for illicit activities. Early spam e-mails, which were used to spread malware and advertisements, were originally sent from centralized server farms. However, with botnets, thousands or millions of computes could be used to send spam. When using a botnet, criminals did not need to pay for server upkeep or find a place to store physical servers. Additionally, if the authorities managed to seize or shut down a server farm, the spamming operation would be shut down. To shut down a botnet, authorities would need to find and apprehend whoever controlled the botnet. Such a task is incredibly difficult. Modern botnets can generate fake names and locations for their controlling server on a daily basis. They can also encrypt these fake credentials, making them appear legitimate and forcing authorities to spend time decoding false information.

In addition to spam, botnets are often used in distributed denial of service (DDoS) attacks. DDoS attacks are used to shut down access to a webpage. Webpages are hosted on servers, and servers can only process a finite amount of requests in any given period. DDoS attacks send more requests to the server than it can process, overloading it. An overloaded server can no longer respond to requests, so no other computers can view websites hosted on the server. Traditionally, powerful computers carried out DDoS attacks. Using botnets, thousands of computers can be ordered to access the same server at once. This massive number of sustained requests overloads the server. Such botnet attacks are more difficult to shut down than a traditional DDoS attack.

Computers that are part of a botnet display several symptoms. For example, some computers connected to a botnet will send large amounts of outgoing messages. Others will activate their fans and cooling systems when the primary user is not running demanding programs. This indicates that the botnet is utilizing a large percentage of the computer's resources. Additionally, some botnets will cause computers to start up and shut down extremely slowly or cause programs on the computer to run at a drastically reduced speed. Lastly, some botnets will stop the computer from updating its operating system or updating any antivirus software. This is because newer variants of the operating system may stop the botnet from operating, and newer versions of antivirus software may be able to remove the virus that created the zombie.

—*Tyler Biscontini*

BIBLIOGRAPHY

"Botnet." *Radware.com*, 2016, security.radware.com/ddos-knowledge-center/ddospedia/botnet/. Accessed 30 Nov. 2016.

Bradely, Tony. "Sub7 Trojan/Backdoor." *Lifewire*, 20 Oct. 2016, www.lifewire.com/sub7-trojan-backdoor-2486800. Accessed 6 Dec. 2016.

Geier, Eric, and Norem, Josh. "How to Remove Malware from Your Windows PC." *PCWorld*, 18 July 2016, www.pcworld.com/article/243818/

security/how-to-remove-malware-from-your-windows-pc.html. Accessed 30 Nov. 2016.

"The History of the Botnet, Part I." *TrendMicro*, 24 Sept. 2010, countermeasures.trendmicro.eu/the-history-of-the-botnet-part-i. Accessed 30 Nov. 2016.

Rouse, Margaret. "Zombie (bot)." *Tech Target*, 2016, searchmidmarketsecurity.techtarget.com/definition/zombie. Accessed 30 Nov. 2016.

Strickland, Jonathan. "How Zombie Computers Work." *HowStuffWorks*, 2016, computer.howstuffworks.com/zombie-computer1.htm. Accessed 30 Nov. 2016.

"Top 10 Signs Your Computer May be Part of a Botnet." *WeLiveSecurity*, 21 Apr. 2010, www.welivesecurity.com/2010/04/21/top-10-signs-your-computer-may-be-part-of-a-botnet. Accessed 30 Nov. 2016.

"What Is a Botnet Attack?" *Kaspersky Lab*, 2016, usa.kaspersky.com/internet-security-center/threats/botnet-attacks#.WEcmCH0koU0. Accessed 30 Nov. 2016.

"What Is Malware and How Can We Prevent It?" *PC-Tools*, 2016, www.pctools.com/security-news/what-is-malware/. Accessed 30 Nov. 2016.

TIME LINE OF MACHINE LEARNING AND ARTIFICIAL INTELLIGENCE

Year	Event
Antiquity	Greek myths of Hephaestus and Pygmalion incorporated the idea of intelligent robots (such as Talos) and artificial beings (such as Galatea and Pandora).
Antiquity	Yan Shi presented King Mu of Zhou with mechanical men.
Antiquity	Sacred mechanical statues built in Egypt and Greece were believed to be capable of wisdom and emotion. Hermes Trismegistus would write "they have sensus and spiritus ... by discovering the true nature of the gods, man has been able to reproduce it." Mosaic law prohibits the use of automatons in religion.
384 BC–322 BC	Aristotle described the syllogism, a method of formal, mechanical thought.
1st century	Heron of Alexandria created mechanical men and other automatons.
260	Porphyry of Tyros wrote Isagogê which categorized knowledge and logic.
~800	Geber develops the Arabic alchemical theory of Takwin, the artificial creation of life in the laboratory, up to and including human life.
1206	Al-Jazari created a programmable orchestra of mechanical human beings.
1275	Ramon Llull, Spanish theologian invents the Ars Magna, a tool for combining concepts mechanically, based on an Arabic astrological tool, the Zairja. The method would be developed further by Gottfried Leibniz in the 17th century.
~1500	Paracelsus claimed to have created an artificial man out of magnetism, sperm and alchemy.
~1580	Rabbi Judah Loew ben Bezalel of Prague is said to have invented the Golem, a clay man brought to life.
Early 17th century	René Descartes proposed that bodies of animals are nothing more than complex machines (but that mental phenomena are of a different "substance").
1623	Wilhelm Schickard drew a calculating clock on a letter to Kepler. This will be the first of five unsuccessful attempts at designing a direct entry calculating clock in the 17th century (including the designs of Tito Burattini, Samuel Morland and René Grillet)).
1641	Thomas Hobbes published Leviathan and presented a mechanical, combinatorial theory of cognition. He wrote "...for reason is nothing but reckoning".
1642	Blaise Pascal invented the mechanical calculator, the first digital calculating machine
1672	Gottfried Leibniz improved the earlier machines, making the Stepped Reckoner to do multiplication and division. He also invented the binary numeral system and envisioned a universal calculus of reasoning (alphabet of human thought) by which arguments could be decided mechanically. Leibniz worked on assigning a specific number to each and every object in the world, as a prelude to an algebraic solution to all possible problems.
1726	Jonathan Swift published Gulliver's Travels, which includes this description of the Engine, a machine on the island of Laputa: "a Project for improving speculative Knowledge by practical and mechanical Operations " by using this "Contrivance", "the most ignorant Person at a reasonable Charge, and with a little bodily Labour, may write Books in Philosophy, Poetry, Politicks, Law, Mathematicks, and Theology, with the least Assistance from Genius or study." The machine is a parody of Ars Magna, one of the inspirations of Gottfried Leibniz' mechanism.

1750	Julien Offray de La Mettrie published L'Homme Machine, which argued that human thought is strictly mechanical.
1763	Thomas Bayes's work An Essay towards solving a Problem in the Doctrine of Chances is published two years after his death, having been amended and edited by a friend of Bayes, Richard Price. The essay presents work which underpins Bayes theorem.
1769	Wolfgang von Kempelen built and toured with his chess-playing automaton, The Turk. The Turk was later shown to be a hoax, involving a human chess player.
1805	Adrien-Marie Legendre describes the "méthode des moindres carrés", known in English as the least squares method. The least squares method is used widely in data fitting.
1812	Pierre-Simon Laplace publishes Théorie Analytique des Probabilités, in which he expands upon the work of Bayes and defines what is now known as Bayes' Theorem.
1818	Mary Shelley published the story of Frankenstein; or the Modern Prometheus, a fictional consideration of the ethics of creating sentient beings.
1822–1859	Charles Babbage & Ada Lovelace worked on programmable mechanical calculating machines.
1837	The mathematician Bernard Bolzano made the first modern attempt to formalize semantics.
1854	George Boole set out to "investigate the fundamental laws of those operations of the mind by which reasoning is performed, to give expression to them in the symbolic language of a calculus", inventing Boolean algebra.
1863	Samuel Butler suggested that Darwinian evolution also applies to machines, and speculates that they will one day become conscious and eventually supplant humanity.
1913	Andrey Markov first describes techniques he used to analyse a poem. The techniques later become known as Markov chains.
1913	Bertrand Russell and Alfred North Whitehead published Principia Mathematica, which revolutionized formal logic.
1915	Leonardo Torres y Quevedo built a chess automaton, El Ajedrecista and published speculation about thinking and automata.
1920s and 1930s	Ludwig Wittgenstein and Rudolf Carnap lead philosophy into logical analysis of knowledge. Alonzo Church develops Lambda Calculus to investigate computability using recursive functional notation.
1923	Karel apek's play R.U.R. (Rossum's Universal Robots) opened in London. This is the first use of the word "robot" in English.
1931	Kurt Gödel showed that sufficiently powerful formal systems, if consistent, permit the formulation of true theorems that are unprovable by any theorem-proving machine deriving all possible theorems from the axioms. To do this he had to build a universal, integer-based programming language, which is the reason why he is sometimes called the "father of theoretical computer science".
1941	Konrad Zuse built the first working program-controlled computers.
1943	Warren Sturgis McCulloch and Walter Pitts publish "A Logical Calculus of the Ideas Immanent in Nervous Activity" (1943), laying foundations for artificial neural networks.
1943	Arturo Rosenblueth, Norbert Wiener and Julian Bigelow coin the term "cybernetics". Wiener's popular book by that name published in 1948.

1945	Game theory which would prove invaluable in the progress of AI was introduced with the 1944 paper, Theory of Games and Economic Behavior by mathematician John von Neumann and economist Oskar Morgenstern.
1945	Vannevar Bush published As We May Think (The Atlantic Monthly, July 1945) a prescient vision of the future in which computers assist humans in many activities.
1948	John von Neumann (quoted by E.T. Jaynes) in response to a comment at a lecture that it was impossible for a machine to think: "You insist that there is something a machine cannot do. If you will tell me precisely what it is that a machine cannot do, then I can always make a machine which will do just that!". Von Neumann was presumably alluding to the Church-Turing thesis which states that any effective procedure can be simulated by a (generalized) computer.
1950	Alan Turing proposes a 'learning machine' that could learn and become artificially intelligent. Turing's specific proposal foreshadows genetic algorithms.
1950	Alan Turing proposes the Turing Test as a measure of machine intelligence.
1950	Claude Shannon published a detailed analysis of chess playing as search.
1950	Isaac Asimov published his Three Laws of Robotics.
1951	Marvin Minsky and Dean Edmonds build the first neural network machine, able to learn, the SNARC.
1951	The first working AI programs were written in 1951 to run on the Ferranti Mark 1 machine of the University of Manchester: a checkers-playing program written by Christopher Strachey and a chess-playing program written by Dietrich Prinz.
1952	Arthur Samuel joins IBM's Poughkeepsie Laboratory and begins working on some of the very first machine learning programs, first creating programs that play checkers.
1952–1962	Arthur Samuel (IBM) wrote the first game-playing program, for checkers (draughts), to achieve sufficient skill to challenge a respectable amateur. His first checkers-playing program was written in 1952, and in 1955 he created a version that learned to play.
1956	The Dartmouth College summer AI conference is organized by John McCarthy, Marvin Minsky, Nathan Rochester of IBM and Claude Shannon. McCarthy coins the term artificial intelligence for the conference.
1956	The first demonstration of the Logic Theorist (LT) written by Allen Newell, J.C. Shaw and Herbert A. Simon (Carnegie Institute of Technology, now Carnegie Mellon University or CMU). This is often called the first AI program, though Samuel's checkers program also has a strong claim.
1957	Frank Rosenblatt invents the perceptron while working at the Cornell Aeronautical Laboratory. The invention of the perceptron generated a great deal of excitement and was widely covered in the media.
1958	John McCarthy (Massachusetts Institute of Technology or MIT) invented the Lisp programming language.
1958	Herbert Gelernter and Nathan Rochester (IBM) described a theorem prover in geometry that exploits a semantic model of the domain in the form of diagrams of "typical" cases.
1958	Teddington Conference on the Mechanization of Thought Processes was held in the UK and among the papers presented were John McCarthy's Programs with Common Sense, Oliver Selfridge's Pandemonium, and Marvin Minsky's Some Methods of Heuristic Programming and Artificial Intelligence.
1959	The General Problem Solver (GPS) was created by Newell, Shaw and Simon while at CMU.

1959	John McCarthy and Marvin Minsky founded the MIT AI Lab.
Late 1950s, early 1960s	Margaret Masterman and colleagues at University of Cambridge design semantic nets for machine translation.
1960s	Ray Solomonoff lays the foundations of a mathematical theory of AI, introducing universal Bayesian methods for inductive inference and prediction.
1960	Man-Computer Symbiosis by J.C.R. Licklider.
1961	James Slagle (PhD dissertation, MIT) wrote (in Lisp) the first symbolic integration program, SAINT, which solved calculus problems at the college freshman level.
1961	In Minds, Machines and Gödel, John Lucas denied the possibility of machine intelligence on logical or philosophical grounds. He referred to Kurt Gödel's result of 1931: sufficiently powerful formal systems are either inconsistent or allow for formulating true theorems unprovable by any theorem-proving AI deriving all provable theorems from the axioms. Since humans are able to "see" the truth of such theorems, machines were deemed inferior.
1961	Unimation's industrial robot Unimate worked on a General Motors automobile assembly line.
1963	Donald Michie creates a 'machine' consisting of 304 match boxes and beads, which uses reinforcement learning to play Tic-tac-toe (also known as noughts and crosses).
1963	Thomas Evans' program, ANALOGY, written as part of his PhD work at MIT, demonstrated that computers can solve the same analogy problems as are given on IQ tests.
1963	Edward Feigenbaum and Julian Feldman published Computers and Thought, the first collection of articles about artificial intelligence.
1963	Leonard Uhr and Charles Vossler published "A Pattern Recognition Program That Generates, Evaluates, and Adjusts Its Own Operators", which described one of the first machine learning programs that could adaptively acquire and modify features and thereby overcome the limitations of simple perceptrons of Rosenblatt.
1964	Danny Bobrow's dissertation at MIT (technical report #1 from MIT's AI group, Project MAC), shows that computers can understand natural language well enough to solve algebra word problems correctly.
1964	Bertram Raphael's MIT dissertation on the SIR program demonstrates the power of a logical representation of knowledge for question-answering systems.
1965	J. Alan Robinson invented a mechanical proof procedure, the Resolution Method, which allowed programs to work efficiently with formal logic as a representation language.
1965	Joseph Weizenbaum (MIT) built ELIZA, an interactive program that carries on a dialogue in English language on any topic. It was a popular toy at AI centers on the ARPANET when a version that "simulated" the dialogue of a psychotherapist was programmed.
1965	Edward Feigenbaum initiated Dendral, a ten-year effort to develop software to deduce the molecular structure of organic compounds using scientific instrument data. It was the first expert system.
1966	Ross Quillian (PhD dissertation, Carnegie Inst. of Technology, now CMU) demonstrated semantic nets.
1966	Machine Intelligence workshop at Edinburgh – the first of an influential annual series organized by Donald Michie and others.

1966	Negative report on machine translation kills much work in Natural language processing (NLP) for many years.
1967	The nearest neighbor algorithm was created, which is the start of basic pattern recognition. The algorithm was used to map routes.
1967	Dendral program (Edward Feigenbaum, Joshua Lederberg, Bruce Buchanan, Georgia Sutherland at Stanford University) demonstrated to interpret mass spectra on organic chemical compounds. First successful knowledge-based program for scientific reasoning.
1968	Joel Moses (PhD work at MIT) demonstrated the power of symbolic reasoning for integration problems in the Macsyma program. First successful knowledge-based program in mathematics.
1968	Richard Greenblatt (programmer) at MIT built a knowledge-based chess-playing program, MacHack, that was good enough to achieve a class-C rating in tournament play.
1968	Wallace and Boulton's program, Snob (Comp.J. 11(2) 1968), for unsupervised classification (clustering) uses the Bayesian Minimum Message Length criterion, a mathematical realisation of Occam's razor.
1969	Marvin Minsky and Seymour Papert publish their book Perceptrons, describing some of the limitations of perceptrons and neural networks. The interpretation that the book shows that neural networks are fundamentally limited is seen as a hindrance for research into neural networks.
1969	Stanford Research Institute (SRI): Shakey the Robot, demonstrated combining animal locomotion, perception and problem solving.
1969	Roger Schank (Stanford) defined conceptual dependency model for natural language understanding. Later developed (in PhD dissertations at Yale University) for use in story understanding by Robert Wilensky and Wendy Lehnert, and for use in understanding memory by Janet Kolodner.
1969	Yorick Wilks (Stanford) developed the semantic coherence view of language called Preference Semantics, embodied in the first semantics-driven machine translation program, and the basis of many PhD dissertations since such as Bran Boguraev and David Carter at Cambridge.
1969	First International Joint Conference on Artificial Intelligence (IJCAI) held at Stanford.
1969	Marvin Minsky and Seymour Papert publish Perceptrons, demonstrating previously unrecognized limits of this feed-forward two-layered structure. This book is considered by some to mark the beginning of the AI winter of the 1970s, a failure of confidence and funding for AI. Nevertheless, significant progress in the field continued (see below).
1969	McCarthy and Hayes started the discussion about the frame problem with their essay, "Some Philosophical Problems from the Standpoint of Artificial Intelligence".
Early 1970s	Jane Robinson and Don Walker established an influential Natural Language Processing group at SRI.
Mid-1970s	Barbara Grosz (SRI) established limits to traditional AI approaches to discourse modeling. Subsequent work by Grosz, Bonnie Webber and Candace Sidner developed the notion of "centering", used in establishing focus of discourse and anaphoric references in Natural language processing.
Mid-1970s	David Marr and MIT colleagues describe the "primal sketch" and its role in visual perception.
Late 1970s	Stanford's SUMEX-AIM resource, headed by Ed Feigenbaum and Joshua Lederberg, demonstrates the power of the ARPAnet for scientific collaboration.

1970	Seppo Linnainmaa publishes the general method for automatic differentiation (AD) of discrete connected networks of nested differentiable functions. This corresponds to the modern version of backpropagation, but is not yet named as such.
1970	Jaime Carbonell (Sr.) developed SCHOLAR, an interactive program for computer assisted instruction based on semantic nets as the representation of knowledge.
1970	Bill Woods described Augmented Transition Networks (ATN's) as a representation for natural language understanding.
1970	Patrick Winston's PhD program, ARCH, at MIT learned concepts from examples in the world of children's blocks.
1971	Terry Winograd's PhD thesis (MIT) demonstrated the ability of computers to understand English sentences in a restricted world of children's blocks, in a coupling of his language understanding program, SHRDLU, with a robot arm that carried out instructions typed in English.
1971	Work on the Boyer-Moore theorem prover started in Edinburgh.
1972	Karen Spärck Jones publishes the concept of TF-IDF, a numerical statistic that is intended to reflect how important a word is to a document in a collection or corpus. 83% of text-based recommender systems in the domain of digital libraries use tf-idf.
1972	Prolog programming language developed by Alain Colmerauer.
1972	Earl Sacerdoti developed one of the first hierarchical planning programs, ABSTRIPS.
1973	The Assembly Robotics Group at University of Edinburgh builds Freddy Robot, capable of using visual perception to locate and assemble models. (See Edinburgh Freddy Assembly Robot: a versatile computer-controlled assembly system.)
1973	The Lighthill report gives a largely negative verdict on AI research in Great Britain and forms the basis for the decision by the British government to discontinue support for AI research in all but two universities.
1974	Ted Shortliffe's PhD dissertation on the MYCIN program (Stanford) demonstrated a very practical rule-based approach to medical diagnoses, even in the presence of uncertainty. While it borrowed from DENDRAL, its own contributions strongly influenced the future of expert system development, especially commercial systems.
1975	Earl Sacerdoti developed techniques of partial-order planning in his NOAH system, replacing the previous paradigm of search among state space descriptions. NOAH was applied at SRI International to interactively diagnose and repair electromechanical systems.
1975	Austin Tate developed the Nonlin hierarchical planning system able to search a space of partial plans characterised as alternative approaches to the underlying goal structure of the plan.
1975	Marvin Minsky published his widely read and influential article on Frames as a representation of knowledge, in which many ideas about schemas and semantic links are brought together.
1975	The Meta-Dendral learning program produced new results in chemistry (some rules of mass spectrometry) the first scientific discoveries by a computer to be published in a refereed journal.
1976	Douglas Lenat's AM program (Stanford PhD dissertation) demonstrated the discovery model (loosely guided search for interesting conjectures).
1976	Randall Davis demonstrated the power of meta-level reasoning in his PhD dissertation at Stanford.

1978	Tom Mitchell, at Stanford, invented the concept of Version spaces for describing the search space of a concept formation program.
1978	Herbert A. Simon wins the Nobel Prize in Economics for his theory of bounded rationality, one of the cornerstones of AI known as "satisficing".
1978	The MOLGEN program, written at Stanford by Mark Stefik and Peter Friedland, demonstrated that an object-oriented programming representation of knowledge can be used to plan gene-cloning experiments.
1979	Students at Stanford University develop a cart that can navigate and avoid obstacles in a room.
1979	Bill VanMelle's PhD dissertation at Stanford demonstrated the generality of MYCIN's representation of knowledge and style of reasoning in his EMYCIN program, the model for many commercial expert system "shells".
1979	Jack Myers and Harry Pople at University of Pittsburgh developed INTERNIST, a knowledge-based medical diagnosis program based on Dr. Myers' clinical knowledge.
1979	Cordell Green, David Barstow, Elaine Kant and others at Stanford demonstrated the CHI system for automatic programming.
1979	The Stanford Cart, built by Hans Moravec, becomes the first computer-controlled, autonomous vehicle when it successfully traverses a chair-filled room and circumnavigates the Stanford AI Lab.
1979	BKG, a backgammon program written by Hans Berliner at CMU, defeats the reigning world champion (in part via luck).
1979	Drew McDermott and Jon Doyle at MIT, and John McCarthy at Stanford begin publishing work on non-monotonic logics and formal aspects of truth maintenance.
1980s	Lisp machines developed and marketed. First expert system shells and commercial applications.
Mid-1980s	Neural Networks become widely used with the Backpropagation algorithm (first described by Paul Werbos in 1974).
1980	Kunihiko Fukushima first publishes his work on the neocognitron, a type of artificial neural network (ANN). Neocognition later inspires convolutional neural networks (CNNs).
1980	First National Conference of the American Association for Artificial Intelligence (AAAI) held at Stanford.
1981	Gerald Dejong introduces Explanation Based Learning, where a computer algorithm analyses data and creates a general rule it can follow and discard unimportant data.
1981	Danny Hillis designs the connection machine, which utilizes Parallel computing to bring new power to AI, and to computation in general. (Later founds Thinking Machines Corporation)
1982	John Hopfield popularizes Hopfield networks, a type of recurrent neural network that can serve as content-addressable memory systems.
1982	The Fifth Generation Computer Systems project (FGCS), an initiative by Japan's Ministry of International Trade and Industry, begun in 1982, to create a "fifth generation computer" (see history of computing hardware) which was supposed to perform much calculation utilizing massive parallelism.
1983	John Laird and Paul Rosenbloom, working with Allen Newell, complete CMU dissertations on Soar (program).

1983	James F. Allen invents the Interval Calculus, the first widely used formalization of temporal events.
1985	A program that learns to pronounce words the same way a baby does, is developed by Terry Sejnowski.
1985	The autonomous drawing program, AARON, created by Harold Cohen, is demonstrated at the AAAI National Conference (based on more than a decade of work, and with subsequent work showing major developments).
1986	The process of backpropagation is described by David Rumelhart, Geoff Hinton and Ronald J. Williams.
1986	The team of Ernst Dickmanns at Bundeswehr University of Munich builds the first robot cars, driving up to 55 mph on empty streets.
1986	Barbara Grosz and Candace Sidner create the first computation model of discourse, establishing the field of research.
1987	Marvin Minsky published The Society of Mind, a theoretical description of the mind as a collection of cooperating agents. He had been lecturing on the idea for years before the book came out (c.f. Doyle 1983).
1987	Around the same time, Rodney Brooks introduced the subsumption architecture and behavior-based robotics as a more minimalist modular model of natural intelligence; Nouvelle AI.
1987	Commercial launch of generation 2.0 of Alacrity by Alacritous Inc./Allstar Advice Inc. Toronto, the first commercial strategic and managerial advisory system. The system was based upon a forward-chaining, self-developed expert system with 3,000 rules about the evolution of markets and competitive strategies and co-authored by Alistair Davidson and Mary Chung, founders of the firm with the underlying engine developed by Paul Tarvydas. The Alacrity system also included a small financial expert system that interpreted financial statements and models.
1989	Christopher Watkins develops Q-learning, which greatly improves the practicality and feasibility of reinforcement learning.
1989	Axcelis, Inc. releases Evolver, the first software package to commercialize the use of genetic algorithms on personal computers.
1989	Dean Pomerleau at CMU creates ALVINN (An Autonomous Land Vehicle in a Neural Network).
Early 1990s	TD-Gammon, a backgammon program written by Gerry Tesauro, demonstrates that reinforcement (learning) is powerful enough to create a championship-level game-playing program by competing favorably with world-class players.
1990s	Major advances in all areas of AI, with significant demonstrations in machine learning, intelligent tutoring, case-based reasoning, multi-agent planning, scheduling, uncertain reasoning, data mining, natural language understanding and translation, vision, virtual reality, games, and other topics.
Late 1990s	Web crawlers and other AI-based information extraction programs become essential in widespread use of the World Wide Web.
Late 1990s	Demonstration of an Intelligent room and Emotional Agents at MIT's AI Lab.
Late 1990s	Initiation of work on the Oxygen architecture, which connects mobile and stationary computers in an adaptive network.
1991	DART scheduling application deployed in the first Gulf War paid back DARPA's investment of 30 years in AI research.

1992	Gerald Tesauro develops TD-Gammon, a computer backgammon program that uses an artificial neural network trained using temporal-difference learning (hence the 'TD' in the name). TD-Gammon is able to rival, but not consistently surpass, the abilities of top human backgammon players.
1993	Ian Horswill extended behavior-based robotics by creating Polly, the first robot to navigate using vision and operate at animal-like speeds (1 meter/second).
1993	Rodney Brooks, Lynn Andrea Stein and Cynthia Breazeal started the widely publicized MIT Cog project with numerous collaborators, in an attempt to build a humanoid robot child in just five years.
1993	ISX corporation wins "DARPA contractor of the year" for the Dynamic Analysis and Replanning Tool (DART) which reportedly repaid the US government's entire investment in AI research since the 1950s.
1994	With passengers on board, the twin robot cars VaMP and VITA-2 of Ernst Dickmanns and Daimler-Benz drive more than one thousand kilometers on a Paris three-lane highway in standard heavy traffic at speeds up to 130 km/h. They demonstrate autonomous driving in free lanes, convoy driving, and lane changes left and right with autonomous passing of other cars.
1994	English draughts (checkers) world champion Tinsley resigned a match against computer program Chinook. Chinook defeated 2nd highest rated player, Lafferty. Chinook won the USA National Tournament by the widest margin ever.
1995	Tin Kam Ho publishes a paper describing random decision forests.
1995	Corinna Cortes and Vladimir Vapnik publish their work on support vector machines.
1995	"No Hands Across America": A semi-autonomous car drove coast-to-coast across the United States with computer-controlled steering for 2,797 miles (4,501 km) of the 2,849 miles (4,585 km). Throttle and brakes were controlled by a human driver.
1995	One of Ernst Dickmanns' robot cars (with robot-controlled throttle and brakes) drove more than 1000 miles from Munich to Copenhagen and back, in traffic, at up to 120 mph, occasionally executing maneuvers to pass other cars (only in a few critical situations a safety driver took over). Active vision was used to deal with rapidly changing street scenes.
1997	IBM's Deep Blue beats the world champion at chess.
1997	Sepp Hochreiter and Jürgen Schmidhuber invent long short-term memory (LSTM) recurrent neural networks, greatly improving the efficiency and practicality of recurrent neural networks.
1997	The Deep Blue chess machine (IBM) defeats the (then) world chess champion, Garry Kasparov.
1997	First official RoboCup football (soccer) match featuring table-top matches with 40 teams of interacting robots and over 5000 spectators.
1997	Computer Othello program Logistello defeated the world champion Takeshi Murakami with a score of 6–0.
1998	A team led by Yann LeCun releases the MNIST database, a dataset comprising a mix of handwritten digits from American Census Bureau employees and American high school students. The MNIST database has since become a benchmark for evaluating handwriting recognition.
1998	Tiger Electronics' Furby is released, and becomes the first successful attempt at producing a type of A.I to reach a domestic environment.
1998	Tim Berners-Lee published his Semantic Web Road map paper.

1998	Leslie P. Kaelbling, Michael Littman, and Anthony Cassandra introduce the first method for solving POMDPs offline, jumpstarting widespread use in robotics and automated planning and scheduling
1999	Sony introduces an improved domestic robot similar to a Furby, the AIBO becomes one of the first artificially intelligent "pets" that is also autonomous.
2000	Interactive robopets ("smart toys") become commercially available, realizing the vision of the 18th century novelty toy makers.
2000	Cynthia Breazeal at MIT publishes her dissertation on Sociable machines, describing Kismet (robot), with a face that expresses emotions.
2000	The Nomad robot explores remote regions of Antarctica looking for meteorite samples.
2002	Torch, a software library for machine learning, is first released.
2002	iRobot's Roomba autonomously vacuums the floor while navigating and avoiding obstacles.
2004	OWL Web Ontology Language W3C Recommendation (10 February 2004).
2004	DARPA introduces the DARPA Grand Challenge requiring competitors to produce autonomous vehicles for prize money.
2004	NASA's robotic exploration rovers Spirit and Opportunity autonomously navigate the surface of Mars.
2005	Honda's ASIMO robot, an artificially intelligent humanoid robot, is able to walk as fast as a human, delivering trays to customers in restaurant settings.
2005	Recommendation technology based on tracking web activity or media usage brings AI to marketing. See TiVo Suggestions.
2005	Blue Brain is born, a project to simulate the brain at molecular detail.
2006	The Netflix Prize competition is launched by Netflix. The aim of the competition was to use machine learning to beat Netflix's own recommendation software's accuracy in predicting a user's rating for a film given their ratings for previous films by at least 10%. The prize was won in 2009.
2006	The Dartmouth Artificial Intelligence Conference: The Next 50 Years (AI@50) AI@50 (14–16 July 2006)
2007	Philosophical Transactions of the Royal Society, B – Biology, one of the world's oldest scientific journals, puts out a special issue on using AI to understand biological intelligence, titled Models of Natural Action Selection
2007	Checkers is solved by a team of researchers at the University of Alberta.
2007	DARPA launches the Urban Challenge for autonomous cars to obey traffic rules and operate in an urban environment.
2009	ImageNet is created. ImageNet is a large visual database envisioned by Fei-Fei Li from Stanford University, who realized that the best machine learning algorithms wouldn't work well if the data didn't reflect the real world. For many, ImageNet was the catalyst for the AI boom of the 21st century.
2009	Google builds autonomous car.
2010	Kaggle, a website that serves as a platform for machine learning competitions, is launched.

2010	Microsoft launched Kinect for Xbox 360, the first gaming device to track human body movement, using just a 3D camera and infra-red detection, enabling users to play their Xbox 360 wirelessly. The award-winning machine learning for human motion capture technology for this device was developed by the Computer Vision group at Microsoft Research, Cambridge.
2011	Using a combination of machine learning, natural language processing and information retrieval techniques, IBM's Watson beats two human champions in a Jeopardy! competition.
2011	IBM's Watson computer defeated television game show Jeopardy! champions Rutter and Jennings.
2011–2014	Apple's Siri (2011), Google's Google Now (2012) and Microsoft's Cortana (2014) are smartphone apps that use natural language to answer questions, make recommendations and perform actions.
2012	The Google Brain team, led by Andrew Ng and Jeff Dean, create a neural network that learns to recognize cats by watching unlabeled images taken from frames of YouTube videos.
2013	Robot HRP-2 built by SCHAFT Inc of Japan, a subsidiary of Google, defeats 15 teams to win DARPA's Robotics Challenge Trials. HRP-2 scored 27 out of 32 points in 8 tasks needed in disaster response. Tasks are drive a vehicle, walk over debris, climb a ladder, remove debris, walk through doors, cut through a wall, close valves and connect a hose.
2013	NEIL, the Never Ending Image Learner, is released at Carnegie Mellon University to constantly compare and analyze relationships between different images.
2014	Facebook researchers publish their work on DeepFace, a system that uses neural networks that identifies faces with 97.35% accuracy. The results are an improvement of more than 27% over previous systems and rivals human performance.
2014	Researchers from Google detail their work on Sibyl, a proprietary platform for massively parallel machine learning used internally by Google to make predictions about user behavior and provide recommendations.
2015	An open letter to ban development and use of autonomous weapons signed by Hawking, Musk, Wozniak and 3,000 researchers in AI and robotics.
2015	Google DeepMind's AlphaGo (version: Fan) defeated 3 time European Go champion 2 dan professional Fan Hui by 5 games to 0.
2016	Google's AlphaGo program becomes the first Computer Go program to beat an unhandicapped professional human player using a combination of machine learning and tree search techniques. Later improved as AlphaGo Zero and then in 2017 generalized to Chess and more two-player games with AlphaZero.
2016	Google DeepMind's AlphaGo (version: Lee) defeated Lee Sedol 4–1. Lee Sedol is a 9 dan professional Korean Go champion who won 27 major tournaments from 2002 to 2016. Before the match with AlphaGo, Lee Sedol was confident in predicting an easy 5–0 or 4–1 victory.
2017	Asilomar Conference on Beneficial AI was held, to discuss AI ethics and how to bring about beneficial AI while avoiding the existential risk from artificial general intelligence.
2017	Poker AI Libratus individually defeated each of its 4 human opponents—among the best players in the world—at an exceptionally high aggregated winrate, over a statistically significant sample. In contrast to Chess and Go, Poker is an imperfect information game.
2017	Google DeepMind's AlphaGo (version: Master) won 60–0 rounds on two public Go websites including 3 wins against world Go champion Ke Jie.

2017	An OpenAI-machined learned bot played at The International 2017 Dota 2 tournament in August 2017. It won during a 1v1 demonstration game against professional Dota 2 player Dendi.
2017	Google DeepMind revealed that AlphaGo Zero—an improved version of AlphaGo—displayed significant performance gains while using far fewer tensor processing units (as compared to AlphaGo Lee; it used same amount of TPU's as AlphaGo Master). Unlike previous versions, which learned the game by observing millions of human moves, AlphaGo Zero learned by playing only against itself. The system then defeated AlphaGo Lee 100 games to zero, and defeated AlphaGo Master 89 to 11. Although unsupervised learning is a step forward, much has yet to be learned about general intelligence. AlphaZero masters chess in 4 hours, defeating the best chess engine, StockFish 8. AlphaZero won 28 out of 100 games, and the remaining 72 games ended in a draw.
2018	Alibaba language processing AI outscores top humans at a Stanford University reading and comprehension test, scoring 82.44 against 82.304 on a set of 100,000 questions.
2018	The European Lab for Learning and Intelligent Systems (aka Ellis) proposed as a pan-European competitor to American AI efforts, with the aim of staving off a brain drain of talent, along the lines of CERN after World War II.
2018	Announcement of Google Duplex, a service to allow an AI assistant to book appointments over the phone. The LA Times judges the AI's voice to be a "nearly flawless" imitation of human-sounding speech.

A.M. TURING AWARD WINNERS

Year of Award	Recipient	Award citation
2017	John L. Hennessy David Patterson	For pioneering a systematic, quantitative approach to the design and evaluation of computer architectures with enduring impact on the microprocessor industry.
2016	Tim Berners-Lee	For inventing the World Wide Web, the first web browser, and the fundamental protocols and algorithms allowing the Web to scale.
2015	Whitfield Diffi Martin Hellman	For inventing and promulgating both asymmetric public-key cryptography, including its application to digital signatures, and a practical cryptographic key-exchange method.
2014	Michael Stonebraker	For fundamental contributions to the concepts and practices underlying modern database systems.
2013	Leslie Lamport	For fundamental contributions to the theory and practice of distributed and concurrent systems, notably the invention of concepts such as causality and logical clocks, safety and liveness, replicated state machines, and sequential consistency.
2012	Shafi Goldwasser Silvio Micali	For transformative work that laid the complexity-theoretic foundations for the science of cryptography, and in the process pioneered new methods for efficient verification of mathematical proofs in complexity theory.
2011	Judea Pearl	For fundamental contributions to artificial intelligence through the development of a calculus for probabilistic and causal reasoning.
2010	Leslie Gabriel Valiant	For transformative contributions to the theory of computation, including the theory of probably approximately correct (PAC) learning, the complexity of enumeration and of algebraic computation, and the theory of parallel and distributed computing.
2009	Charles P. (Chuck) Thacker*	For the pioneering design and realization of the first modern personal computer – the Alto at Xerox PARC – and seminal inventions and contributions to local area networks (including the Ethernet), multiprocessor workstations, snooping cache coherence protocols, and tablet personal computers.
2008	Barbara Liskov	For contributions to practical and theoretical foundations of programming language and system design, especially related to data abstraction, fault tolerance, and distributed computing.
2007	Edmund Melson Clarke E. Allen Emerson Joseph Sifakis	For their role in developing Model-Checking into a highly effective verification technology that is widely adopted in the hardware and software industries.
2006	Frances ("Fran") Elizabeth Allen	For pioneering contributions to the theory and practice of optimizing compiler techniques that laid the foundation for modern optimizing compilers and automatic parallel execution.
2005	Peter Naur*	For fundamental contributions to programming language design and the definition of Algol 60, to compiler design, and to the art and practice of computer programming.
2004	Vinton ("Vint") Gray Cerf Robert ("Bob") Elliot Kahn	For pioneering work on internetworking, including the design and implementation of the Internet's basic communications protocols, TCP/IP, and for inspired leadership in networking.

2003	Alan Kay	For pioneering many of the ideas at the root of contemporary object-oriented programming languages, leading the team that developed Smalltalk, and for fundamental contributions to personal computing.
2002	Leonard (Len) Max Adleman Ronald (Ron) Linn Rivest Adi Shamir	For their ingenious contribution to making public-key cryptography useful in practice.
2001	Ole-Johan Dahl* Kristen Nygaard*	With Kristen Nygaard, for ideas fundamental to the emergence of object oriented programming, through their design of the programming languages Simula I and Simula 67.
2000	Andrew Chi-Chih Yao	In recognition of his fundamental contributions to the theory of computation, including the complexity-based theory of pseudorandom number generation, cryptography, and communication complexity.
1999	Frederick ("Fred") Brooks	For landmark contributions to computer architecture, operating systems, and software engineering.
1998	James ("Jim") Nicholas Gray*	For seminal contributions to database and transaction processing research and technical leadership in system implementation.
1997	Douglas Engelbart*	For an inspiring vision of the future of interactive computing and the invention of key technologies to help realize this vision.
1996	Amir Pnueli*	For seminal work introducing temporal logic into computing science and for outstanding contributions to program and system verification.
1995	Manuel Blum	In recognition of his contributions to the foundations of computational complexity theory and its application to cryptography and program checking.
1994	Edward A. ("Ed") Feigenbaum Dabbala Rajagopal ("Raj") Reddy	For pioneering the design and construction of large scale artificial intelligence systems, demonstrating the practical importance and potential commercial impact of artificial intelligence technology.
1993	Juris Hartmanis Richard ("Dick") Edwin Stearns	In recognition of their seminal paper which established the foundations for the field of computational complexity theory.
1992	Butler W. Lampson	For contributions to the development of distributed, personal computing environments and the technology for their implementation: workstations, networks, operating systems, programming systems, displays, security and document publishing.
1991	Arthur John Robin Gorell ("Robin") Milne*r	For three distinct and complete achievements: 2 LCF, the mechanization of Scott's Logic of Computable Functions, probably the first theoretically based yet practical tool for machine assisted proof construction; 2 ML, the first language to include polymorphic type inference together with a type-safe exception-handling mechanism; (3 CCS, a general theory of concurrency. In addition, he formulated and strongly advanced full abstraction, the study of the relationship between operational and denotational semantics.
1990	Fernando J. ("Corby") Corbato	For his pioneering work organizing the concepts and leading the development of the general-purpose, large-scale, time-sharing and resource-sharing computer systems, CTSS and Multics.

1989	William ("Velvel") Morton Kahan	For his fundamental contributions to numerical analysis. One of the foremost experts on floating-point computations. Kahan has dedicated himself to "making the world safe for numerical computations"!
1988	Ivan Sutherland	For his pioneering and visionary contributions to computer graphics, starting with Sketchpad, and continuing after.
1987	John Cocke*	For significant contributions in the design and theory of compilers, the architecture of large systems and the development of reduced instruction set computers (RISC); for discovering and systematizing many fundamental transformations now used in optimizing compilers including reduction of operator strength, elimination of common subexpressions, register allocation, constant propagation, and dead code elimination.
1986	John E. Hopcroft Robert (Bob) Endre Tarjan	For fundamental achievements in the design and analysis of algorithms and data structures.
1985	Richard ("Dick") Manning Karp	For his continuing contributions to the theory of algorithms including the development of efficient algorithms for network flow and other combinatorial optimization problems, the identification of polynomial-time computability with the intuitive notion of algorithmic efficiency, and, most notably, contributions to the theory of NP-completeness. Karp introduced the now standard methodology for proving problems to be NP-complete which has led to the identification of many theoretical and practical problems as being computationally difficult.
1984	Niklaus E. Wirth	For developing a sequence of innovative computer languages, EULER, ALGOL-W, MODULA and PASCAL. PASCAL has become pedagogically significant and has provided a foundation for future computer language, systems, and architectural research.
1983	Dennis M.* Ritchie* Kenneth Lane Thompson	For their development of generic operating systems theory and specifically for the implementation of the UNIX operating system.
1982	Stephen Arthur Cook	For his advancement of our understanding of the complexity of computation in a significant and profound way. His seminal paper, "The Complexity of Theorem Proving Procedures," presented at the 1971 ACM SIGACT Symposium on the Theory of Computing, laid the foundations for the theory of NP-Completeness. The ensuing exploration of the boundaries and nature of NP-complete class of problems has been one of the most active and important research activities in computer science for the last decade.
1981	Edgar F. ("Ted") Codd*	For his fundamental and continuing contributions to the theory and practice of database management systems.
1980	C. Antony ("Tony") R. Hoare	For his fundamental contributions to the definition and design of programming languages.
1979	Kenneth E. ("Ken") Iverson*	For his pioneering effort in programming languages and mathematical notation resulting in what the computing field now knows as APL, for his contributions to the implementation of interactive systems, to educational uses of APL, and to programming language theory and practice.
1978	Robert (Bob) W. Floyd*	For having a clear influence on methodologies for the creation of efficient and reliable software, and for helping to found the following important subfields of computer science: the theory of parsing, the semantics of programming languages, automatic program verification, automatic program synthesis, and analysis of algorithms.

1977	John Backus*	For profound, influential, and lasting contributions to the design of practical high-level programming systems, notably through his work on FORTRAN, and for seminal publication of formal procedures for the specification of programming languages.
1976	Michael O. Rabin Dana Stewart Scott	For their joint paper "Finite Automata and Their Decision Problem," which introduced the idea of nondeterministic machines, which has proved to be an enormously valuable concept. Their (Scott & Rabin) classic paper has been a continuous source of inspiration for subsequent work in this field.
1975	Allen Newell* Herbert ("Herb") Alexander Simon*	In joint scientific efforts extending over twenty years, initially in collaboration with J. C. Shaw at the RAND Corporation, and subsequentially with numerous faculty and student collegues at Carnegie-Mellon University, Newell and co-recipient Herbert A. Simon made basic contributions to artificial intelligence, the psychology of human cognition, and list processing.
1974	Donald ("Don") Ervin Knuth	For his major contributions to the analysis of algorithms and the design of programming languages, and in particular for his contributions to the "art of computer programming" through his well-known books in a continuous series by this title.
1973	Charles William Bachman*	For his outstanding contributions to database technology.
1972	Edsger Wybe Dijkstra*	For fundamental contributions to programming as a high, intellectual challenge; for eloquent insistence and practical demonstration that programs should be composed correctly, not just debugged into correctness; for illuminating perception of problems at the foundations of program design.
1971	John McCarthy*	Dr. McCarthy's lecture "The Present State of Research on Artificial Intelligence" is a topic that covers the area in which he has achieved considerable recognition for his work.
1970	James Hardy ("Jim") Wilkinson*	For his research in numerical analysis to facilitiate the use of the high-speed digital computer, having received special recognition for his work in computations in linear algebra and "backward" error analysis.
1969	Marvin Minsky*	For his central role in creating, shaping, promoting, and advancing the field of Artificial Intelligence.
1968	Richard W. Hamming*	For his work on numerical methods, automatic coding systems, and error-detecting and error-correcting codes.
1967	Maurice V. Wilkes*	Professor Wilkes is best known as the builder and designer of the EDSAC, the first computer with an internally stored program. Built in 1949, the EDSAC used a mercury delay line memory. He is also known as the author, with Wheeler and Gill, of a volume on "Preparation of Programs for Electronic Digital Computers" in 1951, in which program libraries were effectively introduced.
1966	Alan J Perlis*	For his influence in the area of advanced programming techniques and compiler construction.

GLOSSARY

1-bit watermarking: a type of digital watermark that embeds one bit of binary data in the signal to be transmitted; also called "0-bit watermarking."

3-D rendering: the process of creating a 2-D animation using 3-D models.

3D Touch: a feature that senses the pressure with which users exert upon Apple touch screens.

abstract class: A class that cannot be directly constructed, one that can be constructed only through construction of some of its subclasses.

abstract type: A type in a nominative type system that cannot be instantiated.

abstract: [Mathematical] algebras and logics that describe several different concrete algebras and logics; Software engineering] descriptions that do not swamp you with unnecessary detail; they provide enough information to use something without knowing its detailed construction.

abstraction: a technique used to reduce the structural complexity of programs, making them easier to create, understand, maintain, and use.

accelerometer IC: an integrated circuit that measures acceleration.

access level: in a computer security system, a designation assigned to a user or group of users that allows access a predetermined set of files or functions.

accessor: A method or member function that does not change the object to which it is applied, also known as a "const" function.

actual argument: A value, or reference to a value, passed to a function.

actual parameter: Any parameter in the call of a subprogram.

actuator: a motor designed to control the movement of a device or machine by transforming potential energy into kinetic energy.

Additive White Gaussian Noise (AWGN): a model used to represent imperfections in real communication channels.

address space: the amount of memory allocated for a file or process on a computer.

adware: software that generates advertisements to present to a computer user.

affinity chromatography: a technique for separating a particular biochemical substance from a mixture based on its specific interaction with another substance.

agile software development: an approach to software development that addresses changing requirements as they arise throughout the process, with programming, implementation, and testing occurring simultaneously.

algorithm: A description in precise but natural language plus mathematical notation of how a problem is solved.

algorithm: a set of step-by-step instructions for performing computations.

alias: Two names or identifiers are aliases if they name or identify the same thing.

American National Standards Institute (ANSI): a nonprofit organization that oversees the creation and use of standards and certifications such as those offered by CompTIA.

amplifier: a device that strengthens the power, voltage, or current of a signal.

analog signal: a continuous signal whose values or quantities vary over time.

analytic combinatorics: a method for creating precise quantitative predictions about large sets of objects.

Android Open Source Project: a project undertaken by a coalition of mobile phone manufacturers and other interested parties, under the leadership of Google. The purpose of the project is to develop the Android platform for mobile devices.

animation variables (avars): defined variables used in computer animation to control the movement of an animated figure or object.

anthropomorphic: resembling a human in shape or behavior; from the Greek words *anthropos* (human) and *morphe* (form).

app: An application that executes on a small, handheld device.

application program interface (API): the code that defines how two pieces of software interact, particularly a software application and the operating system on which it runs.

application suite: a set of programs designed to work closely together, such as an office suite that includes a word processor, spreadsheet, presentation creator, and database application.

application: A program or integrated suite of programs that has a defined function.

application-level firewalls: firewalls that serve as proxy servers through which all traffic to and from applications must flow.

application-specific GUI: a graphical interface designed to be used for a specific application.

argument: A value, or reference to a value, passed to a function; an actual argument.

arithmetic operations: Addition, subtraction, multiplication, and division ideally forming an abstract data type (ADT) with the algebraic properties of a ring or field.

array: An ordered sequence of same-typed values whose elements are fast to access by their numerical index in the array.

artificial intelligence: the intelligence exhibited by machines or computers, in contrast to human, organic, or animal intelligence.

ASCII: The original common character code for computers using 8 bits.

assignment: [Statement] A statement with an expression and a variable. The expression is evaluated and the result is stored in the variables.

associativity: Rules for determining which of two identical infix operators should be evaluated first.

asymmetric-key encryption: a process in which data is encrypted using a public encryption key but can only be decrypted using a different, private key.

attenuation: the loss of intensity from a signal being transmitted through a medium.

attributes: the specific features that define an object's properties or characteristics.

audio codec: a program that acts as a "coder-decoder" to allow an audio stream to be encoded for storage or transmission and later decoded for playback.

authentication: the process by which the receiver of encrypted data can verify the identity of the sender or the authenticity of the data.

automatic sequential control system: a mechanism that performs a multistep task by triggering a series of actuators in a particular sequence.

automaton: a machine that mimics a human but is generally considered to be unthinking.

autonomic components: self-contained software or hardware modules with an embedded capacity for self-management, connected via input/outputs to other components in the system.

autonomous agent: a system that acts on behalf of another entity without being directly controlled by that entity.

autonomous: able to operate independently, without external or conscious control.

backdoor: a hidden method of accessing a computer system that is placed there without the knowledge of the system's regular user in order to make it easier to access the system secretly.

base-16: a number system using sixteen symbols, 0 through 9 and A through F.

base-2 system: a number system using the digits 0 and 1.

BASIC: Beginners All-purpose Symbolic Instruction Code, a family of languages developed for teaching programming and given away with early IBM PCs.

BCD-to-seven-segment decoder/driver: a logic gate that converts a four-bit binary-coded decimal (BCD) input to decimal numerals that can be output to a seven-segment digital display.

behavioral marketing: advertising to users based on their habits and previous purchases.

binary: Pertaining to 2. Binary operators have two operands. Binary numbers have base 2 and use 2 symbols.

binder jetting: the use of a liquid binding agent to fuse layers of powder together.

binding: A relationship between two things, typically an identifier and some one of its properties or attributes. For example a variable is an identifier bound to a piece of storage in the main memory of the computer.

bioinformatics: the scientific field focused on developing computer systems and software to analyze and examine biological data.

bioinstrumentation: devices that combine biology and electronics in order to interface with a patient's body and record or monitor various health parameters.

biomarker: short for "biological marker"; a measurable quality or quantity (e.g., internal temperature, amount of iron dissolved in blood) that serves as an indicator of an organism's health, or some other biological phenomenon or state.

biomaterials: natural or synthetic materials that can be used to replace, repair, or modify organic tissues or systems.

biomechanics: the various mechanical processes such as the structure, function, or activity of organisms.

bioMEMS: short for "biomedical micro-electromechanical system"; a microscale or nanoscale self-contained device used for various applications in health care.

biometrics: measurements that can be used to distinguish individual humans, such as a person's height, weight, fingerprints, retinal pattern, or genetic makeup.

bionics: the use of biologically based concepts and techniques to solve mechanical and technological problems.

biosignal processing: the process of capturing the information the body produces, such as heart rate, blood pressure, or levels of electrolytes, and analyzing it to assess a patient's status and to guide treatment decisions.

bit rate: the amount of data encoded for each second of video; often measured in kilobits per second (kbps) or kilobytes per second (Kbps).

bit width: the number of bits used by a computer or other device to store integer values or other data.

black-box testing: a testing technique in which function is analyzed based on output only, without knowledge of structure or coding.

block: [Program structure] A piece of source code that has one or more declarations in it.

blotting: a method of transferring RNA, DNA, and other proteins onto a substrate for analysis.

Boolean: [Adjective] Any data type that follows George Boole's algebraic axioms. The commonest Boolean data has two values {true, false} and the operations of and, or, and not.

bootstrapping: a self-starting process in a computer system, configured to automatically initiate other processes after the booting process has been initiated.

bridge: a connection between two or more networks, or segments of a single network, that allows the computers in each network or segment to communicate with one another.

broadcast: an audio or video transmission sent via a communications medium to anyone with the appropriate receiver.

building information modeling (BIM): the creation of a model of a building or facility that accounts for its function, physical attributes, cost, and other characteristics.

butterfly effect: an effect in which small changes in a system's initial conditions lead to major, unexpected changes as the system develops.

byte: a group of eight bits.

C: Programming Language invented to help develop operating systems.

C++: Hybrid child of C with object oriented features and generic functions and classes.

caching: the storage of data, such as a previously accessed web page, in order to load it faster upon future access.

call by(X): Old-fashioned way of saying

call: To make use of something by writing its name and the correct protocol; A piece of code that transfers control, temporarily, to a subprogram and suspends the original code until the subprogram returns to the following statement etc.

carrier signal: an electromagnetic frequency that has been modulated to carry analog or digital information.

cathode ray tube (CRT): a vacuum tube used to create images in devices such as older television and computer monitors.

central processing unit (CPU): electronic circuitry that provides instructions for how a computer handles processes and manages data from applications and programs.

chain: [Data structure] Any kind of linked list, a set of records where each record identifies the next record in some sequence or other.

channel capacity: the upper limit for the rate at which information transfer can occur without error.

character: a unit of information that represents a single letter, number, punctuation mark, blank space, or other symbol used in written language.

chatterbot: a computer program that mimics human conversation responses in order to interact with people through text; also called "talkbot," "chatbot," or simply "bot."

circular wait: a situation in which two or more processes are running and each one is waiting for a resource that is being used by another; one of the necessary conditions for deadlock.

class: a collection of independent objects that share similar properties and behaviors.

class-based inheritance: a form of code reuse in which attributes are drawn from a preexisting class to create a new class with additional attributes.

clinical engineering: the design of medical devices to assist with the provision of care.

clock speed: the speed at which a microprocessor can execute instructions; also called "clock rate."

CLOS: "Common LISP Object System", a modern LISP.

COBOL: COmmon Business Oriented Language.

code: [Noun] A piece of text that can not be understood without a key, hence the source code for a program

coding theory: the study of codes and their use in certain situations for various applications.

coercion: An implicit type conversion that lets a smart compiler work out the wrong meaning for a programmers typing mistake.

combinatorial design: the study of the creation and properties of finite sets in certain types of designs.

command line: a text-based computer interface that allows the user to input simple commands via a keyboard.

command-line interpreter: an interface that interprets and carries out commands entered by the user.

commodities: consumer products, physical articles of trade or commerce.

communication architecture: the design of computer components and circuitry that facilitates the rapid and efficient transmission of signals between different parts of the computer.

communication devices: devices that allow drones to communicate with users or engineers in remote locations.

compile: [Verb] Translate source code into executable object code.

compiler: A computer program which transforms source code into object code.

compliance: adherence to standards or specifications established by an official body to govern a particular industry, product, or activity.

component: [Technology] A unit of composition with contractually specified interfaces and only explicit context dependencies; components can be deployed and composed by third parties, often a collection of objects with a given set of methods for handling them and abstract classes that can be defined by other people.

component-based development: an approach to software design that uses standardized software components to create new applications and new software.

compound: A single statement or object that can have any number of other statements as its parts.

compressed data: data that has been encoded such that storing or transferring the data requires fewer bits of information.

computational linguistics: a branch of linguistics that uses computer science to analyze and model language and speech.

Computer Fraud and Abuse Act (CFAA): a 1986 legislative amendment that made accessing a protected computer without authorization, or exceeding one's authorized level of access, a federal offense.

computer technician: a professional tasked with the installation, repair, and maintenance of computers and related technology.

conditional: An expression or statement that selects one out of a number of alternative subexpressions.

constant: An identifier that is bound to an invariant value.

constraints: limitations on values in computer programming that collectively identify the solutions to be produced by a programming problem.

constructor: A method or function in a class that creates a new object of that class.

context switch: a multitasking operating system shifting from one task to another; for example, after formatting a print job for one user, the computer might switch to resizing a graphic for another user.

context switching: pausing and recording the progress of a thread or process such that the process or thread can be executed at a later time.

control characters: units of information used to control the manner in which computers and other devices process text and other characters.

control statement: Statements that permit a processor to select the next of several possible computations according to various conditions.

control unit design: describes the part of the CPU that tells the computer how to perform the instructions sent to it by a program.

converter: a device that expands a system's range of reception by bringing in and adapting signals that the system was not originally designed to process.

cookies: small data files that allow websites to track users.

cooperative multitasking: an implementation of multitasking in which the operating system will not initiate a context switch while a process is running in order to allow the process to complete.

core voltage: the amount of power delivered to the processing unit of a computer from the power supply.

counter: a digital sequential logic gate that records how many times a certain event occurs in a given amount of time.

coupling: the degree to which different parts of a program are dependent upon one another.

cracker: a criminal hacker; one who finds and exploits weak points in a computer's security system to gain unauthorized access for malicious purposes.

crippleware: software programs in which key features have been disabled and can only be activated after registration or with the use of a product key.

crosstalk: interference of the signals on one circuit with the signals on another, caused by the two circuits being too close together.

cybercrime: crime that involves targeting a computer or using a computer or computer network to commit a crime, such as computer hacking, digital piracy, and the use of malware or spyware.

data granularity: the level of detail with which data is collected and recorded.

data integrity: the degree to which collected data is and will remain accurate and consistent.

data source: the origin of the information used in a computer model or simulation, such as a database or spreadsheet.

data type: A collection of values together with the operations that use them and produce them, plus the assumptions that can be made about the operations and values.

data width: a measure of the amount of data that can be transmitted at one time through the computer bus, the specific circuits and wires that carry data from one part of a computer to another.

datapath design: describes how data flows through the CPU and at what points instructions will be decoded and executed.

declaration: A piece of source code that adds a name to the program's environment and binds it to a class of meanings, and may also define the name.

declarative language: language that specifies the result desired but not the sequence of operations needed to achieve the desired result.

deep learning: an emerging field of artificial intelligence research that uses neural network algorithms to improve machine learning.

default constructor: Actions carried out to create an object of a given class when no other data is provided about the object.

default: An item provided in place of an omitted item.

definition: A piece of source code or text that binds a name to a precise "definite" meaning. A definition may implicitly also declare the name at the same time or bind more information to an already defined name.

delta debugging: an automated method of debugging intended to identify a bug's root cause while eliminating irrelevant information.

destructor: In object-oriented programming, the command sequence that is launched when the execution of an object is finished.

deterministic algorithm: an algorithm that when given a particular input will always produce the same output.

device fingerprinting: information that uniquely identifies a particular computer, component, or piece of software installed on the computer. This can be used to find out precisely which device accessed a particular online resource.

device: equipment designed to perform a specific function when attached to a computer, such as a scanner, printer, or projector.

dexterity: finesse; skill at performing delicate or precise tasks.

digital commerce: the purchase and sale of goods and services via online vendors or information technology systems.

digital legacy (digital remains): the online accounts and information left behind by a deceased person.

digital literacy: familiarity with the skills, behaviors, and language specific to using digital devices to access, create, and share content through the Internet.

digital native: an individual born during the digital age or raised using digital technology and communication.

direct manipulation interfaces: computer interaction format that allows users to directly manipulate graphical objects or physical shapes that are automatically translated into coding.

direct-access storage: a type of data storage in which the data has a dedicated address and location on the storage device, allowing it to be accessed directly rather than sequentially.

directed energy deposition: a process that deposits wire or powdered material onto an object and then melts it using a laser, electron beam, or plasma arc.

distributed algorithm: an algorithm designed to run across multiple processing centers and so is capable of directing a concentrated action between several computer systems.

domain: the range of values that a variable may take on, such as any even number or all values less than –23.7.

domain-dependent complexity: a complexity that results from factors specific to the context in which the computational problem is set.

DRAKON chart: a flowchart used to model algorithms and programmed in the hybrid DRAKON computer language.

dual-mesh network: a type of mesh network in which all nodes are connected both through wiring and wirelessly, thus increasing the reliability of the system. A node is any communication intersection or endpoint in the network (e.g., computer or terminal).

dump: A formatted listing of the contents of program storage, especially when produced automatically by a failing program

dynamic balance: the ability to maintain balance while in motion.

dynamic binding: A binding that can be made at any time as a program runs and may change as the program runs.

dynamic polymorphism: A kind of polymorphism where the current type of an object determines which of several alternate subprograms is invoked as the program runs.

dynamic random-access memory (DRAM): a form of RAM in which the device's memory must be refreshed on a regular basis, or else the data it contains will disappear.

dynamic scoping: Determining the global environment of a subprogram as that which surrounds its call.

dynamic testing: a testing technique in which software input is tested and output is analyzed by executing the code.

dynamic: Something that is done as the program runs rather than by the compiler before the program runs.

EBNF: Extended BNF. A popular way to define syntax as a dictionary of terms defined by using iteration, options, alternatives, etc. [BNF].

edit decision list (EDL): a list that catalogs the reel or time code data of video frames so that the frames can be accessed during video editing.

electromagnetic spectrum: the complete range of electromagnetic radiation, from the longest wavelength and lowest frequency (radio waves) to the shortest wavelength and highest frequency (gamma rays).

Electronic Communications Privacy Act: a 1986 law that extended restrictions on wiretapping to cover the retrieval or interception of information transmitted electronically between computers or through computer networks.

electronic interference: the disturbance generated by a source of electrical signal that affects an electrical circuit, causing the circuit to degrade or malfunction. Examples include noise, electrostatic discharge, and near-field and far-field interference.

embedded systems: computer systems that are incorporated into larger devices or systems to monitor performance or to regulate system functions.

encapsulated: [Programming] Coding that can be changed with out breaking client code; Being able to place all relevant information in the same piece of code; for example data and the operations that manipulate it in a C++ class.

encapsulation: [Programming] The ability to hide unwanted details inside an interface so that the result works like a black box or vending machine, providing useful services to many clients (programs or people).

entanglement: the phenomenon in which two or more particles' quantum states remain linked even if the particles are later separated and become part of distinct systems.

enumeration: A data type whose values are a set of mutually exclusive named constants.

enumerative combinatorics: a branch of combinatorics that studies the number of ways that certain patterns can be formed using a set of objects.

e-waste: short for "electronic waste"; computers and other digital devices that have been discarded by their owners.

exception: An interruption in normal processing, especially as caused by an error condition.

expression: A shorthand description of a calculation.

false match rate: the probability that a biometric system incorrectly matches an input to a template contained within a database.

fault detection: the monitoring of a system in order to identify when a fault occurs in its operation.

fiber: a small thread of execution using cooperative multitasking.

field programmable gate array: an integrated circuit that can be programmed in the field and can therefore allow engineers or users to alter a machine's programming without returning it to the manufacturer.

filter: in signal processing, a device or procedure that takes in a signal, removes certain unwanted elements, and outputs a processed signal.

firewall: a virtual barrier that filters traffic as it enters and leaves the internal network, protecting internal resources from attack by external sources.

fixed point: a type of digital signal processing in which numerical data is stored and represented with a fixed number of digits after (and sometimes before) the decimal point, using a minimum of sixteen bits.

fixed-point arithmetic: a calculation involving numbers that have a defined number of digits before and after the decimal point.

flash memory: nonvolatile computer memory that can be erased or overwritten solely through electronic signals, i.e. without physical manipulation of the device.

flashing: a process by which the flash memory on a motherboard or an embedded system is updated with a newer version of software.

floating point: a type of digital signal processing in which numerical data is stored and represented in the form of a number (called the mantissa, or significand) multiplied by a base number (such as base-2) raised to an exponent, using a minimum of thirty-two bits.

floating-point arithmetic: a calculation involving numbers that have a decimal point that can be placed anywhere through the use of exponents, as is done in scientific notation.

flooding: sending information to every other node in a network to get the data to its appropriate destination.

flow chart: A schematic representation of the logic that defines the flow of control through a program

force-sensing touch technology: touch display that can sense the location of the touch as well as the amount of pressure the user applies, allowing for a wider variety of system responses to the input.

formal argument: A parameter in a function definition.

formal parameter: The symbol used inside a subprogram in place of the actual parameter provided when the subprogram is called.

FORTRAN: FORmula TRANslation. There have been many FORTRANs. The series includes

four-dimensional building information modeling (4-D BIM): the process of creating a 3-D model that incorporates time-related information to guide the manufacturing process.

Fourier transform: a mathematical operator that decomposes a single function into the sum of multiple sinusoidal (wave) functions.

frames per second (FPS): a measurement of the rate at which individual video frames are displayed.

free software: software developed by programmers for their own use or for public use and distributed without charge; it usually has conditions attached that prevent others from acquiring it and then selling it for their own profit.

function: A subprogram that returns a value but can not change its parameters or have side effects; Any subprogram; [Mathematics] A total many to one relation between a domain and a co-domain; A routine that receives zero or more arguments and may return a result.

functional programming: A programming paradigm that treats computation as the evaluation of mathematical functions, avoids state and mutable data, and makes it easy to construct functions as if they were data objects.

functional requirement: a specific function of a computer system, such as calculating or data processing.

fundamental data type: A type of data that is not defined by a class, struct, or union declaration.

game loop: the main part of a game program that allows the game's physics, artificial intelligence, and graphics to continue to run with or without user input.

garbage: A piece of storage that has been allocated but can no longer be accessed by a program. If not collected and recycled garbage can cause a memory leak.

gateway: a device capable of joining one network to another that has different protocols.

generic: A package or Subprogram that can generate a large number of similar yet different packages or subprograms. See template.

genetic modification: direct manipulation of an organism's genome, often for the purpose of engineering useful microbes or correcting for genetic disease.

genome-wide association study: a type of genetic study that compares the complete genomes of individuals within a population to find which genetic markers, if any, are associated with various traits, most often diseases or other health problems.

gestures: combinations of finger movements used to interact with multitouch displays in order to accomplish various tasks. Examples include tapping the finger on the screen, double-tapping, and swiping the finger along the screen.

global: Something that can be used in all parts of program. Compare with

Glossary:

glyph: a specific representation of a grapheme, such as the letter A rendered in a particular typeface.

goto: A 4 letter word no longer considered correct that is still usable in all practical languages to indicate an unconditional jump.

grammar: [Math] A set of definitions that define the syntax of a language. A grammar generates the strings in the language and so implicitly describes how to recognize and parse strings in the language.

graph theory: the study of graphs, which are diagrams used to model relationships between objects.

grapheme: the smallest unit used by a writing system, such as alphabetic letters, punctuation marks, or Chinese characters.

graphical user interface (GUI): an interface that allows users to control a computer or other device by interacting with graphical elements such as icons and windows.

Green Electronics Council: a US nonprofit organization dedicated to promoting green electronics.

hacking: the use of technical skill to gain unauthorized access to a computer system; also, any kind of advanced tinkering with computers to increase their utility.

hamming distance: a measurement of the difference between two characters or control characters that effects character processing, error detection, and error correction.

hardware interruption: a device attached to a computer sending a message to the operating system to inform it that the device needs attention, thereby "interrupting" the other tasks that the operating system was performing.

hardware: the physical parts that make up a computer. These include the motherboard and processor, as well as input and output devices such as monitors, keyboards, and mice.

Harvard architecture: a computer design that has physically distinct storage locations and signal routes for data and for instructions.

hash function: an algorithm that converts a string of characters into a different, usually smaller, fixed-length string of characters that is ideally impossible either to invert or to replicate.

hashing algorithm: a computing function that converts a string of characters into a different, usually smaller string of characters of a given length, which is ideally impossible to replicate without knowing both the original data and the algorithm used.

header file: [C++] A collection of function headers, class interfaces, constants and definitions that is read by a compiler and changes the interpretation of the rest of the program by (for example) defining operation for handling strings. Contrast with object file.

header: [C++] The first part of a function definition that describes how to call the function but does not describe what it does. The header defines the function's signature. A function header can be separated from its function when the body of the function is replaced by a semicolon. This allows information hiding and separate development.

Health Insurance Portability and Accountability Act (HIPAA): a 1996 law that established national standards for protecting individuals' medical records and other personal health information.

heap: An area of memory reserved for dynamically allocated data objects, contrasted to the stack.

heavy metal: one of several toxic natural substances often used as components in electronic devices.

heterogeneous scalability: the ability of a multiprocessor system to scale up using parts from different vendors.

hexadecimal: a base-16 number system that uses the digits 0 through 9 and the letters A, B, C, D, E, and F as symbols to represent numbers.

hibernation: a power-saving state in which a computer shuts down but retains the contents of its random-access memory.

hidden Markov model: a type of model used to represent a dynamic system in which each state can only be partially seen.

hierarchical file system: a directory with a treelike structure in which files are organized.

high-fidelity wireframe: an image or series of images that represents the visual elements of the website in great detail, as close to the final product as possible, but still does not permit user interaction.

homebrew: software that is developed for a device or platform by individuals not affiliated with the device manufacturer; it is an unofficial or "homemade" version of the software that is developed to provide additional functionality not included or not permitted by the manufacturer.

hopping: the jumping of a data packet from one device or node to another as it moves across the network. Most transmissions require each packet to make multiple hops.

horizontal scaling: the addition of nodes (e.g., computers or servers) to a distributed system.

host-based firewalls: firewalls that protect a specific device, such as a server or personal computer, rather than the network as a whole.

HTML editor: a computer program for editing web pages encoded in hypertext markup language (HTML).

HTML: HyperText Markup Language; used to define pages on the WWW.

Human Brain Project: a project launched in 2013 in an effort at modeling a functioning brain by 2023; also known as HBP.

humanoid: resembling a human.

hybrid cloud: a cloud computing model that combines public cloud services with a private cloud platform linked through an encrypted connection.

identifier: A formal name used in source code to refer to a variable, function, procedure, package, etc.

identifiers: measurable characteristics used to identify individuals.

identity operation: An operation that returns its arguments unchanged.

identity: [Mathematics] An equation that is true for all values of its variables.

imitation game: Alan Turing's name for his proposed test, in which a machine would attempt to respond to questions in such a way as to fool a human judge into thinking it was human.

immersive mode: a full-screen mode in which the status and navigation bars are hidden from view when not in use.

imperative language: language that instructs a computer to perform a particular sequence of operations.

imperative programming: programming that produces code that consists largely of commands issued to the computer, instructing it to perform specific actions.

implementation: The way something is made to work. There are usually many ways to implement something.

in: A way of handling parameters that gives a subprogram access to the value of an actual parameter without permitting the subprogram to change the actual parameter. Often implemented by pass by value.

in-circuit emulator: a device that enables the debugging of a computer system embedded within a larger system.

infix: An operator that is placed between operands. Infix notation dates back to the invention of algebra 3 or 4 hundred years ago.

information hiding: The doctrine that design choices should be hidden by the modules in which they are implemented

information hierarchy: the relative importance of the information presented on a web page.

information technology: the use of computers and related equipment for the purpose of processing and storing data.

infrastructure as a service: a cloud computing platform that provides additional computing resources by linking hardware systems through the Internet; also called "hardware as a service."

inheritance: [Objects] The ability to easily construct new data types or classes by extending existing structures, data types, or classes.

Initiative for Software Choice (ISC): a consortium of technology companies founded by CompTIA, with the goal of encouraging governments to allow competition among software manufacturers.

inout: A way of handling parameters that lets a subprogram both use and change the values of an actual parameter. It can be implemented by pass by reference, pass by name, or pass by value result.

input/output instructions: instructions used by the central processing unit (CPU) of a computer when information is transferred between the CPU and a device such as a hard disk.

input: Data supplied to some program, subprogram, OS, machine, system, or abstraction.

int: [Integer data type] Fixed point data representing a subset of the whole numbers.

integer: A data type for integer values.

integration testing: a process in which multiple units are tested individually and when working in concert.

interaction design: the practice of designing a user interface, such as a website, with a focus on how to facilitate the user's experience.

interface metaphors: linking computer commands, actions, and processes with real-world actions, processes, or objects that have functional similarities.

interface: the function performed by the device driver, which mediates between the hardware of the peripheral and the hardware of the computer.

interference: anything that disrupts a signal as it moves from source to receiver.

interferometry: a technique for studying biochemical substances by superimposing light waves, typically one reflected from the substance and one reflected from a reference point, and analyzing the interference.

internal-use software: software developed by a company for its own use to support general and administrative functions, such as payroll, accounting, or personnel management and maintenance.

Internet of things: a wireless network connecting devices, buildings, vehicles, and other items with network connectivity.

interpolation: a process of estimating intermediate values when nearby values are known; used in image editing to "fill in" gaps by referring to numerical data associated with nearby points.

interpreter: A program which executes another program written in a programming language other than machine code.

interrupt vector table: a chart that lists the addresses of interrupt handlers.

intrusion detection system: a system that uses hardware, software, or both to monitor a computer or network in order to determine when someone attempts to access the system without authorization.

inverter: a logic gate whose output is the inverse of the input; also called a NOT gate.

iOS: Apple's proprietary mobile operating system, installed on Apple devices such as the iPhone, iPad, and iPod touch.

IRL relationships: relationships that occur "in real life," meaning that the relationships are developed or sustained outside of digital communication.

item field: A component in a compound data structure.

iterator: An object that is responsible for tracking progress through a collection of other objects. Often it is implemented as a reference or pointer plus methods for navigating the set of objects. The C++ STL provides many iterators Java has an Enumeration class for iterators.

jailbreak: the removal of restrictions placed on a mobile operating system to give the user greater control over the mobile device.

JavaScript: a flexible programming language that is commonly used in website design.

jQuery: a free, open-source JavaScript library.

kernel: the central component of an operating system.

keyframing: a part of the computer animation process that shows, usually in the form of a drawing, the position and appearance of an object at the beginning of a sequence and at the end.

language interface packs: programs that translate interface elements such as menus and dialog boxes into different languages.

learner-controlled program: software that allows a student to set the pace of instruction, choose which content areas to focus on, decide which areas to explore when, or determine the medium or difficulty level of instruction; also known as a "student-controlled program."

learning strategy: a specific method for acquiring and retaining a particular type of knowledge, such as memorizing a list of concepts by setting the list to music.

learning style: an individual's preferred approach to acquiring knowledge, such as by viewing visual stimuli, reading, listening, or using one's hands to practice what is being taught.

lexicon: the total vocabulary of a person, language, or field of study.

linear predictive coding: a popular tool for digital speech processing that uses both past speech samples and a mathematical approximation of a human vocal tract to predict and then eliminate certain vocal frequencies that do not contribute to meaning. This allows speech to be processed at a faster bit rate without significant loss in comprehensibility.

link editor: Link loader.

link loader: A program that carries out the last stage of compilation by binding together the different uses of identifiers in different files.

link: [Verb] to connect to things together. In computing

LISP: LISt Processing language, The key versions are LISP1.5, CLOS, and Scheme.

livelock: a situation in which two or more processes constantly change their state in response to one another in such a way that neither can complete.

loader: Nowadays a link loader, in the past any program that placed an executable program and placed it into memory.

local area network (LAN): a network that connects electronic devices within a limited physical area.

local: Related to the current instruction rather than a larger context.

logic implementation: the way in which a CPU is designed to use the open or closed state of combinations of circuits to represent information.

logic programming: A style or paradigm of computer programming exemplified by the language Prolog.

logical copy: a copy of a hard drive or disk that captures active data and files in a different configuration from the original, usually excluding free space and artifacts such as file remnants; contrasts with a physical copy, which is an exact copy with the same size and configuration as the original.

logotype: a company or brand name rendered in a unique, distinct style and font; also called a "wordmark."

long: A Fixed point data type that may have more bits than ints.

lossless compression: data compression that allows the original data to be compressed and reconstructed without any loss of accuracy.

lossy compression: a method of decreasing image file size by discarding some data, resulting in some image quality being irreversibly sacrificed.

low-energy connectivity IC: an integrated circuit that enables wireless Bluetooth connectivity while using little power.

LZW compression: a type of lossless compression that uses a table-based algorithm.

machine code: System of instructions and data directly understandable by a computer's central processing unit.

magnetic storage: a device that stores information by magnetizing certain types of material.

main loop: the overarching process being carried out by a computer program, which may then invoke subprocesses.

main memory: the primary memory system of a computer, often called "random access memory" (RAM), accessed by the computer's central processing unit (CPU).

mapping: [Mathematics] A relationship that takes something and turns it into something uniquely determined by the relationship.

mastering: the creation of a master recording that can be used to make other copies for distribution.

Material Design: a comprehensive guide for visual, motion, and interaction design across Google platforms and devices.

material extrusion: a process in which heated filament is extruded through a nozzle and deposited in layers, usually around a removable support.

material jetting: a process in which drops of liquid photopolymer are deposited through a printer head and heated to form a dry, stable solid.

Matrix: [Mathematics] An ADT that can be implemented by rectangular arrays and has many of the arithmetic operations defined on them. A matrix abstracts the structure and behavior of linear maps.

Medical Device Innovation Consortium: a nonprofit organization established to work with the US Food and Drug Administration on behalf of medical device manufacturers to ensure that these devices are both safe and effective.

medical imaging: the use of devices to scan a patient's body and create images of the body's internal structures to aid in diagnosis and treatment planning.

memory dumps: computer memory records from when a particular program crashed, used to pinpoint and address the bug that caused the crash.

memristor: a memory resistor, a circuit that can change its own electrical resistance based on the resistance it has used in the past and can respond to familiar phenomena in a consistent way.

mesh network: a type of network in which each node relays signal through the network. A node is any communication intersection or endpoint in the network (e.g., computer or terminal).

meta-complexity: a complexity that arises when the computational analysis of a problem is compounded by the complex nature of the problem itself.

metadata: data that contains information about other data, such as author information, organizational information, or how and when the data was created.

method: a procedure that describes the behavior of an object and its interactions with other objects.

microcontroller: a tiny computer in which all of the essential parts of a computer are united on a single microchip—input and output channels, memory, and a processor.

micron: a unit of measurement equaling one millionth of a meter; typically used to measure the width of a core in an optical figure or the line width on a microchip.

microwaves: electromagnetic radiation with a frequency higher than that of radio wave but lower than that of visible light.

middle computing: computing that occurs at the application tier and involves intensive processing of data that will subsequently be presented to the user or another, intervening application.

million instructions per second (MIPS): a unit of measurement used to evaluate computer performance or the cost of computing resources.

mixer: a component that converts random input radio frequency (RF) signal into a known, fixed frequency to make processing the signal easier.

mixing: the process of combining different sounds into a single audio recording.

modeling: the process of creating a 2-D or 3-D representation of the structure being designed.

modulator: a device used to regulate or adjust some aspect of an electromagnetic wave.

module: A program that is linked with others to form a functioning application; one method of implementing a subroutine

morphology: a branch of linguistics that studies the forms of words.

multi-agent system: a system consisting of multiple separate agents, either software or hardware systems, that can cooperate and organize to solve problems.

multibit watermarking: a watermarking process that embeds multiple bits of data in the signal to be transmitted.

multicast: a network communications protocol in which a transmission is broadcast to multiple recipients rather than to a single receiver.

multimodal monitoring: the monitoring of several physical parameters at once in order to better evaluate a patient's overall condition, as well as how different parameters affect one another or respond to a given treatment.

multiplexing: combining multiple data signals into one in order to transmit all signals simultaneously through the same medium.

multiplier-accumulator: a piece of computer hardware that performs the mathematical operation of multiplying two numbers and then adding the result to an accumulator.

multiprocessing: the use of more than one central processing unit to handle system tasks; this requires an operating system capable of dividing tasks between multiple processors.

multitasking: in computing, the process of executing multiple tasks concurrently on an operating system (OS); in the mobile phone environment, allowing different apps to run concurrently, much like the ability to work in multiple open windows on a PC.

multitenancy: a software program that allows multiple users to access and use the software from different locations.

multi-terminal configuration: a computer configuration in which several terminals are connected to a single computer, allowing more than one person to use the computer.

multitouch gestures: combinations of finger movements used to interact with touch-screen or other touch-sensitive displays in order to accomplish various tasks. Examples include double-tapping and swiping the finger along the screen.

multiuser: capable of being accessed or used by more than one user at once.

mutator: A method that is permitted to change the state of the object to which it is applied.

mutual exclusion: a rule present in some database systems that prevents a resource from being accessed by more than one operation at a time; one of the necessary conditions for deadlock.

narrowing: A conversion that converts an object to a type that cannot include all the possible values of the object.

natural numbers: The numbers 1,2,3,4...

near-field communication: a method by which two devices can communicate wirelessly when in close proximity to one another.

near-field communications antenna: an antenna that enables a device to communicate wirelessly with a nearby compatible device.

negative-AND (NAND) gate: a logic gate that produces a false output only when both inputs are true

nervous (neural) system: the system of nerve pathways by which an organism senses changes in itself and its environment and transmits electrochemical signals describing these changes to the brain so that the brain can respond.

network firewalls: firewalls that protect an entire network rather than a specific device.

network: two or more computers being linked in a way that allows them to transmit information back and forth.

networking: the use of physical or wireless connections to link together different computers and computer networks so that they can communicate with one another and collaborate on computationally intensive tasks.

neural network: in computing, a model of information processing based on the structure and function of biological neural networks such as the human brain.

neuroplasticity: the capacity of the brain to change as it acquires new information and forms new neural connections.

nibble: a group of four bits.

node: any point on a computer network where communication pathways intersect, are redistributed, or end (i.e., at a computer, terminal, or other device).

noise: interferences or irregular fluctuations affecting electrical signals during transmission.

noise-tolerant signal: a signal that can be easily distinguished from unwanted signal interruptions or fluctuations (i.e., noise).

nondestructive editing: a mode of image editing in which the original content of the image is not destroyed because the edits are made only in the editing software.

nonfunctional requirement: also called extrafunctional requirement; an attribute of a computer system, such as reliability, scalability, testability, security, or usability, that reflects the operational performance of the system.

nongraphical: not featuring graphical elements.

nonlinear editing: a method of editing video in which each frame of video can be accessed, altered, moved, copied, or deleted regardless of the order in which the frames were originally recorded.

nonvolatile memory: computer storage that retains its contents after power to the system is cut off, rather than memory that is erased at system shutdown.

nonvolatile random-access memory (NVRAM): a form of RAM in which data is retained even when the device loses access to power.

normalization: a process that ensures that different code points representing equivalent characters will be recognized as equal when processing text.

nuclear magnetic resonance (NMR) spectroscopy: a technique for studying the properties of atoms or molecules by applying an external magnetic field to atomic nuclei and analyzing the resulting difference in energy levels.

object code: The output of a compiler or assembler, not necessarily executable directly without linking to other modules.

object file: A piece of compiled code that is linked into a compiled program after compilation and either during loading or when the program is running. Do not confuse this use of object with the later use in programming, analysis, and design.

object: An instance of a class; [Analysis] Something that is uniquely identifiable by the user; [Code] A piece identifiable storage that can suffer and/or perform various operations defined by the objects type; [Design] A module that encapsulates a degree of intelligence and know how and has specialized responsibilities.

object-oriented programming: a type of programming in which the source code is organized into objects, which are elements with a unique identity that have a defined set of attributes and behaviors.

object-oriented user interface: an interface that allows users to interact with onscreen objects as they would in real-world situations, rather than selecting objects that are changed through a separate control panel interface.

object-oriented: Using entities called objects that can process data and exchange messages with other objects.

open-source software: software that makes its source code available to the public for free use, study, modification, and distribution.

operating system (OS): a specialized program that manages a computer's functions.

operating system shell: a user interface used to access and control the functions of an operating system.

operation: One of a set of functions with special syntax and semantics that can be used to construct an expression.

operator associativity: Rules that help define the order in which an expression is evaluated when two adjacent infix operators are identical.

operator precedence: Rules that help define the order in which an expression is evaluated when two infix operators can be done next.

operator: [Lexeme] A symbol for an operation. Operators can infix, prefix, or postfix.

optical storage: storage of data by creating marks in a medium that can be read with the aid of a laser or other light source.

optical touchscreens: touchscreens that use optical sensors to locate the point where the user touches before physical contact with the screen has been made.

out: Any mode of passing parameters that permits the subprogram to give a value to an actual parameter without letting the subprogram no what the original value of the subprogram. Only available for general parameters in Ada, it can be implemented by pass by result.

output: A means whereby data or objects are passed from a part to a wider context; for example a program sending data to the operating system so that you can see it on the screen.

overload: To provide multiple context dependent meanings for a symbol in a language.

overloading: Giving multiple meanings to a symbol depending on its context.

packet filters: filters that allow data packets to enter a network or block them on an individual basis.

packet forwarding: the transfer of a packet, or unit of data, from one network node to another until it reaches its destination.

packet sniffers: a program that can intercept data or information as it moves through a network.

packet switching: a method of transmitting data over a network by breaking it up into units called packets, which are sent from node to node along the network until they reach their destination and are reassembled.

paradigm: A fundamental style of computer programming to which the design of a programming language typically has to cater, such as imperative programming, declarative programming, or, on a finer level, functional programming, logic programming or object-oriented programming.

parallel processing: the division of a task among several processors working simultaneously, so that the task is completed more quickly.

parameter passing: The means by which the actual parameters in a call of a subprogram are connected with the formal parameters in the definition of the subprogram.

parameter: A name in a function or subroutine definition that is replaced by, or bound to, the corresponding actual argument when the function or subroutine is called; [Mathematics] A variable constant or perhaps a constant variable; [Programming] Something that is used in a subprogram that can be changed when the subprogram is called.

parametric polymorphism: A piece of code describes a general form of some code by using a parameter. Different instances or special cases are created by replace these parameters by actual parameters. Templates in C++, Generics in Ada, and Functors in SML are particular implementations of this idea.

parse: To convert a sequence of tokens into a data structures (typically a tree and a name table) that can be used to interpret or translate the sequence.

pass by reference: Parameter passing where the parameter is implemented by providing an access path to the actual parameter from the formal parameter. Actions written as if they use or change the formal parameter use or change the actual parameter instead.

pass by value: Parameter passing where The actual parameter is evaluated (if necessary) and the value placed in a location bound to the formal parameter.

PATRIOT Act: a 2001 law that expanded the powers of federal agencies to conduct surveillance and intercept digital information for the purpose of investigating or preventing terrorism.

pedagogy: a philosophy of teaching that addresses the purpose of instruction and the methods by which it can be achieved.

peer-to-peer (P2P) network: a network in which all computers participate equally and share coordination of network operations, as opposed to a client-server network, in which only certain computers coordinate network operations.

pen/trap: short for pen register or trap-and-trace device, devices used to record either all numbers called from a particular telephone (pen register) or all numbers making incoming calls to that phone (trap and trace); also refers to the court order that permits the use of such devices.

peripheral: a device that is connected to a computer and used by the computer but is not part of the computer, such as a printer, scanner, external storage device, and so forth.

personal area network (PAN): a network generated for personal use that allows several devices to connect to one another and, in some cases, to other networks or to the Internet.

personally identifiable information (PII): information that can be used to identify a specific individual.

pharmacogenomics: the study of how an individual's genome influences his or her response to drugs.

phased software development: an approach to software development in which most testing occurs after the system requirements have been implemented.

phishing: the use of online communications in order to trick a person into sharing sensitive personal information, such as credit card numbers or social security numbers.

Phonebloks: a concept devised by Dutch designer Dave Hakkens for a modular mobile phone intended to reduce electronic waste.

phoneme: a sound in a specified language or dialect that is used to compose words or to distinguish between words.

pipelined architecture: a computer design where different processing elements are connected in a series, with the output of one operation being the input of the next.

piracy: in the digital context, unauthorized reproduction or use of copyrighted media in digital form.

PL: Programming Language.

planned obsolescence: a design concept in which consumer products are given an artificially limited lifespan, therefore creating a perpetual market.

platform as a service: a category of cloud computing that provides a virtual machine for users to develop, run, and manage web applications.

platform: the specific hardware or software infrastructure that underlies a computer system; often refers to an operating system, such as Windows, Mac OS, or Linux.

plug-in: an application that is easily installed (plugged in) to add a function to a computer system.

pointer: Data type with values that are addresses of other items of data.

polymerase chain reaction (PCR) machine: a machine that uses polymerase chain reaction to amplify segments of DNA for analysis; also called a thermal cycler.

polymorphism: the ability to maintain the same method name across subclasses even when the method functions differently depending on its class.

port scanning: the use of software to probe a computer server to see if any of its communication ports have been left open or vulnerable to an unauthorized connection that could be used to gain control of the computer.

portlet: an independently developed software component that can be plugged into a website to generate and link to external content.

positional parameter: A parameter that is bound by its position.

postfix: [Operator] An operator that is placed after its single operand.

postproduction: the period after a model has been designed and an image has been rendered, when the architect may manipulate the created image by adding effects or making other aesthetic changes.

powder bed fusion: the use of a laser to heat layers of powdered material in a movable powder bed.

precoding: a technique that uses the diversity of a transmission by weighting an information channel.

predicate: [Logic] A formula that may contain variables, that when evaluated should be either true or false; [Prolog] A procedure that can fail or succeed to achieve a goal. Success is finding an instance of a formula that is true and failure means failing to find such an instance. It is assumed that failing to find a solution is proof that the predicate is false. In fact the definition of the predicate may be incomplete or some infinite instance is needed to fit the predicate.

preemptive multitasking: an implementation of multitasking in which the operating system will initiate a context switch while a process is running, usually on a schedule so that switches between tasks occur at precise time intervals.

prefix: [Operator] An operator that is placed in front of its single operand.

Pretty Good Privacy: a data encryption program created in 1991 that provides both encryption and authentication.

primitive: Something that does not need to be defined.

principle of least privilege: a philosophy of computer security that mandates users of a computer or network be given, by default, the lowest level of privileges that will allow them to perform their jobs. This way, if a user's account is compromised, only a limited amount of data will be vulnerable.

printable characters: characters that can be written, printed, or displayed in a manner that can be read by a human.

printed circuit board: a flat copper sheet shielded by fiberglass insulation in which numerous lines have been etched and holes have been punched, allowing various electronic components to be connected and to communicate with one another and with external components via the exposed copper traces.

Privacy Incorporated Software Agents (PISA): a project that sought to identify and resolve privacy problems related to intelligent software agents.

procedure: A subroutine or function coded to perform a specific task.

process: the execution of instructions in a computer program.

processor coupling: the linking of multiple processors within a computer so that they can work together to perform calculations more rapidly. This can be characterized as loose or tight, depending on the degree to which processors rely on one another.

processor symmetry: multiple processors sharing access to input and output devices on an equal basis and being controlled by a single operating system.

program: A software application, or a collection of software applications, designed to perform a specific task.

programmable oscillator: an electronic device that fluctuates between two states that allows user modifications to determine mode of operation.

programming languages: sets of terms and rules of syntax used by computer programmers to create instructions for computers to follow. This code is then compiled into binary instructions for a computer to execute.

projected capacitive touch: technology that uses layers of glass etched with a grid of conductive material that allows for the distortion of voltage flowing through the grid when the user touches the surface; this distortion is measured and used to determine the location of the touch.

Prolog: PROgramable LOGic.

proprietary software: software owned by an individual or company that places certain restrictions on its use, study, modification, or distribution and typically withholds its source code.

protocol processor: a processor that acts in a secondary capacity to the CPU, relieving it from some of the work of managing communication protocols that are used to encode messages on the network.

protocol: [Networking] Rules for sending and receiving data and commands over the network; [Subprogram] Rules for calling a subprogram.

prototypal inheritance: a form of code reuse in which existing objects are cloned to serve as prototypes.

prototype: [Software engineering] A piece of software that requires more work before it is finished, but is complete enough for the value of the finished product to be evaluated or the currant version improved.

proxy server: a computer through which all traffic flows before reaching the user's computer.

proxy: a dedicated computer server that functions as an intermediary between a user and another server.

pseudocode: a combination of a programming language and a spoken language, such as English, that is used to outline a program's code.

public-key cryptography: a system of encryption that uses two keys, one public and one private, to encrypt and decrypt data.

push technology: a communication protocol in which a messaging server notifies the recipient as soon as the server receives a message, instead of waiting for the user to check for new messages.

quantum bit (qubit): a basic unit of quantum computation that can exist in multiple states at the same time, and can therefore have multiple values simultaneously.

quantum logic gate: a device that alters the behavior or state of a small number of qubits for the purpose of computation.

quicksort: [Algorithm] Split the data into two roughly equal parts with all the lesser elements in one and the greater ones in the other and then sort each part; Prof. C. A. R. Hoare wrote this a young programmer and team leader. His future career started with the publication of this elegant recursion for placing an array of numbers into order.

radio waves: low-frequency electromagnetic radiation, commonly used for communication and navigation.

random access memory (RAM): memory that the computer can access very quickly, without regard to where in the storage media the relevant information is located.

ransomware: malware that encrypts or blocks access to certain files or programs and then asks users to pay to have the encryption or other restrictions removed.

rapid prototyping: the process of creating physical prototype models that are then tested and evaluated.

raster: a means of storing, displaying, and editing image data based on the use of individual pixels.

read-only memory (ROM): type of nonvolatile data storage that is typically either impossible to erase and alter or requires special equipment to do so; it is usually used to store basic operating information needed by a computer.

real: a number containing a decimal point, e.g. the number pi is a real number with a value of approximately 3.14159268

real-time monitoring: a process that grants administrators access to metrics and usage data about a software program or database in real time.

real-time operating system: an operating system that is designed to respond to input within a set amount of time without delays caused by buffering or other processing backlogs.

receiver: a device that reads a particular type of transmission and translates it into audio or video.

recursive: describes a method for problem solving that involves solving multiple smaller instances of the central problem.

relational expression: An infix expression in which two non-Boolean values are compared and a Boolean value returned.

relational operator: An infix operator that returns a Boolean value when given non-Boolean operands.

relational: Pertaining to a relation.

remote monitoring: a platform that reviews the activities on software or systems that are located off-site.

render farm: a cluster of powerful computers that combine their efforts to render graphics for animation applications.

rendering: the process of transforming one or more models into a single image; the production of a computer image from a 2-D or 3-D computer model; the process of selecting and displaying glyphs.

resistive touchscreens: touchscreens that can locate the user's touch because they are made of several layers of conductive material separated by small spaces; when the user touches the screen, the layers touch each other and complete a circuit.

resource allocation: a system for dividing computing resources among multiple, competing requests so that each request is eventually fulfilled.

resource distribution: the locations of resources available to a computing system through various software or hardware components or networked computer systems.

resource holding: a situation in which one process is holding at least one resource and is requesting further resources; one of the necessary conditions for deadlock.

retriggerable single shot: a monostable multivibrator (MMV) electronic circuit that outputs a single pulse when triggered but can identify a new trigger during an output pulse, thus restarting its pulse time and extending its output.

reversible data hiding: techniques used to conceal data that allow the original data to be recovered in its exact form with no loss of quality.

RGB: a color model that uses red, green, and blue to form other colors through various combinations.

routing: selecting the best path for a data packet to take in order to reach its destination on the network.

run time: The time during which a program is executing, as oppose to the compile time.

Sarbanes-Oxley Act (SOX): a 2002 law that requires all business records, including electronic records and electronic messages, to be retained for a certain period of time.

scareware: malware that attempts to trick users into downloading or purchasing software or applications to address a computer problem.

Scheme: A modern statically scoped version of LISP.

Scientific Working Group on Digital Evidence (SWGDE): an American association of various academic and professional organizations interested in the development of digital forensics systems, guidelines, techniques, and standards.

scope: The parts of a program where a particular identifier has a particular meaning (set of bindings).

scoped: Pertains to languages with particular scoping rules.

scoping: The rules used to determine an identifier's scope in a language, see dynamic scoping and static scoping.

screen-reading program: a computer program that converts text and graphics into a format accessible to visually impaired, blind, learning disabled, or illiterate users.

script: a group of written signs, such as Latin or Chinese characters, used to represent textual information in a writing system.

scrubbing: navigating through an audio recording repeatedly in order to locate a specific cue or word.

search engine optimization (SEO): techniques used to increase the likelihood that a website will appear among the top results in certain search engines.

selection: A statement that chooses between several possible executions paths in a program.

self-star properties: a list of component and system properties required for a computing system to be classified as an autonomic system.

semantics: A description of how the meaning of a valid statement or sentence can be worked out from its parsed form.

semiconductor intellectual property (SIP) block: a quantity of microchip layout design that is owned by a person or group; also known as an "IP core."

sensors: devices capable of detecting, measuring, or reacting to external physical properties.

separation of concerns: a principle of software engineering in which the designer separates the computer program's functions into discrete elements.

set: A collection of objects, usually of the same type, described either by enumerating the elements or by stating a common property, or by describing rules for constructing items in the set.

shadow RAM: a form of RAM that copies code stored in read-only memory into RAM so that it can be accessed more quickly.

sheet lamination: a process in which thin layered sheets of material are adhered or fused together and then extra material is removed with cutting implements or lasers.

shell: an interface that allows a user to operate a computer or other device.

Short Message Service (SMS): the technology underlying text messaging used on cell phones.

side effect: A function or expression has a side effect if executing it changes the values of global variables or its arguments.

signal-to-noise ratio (SNR): the power ratio between meaningful information, referred to as "signal," and background noise.

silent monitoring: listening to the exchanges between an incoming caller and an employee, such as a customer service agent.

simulation: a computer model executed by a computer system.

software as a service: a software service system in which software is stored at a provider's data center and accessed by subscribers.

software patches: updates to software that correct bugs or make other improvements.

software: the sets of instructions that a computer follows in order to carry out tasks. Software may be stored on physical media, but the media is not the software.

software-defined antennas: reconfigurable antennas that transmit or receive radio signals according to software instructions.

solid modeling: the process of creating a 3-D representation of a solid object.

solid-state storage: computer memory that stores information in the form of electronic circuits, without the use of disks or other read/write equipment.

sound card: a peripheral circuit board that enables a computer to accept audio input, convert signals between analog and digital formats, and produce sound.

source code: Human-readable instructions in a programming language, to be transformed into machine instructions by a compiler, interpreter, assembler or other such system.

speaker independent: describes speech software that can be used by any speaker of a given language.

special character: a character such as a symbol, emoji, or control character.

spyware: software installed on a computer that allows a third party to gain information about the computer user's activity or the contents of the user's hard drive.

stack: A collection of data items where new items are added and old items retrieved at the same place, so that the last item added is always the first item retrieved, and so on. An important part of compilers, interpreters, processors, and programs.

state: a technical term for all of the stored information, and the configuration thereof, that a program or circuit can access at a given time; a complete description of a physical system at a specific point in time, including such factors as energy, momentum, position, and spin.

stateful filters: filters that assess the state of a connection and allow or disallow data transfers accordingly.

static random-access memory (SRAM): a form of RAM in which the device's memory does not need to be regularly refreshed but data will still be lost if the device loses power.

static testing: a testing technique in which software is tested by analyzing the program and associated documentation, without executing its code.

static: [C] A keyword with too many different meanings pertaining to the life history and scope of variables.

string: A data type for a sequence of characters such as letters of English alphabet.

structure: [Data type] A finite collection of named items of data of different types.

subclass: In object-oriented programming, an object class derived from another class (its superclass) from which it inherits a base set of properties and methods.

subprogram header: The part of a subprogram definition that describes how the subprogram can be called without defining how it works.

subprogram: A piece of code that has been named and can be referred to by that name (called) as many times as is needed. Either a procedure or a function.

subroutine: A section of code that implements a task. While it may be used at more than one point in a program, it need not be.

substitution cipher: a cipher that encodes a message by substituting one character for another.

subtype: A type S is a subtype of type T if every valid operation on an object of type T is also a valid operation of type S.

subtyping: a relation between data types where one type is based on another, but with some limitations imposed.

superclass: A class that passes attributes and methods down the hierarchy to subclasses.

supercomputer: an extremely powerful computer that far outpaces conventional desktop computers.

superposition: the principle that two or more waves, including waves describing quantum states, can be combined to give rise to a new wave state with unique properties. This allows a qubit to potentially be in two states at once.

surface capacitive technology: a glass screen coated with an electrically conductive film that draws current across the screen when it is touched; the flow of current is measured in order to determine the location of the touch.

symmetric-key cryptography: a system of encryption that uses the same private key to encrypt and decrypt data.

syntax: a branch of linguistics that studies how words and phrases are arranged in sentences to create meaning.

system agility: the ability of a system to respond to changes in its environment or inputs without failing altogether.

system identification: the study of a system's inputs and outputs in order to develop a working model of it.

system software: the basic software that manages the computer's resources for use by hardware and other software.

system: a set of interacting or interdependent component parts that form a complex whole; a computer's combination of hardware and software resources that must be managed by the operating system.

tableless web design: the use of style sheets rather than HTML tables to control the layout of a web page.

TCO certification: a credential that affirms the sustainability of computers and related devices.

telecom equipment: hardware that is intended for use in telecommunications, such as cables, switches, and routers.

telemedicine: health care provided from a distance using communications technology, such as video chats, networked medical equipment, smartphones, and so on.

telemetry: automated communication process that allows a machine to identify its position relative to external environmental cues.

temporal synchronization: the alignment of signals from multiple devices to a single time standard, so that, for example, two different devices that record the same event will show the event happening at the exact same time.

terminals: a set of basic input devices, such as a keyboard, mouse, and monitor, that are used to connect to a computer running a multi-user operating system.

ternary: Pertaining to 3. Ternary operators have two operands. Ternary numbers have base 3 and use 3 symbols.

third-party data center: a data center service provided by a separate company that is responsible for maintaining its infrastructure.

thread: the smallest part of a programmed sequence that can be managed by a scheduler in an OS.

time-sharing: the use of a single computing resource by multiple users at the same time, made possible by the computer's ability to switch rapidly between users and their needs.

token: A particular representation of lexemes.

topology: the way a network is organized, including nodes and the links that connect nodes.

toxgnostics: a subfield of personalized medicine and pharmacogenomics that is concerned with whether an individual patient is likely to suffer a toxic reaction to a specific medication.

trace impedance: a measure of the inherent resistance to electrical signals passing through the traces etched on a circuit board.

transactional database: a database management system that allows a transaction— a sequence of operations to achieve a single, self-contained task—to be undone, or "rolled back," if it fails to complete properly.

transistor: a computing component generally made of silicon that can amplify electronic signals or work as a switch to direct electronic signals within a computer system.

transmission medium: the material through which a signal can travel.

transmitter: a device that sends a signal through a medium to be picked up by a receiver.

transparent monitoring: a system that enables employees to see everything that the managers monitoring them can see.

transposition cipher: a cipher that encodes a message by changing the order of the characters within the message.

tree: A collection of connected objects called nodes with all nodes connected indirectly by precisely one path. An ordered tree has a root and the connections lead from this root to all other nodes. Nodes at the end of the paths are called leaves. The connections are called branches. All computer science tress are drawn upside-down with the root at the top and the leaves at the bottom.

trusted platform module (TPM): a standard used for designing cryptoprocessors, which are special chips that enable devices to translate plain text into cipher text and vice versa.

tuning: the process of making minute adjustments to a computer's settings in order to improve its performance.

Turing complete: a programming language that can perform all possible computations.

type: A tag attached to variables and values used in determining what values may be assigned to what variables; A collection of similar objects, See ADT and data type. Objects can be fundamental, pointers, or have their type determined by their class.

typography: the art and technique of arranging type to make language readable and appealing.

ubiquitous computing: an approach to computing in which computing activity is not isolated in a desktop, laptop, or server, but can occur everywhere and at any time through the use of microprocessors embedded in everyday devices.

UML: Unified Modeling Language.

unary: Pertaining to 1. Unary operators have one operand, unary numbers use base 1 and one symbol.

undervolting: reducing the voltage of a computer system's central processing unit to decrease power usage.

UNICODE: A new 16 bit International code for characters. Used in Java.

unmanned aerial vehicle (UAV): an aircraft that does not have a pilot onboard but typically operates through remote control, automated flight systems, or preprogrammed computer instructions.

user-centered design: design based on a perceived understanding of user preferences, needs, tendencies, and capabilities.

utility program: a type of system software that performs one or more routine functions, such as disk partitioning and maintenance, software installation and removal, or virus protection.

variable: A named memory location in which a program can store intermediate results and from which it can read them.

vat photopolymerization: a process in which a laser hardens layers of light-sensitive material in a vat.

vector: a means of storing, displaying, and editing image data based on the use of defined points and lines.

vertical scaling: the addition of resources to a single node (e.g., a computer or server) in a system.

vibrator: an electronic component that vibrates.

video scratching: a technique in which a video sequence is manipulated to match the rhythm of a piece of music.

virtual device driver: a type of device driver used by the Windows operating system that handles communications between emulated hardware and other devices.

virtual memory: memory used when a computer configures part of its physical storage (on a hard drive, for example) to be available for use as additional RAM. Information is copied from RAM and moved into virtual memory whenever memory resources are running low.

virtual reality: the use of technology to create a simulated world into which a user may be immersed through visual and auditory input.

virtual: [C++] A member function or method is virtual if when applied to a pointer the class of the object pointed at is used rather than the class of the pointer. Virtual inheritance means that when a class in inherited by two different path only one single parent object is stored for both paths.

vision mixing: the process of selecting and combining multiple video sources into a single video.

visual programming: a form of programming that allows a programmer to create a program by dragging and dropping visual elements with a mouse instead of having to type in text instructions.

voice over Internet Protocol (VoIP): a set of parameters that make it possible for telephone calls to be transmitted digitally over the Internet, rather than as analog signals through telephone wires.

void pointer: [C] A pointer to an object of unknown type and size.

volatile memory: memory that stores information in a computer only while the computer has power; when the computer shuts down or power is cut, the information is lost.

wardriving: driving around with a device such as a laptop that can scan for wireless networks that may be vulnerable to hacking.

web application: an application that is downloaded either wholly or in part from the Internet each time it is used.

white-box testing: a testing technique that includes complete knowledge of an application's coding and structure.

widening: A conversion that places an object in a type that includes all the possible values of the type.

widget: an independently developed segment of code that can be plugged into a website to display information retrieved from other sites.

wireframe: a schematic or blueprint that represents the visual layout of a web page, without any interactive elements.

word prediction: a software feature that recognizes words that the user has typed previously and offers to automatically complete them each time the user begins typing.

worm: a type of malware that can replicate itself and spread to other computers independently; unlike a computer virus, it does not have to be attached to a specific program.

XBNF: [MATHS] An extension to EBNF invented by Dr. Botting so that ASCII can be used to describe formal syntax and semantics.

zombie computer: a computer that is connected to the Internet or a local network and has been compromised such that it can be used to launch malware or virus attacks against other computers on the same network.

BIBLIOGRAPHY

"10 Reasons Why GMO Smart Label Isn't 'Smart' at All." *EcoWatch*, www.ecowatch.com. Accessed 21 Dec. 2016.

"10 Surprising Things You Might Do with a Mechatronics Degree." *East Coast Polytechnic Institute Blog*, www.ecpi.edu. Accessed 19 Jan. 2017.

"A 10 Gigabit Programmable Network Interface Card – An Overview. *Rice University*, www.ece.rice.edu. Accessed 28 Jan. 2017.

Abbany, Zulfikar. "Konrad Zuse and the Digital Revolution He Started with the Z3 Computer 75 Years Ago." *Deutsche Welle*, 5 Nov. 2016, www.dw.com. Accessed 10 Nov. 2016.

Abelson, Harold, Gerald Jay Sussman, and Julie Sussman. *Structure and Interpretation of Computer Programs.* 2nd ed, Cambridge: MIT P, 1996. Print.

"About Cognitive Linguistics." *ICLA*, www.cognitivelinguistics.org. Accessed 19 Dec. 2016.

"About Scratch." *Scratch.* MIT Media Lab, 2016. Web. 23 Aug. 2016.

"About Us." *The Open Group.* The Open Group. Web. 18 Feb. 2016.

"About Us: SmartLabel™ Working for You." *Smartlabel.org*, www.smartlabel.org. Accessed 21 Dec. 2016.

"About." *Cognitive Science Society*, www.cognitivesciencesociety.org. Accessed 18 Dec. 2016.

Abrash, Michael. *Michael Abrash's Graphics Programming Black Book.* Albany, N.Y.: Coriolis Group, 1997. Consisting of seventy short chapters, the book covers graphics programming up to the time of the Pentium processor.

Ackermann, Gerhard K., and Jürgen Eichler. *Holography: A Practical Approach.* Wiley, 2007.

Ackermann, Robert John. *Data, Instruments, and Theory: A Dialectical Approach to Understanding Science.* Princeton: Princeton UP, 1985. Print.

"Ada Lovelace." *History Mesh.* History Mesh, n.d. Web. 15 June 2016.

Adam, John A. *Mathematics in Nature: Modeling Patterns in the Natural World.* Princeton, NJ: Princeton UP, 2003.

Adams, Ernest. *Break into the Game Industry: How to Get a Job Making Video Games.* McGraw-Hill/Osborne, 2003.

______. *Fundamentals of Game Design.* 2nd ed., New Riders, 2010. Print.

Adams, William J. *The Life and Times of the Central Limit Theorem.* 2nd ed., American Mathematical Society, 2009.

"Advanced Encryption Standard (AES)." *Techopedia.com.* Janalta Interactive Inc.Web. 31 July 2015.

"Advantages of Multitasking." *Microsoft.* Microsoft. Web. 13 Aug. 2015.

"AES." *Webopedia.* QuinStreet Inc. Web. 31 July 2015.

Agazzi, Evandro. *Scientific Objectivity and Its Contexts.* Cham: Springer, 2014. Print.

"Agua Caliente Solar Project." *First Solar*, www.firstsolar.com. Accessed 10 Nov. 2016.

Aichinger, H., et al. *Radiation Exposure and Image Quality in X-ray Diagnostic Radiology: Physical Principles and Clinical Applications.* Springer, 2003.

Aigner, Martin, and Gunter M. Ziegler. *Proofs from the Book.* New York: Springer, 2014.

Albino, Vito, et. al. "Smart Cities: Definitions, Dimensions, and Performance." *International Forum on Knowledge Asset Dynamics*, 2013. www.ifkad.org. Accessed 23 Dec. 2016.

Alfred, Randy. "April 4, 1975: Bill Gates, Paul Allen Form a Little Partnership." *Wired.* Conde Nast. 4 Apr. 2008. Web. 27 Aug. 2015.

Al-Jumaily, A., and R. A. Olivares. "Bio-Driven System-Based Virtual Reality for Prosthetic and Rehabilitation Systems." *Signal, Image and Video Processing* 6.1 (2012): 71–84. *Inspec.* Web. 10 Mar. 2015.

"Altair 8800 Microcomputer." *The National Museum of American History.* Smithsonian. Web. 27 Aug. 2015.

Altavilla, Dave. "Apple Further Legitimizes Augmented Reality Tech With Acquisition of Metaio." *Forbes.* Forbes.com, LLC. 30 May 2015. Web. 13 Aug. 2015.

Amir, Yaniv, et al. "Universal Computing by DNA Origami Robots in a Living Animal." *Nature Nanotechnology*, vol. 9, 2014, pp. 353–357. *PubMed Central*, doi: 10.1038/nnano.2014.58. Accessed 20 Nov. 2016.

"Analysis of Variance." *Khan Academy.* Khan Acad., n.d. Web. 11 July 2016.

"The Analytical Engine of Charles Babbage. *History-computer.com.* History-computer.com, n.d. Web. 15 June 2016.

"The Analytical Engine. *History Mesh.* History Mesh, n.d. Web. 15 June 2016.

Anderson, James M. et al. "Autonomous Vehicle Technology: A Guide for Policymakers." *RAND.* RAND, 2014. Web. 16 Nov. 2014. PDF file.

Angel, Allen R., and Dennis C. Runde. *Elementary Algebra for College Students.* Boston: Pearson, 2011.

Annis, Charles. "Central Limit Theorem." *Statistical Engineering,* www.statisticalengineering.com. Accessed 24 Jan. 2017.

Anthony, Sebastian. "A History of Supercomputers." *Extreme Tech,* 10 Apr. 2012, www.extremetech.com. Accessed 5 Dec. 2016.

Anton, Howard, Irl Bivens, and Stephen Davis. *Calculus.* Hoboken, NJ: Wiley, 2012.

Archimedes, and Thomas L. Heath. *The Works of Archimedes.* Mineola, NY: Dover, 2002.

"An Architectural Blueprint for Autonomic Computing." 3rd ed. IBM, June 2005. PDF file.

"Assembly Language." *Techopedia.* Techopedia Inc. Web. 10 Mar. 2016.

Atherton, Kelsey D. "NASA is Testing a Drone for Mars." *Popular Science,* 6 July 2015, www.popsci.com. Accessed 10 Dec. 2016.

"Augmented Reality." *Webopedia.* Quinstreet Enterprise. Web. 13 Aug. 2051.

"The Babbage Engine. A Brief History. *Computer History Museum.* Computer History Museum, n.d. Web. 15 June 2016.

Bagaria, Joan. "Set Theory." Palo Alto, CA: Stanford University, 8 Oct. 2014. Web. 3 Jan. 2015.

Bailey, David Evans. "Ten Cool Applications for Virtual Reality That Aren't Just Games." *The Conversation.* The Conversation U.S., 22 Mar. 2016. Web. 24 June 2016.

Bailey, James E., and David F. Ollis. *Biochemical Engineering Fundamentals.* 2d cd. New York: McGraw-Hill, 2006. Covers all aspects of biochemical engineering in an understandable manner.

Bainbridge, William Sims, editors. *Berkshire Encyclopedia of Human-Computer Interaction: When Science Fiction Becomes Fact.* Berkshire, 2004. 2 vols.

Bakry, Rania, et al. "Medicinal Applications of Fullerenes." *International Journal of Nanomedicine,* vol. 2, no. 4, pp. 639–49, Dec. 2007, www.ncbi.nlm.nih.gov. Accessed 14 Dec. 2016.

Banner, Adrian D. *The Calculus Lifesaver: All the Tools You Need to Excel at Calculus.* Princeton UP, 2007.

Barbour, Virginia. "How the Insights of the Large Hadron Collider Are Being Made Open to Everyone." *Phys Org,* 13 Jan. 2017, phys.org. Accessed 14 Jan. 2017.

Bar-Cohen, Yoseph. "Biomimetics—Using Nature to Inspire Human Innovation." *Bioinspiration & Biomimetics,* vol. 1, 27 Apr. 2006, pp. 1–12, biomimetic.pbworks.com. Accessed 12 Jan. 2017.

Barr, Jeff. "API Gateway Update – New Features Simplify API Development." *Amazon Web Services,* 20 Sept. 2016, aws.amazon.com. Accessed 29 Dec. 2016.

Basha, Habib A. "Nonlinear Reservoir Routing: Particular Analytical Solution." *Journal of Hydraulic Engineering* 120, no. 5 (May, 1994): 624-632.

Basha, Habib A., and W. El-Asmar. "The Fracture Flow Equation and Its Perturbation Solution." *Water Resources Research* 39, no. 12 (December, 2003): 1365.

"The Basics: How Programming Works. *Microsoft Developer Network.* Microsoft. Web. 11 Aug. 2015.

Basl, John. "The Ethics of Creating Artificial Consciousness." *American Philosophical Association Newsletters: Philosophy and Computers* 13.1 (2013): 25–30. *Philosophers Index with Full Text.* Web. 25 Feb. 2015.

Batty, M., et. al. "Smart Cities of the Future." *The European Physical Journal: Special Topics,* vol. 214, no. 1, Nov. 2012, pp. 481–518.

Bay-Williams, Jennifer M, and William R. Speer. *Professional Collaborations in Mathematics Teaching and Learning: Seeking Success for All.* Reston, VA: National Council of Teachers of Mathematics, 2012.

Beal, Vangie. "Client-Server Architecture." *Webopedia.* Quinstreet Enterprise, n.d. Web. 7 Oct. 2015.

______. "Machine Language." *Webopedia.* QuinStreet Inc. Web. 10 Mar. 2016.

______. "OOP—Object Oriented Programming." *Webopedia,* www.webopedia.com. Accessed 18 Jan. 2017.

Bell, Gordon. "Supercomputers: The Amazing Race (A History of Supercomputing, 1960–2020)." *Microsoft*, Nov. 2014, www.research.microsoft.com. Accessed 5 Dec. 2016.

Bender, Oliver, et al., eds. *Geoinformation Technologies for Geocultural Landscapes: European Perspectives.* Leiden, the Netherlands: CRC Press, 2009.

"Benefits of Open Access Journals." *PLOS*, www.plos.org. Accessed 14 Jan. 2017.

Bengio, Yoshua. "Machines Who Learn." *Scientific American*, vol. 314, no. 6, 2016, pp. 46–51.

Benjamin, Arthur T., and Jennifer J. Quinn. *Proofs That Really Count—The Art of Combinatorial Proof.* Washington, DC: Mathematical Assoc. of America, 2003.

Benson, Pete. *Scratch.* Ann Arbor: Cherry Lake, 2015. Print.

Berg, Hugo. *Mathematical Models of Biological Systems.* New York: Oxford UP, 2011.

Berlatsky, Noah. *Artificial Intelligence.* Detroit: Greenhaven, 2011. Print.

______. *Nanotechnology.* Greenhaven, 2014.

Berlinski, David. "Sets." *Infinite Ascent: A Short History of Mathematics.* New York, NY: Modern Library, 2005. Print.

Berne, Olivier, et al. "30 Years of Cosmic Fullerenes." *Cornell University Library,* 27 Oct. 2015, arxiv.org. Accessed 14 Dec. 2016.

Berners-Lee, Tim, James Hendler, and Ora Lassila. "The Semantic Web." *Scientific American.* Scientific American, a Division of Nature America, Inc. May 2001. Web. 4 Aug. 2015.

Bertram, H. Neal. *Theory of Magnetic Recording.* Cambridge, England: Cambridge University Press, 1994.

Bibby, Joe. "Robonaut: Home." *Robonaut.* NASA, 31 May 2013. Web. 21 Jan. 2016.

Bickle, John, ed. *The Oxford Handbook of Philosophy and Neuroscience.* New York: Oxford UP, 2009. Print.

Bilger, Burkhard. "Auto Correct: Has the Self-Driving Car at Last Arrived?" *New Yorker.* Condé Nast, 25 Nov. 2013. Web. 16 Nov. 2014.

Binmore, K. G. *Game Theory: A Very Short Introduction.* New York: Oxford University Press, 2007.

Binns, Chris. *Introduction to Nanoscience and Nanotechnology.* Hoboken: Wiley, 2010. Print.

"Biomechanics." *Engineering in Medicine and Biology Society,* IEEE, 2018, www.embs.org. Accessed 19 Mar. 2018.

"Biomedical Engineers." *Occupational Outlook Handbook.* Bureau of Labor Statistics, 17 Dec. 2015, www.bls.gov. Accessed 27 Oct. 2016.

"Bioplastics Made From Egg Whites May Soon Make Conventional Plastic History." *Business Standard.* Business Standard Private Ltd. 19 Apr. 2015. Web. 6 Feb. 2016.

Biswas, Abhijit, et al. "Advances in Top-Down and Bottom-Up Surface Nanofabrication: Techniques, Applications & Future Prospects." *Advances in Colloid and Interface Science* 170.1–2 (2012): 2–27. Print.

Black, David A. *The Well-Grounded Rubyist.* Shelter Island: Manning, 2014. Print.

Black, Ken. *Business Statistics: For Contemporary Decisionmaking.* 8th ed. Malden: Wiley, 2014. Print.

Blanchard, Philippe, Felix Bühlmann, and Jacques-Antoine Gauthier, eds. *Advances in Sequence Analysis: Theory, Method, Applications.* New York: Springer, 2014. Print.

Blitzstein, J. K., and J. Hwant. *Introduction to Probability.* Boca Raton, FL: CRC, 2015.

Bloomfield, Victor. *Using R for Numerical Analysis in Science and Engineering.* Boca Raton: CRC, 2014. Print.

Blume, Stuart. *The Artificial Ear: Cochlear Implants and the Culture of Deafness.* New Brunswick, N.J.: Rutgers University Press, 2010.

Blunsom, Phil. "Hidden Markov Models." *Utah State University Computer Science Department.* PDF. Web. 8 Mar. 2016.

Bogost, Ian. "The Secret History of the Robot Car." *Atlantic.* Atlantic Monthly Group, 14 Oct. 2014. Web. 16 Nov. 2014.

Bolton, David. "Is C Still Relevant in the 21st Century?" *Dice.* Dice. 6 Dec. 2014. Web. 11 Aug. 2015.

Borovik, Alexandre, and Mikhail Katz. "Who Gave You the Cauchy-Weierstrass Tale? The Dual History of Rigorous Calculus." *Foundations of Science* 17.3 (2012): 245–76.

Bosker, Bianca. "Siri Rising: The Inside Story of Siri's Origins—And Why She Could Overshadow the iPhone." *Huffington Post,* 22 Jan. 2013, www.huffingtonpost.com. Accessed 9 Dec. 2016.

Boslaugh, Sarah. *Statistics in a Nutshell.* 2nd ed. Sebastopol: O'Reilly, 2013. Print.

Bossi, R. H., et al. "Computed Tomography." *Non-destructive Testing of Fibre-reinforced Plastics Composites.* Edited by John Summerscales, Elsevier Science, 1990.

Bostom, Nick, and Eliezer Yudkowsky. "The Ethics of Artificial Intelligence." *Machine Intelligence Research Institute.* MIRI, n.d. Web. 23 Sept. 2016.

Bostrom, Nick. "Ethical Issues in Advanced Artificial Intelligence." *NickBostrom.com.* Nick Bostrom, 2003. Web. 23 Sept. 2016.

"Botnet." *Radware.com,* 2016, security.radware.com. Accessed 30 Nov. 2016.

Boudreaux, Ryan. "HTML 5 Trends and What They Mean for Developers. Web Designer. 9 Feb. 2012. Web. 28 Aug, 2015.

Bougaze, David, Thomas R. Jewell, and Rodolfo G. Buiser. *Biotechnology. Demystifying the Concepts.* San Francisco: Benjamin/Cummings, 2000.

Boulos, Paul F. "H2ONET Hydraulic Modeling." *Journal of Water Supply Quarterly, Water Works Association of the Republic of China (Taiwan)* 16, no. 1 (February, 1997): 17–29.

Boulos, Paul F., Kevin E. Lansey, and Bryan W. Karney. *Comprehensive Water Distribution Systems Analysis Handbook for Engineers and Planners.* 2d ed. Pasadena, Calif.: MWH Soft Press, 2006.

Bouwkamp, Katie. "The 9 Most In-Demand Programming Languages of 2016." *Coding Dojo Blog,* 27 Jan. 2016, www.codingdojo.com. Accessed 18 Jan. 2017.

Boyer, Carl B. "The Fibonacci Sequence." *A History of Mathematics.* 2nd ed. New York: Wiley, 1991. 255–56. Print.

Bradbury, Danny. "Remote Replication: Comparing Data Replication Methods." *ComputerWeekly.com.* TechTarget. Web. 29 Dec. 2014.

Bradely, Tony. "Sub7 Trojan/Backdoor." *Lifewire,* 20 Oct. 2016, www.lifewire.com. Accessed 6 Dec. 2016.

Bradford, Alina. "Empirical Evidence: A Definition." *Live Science,* 24 Mar. 2015, www.livescience.com. Accessed 19 Jan. 2017.

______. "Science & the Scientific Method: A Definition." *Live Science,* 30 Mar. 2015, www.livescience.com. Accessed 19 Jan. 2017.

Bradley, David. "A Chemical History of Graphene." *MaterialsToday.* Elsevier Ltd. 10 June 2014. Web. 22 June 2016.

Braga, Newton C. *Bionics for the Evil Genius: Twenty-five Build-It-Yourself Projects.* McGraw-Hill, 2006.

Brenner, David, and Eric Hall. "Computed Tomography: An Increasing Source of Radiation Exposure." *New England Journal of Medicine,* vol. 357, no. 22, 2007, pp. 2277–284.

"A Brief Description. *cplusplus.com,* www.cplusplus.com. Accessed 18 Jan. 2017.

Bright, Peter. "Locking the Bad Guys Out with Asymmetric Encryption." *ARSTechnica.* Condé Nast. 12 Feb. 2013. Web. 4 March 2016.

Brighton, Henry, and Howard Selina. *Introducing Artificial Intelligence.* Edited by Richard Appignanesi. 2003. Reprint. Cambridge, England: Totem Books, 2007.

Bronzino, Joseph D., and Donald R. Peterson. *Biomechanics: Principles and Practices.* Boca Raton: CRC, 2014. *eBook Collection (EBSCOhost).* Web. 25 Feb. 2015.

Brooks, Frederick P., Jr. *The Mythical Man-Month: Essays on Software Engineering.* Anniv. ed. Reading: Addison, 1995. Print.

Brousseau, Guy, Virginia Warfield. *Teaching Fractions Through Situations: A Fundamental Experiment.* New York: Springer, 2013.

Brown, Alan S. "Mechatronics and the Role of Engineers." *American Society of Mechanical Engineers,* Aug. 2011, www.asme.org. Accessed 19 Jan. 2017.

Brown, Eric, et al. "Universal Robotic Gripper Based on the Jamming of Granular Material." *Proceedings of the National Academy of Sciences of the United States of America,* 2 Nov. 2010, www.ncbi.nlm.nih.gov. Accessed 12 Jan. 2017.

Brown, Eric. "True Physics." *Game Developer* 17, no. 5 (May, 2010): 13-18.

Brown, Julian. *Minds, Machines, and the Multiverse: The Quest for the Quantum Computer.* Simon & Schuster, 2000.

Brown, Mike. "Konrad Zuse's Z3, the World's First Programmable Computer, Was Unveiled 75 Years Ago." *Inverse,* 12 May 2016, www.inverse.com. Accessed 10 Nov. 2016.

Bryan, L. A., and E. A. Bryan. *Programmable Controllers: Theory and Application.* Industrial Text, 1988.

Bryan, Luis A., and E. A. Bryan. *Programmable Controllers: Theory and Implementation.* 2nd ed. Atlanta: Industrial Text, 1997. Print.

Bueno de Mesquita, Bruce. *The Predictioneer's Game: Using the Logic of Brazen Self-Interest to See and Shape the Future.* New York: Random House, 2009.

Bunzel, Tom. *Easy Creating CDs and DVDs.* 2d ed. Indianapolis: Que, 2005.

Burdea, Grigore C., and Philippe Coiffet. *Virtual Reality Technology.* 2nd ed. Hoboken: Wiley-Interscience, 2003. Print.

Burden, Richard L., and J. Douglas Faires. *Numerical Analysis.* 9th ed. Boston: Brooks, 2011. Print

Bureau of Labor Statistics. Office of Survey Methods Research. "Wage Estimates by Job Characteristic: NCS and OES Program Data." By Michael Lettau and Dee Zamora. N.p.: n.p., 2013.

Burgess, Mark. "C Programming Tutorial." *Mark Burgess.* Mark Burgess. Web. 11 Aug. 2015.

Burrows, Leah. "The First Autonomous, Entirely Soft Robot." *Harvard Gazette,* 24 Aug. 2016, news.harvard.edu. Accessed 12 Jan. 2017.

"Business Process Interoperability Framework." *Australian Government Department of Finance,* www.finance.gov.au. Accessed 23 Jan. 2017.

"C Language Reference." *Microsoft Developer Network.* Microsoft. Web. 11 Aug. 2015.

"The C Programming Language. *College of Engineering & Computer Science: University of Michigan, Dearborn.* The Regents of the University of Michigan. Web. 11 Aug. 2015.

"C Programming Language." *DI Management Services.* DI Management Services Pty Limited. Web. 11 Aug. 2015.

"C/C++." *TechTerms,* techterms.com. Accessed 18 Jan. 2017.

"C++ Programming Language." *Techopedia,* www.techopedia.com. Accessed 18 Jan. 2017.

Cannon, Leah. "What Can DNA-Based Computers Do?" *MIT Technology Review,* 4 Feb. 2015, www.technologyreview.com. Accessed 20 Nov. 2016.

"Carbon Nanotubes and Buckyballs." *University of Wisconsin-Madison,* education.mrsec.wisc.edu. Accessed 14 Dec. 2016.

Carlson, Lucas, and Leonard Richardson. *Ruby Cookbook.* Sebastopol: O'Reilly Media, 2015. Print.

Carlson, Neil R. *Foundations of Behavioral Neuroscience.* 9th ed. Boston: Pearson, 2014. Print.

Carraher, Charles E., Jr. *Giant Molecules: Essential Materials for Everyday Living and Problem Solving.* 2d ed. Hoboken, N.J.: John Wiley & Sons, 2003.

______. *Introduction to Polymer Chemistry.* 2d ed. Boca Raton, Fla.: CRC Press, 2010.

Ceccato, Cristiano, Lars Hesselgren, Mark Pauly, Helmut Pottmann, and Johannes Wallner. *Advances in Architectural Geometry.* Wien: Springer, 2010.

"Central Limit Theorem." *Boston University School of Public Health,* 24 July 2016, sphweb.bumc.bu.edu. Accessed 25 Jan. 2017.

Cerf, V. G., and R. E. Kahn. "A Protocol for Packet Network Interconnection." *IEEE Transactions on Communication Technology* 22 (May, 1974): 627-641.

Chakravarty, Satya R., Manipushpak Mitra, and Palash Sarkar. *A Course on Cooperative Game Theory.* Cambridge UP, 2015.

Chan, Melanie. *Virtual Reality: Representations in Contemporary Media.* New York: Bloomsbury, 2014. Print.

Chang, Sau Sheong. *Go Web Programming.* Shelter Island: Manning, 2016. Print.

"Charles Babbage (1791-1871)." *BBC.* BBC, 2014. Web. 15 June 2016.

"Charles Babbage." *History Mesh.* History Mesh, n.d. Web. 15 June 2016.

"Charles Babbage's Analytical Engine." Online video clip. *YouTube.* YouTube, 12 Aug. 2012. Web. 15 June 2016.

"Chemists Find Key to Manufacturing More Efficient Solar Cells." *Phys.org,* 23 Sept. 2016, phys.org. Accessed 10 Nov. 2016.

Chen, Ying Jian. "Bioplastics and Their Role in Achieving Global Sustainability." *Journal of Chemical and Pharmaceutical Research* 6.1(2014): 226–231. Web. 6 Feb. 2016.

Chow, Ven Te. *Open-Channel Hydraulics.* 1959. Reprint. Caldwell, N.J.: Blackburn, 2008. Contains chapters on uniform flow, varied flow, and unsteady flow.

______. *Neurophilosophy: Toward a Unified Science of the Mind-Brain.* Cambridge: MIT P, 1986. Print.

Churchland, Patricia S. *Touching a Nerve: The Self as Brain.* New York: Norton, 2013. Print.

Cintula, Petr, et al. "Fuzzy Logic." *Stanford Encyclopedia of Philosophy*, 15 Nov. 2016, plato.stanford.edu. Accessed 24 Jan. 2017.

Claerr, Jennifer. "What Are the Four Basic Functions of a Computer?" *Techwalla*, www.techwalla.com. Accessed 17 Nov. 2016.

Clark, Don. "Computer Chips Evolve to Keep Up with Deep Learning." *Wall Street Journal*, 11 Jan. 2017, www.wsj.com. Accessed 16 Jan. 2017.

______. "Smart-Home Gadgets Still a Hard Sell." *Wall Street Journal*. Dow Jones, 5 Jan. 2016. Web. 12 Mar. 2016.

"Client/Server (Client/Server Model, Client/Server Architecture) Definition." *TechTarget*. TechTarget, 2015. Web. 7 Oct. 2015.

"Client/Server Architecture." *Techopedia*. Janalta Interactive, 2015. Web. 7 Oct. 2015.

"Cognitive Behavioral Therapy." *Mayo Clinic*, 23 Feb. 2016, www.mayoclinic.org. Accessed 17 Dec. 2016.

Cohen, Michael. "Classical Mechanics: A Critical Introduction." *Physics & Astronomy*. The Trustees of the University of Pennsylvania. PDF. Web. 14 Mar. 2016.

Colapinto, John. "Graphene May Be the Most Remarkable Substance Ever Discovered. But What's It for?" *The New Yorker*. Conde Nast, 22 and 29 Dec., 2014. Web. 22 June 2016.

Coldewey, Devin. "Scientists Aspire to Nature's Genius with 'Biomimetic' Research." *NBCNews*, 6 June 2015, www.nbcnews.com. Accessed 12 Jan. 2017.

"Colussus: Birth of the Digital Computer." *Crypto Museum*, www.cryptomuseum.com. Accessed 5 Dec. 2016.

Comer, Douglas. *Computer Networks and Internets*. Pearson Education, 2015.

"Computer – Introduction to Memory." *CCM Benchmark Group*, ccm.net. Accessed 17 Nov. 2016.

"Computer AI Passes Turing Test in 'World First.'" *BBC*. BBC. 9 June 2014. Web. 30 July 2015.

"Computer Bus—What Is It?" *CCM*. CCM Benchmark Group. Feb. 2016. Web. 23 Feb. 2016.

"A Computer Called Watson. *IBM 100*, www-03.ibm.com. Accessed 5 Dec. 2016.

Cong-Vinh, Phan. *Formal and Practical Aspects of Autonomic Computing and Networking: Specification, Development, and Verification*. Hershey: Information Science Reference, 2012. Print.

Cook, William. *In Pursuit of the Traveling Salesman: Mathematics at the Limits of Computation*. Princeton, NJ: Princeton UP, 2012.

Copeland, Jack. "What Is Artificial Intelligence?" *AlanTuring*. AlanTuring, 2000. Web. 5 Oct. 2015.

Cormen, Thomas H. *Algorithms Unlocked*. Cambridge, MA: MIT P, 2013.

______. *Introduction to Algorithms*. Cambridge, MA: MIT P, 2009.

Cothron, Julia H., et al. *Science Experiments by the Hundreds*. Kendall/Hunt Publishing Company, 1996, 12–13.

Courant, Richard. *Differential and Integral Calculus*. Hoboken: Wiley, 2011.

Couvalis, George. *The Philosophy of Science: Science and Objectivity*. London: Sage, 1997. Print.

Cox, Kate. "It's Time to Start Treating Video Game Industry Like the $21 Billion Business It Is." *Consumerist*, 9 June 2014, consumerist.com. Accessed 22 Dec. 2014.

Craig, Alan B., William R. Sherman, and Jeffrey D. Will. *Developing Virtual Reality Applications. Foundations of Effective Design*. Burlington: Morgan Kaufmann, 2009. Print.

Crawley, Michael, J. *Statistics: An Introduction using R*. Medford: Wiley, 2014. Print.

"Create the Internet of Your Things." *Microsoft*. Microsoft. Web. 3 Feb. 2015.

Crilly, Rob. "Top-Secret U.S. Space Drone Returns to Earth after Two-Year Orbit." *The Telegraph*, 18 Oct. 2014, www.telegraph.co.uk. Accessed 10 Dec. 2016.

Crowder, C.D. "Explaining the Control in a Science Experiment." *Bright Hub Education*, 1 Jan. 2013, www.brighthubeducation.com. Accessed 19 Jan. 2017.

"Cryptography Defined/Brief History." *The College of Liberal Arts*. University of Texas at Austin. Web. 4 March 2016.

Culler, David, Jaswinder Pal Singh, and Anoot Gupta. *Parallel Computer Architecture: A Hardware/Software Approach*. San Francisco: Kaufmann, 1999. Print.

Cundiff, John S. *Fluid Power Circuits and Controls: Fundamentals and Applications*. Boca Raton: CRC, 2001. Print.

"Cyber Security Primer." Cybersecurity. University of Maryland University College, n.d. Web. 2 June 2014.

Dahm, Donald J., and Eric A. Nelson. *Calculations in Chemistry: An Introduction.* New York: Norton, 2013.

Daines, James R. *Fluid Power: Hydraulics and Pneumatics.* 2nd ed. Tinley Park: Goodheart, 2013. Print.

D'Almeida, Carolyn M., and Bradley J. Roth. "Medical Applications of Nanoparticles." *University of Michigan–Flint,* www.umflint.edu. Accessed 2 Dec. 2016.

Damien, Jose. *Introduction to Computers and Application Software.* Sudbury: Jones & Bartlett Learning, 2011. 4–7, 25. Print.

Daniels, Peter T., and William Bright, eds. *The World's Writing Systems.* New York: Oxford University Press, 1996. These essays continue and refine the work done by Gelb.

Dantzig, George B. "Linear Programming." *Operations Research,* vol. 50, no. 1, 2002, pp. 42–47, pubsonline.informs.org. Accessed 27 Jan. 2017.

Dargie, Waltenegu. *Context-Aware Computing and Self-Managing Systems.* Boca Raton: CRC, 2009. Print.

Das, Sumitabha. *Your UNIX: The Ultimate Guide.* 2d ed. Boston: McGraw-Hill, 2006.

Daston, Lorraine, and Peter Galison. *Objectivity.* 2007. New York: Zone, 2010. Print.

"Data Measurement Chart." *University of Florida,* www.wu.ece.ufl.edu. Accessed 17 Nov. 2016.

"Data Replication." *Documentation.commvault.com.* CommVault Systems Inc. Web. 29 Dec. 2014.

"Database Replication." *UMBC Computer Science and Engineering.* University of Maryland, Baltimore County. Web. 29 Dec. 2014.

Davenport, Thomas H. *Analytics at Work: Smarter Decisions, Better Results.* Cambridge: Harvard Business Review, 2010. Print.

______. *Keeping Up with the Quants: Your Guide to Understanding and Using Analytics.* Cambridge: Harvard Business Review, 2013. Print.

Davies, Alex. "In 20 Years, Most New Cars Won't Have Steering Wheels or Pedals." *Wired.* Condé Nast, 21 July 2014. Web. 16 Nov. 2014.

Dayal, Vikram. *An Introduction to R for Quantitative Economics: Graphing, Simulating and Computing.* New York: Springer, 2015. Print.

De Chant, Tim. "A Drug-Delivering DNA Nanobot Computer, Built Inside a Cockroach." *Nova Next,* 10 Apr. 2014, www.pbs.org. Accessed 20 Nov. 2016.

De Waal, Frans. "To Each Animal Its Own Cognition." *The Scientist,* 1 May 2016, www.the-scientist.com. Accessed 20 Dec. 2016.

DeAngelis, Stephen F. "Closing In on Quantum Computing." *Wired.* Conde Nast. Web. 6 Aug. 2015.

"Definition of: Non-Preemptive Multitasking." *PC Mag.* PC Mag. Web. 14 Aug. 2015.

"Definition of: Preemptive Multitasking." *PC Mag.* PC Mag. Web. 14 Aug. 2015.

"Definition: Expert System." *PCMag.* PCMag Digital Group. Web. 9 Mar 2016.

DeGroot, Morris H., and Mark J. Schervish. *Probability and Statistics.* 4th ed. Upper Saddle River, NJ: Pearson, 2011.

Delaney, Frank. "History of the Microcomputer Revolution." *Micro Technology Associates.* Micro Technology Associates. Web. 27 Aug. 2015.

Dell, Kristina. "The Promise and Pitfalls of Bioplastic." *Time.* Time Inc. 3 May 2010. Web. 6 Feb. 2016.

"Description of Motion in One Dimension." *HyperPhysics,* hyperphysics.phy-astr.gsu.edu. Accessed 13 Dec. 2016.

Deviant, S. "Central Limit Theorem." *The Practically Cheating Statistics Handbook.* 3rd ed., CreateSpace Independent Publishing, 2010, pp. 88–97.

Devlin, Keith. J. *The Joy of Sets: Fundamentals of Contemporary Set Theory.* 2nd ed. New York: Springer, 1993. Print.

Dhillon, Gupreet. "Dimensions of Power and IS Implementation." *Information and Management* 41, no. 5 (May, 2004): 635-644.

Dickerson, Kelly. "Here's Why We Should Be Really Excited about Quantum Computers." *Business Insider.* Business Insider, Inc. 17 Apr. 2015. Web. 6 Aug. 2015.

"Differences between Multithreading and Multitasking for Programmers." *National Instruments.* National Instruments Corporation. 20 Jan. 2014. Web. 14 Aug. 2015.

"Different Types of Software." *Introduction to IT English.* University of Victoria. Web. 9 July 2015.

DiGregorio, Barry E. "Biobased Performance Bioplastic: Mirel." *Chemistry & Biology* 16.1. (2009): 1–2. *ScienceDirect.* Web. 6 Feb. 2016.

DiLorenzo, Daniel J., and Joseph D. Bronzino, eds. *Neuroengineering.* Boca Raton, Fla.: CRC Press, 2008.

Dingle, Norm. "Artificial Intelligence: Fuzzy Logic Explained." *Control Engineering,* 4 Nov. 2011, www.controleng.com. Accessed 24 Jan. 2017.

Dinwiddie, Keith. *Basic Robotics.* Cengage Learning, 2016.

"Discovery of Fullerenes National Historic Chemical Landmark." *American Chemical Society,* www.acs.org. Accessed 14 Dec. 2016.

"The Discovery of Fullerenes, *American Chemical Society,* www.acs.org. Accessed 14 Dec. 2016.

"Do You Really Need a Scan?" *Consumer Reports on Health,* vol. 27, no. 3, 2015, pp., 1–5.

Doehring, James. "What Is Mechatronics Engineering?" *Wisegeek,* 20 Dec. 2016, www.wisegeek.com. Accessed 19 Jan. 2017.

Doncaster, P., and A. Davey. *Analysis of Variance and Covariance: How to Choose and Construct Models for the Life Sciences.* Cambridge: Cambridge UP, 2007. Print.

Donovan, Alan A. A., and Brian W. Kernighan. *The Go Programming Language.* New York: Addison-Wesley, 2015. Print.

Doran, Pauline M. *Bioprocess Engineering Principles.* London: Academic Press, 2009. A solid, basic textbook for students entering the field.

Dordal, Peter L. "An Introduction to Computer Networks." Loyola University Chicago, 2015, intronetworks.cs.luc.edu/current/html/.

Doxsey, Caleb. *Introducing Go: Build Reliable, Scalable Programs.* Sebastopol: O'Reilly, 2016. Print.

"Dr. Otto H. Schmitt—1978 Inductee." *Minnesota Inventors Hall of Fame,* www.minnesotainventors.org. Accessed 12 Jan. 2017.

Drexler, K. Eric. *Engines of Creation: The Coming Era of Nanotechnology.* New York: Anchor, 1986. Print.

______. *Radical Abundance: How a Revolution in Nanotechnology Will Change Civilization.* New York: PublicAffairs, 2013. Print.

Dshalalow, Eugene. *Foundations of Abstract Analysis.* New York: Springer, 2014.

Duggal, Vijay, Sella Rush, and Al Zoli, eds. *CADD Primer: A General Guide to Computer-Aided Design and Drafting.* Elmhurst: Mailmax, 2000. Print.

Dumas, M. Barry, and Morris Schwartz. *Principles of Computer Networks and Communications.* Upper Saddle River, N.J.: Pearson/Prentice Hall, 2009.

Dunn, Casey. "As 'Normal' as Rabbits' Weights and Dragons' Wings." *New York Times,* 23 Sept. 2013, www.nytimes.com. Accessed 24 Jan. 2017.

Dunn, Fletcher, and Ian Parberry. *3D Math Primer for Graphics and Game Development.* Boca Raton, FL: CRC, 2011.

Duraiswami, Ramani. "Projective Geometry." University of Maryland Institute of Advanced Computer Studies, University of Maryland. Web. 3 Mar 2016.

Durand, Dominique M. "What Is Neural Engineering?" *Journal of Neural Engineering* 4, no. 4 (September, 2006).

Dwoskin, Elizabeth. "Siri's Creators Say They've Made Something Better That Will Take Care of Everything for You." *Washington Post,* 4 May 2016, www.washingtonpost.com. Accessed 9 Dec. 2016.

Dyson, George. *Turing's Cathedral: The Origins of the Digital Universe.* London: Penguin Books, 2013. Print.

Ebewele, Robert O. *Polymer Science and Technology.* Boca Raton, Fla.: CRC Press, 2000.

Elwes, Richard. "Cardinal Numbers." *Math in 100 Key Breakthroughs.* New York: Quercus, 2013. Print.

Emmio, Nicolette. "Why Do Engineers Care So Much About Graphene?" *Electronics 360.* Electronics 360, 22 June 2016. Web. 23 June 2016.

Enders, Walter, and Todd Sandler. *The Political Economy of Terrorism.* New York: Cambridge UP, 2005. Print.

"The Engines. *Computer History Museum.* Computer History Museum, n.d. Web. 15 June 2016.

Entertainment Software Assoc. *2016 Sales, Demographic, and Usage Data: Essential Facts about the Computer and Video Game Industry.* Entertainment Software Association, 2016, essentialfacts.theesa.com. Accessed 31 Oct. 2016.

Epp, Susana S. *Discrete Mathematics and Applications.* Boston: Cengage, 2011. Print.

Ercole, Leslie K., Manny Frantz, and George Ashline. "Multiple Ways to Solve Proportions." *Mathematics Teaching in the Middle School* 16.8 (2011): 482-490.

Estienne, Sophie. "Artificial Intelligence Creeps into Daily Life." *Phys. Org,* 15 Dec. 2016, phys.org. Accessed 21 Dec. 2016.

"Ethernet Protocols and Packets." *Savvius,* www.wildpackets.com. Accessed 28 Jan. 2017.

Ethier, C. Ross, and Craig A. Simmons. *Introductory Biomechanics: From Cells to Organisms.* Cambridge, England: Cambridge University Press, 2007. Provides an introduction to biomechanics and also discusses clinical specialties, such as cardiovascular, musculoskeletal, and ophthalmology.

"European Smart Cities." *Vienna University of Technology,* 2015, www.smart-cities.eu. Accessed 23 Dec. 2016.

"The Evolution of AI: Can Morality Be Programmed? *Future of Life,* 6 July 2016, futureoflife.org. Accessed 20 Dec. 2016.

"Evolution of Computer Networks." *University of Notre Dame,* www3.nd.edu. Accessed 28 Jan. 2017.

"Expert System." *TechTarget.* SearchHealthIT. 2014 Nov. Web. 9 Mar 2016.

"Factbox: What Is Stuxnet?" *Reuters.* Thomson Reuters. 24 Sept. 2010. Web. 31 July 2015.

Farber, Dan. "The Next Big Thing in Tech: Augmented Reality." *CNET.* CBS Interactive Inc. 7 June 2013. Web. 13 Aug. 2015.

Faust, Russell A., ed. *Robotics in Surgery: History, Current and Future Applications.* New York: Nova Science, 2007. Print.

"Features of C++." *Sitesbay,* www.sitesbay.com. Accessed 18 Jan. 2017.

Fenichell, Stephen. *Plastic: The Making of a Synthetic Century.* New York: HarperCollins, 1996.

Ferguson, Thomas S. "Linear Programming: A Concise Introduction." *University of California Los Angeles,* www.math.ucla.edu. Accessed 27 Jan. 2017.

"Fiber Optic Network Topologies for ITS and Other Systems." *Fiber-Optics Info.* Fiber-Optics Info. Web. 11 Feb. 2016.

Finkle, Jim, Soham Chatterjee, and Lehar Maan. "eBay Asks 145 Million Users to Change Passwords after Cyber Attack." Reuters. Thomson Reuters, 21 May 2014. Web. 2 June 2014.

Finnegan, Matthew. "Google Go: Why Google's Programming Language Can Rival Java in the Enterprise." *TechWorld.* IDG UK, 25 Sept.2015. Web. 19 Aug. 2016.

Finnemore, E. John, and Joseph B. Franzini. *Fluid Mechanics With Engineering Applications.* 10th international ed. Boston: McGraw-Hill, 2009.

Fischler, Martin A, and Oscar Firschein. *Intelligence: The Eye, the Brain, and the Computer.* Reading (MA): Addison-Wesley, 1987. Print.

Fischman, Josh. "Merging Man and Machine: The Bionic Age." *National Geographic,* vol. 217, no. 1, 2010, pp. 34–53.

Fish, Eric. "Smart Labels—The Next Big Thing in Packaging?" *Flexible Packaging Magazine,* 11 Jan. 2016, www.flexpackmag.com. Accessed 21 Dec. 2016.

Fisher, Len. *Rock, Paper, Scissors: Game Theory in Everyday Life.* New York: Basic Books, 2008.

Fitz-Gibbon, Carol T., and Lynn Lyons Morris. *How to Design a Program Evaluation.* Sage Publications, 1987, 25–27.

"Five ICT Essentials for Smart Cities." *Escher Group,* www.eschergroup.com. Accessed 23 Dec. 2016.

Flanagan, David, and Yukihiro Matsumoto. *The Ruby Programming Language.* Sebastopol: O'Reilly Media, 2008. Print.

Fleischer, Henry. *Manual of Pneumatic Systems Optimization.* New York: McGraw, 1995. Print.

Floreano, Dario, and Claudio Mattuissi. *Bio-Inspired Artificial Intelligence: Theories, Methods, and Technologies.* Cambridge: MIT P, 2008. Print.

Folger, Tim. "Revealed World." *National Geographic.* National Geographic Society. Web. 13 Aug. 2015.

Forbes, Catherine, Merran Evans, Nicholas Hastings, and Brian Peacock. *Statistical Distributions.* Hoboken, NJ: Wiley, 2011.

Foreman, John W. *Data Smart: Using Data Science to Transform Information into Insight.* Indianapolis: Wiley, 2013. Print.

Forouzan, Behrouz. *Data Communications and Networking.* 5th ed. New York: McGraw-Hill, 2013.

"Foundations of Fuzzy Logic." *MathWorks,* www.mathworks.com. Accessed 24 Jan. 2017.

Fowler, Martin. *Domain-Specific Languages.* Upper Saddle River: Addison, 2011.

Fox, J. *Applied Regression Analysis and Generalized Linear Models.* 3rd ed. Thousand Oaks: Sage, 2016. Print.

Franceschetti, Donald R. *Biographical Encyclopedia of Mathematicians.* New York: Marshall Cavendish, 1999. Print.

Franklin, Stan. *Artificial Minds.* Cambridge, Mass: MIT Press, 2001. Print.

Freedman, David, Robert Pisani, and Roger Purves. *Statistics*. 4th ed. London: Norton, 2011.

Frenay, Robert. *Pulse: The Coming Age of Systems and Machines Inspired by Living Things*. 2006. Reprint. Lincoln: University of Nebraska Press, 2008.

"Fullerenes and Their Applications in Science and Technology." *Florida International University College of Engineering and Computing,* Spring 2013, web.eng.fiu.edu. Accessed 14 Dec. 2016.

"Fullerenes." *Stanford University,* web.stanford.edu. Accessed 14 Dec. 2016.

Fulton, Hal, and André Arko. *The Ruby Way*. 3rd ed. Boston: Pearson Education, 2015. Print.

Fung, Kaiser. *Numbers Rule Your World: The Hidden Influence of Probabilities and Statistics on Everything You Do*. New York: McGraw, 2010. Print.

"Fuzzy Logic Introduction." *Imperial College London,* www.doc.ic.ac.uk. Accessed 24 Jan. 2017.

Galati, Steven R. *Geographic Information Systems Demystified*. Boston: Artech House, 2006. Introductory textbook provides graphic examples of the concepts presented and includes a very useful glossary of terms.

Galchen, Rivka. "Dream Machine." *New Yorker*. Conde Nast. 2 May 2011. Web. 6 Aug. 2015.

Gauch, Hugh G., Jr. *Scientific Method in Brief*. New York: Cambridge UP, 2012. Print.

Gaudiosi, John. "Virtual Reality Video Game Industry to Generate $5.1 Billion in 2016." *Fortune*. Time, 5 Jan. 2016. Web. 26 June 2016.

Gee, James Paul. *Unified Discourse Analysis: Language, Reality, Virtual Worlds, and Video Games*. New York: Routledge, 2015. Print.

Geier, Eric, and Norem, Josh. "How to Remove Malware from Your Windows PC." *PCWorld,* 18 July 2016, www.pcworld.com. Accessed 30 Nov. 2016.

Geim, A.K., and K.S. Novoselov. "The Rise of Graphene." *Nature Materials*. Macmillan Publishers Limited 6 (2007): 183-91. Web. 22 June 2016.

Gelb, Ignace. *A Study of Writing: The Foundations of Grammatology*. 1952. Reprint. Chicago: University of Chicago Press, 1989. The foundational text for grammatology.

"George Dantzig, 90; Created Linear Programming." *Los Angeles Times,* 22 May 2005, articles.latimes.com. Accessed 27 Jan. 2017.

"'Getting Started' Guide to Cybernetics." *Pangaro.com*. Paul Pangaro. 2013. Web. 14 Mar. 2016.

Ghosh, Debasish. *DSLs in Action*. New York: Manning, 2010.

Giarratano, Joseph, and Peter Riley. *Expert Systems: Principles and Programming*. 4th ed. Boston: Thomson, 2005. Print.

Gibbs, Samuel, and Alex Hern. "Apple Watch 2 Brings GPS, Waterproofing and Faster Processing." *The Guardian,* 8 Sept. 2016, www.theguardian.com. Accessed 28 Oct. 2016.

Gibbs, Samuel. "Google's Massive Humanoid Robot Can Now Walk and Move without Wires." *Guardian*. Guardian News and Media, 21 Jan. 2015. Web. 21 Jan. 2016.

Gintis, Herbert. *Game Theory Evolving: A Problem-Centered Introduction to Modeling Strategic Behavior*. 2d ed. Princeton, N.J.: Princeton University Press, 2009.

Glantz, Stanton A. *Primer of Biostatistics*. 7th ed. New York,: McGraw, 2011.

Glazer, Alexander N., and Hiroshi Nikaido. *Microbial Biotechnology: Fundamentals of Applied Microbiology*. New York: Cambridge University Press, 2007. In-depth analysis of the application of microorganisms in bioprocessing.

Glink, Ilyce. "10 Smart Home Features Buyers Actually Want." *CBS News*. CBS Interactive, 11 Apr. 2015. Web. 12 Mar. 2016.

"Global Open Access Portal." *United Nations Educational, Scientific and Cultural Organization,* www.unesco.org. Accessed 14 Jan. 2017.

Glydon, Natasha. "Linear Programming." *Regina University,* mathcentral.uregina.ca. Accessed 27 Jan. 2017.

Golang.org. Go Programming Language, 2016. Web. 19 Aug. 2016.

Gordon, Whitson. "How to Build a Computer, Lesson 1: Hardware Basics." *Lifehacker*. Gawker Media Group. Web. 23 Feb. 2016.

Goriunova, Olga, ed. *Fun and Software: Exploring Pleasure, Paradox, and Pain in Computing*. New York: Bloomsbury, 2014. Print.

Grabiner, Judith V. *The Origins of Cauchy's Rigorous Calculus*. Mineola: Dover, 2011. Print.

Graham, Ronald L., Donald E. Knuth, and Oren Patashnik. *Concrete Mathematics: A Foundation for Computer Science*. 2nd ed. Reading: Addison, 1994. Print.

Graham, Susan L., et al., editors. *Getting Up to Speed: The Future of Supercomputing.* National Academies Press, 2005.

"Graphene-Based Material Illuminates Bright New Future for Flexible Lighting Devices." *Nanowerk News*: Nanowerk, June 2016. Web. 23 June 2016.

Greenemeier, Larry. "Soft Touch: Squishy Robots Could Lead to Cheaper, Safer Medical Devices." *Scientific American,* 24 Sept. 2013, www.scientificamerican.com. Accessed 12 Jan. 2017.

Griffiths, Devin C. *Virtual Ascendance: Video Games and the Remaking of Reality.* Lanham: Rowman, 2013. Print.

Grunbaum, Branko, and G. C. Shephard. *Tilings and Patterns.* New York: Dover, 2015.

Grus, Joel. *Data Science from Scratch: First Principles with Python.* Sebastopol: O'Reilly, 2015. Print.

Guelich, Scott, Shishir Gundavaram, and Gunther Birznieks. *CGI Programming with Perl.* 2d ed. Cambridge, Mass.: O'Reilly, 2000.

"Guide to NoSQL Databases: How They Can Help Users Meet Big Data Needs." *TechTarget.* TechTarget. Web. 6 Aug. 2015.

Gupta, Shalene. "For the Disabled, Smart Homes Are Home Sweet Home." *Fortune.* Fortune, 1 Feb. 2015. Web. 15 Mar. 2016.

Gur-Arie, Margalit. "The History of Healthcare Interoperability." *HIT,* 11 Apr. 2013, hitconsultant.net. Accessed 23 Jan. 2017.

Gutkind, Lee. *Almost Human: Making Robots Think.* 2006. Reprint. New York: Norton, 2009. Print.

Hadjipanayis, George C, ed. *Magnetic Storage Systems Beyond 2000.* Dordrecht, The Netherlands: Kluwer Academic, 2001.

Hall, Susan J. *Basic Biomechanics.* 5th ed. New York: McGraw-Hill, 2006. A good introduction to biomechanics, regardless of one's math skills.

Hamill, Joseph, and Kathleen Knutzen. *Biomechanical Basis of Human Movement.* 4th ed. Philadelphia: Lippincott, 2015. Print.

Hanna, Gila. *Explanation and Proof in Mathematics.* New York: Springer, 2014.

"Hard Drive." *Computer Hope,* www.computerhope.com. Accessed 17 Nov. 2016.

Hariharan, P. *Basics of Holography.* Cambridge UP, 2002.

Harmon, Elliot. "Open Access Rewards Passionate Curiosity: 2016 in Review." *Electronic Frontier Foundation,* 24 Dec. 2016, www.eff.org. Accessed 14 Jan. 2017.

______. "What If Elsevier and Researchers Quit Playing Hide-and-Seek?" *Electronic Frontier Foundation,* 16 Dec. 2015, www.eff.org. Accessed 14 Jan. 2017.

Harper, Gavin D. J. *Holography Projects for the Evil Genius.* McGraw-Hill, 2010.

Harris, William. "How the Scientific Method Works." *HowStuffWorks.com,* science.howstuffworks.com. Accessed 19 Jan. 2017.

Hart, Archibald D., and Sylvia Hart Frejd. *The Digital Invasion: How Technology Is Shaping You and Your Relationships.* Grand Rapids: Baker, 2013. Print.

Harvey, Francis. *A Primer of GIS: Fundamental Geographic and Cartographic Concepts.* New York: Guildford Press. 2008.

Hass, Joel, Maurice Weir, and George B. Thomas. *Thomas' Calculus.* 13th ed. Boston: Pearson, 2014. Print.

Hatze, H. "The Meaning of the Term 'Biomechanics.'" *Journal of Biomechanics* 7.2 (1974): 89–90. Print.

Hay, James G. *The Biomechanics of Sports Techniques.* 4th ed. Englewood Cliffs: Prentice, 1993. Print.

Hay, James G., and J. Gavin Reid. *Anatomy, Mechanics, and Human Motion.* 2d ed. Englewood Cliffs, N.J.: Prentice Hall, 1988. A good resource for upper high school students, this text covers basic kinesiology.

Hayenga, Heather N., and Helim Aranda-Espinoza. *Biomaterials Mechanics.* CRC Press, 2017.

He, Bin, ed. *Neural Engineering.* New York: Kluwer Academic/Plenum Publishers, 2005.

Heiligtag, Florian J., and Markus Niederberger. "The Fascinating World of Nanoparticle Research." *Materials Today,* vol. 16, no. 7–8, July–Aug. 2013, pp. 262–71, www.sciencedirect.com. Accessed 2 Dec. 2016.

Heim, Michael. *Virtual Realism.* New York: Oxford UP, 2000. Print.

Heinzle, Elmar, Arno P. Biwer, and Charles L. Cooney. *Development of Sustainable Bioprocesses: Modeling and Assessment.* Hoboken, N.J.: John Wiley & Sons, 2007.

Helander, Martin. *A Guide to Human Factors and Ergonomics.* 2nd ed., CRC Press, 2006.

"Help NASA Create Better Vision for Robonaut." *National Aeronautics and Space Administration,* robotics.nasa.gov. Accessed 12 Jan. 2017.

Hern, Alex. "What Is the Turing Test? And Are We All Doomed Now?" *The Guardian.* Guardian News and Media Limited. 9 June 2014. Web. 30 July 2015.

Higginbotham, Stacey. "5 Reasons Why the 'Smart Home' Is Still Stupid." *Fortune.* Fortune, 19 Aug. 2015. Web. 12 Mar. 2016.

Hildebrand, F. B. *Introduction to Numerical Analysis.* 2nd ed. New York: McGraw, 1974. Print.

Hirschman, Isadore Isaac. *Infinite Series.* Belmont: Cengage, 2014. Print.

"History & Timeline." *Brain & Cognitive Sciences,* bcs.mit.edu. Accessed 18 Dec. 2016.

"History and Timeline." *The Open Group.* The Open Group. Web. 16 Feb. 2016.

"History of APIs." *API Evangelist,* 20 Dec. 2012, apievangelist.com. Accessed 29 Dec. 2016.

"History of C++." *cplusplus.com,* www.cplusplus.com. Accessed 18 Jan. 2017.

"History of Computers." *University of Rhode Island,* homepage.cs.uri.edu. Accessed 29 Dec. 2016.

"History of Cybernetics." *American Society for Cybernetics.* American Society for Cybernetics. Web. 14 Mar. 2016.

"The History of HTML. Ironspider. Iron Spider. Web. 20 Aug. 2015.

"The History of HTML. Landofcode.com. Web. 20 Aug. 2015.

"History of Photogrammetry." The Center for Photogrammetric Training. Department of Spatial Sciences, Curtain University. Web. 3 Mar 2016.

"The History of Solar. *U.S. Department of Energy,* www1.eere.energy.gov. Accessed 10 Nov. 2016.

"The History of the Botnet, Part I. *TrendMicro,* 24 Sept. 2010, countermeasures.trendmicro.eu. Accessed 30 Nov. 2016.

"History of the C++ Language." *CodingUnit Programming Tutorials,* www.codingunit.com. Accessed 18 Jan. 2017.

"History of the Microprocessor." *Meeting Tomorrow.* Meeting Tomorrow, Inc. Web. 23 Feb. 2016.

Hitzler, Pascal, Markus Krotzsch, and Sebastian Rudolph. *Foundations of Semantic Web Technologies.* Boca Raton, FL: Taylor & Francis Group, 2010. Print.

Hoare, C. A. R. *Communicating Sequential Processes.* Upper Saddle River: Prentice, 1985. Print.

Hock-Chuan, Chua. "Java Programming Tutorial: Object-Oriented Programming (OOP) Basics." Nanyang Technological University. Nanyang Technological University. Apr. 2013. Web. 11 Aug. 2015.

Hodges, Andrew. "Alan Turing and the Turing Test." *Alan Turing: The Enigma.* Andrew Hodges. Web. 30 July 2015.

Hof, Robert D. "Deep Learning." *MIT Technology Review,* www.technologyreview.com. Accessed 16 Jan. 2017.

Hofmann, Angelika H. *Scientific Writing and Communication: Papers, Proposals, and Presentations.* 2nd ed. New York: Oxford UP, 2014. Print.

Holdsworth, Brian, and R. Clive Woods. *Digital Logic Design.* 4th ed., Newnes, 2002.

Hollos, Stefan, J. Richard Hollos. *Combinatorics Problems and Solutions.* Longmont, CO:Abrazol, 2013.

Horikoshi, Satoshi, and Nick Serpone. "Introduction to Nanoparticles." *Microwaves in Nanoparticle Synthesis: Fundamentals and Applications.* Wiley-VCH, 2013, pp. 1–24.

Hornsby, John E., Margaret L. Lial, Gary K. Rockswold. A *Graphical Approach to Precalculus with Limits.* 6th ed. London: Pearson, 2014.

Horstmann, Cay. *Big Java.* 4th ed. Hoboken, N.J.: John Wiley & Sons, 2010.

Hothorn, Torsten, and B. S. Everitt. *A Handbook of Statistical Analyses Using R.* 3rd ed. Boca Raton: CRC, 2014. Print.

"How Apple's Siri Got Her Name." *The Week,* 29 Mar. 2012, theweek.com. Accessed 9 Dec. 2016.

"How Computers Work: The CPU and Memory." *University of Rhode Island,* homepage.cs.uri.edu. Accessed 17 Nov. 2016.

"How Did the Human Genome Project Make Science More Accessible?" *YourGenome.org,* 13 June 2016, www.yourgenome.org. Accessed 23 Jan. 2017.

"How Do Solar Cells Work?" *Physics.org,* www.physics.org. Accessed 10 Nov. 2016.

"How Speech Recognition Works." *Microsoft.* Microsoft. Web. 8 Mar. 2016.

Howell, Elizabeth. "What is a Drone?" *Space.com,* 2 June 2015, www.space.com. Accessed 10 Dec. 2016.

Hsieh, Jiang. *Computed Tomography: Principles, Design, Artifacts, and Recent Advances.* 2nd ed., International Society for Optical Engineering, 2009.

"HTML Basics." MediaCollege.com. Web 20 Aug. 2015.

"HTML Introduction." W3Schools.com. W3 Schools. Web. 28 Aug. 2015.

"HTML Tutorial." Refsnes Data. Web. 20 Aug. 2015.

Hu, Xiaoling. "Advances in Neural Engineering for Rehabilitation." *Behavioural Neurology*, vol. 2017, 2017, doi: 10.1155/2017/9240921. Accessed 28 Feb. 2018.

Hughes, Brian. "Technology Becomes Us: The Age of Human-Computer Interaction." *The Huffington Post,* 20 Apr. 2016, www.huffingtonpost.com. Accessed 28 Oct. 2016.

Hung, George K. *Biomedical Engineering: Principles of the Bionic Man.* World Scientific, 2010.

Hwang, Jangsun, et al. "Biomimetics: Forecasting the Future of Science, Engineering, and Medicine." *International Journal of Nanomedicine,* vol. 10, Oct. 2015, pp. 5701–13, www.ncbi.nlm.nih.gov. Accessed 12 Jan. 2017.

Hwang, Kai, and Doug Degroot, eds. *Parallel Processing for Supercomputers and Artificial Intelligence.* New York: McGraw, 1989. Print.

"IBM SPSS Statistics–Monte Carlo Simulation." *IBM.* IBM. Web. 4 Aug. 2015.

Igarashi, Yoshihide, et al. *Computing: A Historical and Technical Perspective.* CRC Press, 2014.

"Impact of eHealth Exchange on Nationwide Interoperability." *HealthIT Interoperability,* 10 Jan. 2017, healthitinteroperability.com. Accessed 23 Jan. 2017.

"Intel's First Microprocessor." *Intel Corporation.* Intel Corporation. Web. 27 Aug. 2015.

"Intel's First Microprocessor: Its Invention, Introduction, and Lasting Influence." *Intel.* Intel Corporation. Web. 23 Feb. 2016.

"Internet Hall of Fame Honors UT Austin Professor Bob Metcalfe and Alum Robert Taylor." *University of Texas at Austin,* 3 Aug. 2013, www.engr.utexas.edu. Accessed 28 Jan. 2017.

"The Internet of Things Research Study. *HP.* Hewlett-Packard Development Company, L.P. Sept. 2014. Web. 3 Feb. 2015.

"Introduction to PROLOG." *Mind.* Consortium on Cognitive Science Instruction, 2006. Web. 5 Oct. 2015.

"Introduction to Structured Query Language (SQL) – Part 1." *University of Delaware.* University of Delaware. Web. 6 Aug. 2015.

"Introduction to the Semantic Web." *Cambridge Semantics.* Cambridge Semantics. Web. 4 Aug. 2015.

Irvine, Kip R. *Assembly Language for Intel-Based Computers.* 5th ed. Upper Saddle River: Prentice, 2006. Print.

Isbister, Katherine. *How Games Move Us: Emotion by Design.* MIT Press, 2016.

Iserles, Arieh. *A First Course in the Numerical Analysis of Differential Equations.* 2nd ed. New York: Cambridge UP, 2009. Print.

Jain, Roopesh, and Archana Tiwari. "Biosynthesis of Planet Friendly Bioplastics Using Renewable Carbon Source." *Journal of Environmental Health Science and Engineering* 13.11 (2015). *PMC.* Web. 6 Feb. 2016.

James, Hubert M. *Theory of Servomechanisms.* New York: McGraw, 1947. Print.

Jamieson, Alastair, and Eric McClam. "Millions of Target Customers' Credit, Debit Card Accounts May Be Hit by Data Breach." NBC News. NBCNews, 19 Dec. 2013. Web. 2 June 2014.

Janusz, Stefan. "The Known and Unknown Pioneers of Modern Computing." *Science Node,* 20 June 2012, sciencenode.org. Accessed 10 Nov. 2016.

Jeffress, D. "What Does a Mechatronics Engineer Do?" *Wisegeek,* 12 Jan. 2017, www.wisegeek.com. Accessed 19 Jan. 2017.

Jimenez, Jorge, et al. "Destroy All Jaggies." *Game Developer* 18, no. 6 (June/July, 2011): 13-20. This article describes the method of anti-aliasing in animation graphics.

Johnson, Olaf A. *Fluid Power: Pneumatics.* Chicago: Amer. Technical Soc., 1975. Print.

Johnston, Sean F. "Absorbing New Subjects: Holography as an Analog of Photography." *Physics in Perspective,* vol. 8, no. 2, 2006, pp. 164–88.

Jokinen, Jussi P. P. "Emotional User Experience: Traits, Events, and States." *International Journal of Human-Computer Studies,* vol. 76, 2015, pp. 67–77.

Jones, James. "Stats: One-Way ANOVA." *Statistics: Lecture Notes.* Richland Community Coll., n.d. Web. 11 July 2016.

Jones, Joseph L. *Robot Programming: A Practical Guide to Behavior-Based Robotics.* New York: McGraw-Hill, 2004. Print.

Jonscher, Charles. *Wired Life: Who Are We in the Digital Age?* Bantam Press, 1999.

Joshi, Kailash. "Chapter 11: Expert Systems and Applied Artificial Intelligence." Management Information Systems, College of Business Administration. University of Missouri, St. Louis. Web. 9 Mar 2016.

Juang, B.H. and Lawrence R. Rabiner. "Automatic Speech Recognition – A Brief History of the Technology Development." *University of California, Santa Barbara.* The Regents of the University of California. PDF. Web. 8 Mar. 2016.

Kabacoff, R. *R in Action: Data Analysis and Graphics with R.* Greenwich: Manning, 2015. Print.

Kaehler, Steven D. "Fuzzy Logic – An Introduction." *Seattle Robotics Society,* www.seattlerobotics.org. Accessed 24 Jan. 2017.

Kafai, Yasmin B., and Quinn Burke. *Connected Code: Why Children Need to Learn Programming.* Cambridge: MIT P, 2014. Print.

Kamenetz, Anya. "Why Video Games Succeed Where the Movie and Music Industries Fail." *Fast Company,* 7 Nov. 2013, www.fastcompany.com. Accessed 22 Dec. 2014.

Katz, Bruce F. *Neuroengineering the Future: Virtual Minds and the Creation of Immortality.* Hingham, Mass.: Infinity Science Press, 2008.

Kavis, Mike. "The Smart Labels That Will Power the Internet of Things." *Forbes,* 17 Feb. 2015, www.forbes.com. Accessed 21 Dec. 2016.

Kayne, R. "What Is P2P?" *WiseGeek.* Conjecture, 2015. Web. 7 Oct. 2015.

Kean, Sam. *The Tale of the Dueling Neurosurgeons: The History of the Human Brain as Revealed by True Stories of Trauma, Madness, and Recovery.* New York: Little, 2014. Print.

Kelley, Michael B. "The Stuxnet Attack on Iran's Nuclear Plant Was 'Far More Dangerous' than Previously Thought." *Business Insider.* Business Insider Inc. 20 Nov. 2013. Web. 31 July 2015.

Kennedy, William, Brian Ketelsen, and Erik St. Martin. *Go in Action.* Shelter Island: Manning, 2016. Print.

Kephart, Jeffrey O., and David M. Chess. "The Vision of Autonomic Computing." IEEE Computer Society, Jan. 2003. PDF file.

Kerlow, Issac. *The Art of 3D Computer Animation and Effects.* 4th ed. Hoboken: Wiley, 2009. Print.

Kerns, David V., Jr., and J. David Irwin. *Essentials of Electrical and Computer Engineering* 2nd ed. Upper Saddle River: Prentice, 2004. Print.

Kerr, Andrew. *Introductory Biomechanics.* London: Elsevier, 2010. Print.

Kershner, Kate. "What Are Supercomputers Currently Used For?" *HowStuffWorks,* computer. howstuffworks.com. Accessed 5 Dec. 2016.

Kessler, Sarah. "For a Glimpse of the Future, Try Reading a 3.5-Inch Floppy Disk." *Fast Company,* January 7, 2013.

Khudyakov, Yury E., and Paul Pumpens. *Viral Nanotechnology.* CRC Press, 2016.

Kincaid, Jason. "Google's Go: A New Programming Language That's Python Meets C++." *TechCrunch.* AOL, 10 Nov. 2009. Web. 19 Aug. 2016.

"The Kinematic Equations." *The Physics Classroom.* The Physics Classroom. Web. 14 Mar. 2016.

Kirchmer, Mathias. *High Performance through Process Excellence: From Strategy to Execution with Business Process Management.* 2nd ed. Heidelberg: Springer, 2011. Print.

Kirk, David, and Wen-mei W. Hwu. *Programming Massively Parallel Processors: A Hands-On Approach.* Burlington: Kaufmann, 2010. Print.

Kizza, Joseph Migga. *Ethical and Social Issues in the Information Age.* 5th ed. London: Springer, 2013. Print.

Kjaerulff, Uffe, and Anders Madsen. *Bayesian Networks and Influence Diagrams.* New York: Springer, 2014.

Kline, Ronald R. *The Cybernetics Movement, or Why We Call Our Age the Information Age.* Baltimore, MD: John Hopkins University Press, 2015. Print.

Knight, Will. "AI's Unspoken Problem." *MIT Technology Review,* vol. 119, no. 5, 2016, pp. 28–37.

______. "Kindergarten for Computers." *MIT Technology Review,* vol. 119, no. 1, 2016, pp. 52–58.

Knopp, Konrad. *Theory and Application of Infinite Series.* Mineola: Dover, 1990. Print.

Kofman, Ava. "Dueling Realities." *The Atlantic.* The Atlantic Monthly Group. 9 June 2015. Web. 13 Aug. 2015.

Kolata, Gina Bari. "Geodesy: Dealing with an Enormous Computer Task." *Science* vol. 200, no. 4340 (April 28, 1978): pp. 421-466.

Kolo, Brian. *Binary and Multiclass Classification.* New York: Weatherford, 2010.

Koster, Raph. *A Theory of Fun for Game Design.* Paraglyph, 2004.

Kubitz, Alan A. *The Elusive Notion of Motion: The Genius of Kepler, Galileo, Newton, and Einstein.* Dog Ear Publishing, 2010.

Kulkarni, Ashish. "New Nanoparticle Reveals Cancer Treatment Effectiveness in Real Time." *Brigham and Women's Hospital,* 28 Mar. 2016, www.sciencedaily.com. Accessed 2 Dec. 2016.

Kulkarni-Thaker, Shefali. "The Diet Problem." *The OR Café, University of Toronto,* 14 Jan. 2013, org.mie.utoronto.ca. Accessed 27 Jan. 2017.

Kushner, David. "The Real Story of Stuxnet." *IEEE Spectrum.* IEEE Spectrum. 26 Feb. 2013. Web. 31 July 2015.

Kutz, Jose N. *Data-Driven Modeling & Scientific Computation: Methods for Complex Systems & Big Data.* Oxford: Oxford UP, 2013.

Kwon, Jaerock. "Lecture 1: Introduction to Microcomputers." *Mobile: Intelligent Robotics Lab.* Kettering University. 1 Jan. 2010. Web. 27 Aug. 2015.

Lafrance, Adrienne. "Why Do So Many Digital Assistants Have Feminine Names?" *Atlantic,* 30 Mar. 2016, www.theatlantic.com. Accessed 9 Dec. 2016.

Lalanda, Philippe, Julie McCann, and Ada Diaconescu. *Autonomic Computing: Principles, Design and Implementation.* London: Springer, 2014. Print.

Lang, Helen S. *Aristotle's Physics and Its Medieval Varieties.* State U of New York P, 1992.

"LANs – Local Area Networks." *Revision World.* Revision World Networks, Ltd. Web. 11 Feb. 2016.

Larcher, Gerhard, and Friedrich Pilichshammer, eds. *Applied Algebra and Number Theory.* New York: Cambridge UP, 2015.

Larson, Ron. *Algebra and Trigonometry.* Boston: Cengage, 2015.

______. *Precalculus with Limits.* Boston: Cengage, 2014.

Laschi, Cecilia, and Matteo Cianchetti. "Soft Robotics: New Perspectives for Robot Bodyware and Control." *Frontiers in Bioengineering and Biotechnology,* 30 Jan. 2014, journal.frontiersin.org. Accessed 12 Jan. 2017.

LaTorre, D. R. *Calculus Concepts : An Informal Approach to the Mathematics of Change.* Boston: Cengage, 2012.

Lax, Peter D., and Maria S. Terrell. *Calculus with Applications.* New York: Springer, 2013.

"Learning to Live with the Laws of Motion." *European Space Agency,* 12 Nov. 2012, www.esa.int. Accessed 13 Dec. 2016.

Lebert, Marie. "eBooks: 1998 – The First eBook Readers." Project Gutenberg News. Project Gutenberg News, 16 July 2011. Web. 3 June 2014.

Lecky-Thompson, Guy W. *Video Game Design Revealed.* Charles River Media, 2007.

Lee, Kent D. *Foundations of Programming Languages.* Springer, 2014.

Lee, Kunwoo. *Principles of CAD/CAM/CAE Systems.* Reading: Addison, 1999. Print.

Lee, Timothy B. "Why It's Time for Uber to Get Out of the Self-Driving Car Business." *Ars Technica,* Condé Nast, 27 Mar. 2018, arstechnica.com. Accessed 27 Mar. 2018.

Lemley, Linda. "Chapter 8: Operating Systems and Utility Programs." *University of West Florida.* University of West Florida. Web. 13 Aug. 2015.

Lerner, K. Lee, and Brenda Wilmoth Lerner, editors. "Zuse, Konrad." *Computer Sciences.* Macmillan Reference U.S.A, 2013.

"Lesson: Object-Oriented Programming Concepts." *Oracle.* Oracle. Web. 11 Aug. 2015.

Leszczynski, Jerzy. *Handbook of Computational Chemistry.* Dordrecht: Springer, 2012.

Levin, Mindy F., Emily A. Keshner, and Patrice L. (Tamar) Weiss. *Virtual Reality for Physical and Motor Rehabilitation.* New York: Springer, 2014. *eBook Collection (EBSCOhost).* Web. 10 Mar. 2015.

Lewin, Sarah. "Drones in Space! NASA's Wild Idea to Explore Mars." *Space.com,* 5 Aug. 2015, .com/30155-nasa-drones-on-mars-video.html" www.space.com. Accessed 10 Dec. 2016.

Lial, Margaret L., E. J. Hornsby, and Terry McGinnis. *Intermediate Algebra.* Boston: Pearson, 2012.

Lien, Tracey. "Virtual Reality Isn't Just for Video Games." *Los Angeles Times.* Tribune, 8 Jan. 2015. Web. 23 Mar. 2016.

"Lifecycle of a Plastic Product." *American Chemistry Council.* American Chemistry Council Inc. Web. 11 Feb. 2016.

Limoncelli, Tom, Strata R. Chalup, and Christina J. Hogan. *The Practice of Cloud System Administration: Designing and Operating Large Distributed Systems.* Upper Saddle River: Addison, 2015. Print.

Lind, Michael. "Stop Pretending Cyberspace Exists." *Salon.* Salon Media Group, Inc. 12 Feb. 2013. Web. 2 Mar. 2016.

Lindsey, Victor. "Konrad Zuse." *Great Lives from History: Inventors & Inventions.* Edited by Alvin K. Benson, Salem Press, 2010.

"Linear Programming." *High School Operations Research,* www.hsor.org. Accessed 27 Jan. 2017.

"Linear Programming." *IBM,* www-01.ibm.com. Accessed 27 Jan. 2017.

"Linear Programming." *Richland University,* people.richland.edu. Accessed 27 Jan. 2017.

"A Little Bit, Better. *The Economist.* The Economist Newspaper. 20 June 2015. Web. 6 Aug. 2015.

Liu, Baoding. "Extreme Value Theorems of Uncertain Process with Application to Insurance Risk Model." *Soft Computing* 17.4 (2013): 549-556.

Lobato, Joanne, and Amy B. Ellis. *Developing Essential Understanding of Ratios, Proportions, and Proportional Reasoning for Teaching Mathematics: Grades 6–8.* Reston, VA: National Council of Teachers of Mathematics, 2010.

Love, Paul, Joe Merlino, Jeremy C. Reed, Craig Zimmerman, and Paul Weinstein. *Beginning Unix.* Indianapolis, IN: Wiley Publishing, Inc., 2005. Print.

Lubanovic, Bill. *Introducing Python: Modern Computing in Simple Packages.* Sebastopol: O'Reilly, 2014. Print.

Lucas, Jim. "Newton's Laws of Motion." *Live Science,* 26 June 2014, www.livescience.com. Accessed 13 Dec. 2016.

Lunney, G. H. "Using Analysis of Variance with a Dichotomous Dependent Variable: An Empirical Study." *Journal of Educational Measurement* 7 (1970): 263–69. Print.

Lutz, Mark. *Learning Python.* Sebastopol: O'Reilly, 2015. Print.

MacCormick, John. *Nine Algorithms That Changed the Future: The Ingenious Ideas That Drive Today's Computers.* Princeton: Princeton UP, 2012.

"Machine Code (Machine Language)." *TechTarget.* TechTarget. Web. 10 Mar. 2016.

"Machine Code." *Techopedia.* Techopedia Inc. Web. 10 Mar. 2016.

"Machine Language vs High-Level Languages." *Bright Hub Engineering.* Bright Hub Inc. Web. 10 Mar. 2016.

Machlis, Sharon. "Beginner's Guide to R." *Computerworld.* Computerworld, 6 June 2013. Web. 22 Aug. 2016.

Mack, Pamela E. "The Microcomputer Revolution." *Clemson University.* Clemson University. 30 Nov. 2005. Web. 27 Aug. 2015.

Maehlum, Mathias Aarre. "Solar Energy Pros and Cons." *Energy Informative,* 12 May 2014, energyinformative.org. Accessed 10 Nov. 2016.

Magee, Jeff, and Jeff Kramer. *Concurrency: State Models and Java Programming.* Hoboken: Wiley, 2006. Print.

"Major Advance in Solar Cells Made from Cheap, Easy-to-Use Perovskite." *Phys.org,* 8 Nov. 2016, phys.org. Accessed 10 Nov. 2016.

Makarov, Mikhail. "Applications of Exact Extreme Value Theorem." *Journal of Operational Risk* 2.1 (2007): 115-120.

"Making Technology Talk: How Interoperability Can Improve Care, Drive Efficiency, and Reduce Waste." *Center for Medical Interoperability,* 27 Apr. 2016, medicalinteroperability.org. Accessed 23 Jan. 2017.

Mann, Charles C. "A Primer in Public-Key Encryption." *The Atlantic.* The Atlantic Monthly Group. Sept. 2002. Web. 4 March 2016.

"The Many Uses of Soft and Flexible Robots. *Association for Advancing Automation,* 31 Aug. 2016, www.a3automate.org. Accessed 12 Jan. 2017.

Marchand-Maillet, S. *Binary Digital Image Processing.* Waltham, MA: Academic P, 1999.

Marcovitz, Hal. *Video Games.* Lucent, 2010.

Marji, Majed. *Learn to Program with Scratch: A Visual Introduction to Programming with Games, Art, Science, and Math.* San Francisco: No Starch, 2014. Print.

Marks, Myles H. *Basic Integrated Circuits.* TAB Books, 1986.

Marques de Sá, J. P. *Pattern Recognition: Concepts, Methods and Applications.* New York: Springer Books, 2001.

Martin, George E. *Counting: The Art of Enumerative Combinatorics.* New York, NY: Springer, 2001.

Martin, James A. "Virtual Assistant Faceoff: Alexa, Cortana, Google Assistant and Siri." *CIO*, 5 Oct. 2016, www.cio.com. Accessed 9 Dec. 2016.

Martz, Eston. "How the Central Limit Theorem Works." *The Minitab Blog*, 15 Apr. 2011, blog.minitab.com. Accessed 24 Jan. 2017.

Mathews, J. D., et al. "Cancer Risk in 680,000 People Exposed to Computed Tomography Scans in Childhood or Adolescence: Data Linkage Study of 11 Million Australians." *BMJ*, vol. 346, 2013, p. f2360.

Mazalov, Vladimir. *Mathematical Game Theory and Applications.* John Wiley & Sons, 2014.

McConnell, Steve. *Code Complete: A Practical Handbook of Software Construction.* 2nd ed. Redmond: Microsoft, 2004. Print.

McCorduck, Pamela. *Machines Who Think: A Personal Inquiry into the History and Prospects of Artificial Intelligence.* 2d ed. Natick, Mass.: A. K. Peters, 2004.

McDonald, Paul. *Video and DVD Industries.* London: British Film Institute, 2008.

McGrayne, Sharon Bertsch. *The Theory That Would Not Die: How Bayes' Rule Cracked the Enigma Code, Hunted Down Russian Submarines, and Emerged Triumphant from Two Centuries of Controversy.* New Haven: Yale UP, 2012. Print.

McHaney, Roger. *Computer Simulation: A Practical Perspective.* San Diego: Academic Press, Inc., 1991. 1–9. Print.

______. *Understanding Computer Simulation.* StudyGuide.pk. Web. 4 Aug. 2015.

McKalin, Vamien. "Augmented Reality vs. Virtual Reality: What are the differences and similarities?" 6 April 2014. Web. 13 Aug. 2015. TechTimes.com.

McKellar, Danica. *Hot X: Algebra Exposed!* New York: Plume, 2011.

McMillan, Robert. "Xerox: Uh, We Didn't Invent the Internet." *Wired*, 23 July 2012, www.wired.com. Accessed 28 Jan. 2017.

McNeill, Daniel, and Paul Freiberger. *Fuzzy Logic: The Revolutionary Computer Technology That Is Changing Our World.* Touchstone, 1993.

"Mechatronics History." *Bright Hub Engineering*, 13 May 2010, www.brighthubengineering.com. Accessed 19 Jan. 2017.

Mee, C. Denis, and Eric D. Daniel. *Magnetic Recording Technology.* 2d ed. New York: McGraw-Hill, 1996.

Mele, Alfred R. *Free: Why Science Hasn't Disproved Free Will.* New York: Oxford UP, 2014. Print.

Menabrea, L.F. "Sketch of The Analytical Engine Invented by Charles Babbage." *Fourmilab.* Fourmilab, n.d. Web. 15 June 2016

Mendenhall, William, Robert J. Beaver, Barbara M. Beaver. *Introduction to Probability and Statistics.* Boston: Cengage, 2013.

Merrin, George. "The Computer Age." *University Observer*, 8 Nov. 2016, www.universityobserver.ie. Accessed 10 Nov. 2016.

Mertens, Ron. *The Graphene Handbook.* Graphene-Info, 2016. Print and online.

Metcalfe, Robert, and David Boggs. "Ethernet: Distributed Packet Switching for Local Computer Networks." *Communications of the ACM* 19, no. 7 (July, 1976): 395-404.

Metz, Cade. "2016: The Year That Deep Learning Took Over the Internet." *Wired*, 25 Dec. 2016, www.wired.com. Accessed 16 Jan. 2017.

______. "Finally, Neural Networks That Actually Work." *Wired*, 21 Apr. 2015, www.wired.com. Accessed 16 Jan. 2017.

Meyer, David. "'Cyberspace' Must Die. Here's Why." *Gigaom.* Knowingly, Inc. 7 Feb. 2015. Web. 2 Mar. 2016.

Michalski, Ryszard S., Jaime G. Carbonell, and Tom M. Mitchell. *Machine Learning: An Artificial Intelligence Approach.* New York: Springer, 2013. Print.

Michie, Donald. *Expert Systems in the Micro-Electric Age: Proceedings of the 1979 Aisb Summer School.* Edinburgh: Edinburgh University Press, 1979. Print.

"Microprocessors in the Home." *Teach-ICT.com.* Teach-ICT.com. Web. 23 Feb. 2016.

Miczo, Alexander. *Digital Logic Testing and Simulation.* 2nd ed., John Wiley & Sons, 2003.

Miller, George. "The Cognitive Revolution: A Historical Perspective." *TRENDS in Cognitive Sciences*, vol. 7, no.3, Mar. 2003, pp. 141–144.

Miller, James. *Game Theory at Work: How to Use Game Theory to Outthink and Outmaneuver Your Competition.* New York: McGraw-Hill, 2003.

Miller, Julie. *College Algebra Essentials.* New York: McGraw, 2013.

Miller, Zach. "Move Over Internet of Things, Here Is Pixie's Location of Things." *Forbes.* Forbes.com LLC. 3 Feb. 2015. Web. 3 Feb. 2015.

Millman, Richard, Peter Shiue, and Eric Brendan Kahn. *Problems and Proofs in Numbers and Algebra.* New York: Springer, 2015.

Minich, Curt. "Types of Software Applications." *INSYS 400 Home Page.* Penn State University, Berks Campus. Web. 9 July 2015.

Minsky, Marvin, and Seymour Papert. *Perceptrons: An Introduction to Computational Geometry.* Rev. ed. Boston: MIT P, 1990. Print.

Mishkoff, Henry C. *Understanding Artificial Intelligence.* Indianapolis, Indiana: Howard W. Sams & Company, 1999. Print.

Mlodinow, Leonard. *The Drunkard's Walk: How Randomness Rules Our Lives.* New York: Vintage, 2009. Print.

Mobbs, Richard. "HTML Developments." University of Leicester. Oct. 2009. Web. 28 Aug. 2015.

Moler, Cleve B. *Numerical Computing with MATLAB.* Philadelphia: Soc. for Industrial and Applied Mathematics, 2008. Print.

Monk, Paul, and Lindsey J. Munro. *Maths for Chemistry: A Chemist's Toolkit of Calculations.* New York: Oxford UP, 2010.

Montaigne, Fen. *Medicine By Design: The Practice and Promise of Biomedical Engineering.* Baltimore: The Johns Hopkins University Press, 2006.

Moon, Mariella. "Now California's DMV Can Allow Fully Driverless Car Testing." *Engadget,* 3 Apr. 2018, www.engadget.com. Accessed 3 Apr. 2018.

Morawetz, Herbert. *Polymers: The Origins and Growth of a Science.* Mineola, N.Y.: Dover, 2002.

Morgan, Jacob. "A Simple Explanation of 'The Internet of Things.'" *Forbes.* Forbes.com LLC. 13 May 2015. Web. 3 Feb. 2015.

______. "Everything You Need to Know about the Internet of Things." *Forbes.* Forbes.com LLC. 30 Oct. 2015. Web. 3 Feb. 2015.

Morrison, Robert Thornton, and Robert Nielson Boyd. *Organic Chemistry.* 5th ed. Newton, Mass.: Allyn & Bacon, 1987.

Mostashari, Farzad. "Mostashari: Value-Based Care Demands Free-Flowing Data." *Healthcare IT News,* 20 Jan. 2017, www.healthcareitnews.com. Accessed 23 Jan. 2017.

Mueller, Tom. "Biomimetics: Design by Nature." *National Geographic,* Apr. 2008, ngm.nationalgeographic.com. Accessed 12 Jan. 2017.

"Multitasking." *WhatIs.Com.* TechTarget. Web 15 Aug. 2015.

Murdoc, Shelby. "The Air Force's Space Drone Has Been in Orbit for Over 500 Days — And Its Mission Is Classified." *Business Insider,* 31 Oct. 2016, www.businessinsider.com. Accessed 10 Dec. 2016.

Murphy, Michael P., and Metin Sitti. "Waalbot: Agile Climbing with Synthetic Fibrillar Dry Adhesives." *2009 IEEE International Conference on Robotics and Automation.* Piscataway: IEEE, 2009. *IEEE Xplore.* Web. 21 Jan. 2016.

Murphy, Mike. "NASA Is Working on Drones That Can Fly in Space." *Quartz,* 31 July 2015. qz.com. Accessed 10 Dec. 2016.

Myers, Rusty L. *The Basics of Physics.* Greenwood Press, 2006.

Nakov, Svetlin. "Chapter 20. Object-Oriented Programming Principles (OOP)." *Introduction to Programming with C#/Java Books.* Svetlin Nakov and Team.

"Nano-Biomimetics." *Stanford University,* web. stanford.edu. Accessed 12 Jan. 2017.

"Nanotechnology Timeline." *National Nanotechnology Initiative,* www.nano.gov. Accessed 2 Dec. 2016.

"NASA Creating Robotic Drones for Future Space Exploration." *RT,* 5 Aug. 2015, www.rt.com/usa. Accessed 10 Dec. 2016.

Nash, John C. *Nonlinear Parameter Optimization Using R Tools.* Medford: Wiley, 2014. Print.

Nassiff, Peter, and Wendy A. Czerwinski. "Using Paperclips to Explain Empirical Formulas to Students." *Journal of Chemical Education* 91.11 (2014): 1934–38.

National Institute for Standards and Technology. "Announcing the Advanced Encryption Standard (AES): Federal Information Processing Standards Publication 197." NIST, 2001. Web. 31 July 2015.

______. "Fact Sheet: CNSS Policy No. 15, Fact Sheet No. 1, National Policy on the Use of the Advanced Encryption Standard (AES) to Protect National Security Systems and National Security Information." NIST, 2003. Web. 31 July 2015.

National Research Council of the National Academies. *Innovation in Information Technology.*

Washington, D.C.: National Academies Press, 2003.

Nebel, Bernard J., and Richard T. Wright. *Environmental Science: Towards a Sustainable Future.* 10th ed. Englewood Cliffs: Prentice Hall, 2008. Describes several bioprocesses used in waste treatment and pollution control.

"NEC Pushes Improved Interoperability for Smart City Solutions." *Enterprise Innovation,* 18 Jan. 2017, www.enterpriseinnovation.net. Accessed 23 Jan. 2017.

Nedrich, Matt. "An Introduction to the Central Limit Theorem." *Atomic Object,* 15 Feb. 2015, spin.atomicobject.com. Accessed 24 Jan. 2017.

"Network Interface Card (NIC)." *Scottish Qualifications Authority,* 21 Mar. 2010, www.sqa.org.uk. Accessed 28 Jan. 2017.

"The Network Interface Card. *Knowledge Systems Institute Graduate School,* pluto.ksi.edu. Accessed 28 Jan. 2017.

"Network Interface Cards (NIC)." *Teach-ICT,* www.teach-ict.com. Accessed 28 Jan. 2017.

"Network Topologies." *Pace University,* webpage.pace.edu. Accessed 28 Jan. 2017.

"Network Topology Definition." *Linux Information Project,* 2 Nov. 2005, www.linfo.org. Accessed 28 Jan. 2017.

"New Chinese Supercomputer Named World's Fastest System on Latest TOP500 List." *Top 500,* 20 June 2016, www.top500.org. Accessed 5 Dec. 2016.

"New Research Reveals How DNA Could Power Computers." *Live Science,* 17 May 2010, www.livescience.com. Accessed 20 Nov. 2016.

"New SmartLabel™ Initiative Gives Consumers Easy Access to Detailed Product Ingredient Information." *Grocery Manufacturers Association,* 2 Dec. 2015, www.gmaonline.org. Accessed 21 Dec. 2016.

Niccolai, James. "As Encryption Debate Rages, Inventors of Public Key Encryption Win Prestigious Turing Award." *PCWorld.* IDG Consumer & SMB. 2 March 2016. Web. 4 March 2016.

Nicholas, Patrick. *Scala for Machine Learning.* New York: Packt, 2014.

Nilsson, Nils J. *Principles of Artificial Intelligence.* San Francisco: Morgan Kaufmann, 2014. Print.

Nisbet, Robert, John Elder, and Gary Miner. *Handbook of Statistical Analysis and Data Mining Applications.* Burlington: Elsevier, 2009. Print.

Norman, Jeremy M., editor. *From Gutenberg to the Internet: A Sourcebook of the History of Information Technology.* HistoryofScience.com, 2005.

Novikov, D.A. *Cybernetics: From Past to Future.* Switzerland: Springer International Publishing, 2016. Print.

Nyholm, Sven, and Jilles Smids. "The Ethics of Accident-Algorithms for Self-Driving Cars: An Applied Trolley Problem?" *Ethical Theory & Moral Practice,* vol. 19, no. 5, pp. 1275–1289. doi:10.1007/s10677-016-9745-2. Accessed 27 Mar. 2018.

Obama, Barack. "Taking the Cyberattack Threat Seriously." Wall Street Journal. Dow Jones, 19 July 2012. Web. 2 June 2014.

"Object-Oriented Programming (C# and Visual Basic)." *Microsoft Developer Network.* Microsoft. Web. 11 Aug. 2015.

"Object-Oriented Programming (OOP) Definition." *TechTarget.* TechTarget. Web. 11 Aug. 2015.

O'Leary, Mike. *Cyber Operations: Building, Defending, and Attacking Modern Computer Networks.* Apress, 2015.

Oliver, Dick, et al. *Tricks of the Graphics Gurus.* Carmel, Ind.: Sams, 1993.

Olsen, Russ. *Eloquent Ruby.* Boston: Pearson Education, 2011. Print.

Olson, David R., and Michael Cole, eds. *Technology, Literacy and the Evolution of Society: Implications of the Work of Jack Goody.* Mahwah, N.J.: Lawrence Erlbaum Associates, 2006.

Ong, Walter J. *Orality and Literacy: The Technologizing of the Word.* 2d ed. Reprint. New York: Routledge, 2009.

"Open Access Week." *Open Access Week,* www.openaccessweek.org. Accessed 14 Jan. 2017.

"The Open Group Membership Page. *The Open Group.* The Open Group. Web. 18 Feb. 2016.

Oremus, Will. "I Didn't Type This Article." *Slate.* The Slate Group. 23 Apr. 2014. Web. 8 Mar. 2016.

Orenstein, David. "Application Programming Interface." *Computerworld,* 10 Jan. 2000, www.computerworld.com. Accessed 29 Dec. 2016.

"Orthophoto Imagery." Pima County, Arizona. Web. 3 Mar 2016.

Otani, Jun, and Yuzo Obara, editors. *X-ray CT for Geomaterials: Soils, Concrete, Rocks.* Balkema, 2004.

Ott, R. Lyman, and Michael T. Longnecker. *An Introduction to Statistical Methods and Data Analysis.* Belmont: Brooks, 2010. Print.

Ouellette, Jennifer. *The Calculus Diaries: How Math Can Help You Lose Weight, Win in Vegas, and Survive a Zombie Apocalypse.* New York: Penguin, 2010.

"Our Friends Electric." *Economist.* Economist Newspaper, 7 Sept. 2013. Web. 14 Nov. 2014.

Overton, Michael L. *Numerical Computing with IEEE Floating Point Arithmetic.* Philadelphia: Soc. for Industrial and Applied Mathematics, 2001. Print.

Painter, Paul C., and Michael M. Coleman. *Essentials of Polymer Science and Engineering.* Lancaster, Pa.: DEStech Publications, 2009.

Paret, Michelle, and Eston Martz. "Tumbling Dice & Birthdays: Understanding the Central Limit Theorem." *Minitab,* Aug. 2009, www.minitab.com. Accessed 25 Jan. 2017.

Parisi, Tony. *Learning Virtual Reality: Developing Immersive Experiences and Applications for Desktop, Web, and Mobile.* Sebastopol: O'Reilly, 2015. Print.

Park, Edwards. "What a Difference the Difference Engine Made From Charles Babbage's Calculator Emerged Today's Computer." *Smithsonian.com.* Smithsonian Magazine, Feb. 1996. Web. 15 June 2016.

Park, John Edgar. *Understanding 3D Animation Using Maya.* New York: Springer, 2005. Print.

Parker, Jack. "Computing with DNA." *EMBO Reports,* vol. 4, no.1, Jan 2003, pp. 7–10. *PubMed Central,* doi:10.1038/sj.embor.embor719. Accessed 20 Nov. 2016.

Parker, Matt. *Things to Make and Do in the Fourth Dimension: A Mathematician's Journey Through Narcissistic Numbers, Optimal Dating Algorithms, at Least Two Kinds of Infinity, and More.* New York: Farrar, 2014.

Parloff, Roger. "Why Deep Learning Is Suddenly Changing Your Life." *Fortune,* 28 Sept. 2016, fortune.com. Accessed 16 Jan. 2017.

Parr, Andrew. *Hydraulics and Pneumatics: A Technician's and Engineer's Guide.* 3rd ed. Burlington: Butterworth, 2011. Print.

Parr, Terence. *The Definitive ANTLR Reference: Building Domain-Specific Languages.* Raleigh: Pragmatic, 2013.

Patterson, David A. and John L. Hennessy. *Computer Organization and Design: The Hardware/Software Interface.* 5th ed. Waltham: Elsevier, Inc., 2014. 3–9, 13–17. Print.

Patterson, Michael. "What Is an API, and Why Does It Matter?" *SproutSocial,* 3 Apr. 2015, sproutsocial.com. Accessed 29 Dec. 2016.

Pennington, James. "4 Ways Smart Cities Will Make Our Lives Better." *World Economic Forum,* 10 Feb. 2016, www.weforum.org. Accessed 23 Dec. 2016.

Penrose, Roger. *The Emperor's New Mind: Concerning Computers, Minds and the Laws of Physics.* Oxford University Press, 2016. Print.

Perlin, John. "The Invention of the Solar Cell." *Popular Science,* 22 Apr. 2014, www.popsci.com. Accessed 10 Nov. 2016.

Peterson, Donald, and Joseph Bronzino. *Biomechanics: Principles and Applications.* Boca Raton: CRC, 2008. Print.

Petty, Luke. "What is Computer Aided Manufacturing? And How Is It Leading the Digital Site Revolution?" *Chapman Taylor,* 19 May 2016, www.chapmantaylor.com. Accessed 24 Oct. 2016.

Philips, Winfred. "Introduction to Logic for ProtoThinker." *Mind.* Consortium on Cognitive Science Instruction, 2006. Web. 6 Oct. 2015.

Phillips, Anna Lena. "Bioplastics Boom." *American Scientist.* The Scientific Research Society. Web. 6 Feb. 2016.

"Photogrammetry Surveys." CalTrans Survey Manual 2006. California Department of Transportation. Web. 3 Mar 2016.

Pillai, Anil Narendran. "A Brief Introduction to Photogrammetry and Remote Sensing." GIS Lounge. 12 July 2015. Web. 3 Mar 2016.

Pinola, Melanie. "Speech Recognition through the Decades: How We Ended Up with Siri." *PCWorld.* IDG Consumer & SMB. Web. 8 Mar. 2016.

Pishro-Nik, Hossein. *Introduction to Probability, Statistics, and Random Processes.* Kappa Research, 2014.

Pollack, Andrew. "A Glance at the Future of DNA: MD's Inside the Body." *New York Times,* 29 Apr. 2004, www.nytimes.com. Accessed 20 Nov. 2016.

Pólya, George. *How to Solve It: A New Aspect of Mathematical Method.* Expanded Princeton Science Lib. ed. Fwd. John H. Conway. 2004. Princeton: Princeton UP, 2014. Print.

Popovic, Marko B. *Biomechanics and Robotics.* Boca Raton: CRC, 2013. Print.

Poulter, Sean. "Bendable Smartphones Are Coming! Devices With Screens Made From Graphene Are So Flexible They Can Be Worn Like a Bracelet." *Mail Online.* The Daily Mail. Associated Newspapers Ltd, 24 May 2016. Web. 22 June 2016.

Pourciau, Bruce. "Newton and the Notion of Limit." *Historia Mathematica* 28.1 (2001): 18–30.

Prendergast, Patrick, ed. *Biomechanical Engineering: From Biosystems to Implant Technology.* London: Elsevier, 2007. One of the first comprehensive books for biomechanical engineers, written with the student in mind.

Prince, Betty. *Emerging Memories: Technologies and Trends.* Norwell, Mass.: Kluwer Academic, 2002.

Privitera, Gregory J. *Research Methods for the Behavioral Sciences.* 2nd ed., Sage Publications, 2016, 264–265.

"Processing a SQL Statement." *Microsoft: Developer Network.* Microsoft. Web. 6 Aug. 2015.

"Programming Language Generations." *TechTarget.* TechTarget. Web. 10 Mar. 2016.s

"Project Argus: Network Topology Discovery, Monitoring, History, and Visualization." *Cornell University,* www.cs.cornell.edu. Accessed 28 Jan. 2017.

Provost, Foster, and Tom Fawcett. *Data Science for Business: What You Need to Know about Data Mining and Data-Analytic Thinking.* Sebastopol: O'Reilly, 2013. Print.

"Public Key Cryptography." *PC Magazine Encyclopedia.* Ziff Davis, LLC. Web. 4 March 2016.

"Punch Cards." *History Mesh.* History Mesh, n.d. Web. 15 June 2016.

Purchase, Helen C. *Experimental Human-Computer Interaction: A Practical Guide with Visual Examples.* Cambridge UP, 2012.

Putrino, David, et al. "A Training Platform for Many-Dimensional Prosthetic Devices Using a Virtual Reality Environment." *Jour. of Neuroscience Methods* (2014). *ScienceDirect.* Web. 10 Mar. 2015.

Rabie, M. Galal. *Fluid Power Engineering.* New York: McGraw 2009. Print.

"Radiation-Emitting Products: Computed Tomography (CT)." *U.S. Food and Drug Administration.* U.S. Food and Drug Administration, 7 Aug. 2014, www.fda.gov. Accessed 3 Mar. 2015.

"Radio Frequency Identification (RFID): What Is It?" *U.S. Department of Homeland Security,* 20 Aug. 2015, www.dhs.gov. Accessed 21 Dec. 2016.

Ralston, Anthony, and Philip Rabinowitz. *A First Course in Numerical Analysis.* 2nd ed. Mineola: Dover, 2001. Print.

"RAM, ROM, and Flash Memory." *For Dummies.* John Wiley & Sons, Inc. Web. 27 Aug. 2015.

Ramachandran, V. S. *The Tell-Tale Brain: A Neuroscientist's Quest for What Makes Us Human.* New York: Norton, 2011. Print.

Ramalho, Luciano. *Fluent Python.* Sebastopol: O'Reilly, 2015. Print.

Ramsden, Jeremy. *Nanotechnology.* Elsevier, 2016.

Ramsey, Mike. "Tesla CEO Musk Sees Fully Autonomous Car Ready in Five or Six Years." *Wall Street Journal.* Dow Jones, 17 Sept. 2014. Web. 16 Nov. 2014.

Rana, Sanjay, and Jayant Sharma, eds. *Frontiers of Geographic Information Technology.* New York: Springer, 2006.

Ratner, Daniel, and Mark A. Ratner. *Nanotechnology and Homeland Security: New Weapons for New Wars.* Upper Saddle River: Prentice, 2004. Print.

______. *Nanotechnology: A Gentle Introduction to the Next Big Idea.* Upper Saddle River: Prentice, 2003. Print.

Rauber, Thomas, and Gudula Rünger. *Parallel Programming: For Multicore and Cluster Systems.* New York: Springer, 2010. Print.

"Readings on Smart Cities." *IEEE Smart Cities,* smartcities.ieee.org. Accessed 23 Dec. 2016.

Reba, Marilyn, and Douglas R. Shier. *Puzzles, Paradoxes, and Problem Solving.* New York: Chapman, 2014.

Reingold, Eyal and Jonathan Nightingale. "Expert Systems." *Artificial Intelligence.* Department of Psychology, University of Toronto. Web. 9 Mar 2016.

Reingold, Eyal. "The Turing Test." *University of Toronto.* University of Toronto. Web. 30 July 2015.

"Replication." *Webopedia.com.* Quinstreet Enterprise. Web. 29 Dec. 2014.

Rettner, Rachael. "DNA: Definition, Structure & Discovery." *Live Science*, 6 June 2013, www.livescience.com. Accessed 20 Nov. 2016.

Reuters. "Apple Enhances Siri but Still Trails in Artificial Intelligence Race." *NBC News*, 14 June 2016, www.nbcnews.com. Accessed 9 Dec. 2016.

Rey, PJ. "The Myth of Cyberspace." *The New Inquiry*. The New Inquiry. 13 Apr. 2012. Web. 2 Mar. 2016.

Richards-Kortum, Rebecca. *Biomedical Engineering for Global Health*. Cambridge UP, 2010.

Richardson, Leonard F. *Advanced Calculus: An Introduction to Linear Analysis*. Chicester: Wiley, 2011.

Roberts, Eric S. *Programming Abstractions in Java*. Boston: Pearson, 2017. Print.

Rockwell, Geoffrey. "Types of Computers: The Microcomputer." *McMaster University*. McMaster University. 15 June 1995. Web. 27 Aug. 2015.

Roessler, Johannes, Hemdat Lerman, and Naomi Eilan, eds. *Perception, Causation, and Objectivity*. New York: Oxford UP, 2011. Print.

Rogers, Ben, Jesse Adams, and Sumita Pennathur. *Nanotechnology: The Whole Story*. Boca Raton: CRC, 2013. Print.

Rogers, Henry. *Writing Systems: A Linguistic Approach*. Malden, Mass.: Wiley-Blackwell, 2005. A very detailed presentation of world languages and their writing systems that describes the classification of writing systems.

Rogers, Scott. *Level Up!: The Guide to Great Video Game Design*. 2nd ed., Wiley, 2014.

Roos, Dave. "How to Leverage an API for Conferencing." *HowStuffWorks*, money.howstuffworks.com. Accessed 29 Dec. 2016.

Roscoe, A.W. *Theory and Practice of Concurrency*. Upper Saddle River: Prentice, 1997. Print.

Ross, Kenneth A. *Elementary Analysis: The Theory of Calculus*. New York: Springer, 2013.

Ross, Timothy J. *Fuzzy Logic with Engineering Applications*. 3rd ed., Wiley, 2010.

Roth, Richard, and Stephen Pentak. *Design Basics*. Boston: Cengage, 2013.

Rouse, Margaret. "Advanced Encryption Standard (AES)." *TechTarget*. TechTarget. Web. 31 July 2015.

______. "Definition: FDDI (Fiber Distributed Data Interface)." *Search Networking*. TechTarget. Web. 11 Feb. 2016.

______. "Definition: Local Area Network (LAN)." *Search Networking*. TechTarget. Web. 11 Feb. 2016.

______. "Definition: Token Ring." *Search Networking*. TechTarget. Web. 11 Feb. 2016.

______. "Smart Label." *IoT Agenda*, internetofthingsagenda.techtarget.com. Accessed 21 Dec. 2016

______. "Zombie (bot)." *Tech Target*, 2016, searchmidmarketsecurity.techtarget.com. Accessed 30 Nov. 2016.

Rumelhart, David E., James L. McClelland, and the PDP Research Group. *Parallel Distributed Processing: Explorations in the Microstructure of Cognition*. 1986. Rpt. 2 vols. Boston: MIT P, 1989. Print.

Russell, Stuart, and Peter Norvig. *Artificial Intelligence: A Modern Approach*. 3rd ed. Upper Saddle River: Prentice, 2010. Print.

Ryan, Dan. *History of Computer Graphics: DLR Associates Series*. Bloomington, Ind.: AuthorHouse, 2011.

Sabbatini, Renato M. E. "Imitation of Life: A History of the First Robots." *Brain & Mind* 9 (1999): n. pag. Web. 21 Jan. 2016.

Sabin, Dyani. "Graphene-Based Computers Could Turn Electricity Into Light, Speeding Processing." *Inverse*, n.p., 23 June 2016. Web. 23 June 2016.

Sadalage, Pramod, and Martin Fowler. *NoSQL Distilled: A Brief Guide to the Emerging World of Polyglot Persistence*. Upper Saddle River: Addison, 2012.

Salz Trautman, Peggy. "A Computer Pioneer Rediscovered, 50 Years On." *New York Times*, 20 Apr. 1994, www.nytimes.com. Accessed 10 Nov. 2016.

Samani, Hooman. *Cognitive Robotics*. CRC Press, 2016.

Sandhu, Robin. "5 Examples of Biomimetic Technology." *Livewire*, 19 Oc. 2016, www.lifewire.com. Accessed 12 Jan. 2017.

Santo Domingo, Joel. "SSD vs. HDD: What's the Difference?" *PC*, 9 June 2017, www.pcmag.com. Accessed 20 Mar. 2018.

Santos, Kay. "What Components Are Necessary for an Experiment to Be Valid?" *Seattle Post-Intelligencer*, education.seattlepi.com. Accessed 19 Jan. 2017.

Sarkar, Jayanta. Computer Aided Design: A Conceptual Approach. CRC Press, 2015.

Saxby, Graham, and Stanislovas Zacharovas. *Practical Holography*. 4th ed., CRC Press, 2016.

Schapire, Robert E., and Yoav Freund. *Boosting: Foundations and Algorithms.* Cambridge, MA: MIT P, 2012.

Schell, Jesse. *The Art of Game Design: A Book of Lenses.* 2nd ed. Boca Raton: CRC, 2015. Print.

Schneier, Bruce. "The Story Behind the Stuxnet Virus." *Forbes.* Forbes.com LLC. 7 Oct. 2010. Web. 30 July 2015.

Schodeck, Daniel, et al. *Digital Design and Manufacturing: CAD/CAM Applications in Architecture and Design.* Hoboken: Wiley, 2005. Print.

Scholten, Henk J., Rob van de Velde, and Niels van Manen, eds. *Geospatial Technology and the Role of Location in Science.* New York: Springer, 2009.

Schuette, Paul. "A Question of Limits." *Mathematics Magazine* 77.1 (2004): 61-68.

Schwartz, John. "In the Lab: Robots That Slink and Squirm." *New York Times.* New York Times, 27 Mar. 2007. Web. 21 Jan. 2016.

"Scientists Achieve Critical Steps to Building First Practical Quantum Computer." *Phys.org.* Phys. org. 29 Apr. 2015. Web. 6 Aug. 2015.

Scott, Clare. "TU Delft Students Develop New Technique for 3D Printing Soft Robotics." *3DPrint.com,* 9 Nov. 2016. Accessed 12 Jan. 2017.

Scott, Gerald. *Polymers and the Environment.* Cambridge, England: Royal Society of Chemistry, 1999.

Scott, Michael L. *Programming Language Pragmatics.* 4th ed., Morgan Kaufmann, 2016.

Seal, Anthony M. *Practical Process Control.* Oxford: Butterworth, 1998. Print.

Seames, Warren S. *Computer Numerical Control Concepts and Programming.* 4th ed. Albany: Delmar, 2002. Print.

Sears, Andrew, and Julie A. Jacko, editors. *The Human-Computer Interaction Handbook: Fundamentals, Evolving Technologies, and Emerging Applications.* 2nd ed., Erlbaum, 2008.

Selinger, Ben. *Chemistry in the Marketplace.* 5th ed. Sydney: Allen & Unwin, 2002. The seventh chapter of this book provides a concise overview of many fiber materials and their common uses and properties.

"Semantic Web." *W3C.* W3C. Web. 6 Aug. 2015.

Serway, Raymond A., John W. Jewett, and Vahé Peroomian. *Physics for Scientists and Engineers with Modern Physics.* Boston: Cengage, 2014.

Seymour, Raymond B., ed. *Pioneers in Polymer Science: Chemists and Chemistry.* Boston: Kluwer Academic Publishers, 1989.

Shankar. "Evolution of Microprocessors." *Bright Hub Engineering.* Bright Hub Inc. Web. 23 Feb. 2016.

Shapiro, Ehud, and Yaakov Benenson. "Bringing DNA Computers to Life." *Scientific American,* vol. 294, no.5, 2006, pp. 44–51, www.scientificamerican. com. Accessed 20 Nov. 2016.

Shapiro, Stewart, ed. *Encyclopedia of Artificial Intelligence.* 2nd ed. New York: Wiley, 1992. Print.

Sharkey, Noel. "Alan Turing: The Experiment That Shaped Artificial Intelligence." *BBC.* BBC. 21 June 2012. Web. 30 July 2015.

Sharp, Heken, et al. *Interaction Design: Beyond Human-Computer Interaction.* 2nd ed., John Wiley & Sons, 2007.

Sharpe, Norean R., Richard De Veaux, and Paul F. Velleman. *Business Statistics: A First Course.* 2nd ed. New York: Pearson, 2013. Print.

Shaw, Zed A. *Learn Ruby the Hard Way.* 3rd ed. Upper Saddle River: Pearson Education, 2015. Print.

Shen, Helen. "Meet the Soft, Cuddly Robots of the Future." *Nature,* 3 Feb. 2016, www.nature.com. Accessed 12 Jan. 2017.

Sherman, William R., and Alan B. Craig. *Understanding Virtual Reality: Interface, Application and Design.* San Francisco: Morgan Kaufmann, 2003. Print.

"A Short History of HTML. W3C-HTML.com. W3C Foundation. Web. 28 Aug. 2015.

Siciliano, Bruno, and Oussama Khatib, eds. *The Springer Handbook of Robotics.* Springer, 2016.

Siciliano, Bruno, et al. *Robotics: Modeling, Planning, and Control.* London: Springer, 2009. Print.

Sifton, John. "A Brief History of Drones." *The Nation,* 7 Feb. 2012, www.thenation.com. Accessed 10 Dec. 2016.

Silver, Nate. *The Signal and the Noise: Why So Many Predictions Fail—But Some Don't.* New York: Penguin, 2015. Print.

Silvester, P. P., and D. A. Lowther. *Computer Engineering Circuits, Programs, and Data.* New York: Oxford UP, 1989. Print.

Simonite, Tom. "Teaching Machines to Understand Us." *MIT Technology Review,* 6 Aug. 2015, www. technologyreview.com. Accessed 16 Jan. 2017.

Singer, Graham. "The History of the Microprocessor and the Personal Computer." *TechSpot.* TechSpot, Inc. 17 Sep. 2014. Web. 23 Feb. 2016.

Singer, Peter Warren. *Wired for War: The Robotics Revolution and Conflict in the Twenty-first Century.* New York: Penguin Press, 2009.

Singh, Ankit Kumar. "Evolution of Microprocessor." *Scanftree.com.* Scanftree. Web. 23 Feb. 2016.

"Siri." *Apple,* www.apple.com. Accessed 9 Dec. 2016.

Smith, Aaron. "U.S. Views of Technology and the Future." *Pew Research.* Pew Foundation, 17 Apr. 2014. Web. 16 Nov. 2014.

Smith, Carlos A. *Automated Continuous Process Control.* New York: Wiley, 2002. Print.

Smith, Marquard, and Joanne Morra, editors. *The Prosthetic Impulse: From a Posthuman Present to a Biocultural Future.* MIT, 2007.

Snow, Colin. "Embrace the Role and Value of Master Data Management," *Manufacturing Business Technology,* 26, no. 2 (February, 2008): 92-95.

Soegaard, Mads, and Rikke Friis Dam, editors. *The Encyclopedia of Human-Computer Interaction.* 2nd ed., Interaction Design Foundation, 2014.

"Soft Robotics." *Harvard Biodesign Lab,* biodesign.seas.harvard.edu. Accessed 12 Jan. 2017.

"Soft Robotics." *Jaeger Lab,* jfi.uchicago.edu. Accessed 12 Jan. 2017.

"Solar Arrays." *NASA,* www.nasa.gov. Accessed 10 Nov. 2016.

Sommerville, Ian. *Software Engineering.* 9th ed. Boston: Addison, 2010. Print.

Spanbauer, Scott. "Wired or Wireless? Choose Your Network." *PC World,* 30 Sept. 2003, www.pcworld.com. Accessed 28 Jan. 2017.

"Speaker Recognition." *U.S. National Science and Technology Council Subcommittee on Biometrics and Identity Management.* PDF. Web. 8 Mar. 2016.

Spiegel, Rob. "Mechatronics: Blended Engineering for the Robotic Future." *DesignNews,* 23 Nov. 2016, www.designnews.com. Accessed 19 Jan. 2017.

Springer, Paul J. *Military Robots and Drones: A Reference Handbook.* Santa Barbara: ABC-CLIO, 2013. Print.

"SQL Modules." *Microsoft: Developer Network.* Microsoft. Web. 6 Aug. 2015.

Stakhov, Alexey, and Scott Anthony Olsen. *The Mathematics of Harmony.* Hackensack: World Scientific, 2009.

Stallings, William. *Data and Computer Communications.* 10th ed. Upper Saddle River, N.J.: Prentice Hall, 2014.

Steiner, Christopher. *Automate This: How Algorithms Came to Rule Our World.* New York: Penguin, 2012.

Stenquist, Paul. "In Self-Driving Cars, a Potential Lifeline for the Disabled." *New York Times.* New York Times, 7 Nov. 2014. Web. 16 Nov. 2014.

Stern, Joanna. "Apple's Siri: A Lot Smarter, but Still Kind of Dumb." *Wall Street Journal,* 20 Sept. 2016, www.wsj.com. Accessed 9 Dec. 2016.

Stewart, James. *Calculus : Early Transcendentals.* Belmont, CA: Cengage, 2012.

Stine, Keith J. *Carbohydrate Nanotechnology.* Wiley, 2016.

Stokes, Jon. "Introduction to Multithreading, Superthreading and Hyperthreading." *ARS Technica.* Conde Nast. 3 Oct. 2002. Web. 14 Aug. 2015.

Stowell, Sarah. *Using R for Statistics.* New York: Apress, 2014. Print.

Strauss, Walter A. *Partial Differential Equations: An Introduction.* 2nd ed. Hoboken: Wiley, 2008. Print.

Streiner, D. L., G. R. Norman, and J. Cairney. *Health Measurement Scales: A Practical Guide to Their Development and Use.* New York: Oxford UP, 2014. Print.

Strickland, Jonathan. "How Zombie Computers Work." *HowStuffWorks,* 2016, computer.howstuffworks.com. Accessed 30 Nov. 2016.

Strogatz, Steven. "The Hilbert Hotel." *The Joy of X.* Boston: Houghton, 2012. Print.

"Structured Query Language (SQL)." *BusinessDictionary.com.* WebFinance, Inc. Web. 6 Aug. 2015.

"Structured Query Language (SQL)." *Microsoft: Developer Network.* Microsoft. Web. 6 Aug. 2015.

"The Stuxnet Worm. *Norton by Symantec.* Symantec Corporation. Web. 31 July 2015.

Suber, Peter. "Open Access Overview." *Earlham College,* 5 Dec. 2015, legacy.earlham.edu. Accessed 14 Jan. 2017.

Subhash, D. "Types of Memory." *IT4NextGen,* 27 Aug. 2016, www.it4nextgen.com. Accessed 17 Nov. 2016.

Suddath, Claire. "A Brief History of: Velcro." *Time,* 15 June 2010, content.time.com. Accessed 12 Jan. 2017.

Sun, Changyou. *Empirical Research in Economics: Growing up with R.* Starkville: Pine Square, 2015. Print.

Sutner, Shaun. "Marc Probst Calls for Government Healthcare Interoperability Standards." *TechTarget,* 29 Dec. 2016, searchhealthit.techtarget.com. Accessed 23 Jan. 2017.

Sweigart, Al. *Scratch Programming Playground: Learn to Program by Making Cool Games.* San Francisco: No Starch, 2016. Print.

Takahashi, Shin. *The Manga Guide to Statistics.* San Francisco: No Starch, 2009. Print.

Tanenbaum, Andrew. *Computer Networks.* 5th ed. Boston: Pearson, 2011.

Taylor, Harriet. "How Your Home Will Know What You Need Before You Do." *CNBC.* CNBC, Jan 6 2016. Web. 11 Mar. 2016.

Taylor, Jim, et al. *Blu-ray Disc Demystified.* New York: McGraw-Hill, 2009.

Taylor, Jim, Mark R. Johnson, and Charles G. Crawford. *DVD Demystified.* 3d ed. New York: McGraw-Hill, 2006.

"Technology Not Enough for Interoperability." *IT-Online,* 17 Jan. 2017, it-online.co.za. Accessed 23 Jan. 2017.

Thatcher, Jim, et al. *Web Accessibility: Web Standards and Regulatory Compliance.* Springer, 2006.

Theodoridis, Sergio, and Konstantinos Koutroumbas. *Pattern Recognition.* 4th ed. Boston: Academic Press, 2009.

Thompson, Jim, Barnaby Berbank-Green, and Nic Cusworth. *Game Design: Principles, Practice, and Techniques.* Quarto, 2007.

Thrower, Norman J. W. *Maps and Civilization: Cartography in Culture and Society.* 3d ed. Chicago: University of Chicago Press. 2007.

"Timeline of Computer History." *Computer History Museum,* www.computerhistory.org. Accessed 17 Nov. 2016.

"Tips & Advice." Stay Safe Online. National Cyber Security Alliance, n.d. Web. 2 June 2014.

Toal, Ray. "What Is Systems Programming?" *Computer Science Division.* College of Science and Engineering, Loyola Marymount University. Web. 9 July 2015.

Toomey, Warren. "The Strange Birth and Long Life of Unix." *IEEE Spectrum.* IEEE Spectrum. 28 Nov. 2011. Web. 16 Feb. 2016.

Toothman, Jessica, and Scott Aldous. "How Solar Cells Work." *HowStuffWorks,* 1 Apr. 2000, science.howstuffworks.com. Accessed 10 Nov. 2016.

"Top 10 Signs Your Computer May be Part of a Botnet." *WeLiveSecurity,* 21 Apr. 2010, www.welivesecurity.com. Accessed 30 Nov. 2016.

"Topology." *College of Education, University of South Florida,* fcit.usf.edu. Accessed 28 Jan. 2017.

Tucker, Alan. *Applied Combinatorics.* Hoboken, NJ: Wiley, 2012.

Tufte, Edward R. *The Visual Display of Quantitative Information.* 2nd ed., Graphics, 2007.

Tukey, J. W. *Exploratory Data Analysis: Past, Present, and Future.* Technical Report No. 302 (Series 2), Department of Statistics. Princeton: Princeton UP, 1993.

"Turing Test." *Webopedia.* QuinStreet Inc. Web. 30 July 2015.

Tyson, Jeff. "How Internet Infrastructure Works." *HowStuffWorks.* InfoSpace, 2015. Web. 7 Oct. 2015.

"U.S. Department of Transportation Releases Policy on Autonomous Vehicle Development." *NHTSA.gov.* National Highway Traffic Safety Administration, 30 May 2013. Web. 16 Nov. 2014.

Ulmer, Gregory. *Applied Grammatology: Post(e)-pedagogy from Jacques Derrida to Joseph Beuys.* 1985. Reprint. Baltimore: The Johns Hopkins University Press, 1992.

______. *Electronic Monuments.* Minneapolis: University of Minnesota Press, 2005.

"Understanding Public Key Cryptography." *Microsoft Tech Net.* Microsoft. Web. 4 March 2016.

"Unimate: The First Industrial Robot." *Robotic Industries Association,* www.robotics.org. Accessed 12 Jan. 2017.

Utt, Jessica M., and Robert F. Heckard. *Mind on Statistics.* Stamford, CT: Cengage, 2015.

Valiant, Leslie. *Probably Approximately Correct: Nature's Algorithms for Learning and Prospering in a Complex World.* New York: Basic, 2013.

Van Helden, Al. "On Motion." *The Galileo Project,* 1995, galileo.rice.edu. Accessed 13 Dec. 2016.

Van Pul, Sergio, and Jessica Chiang. *Scratch 2.0 Game Development.* Birmingham: Packt, 2014. Print.

Vanderbilt, Tom. "Autonomous Cars through the Ages." *Wired.* Condé Nast, Feb. 2012. Web. 16 Nov. 2014.

Vanhemert, Kyle. "Leap Motion's Augmented-Reality Computing Looks Stupid Cool." *Wired.* Conde Nast. 7 July 2015. Web. 13 Aug. 2015.

Varoglu, Sevin, and Stephen Jenks. "Architectural support for thread communications in multi-core processors." *Parallel Computing* 37.1 (2011): 26–41. Print.

Vaughan, Simon. *Scientific Inference: Learning from Data.* Cambridge: Cambridge UP, 2013. Print.

Veit, Stan. "Pre-IBM PC Computers." *PC-History.org.* PC-History.org. Web. 27 Aug. 2015.

Vella, Matt. "Nest CEO Tony Fadell on the Future of the Smart Home." *Time.* Time, 26 June 2014. Web. 12 Mar. 2016.

Vernon, Vaughn. *Implementing Domain-Driven Design.* Upper Saddle River: Addison, 2013.

Vidal, John. "'Sustainable' Bio-plastic Can Damage Environment." *Guardian.* Guardian News and Media Limited. 25 Apr. 2008. Web. 6 Feb 2016.

Vince, John. *Introduction to Virtual Reality.* New York: Springer, 2004. Print.

Vinod Kumar, T.M., editor. *Smart Economy in Smart Cities.* Springer Singapore, 2017.

"VR Futures: Where Will Virtual Reality Take You?" *EandT.* Institution of Engineering and Technology, 15 Mar. 2016. Web. 26 June 2016.

"W3C Data Activity: Building the Web of Data." *W3C.* W3C. Web. 6 Aug. 2015.

"W3C Semantic Web Frequently Asked Questions." *W3C.* W3C. Web. 6 Aug 2015.

Wade, William R. *An Introduction to Analysis.* 4th ed. Boston: Pearson, 2009. Print.

Wadhwa, Vivek. "Quantum Computing Is About to Overturn Cybersecurity's Balance of Power." *The Washington Post.* The Washington Post. 11 May 2015. Web. 6 Aug. 2015.

Wakabayashi, Daisuke. "Uber's Self-Driving Cars Were Struggling before Arizona Crash." *The New York Times,* 23 Mar. 2018, www.nytimes.com. Accessed 3 Apr. 2018.

Walker, John. "The Analytical Engine. The First Computer." *Fourmilab.* Fourmilab, n.d. Web. 15 June 2016.

Wall, Jennifer J., and Christine C. Benson. "So Many Graphs, So Little Time." *Mathematics Teaching in the Middle School* 15.2 (2009): 82-91.

Wallace, David Foster. *Everything and More: A Compact History of Infinity.* New York: Norton, 2010.

Wallberg, Ben. "A Brief Introduction to APIs." *University of Maryland Libraries,* 24 Apr. 2014, dssumd.wordpress.com. Accessed 29 Dec. 2016.

Walski, Thomas M. *Advanced Water Distribution Modeling and Management.* Waterbury, Conn.: Haestad Methods, 2003.

Wang, Shan X., and Alexander Markovich Taratorin. *Magnetic Information Storage Technology.* San Diego: Academic Press, 1999.

Watkins, James. *Introduction to Biomechanics of Sport and Exercise.* London: Elsevier, 2007. Print.

Watson, Elaine. "80% of Packaged Groceries Will Feature SmartLabel within Five Years, Predicts GMA." *Food Navigator U.S.A,* 2 Dec. 2015, www.foodnavigator-usa.com. Accessed 21 Dec. 2016.

Weightman, Gavin. "The History of the Bar Code." *Smithsonian.com,* 23 Sept. 2015, www.smithsonianmag.com. Accessed 21 Dec. 2016.

Weikum, Gerhard, and Gottfried Vossen. *Transactional Information Systems: Theory, Algorithms, and the Practice of Concurrency Control and Recovery.* Burlington: Morgan, 2001. Print.

Weinberger, Charles B. *"Instructional Module on Synthetic Fiber Manufacturing."* Gateway Engineering Education Coalition: 30 Aug. 1996.

"What Is a Botnet Attack?" *Kaspersky Lab,* 2016, usa.kaspersky.com. Accessed 30 Nov. 2016.

"What Is a Nanoparticle?" *Horiba Scientific,* www.horiba.com. Accessed 2 Dec. 2016.

"What Is a Projectile?" *The Physics Classroom.* The Physics Classroom. Web. 14 Mar. 2016.

"What Is an Application?" *Goodwill Community Foundation,* www.gcflearnfree.org. Accessed 29 Dec. 2016.

"What Is Cognitive Science?" *ICOGSCI,* 2012, www.cogs.indiana.edu. Accessed 18 Dec. 2016.

"What Is 'Fuzzy Logic'? Are There Computers That Are Inherently Fuzzy and Do Not Apply the Usual Binary Logic?" *Scientific American,* www.scientificamerican.com. Accessed 24 Jan. 2017.

"What Is Interoperability?" *HIMSS,* www.himss.org. Accessed 23 Jan. 2017.

"What Is Malware and How Can We Prevent It?" *PCTools,* 2016, www.pctools.com. Accessed 30 Nov. 2016.

"What Is Mechatronic Engineering?" *Brightside,* www.brightknowledge.org. Accessed 19 Jan. 2017.

"What Is Mechatronics?" *Institution of Mechanical Engineers,* www.imeche.org. Accessed 19 Jan. 2017.

"What Is NIC (Network Interface Card)?" *OmniSecu.com,* www.omnisecu.com. Accessed 28 Jan. 2017.

"What Is Peer-to-Peer (P2P)?" *GigaTribe.* GigaTribe, n.d. Web. 7 Oct. 2015.

"What Is Unix?" *Indiana University Knowledge Base.* The Trustees of Indiana University. 14 Dec. 2015. Web. 16 Feb. 2016.

"What's So Special about the Nanoscale?" *National Nanotechnology Initiative,* www.nano.gov. Accessed 2 Dec. 2016.

Wheater, Carolyn C. *Practice Makes Perfect Algebra.* New York: McGraw, 2010.

Wheelan, Charles. *Naked Statistics: Stripping the Dread from the Data.* New York: Norton, 2014. Print.

"When to Use Multitasking." *Microsoft.* Microsoft. Web. 13 Aug. 2015.

"Who Are We & Why This Website?" *EnablingOpenScholarship,* www.openscholarship.org. Accessed 14 Jan. 2017.

"Who Was Charles Babbage?" *University of Minnesota.* Regents of the University of Minnesota, 9 Feb. 2016. Web. 15 June 2016.

Wieber, Pierre-Brice, Russ Tedrake, and Scott Kuindersma. "Modeling and Control of Legged Robots." *Handbook of Robotics.* Ed. Bruno Siciliano and Oussama Khatib. 2nd ed. N.p.: Springer, n.d. (forthcoming). *Scott Kuindersma—Harvard University.* Web. 6 Jan. 2016

Wilder, Richard, and Melissa Levine. "Let's Speed Up Science by Embracing Open Access Publishing." *STAT,* 19 Dec. 2016, www.statnews.com. Accessed 14 Jan. 2017.

Williams, Brett. "Cyberspace: What Is It, Where Is It, and Who Cares?" *Armed Forces Journal.* Sightline Media Group. 13 Mar. 2014. Web. 2 Mar. 2016.

Wilson, Bill. "Introduction to PROLOG Programming." *CSE.* Bill Wilson/U of New South Wales, 26 Feb. 2012. Web 5 Oct. 2015.

Winsberg, Eric. "Computer Simulations in Science." *Stanford Encyclopedia of Philosophy.* Center for the Study of Language and Information (CSLI), Stanford University. 23 Apr. 2015. Web. 4 Aug. 2015.

______. *Science in the Age of Computer Simulation.* Chicago: University of Chicago Press, 2010. 1–6. Print.

Wonder, Dan. "The Function of the NIC." *Houston Chronicle,* smallbusiness.chron.com. Accessed 28 Jan. 2017.

Wood, Lamont. "The 8080 Chip at 40: What's Next for the Mighty Microprocessor?" *ComputerWorld.* ComputerWorld, Inc. 8 Jan. 2015. Web. 27 Aug. 2015.

______. "The Clock Is Ticking for Encryption." *Computerworld.* Computerworld, Inc. 21 Mar. 2011. Web. 31 July 2015.

Woodcock, Jon. *Coding Games in Scratch.* New York: DK, 2016. Print.

Wright, Rob. "The Evolution of the Smartphone in 7 Releases." CRN. Channel Company, 3 Apr. 2013. Web. 3 June 2014.

Yager, Ronald, and Ali Abbasov, eds. *Soft Computing.* New York: Springer, 2014.

Yang, Shang-Tian. *Bioprocessing for Value-Added Products from Renewable Resources: New Technologies and Applications.* Amsterdam: Elsevier, 2007.

Yaroslavsky, Leonid. *Digital Holography and Digital Image Processing: Principles, Methods, Algorithms.* Kluwer Academic, 2004.

Young, Cynthia Y. *College Algebra.* Hoboken, NJ: Wiley, 2012.

Young, Peyton, and Shmuel Zamir. *Handbook of Game Theory.* Elsevier, 2015.

Yukihiro Matsumoto. *Ruby in a Nutshell.* Trans. David L. Reynolds Jr. Sebastopol: O'Reilly Media, 2002. Print.

Zanella, A., et. al. "Internet of Things for Smart Cities." *IEEE Internet of Things Journal,* vol. 1, no. 1, 14 Feb. 2014, pp. 22–32.

Zetter, Kim. "An Unprecedented Look at Stuxnet, the World's First Digital Weapon." *Wired.* Condé Nast. 3 Nov. 2014. Web. 31 July 2015.

Zhang, J., and X. Liang. "One-Way ANOVA for Functional Data via Globalizing the Pointwise F-test." *Scandinavian Journal of Statistics* 41 (2014): 51–74. Print.

Zimba, Jason. *Force and Motion: An Illustrated Guide to Newton's Laws.* Johns Hopkins UP, 2009.

Zimmer, Ben. "Is It Time to Welcome Our New Computer Overlords?" *Atlantic.* Atlantic Monthly Group, 17 Feb. 2011. Web. 23 Aug. 2016.

Zimmerman, Kim Ann. "History of Computers: A Brief Timeline." *Live Science,* 8 Sept. 2015, www.livescience.com. Accessed 17 Nov. 2016.

INDEX